New Governance and Management in Touristic Destinations

Marco Valeri
Niccolò Cusano University, Italy

IGI Global
PUBLISHER of TIMELY KNOWLEDGE

A volume in the Advances in Hospitality, Tourism, and the Services Industry (AHTSI) Book Series

Published in the United States of America by
IGI Global
Business Science Reference (an imprint of IGI Global)
701 E. Chocolate Avenue
Hershey PA, USA 17033
Tel: 717-533-8845
Fax: 717-533-8661
E-mail: cust@igi-global.com
Web site: http://www.igi-global.com

Library of Congress Cataloging-in-Publication Data

Names: Valeri, Marco, editor.
Title: New governance and management in touristic destinations / Marco Valeri, editor.
Description: Hershey, PA : Business Science Reference, 2022. | Includes bibliographical references and index. | Summary: "This book offers new insights for traditional paradigms, approaches and methods, as well as more recent developments in research methodology in new approaches to governance and management of destination, outlining how and why tourist destinations have to act more like businesses than they have done in the past"-- Provided by publisher.
Identifiers: LCCN 2021045729 (print) | LCCN 2021045730 (ebook) | ISBN 9781668438893 (hardcover) | ISBN 9781668438909 (paperback) | ISBN 9781668438916 (ebook)
Subjects: LCSH: Tourism--Management--Case studies. | Place marketing--Case studies.
Classification: LCC G155.A1 N428 2022 (print) | LCC G155.A1 (ebook) | DDC 910.68--dc23/eng/20211105
LC record available at https://lccn.loc.gov/2021045729
LC ebook record available at https://lccn.loc.gov/2021045730

This book is published in the IGI Global book series Advances in Hospitality, Tourism, and the Services Industry (AHTSI) (ISSN: 2475-6547; eISSN: 2475-6555)

British Cataloguing in Publication Data
A Cataloguing in Publication record for this book is available from the British Library.

For electronic access to this publication, please contact: eresources@igi-global.com.

New Governance and Management in Touristic Destinations

Marco Valeri
Niccolò Cusano University, Italy

IGI Global
PUBLISHER of TIMELY KNOWLEDGE

A volume in the Advances in Hospitality, Tourism, and the Services Industry (AHTSI) Book Series

Published in the United States of America by
IGI Global
Business Science Reference (an imprint of IGI Global)
701 E. Chocolate Avenue
Hershey PA, USA 17033
Tel: 717-533-8845
Fax: 717-533-8661
E-mail: cust@igi-global.com
Web site: http://www.igi-global.com

Library of Congress Cataloging-in-Publication Data

Names: Valeri, Marco, editor.
Title: New governance and management in touristic destinations / Marco Valeri, editor.
Description: Hershey, PA : Business Science Reference, 2022. | Includes bibliographical references and index. | Summary: "This book offers new insights for traditional paradigms, approaches and methods, as well as more recent developments in research methodology in new approaches to governance and management of destination, outlining how and why tourist destinations have to act more like businesses than they have done in the past"-- Provided by publisher.
Identifiers: LCCN 2021045729 (print) | LCCN 2021045730 (ebook) | ISBN 9781668438893 (hardcover) | ISBN 9781668438909 (paperback) | ISBN 9781668438916 (ebook)
Subjects: LCSH: Tourism--Management--Case studies. | Place marketing--Case studies.
Classification: LCC G155.A1 N428 2022 (print) | LCC G155.A1 (ebook) | DDC 910.68--dc23/eng/20211105
LC record available at https://lccn.loc.gov/2021045729
LC ebook record available at https://lccn.loc.gov/2021045730

This book is published in the IGI Global book series Advances in Hospitality, Tourism, and the Services Industry (AHTSI) (ISSN: 2475-6547; eISSN: 2475-6555)

British Cataloguing in Publication Data
A Cataloguing in Publication record for this book is available from the British Library.

All work contributed to this book is new, previously-unpublished material. The views expressed in this book are those of the authors, but not necessarily of the publisher.

For electronic access to this publication, please contact: eresources@igi-global.com.

Advances in Hospitality, Tourism, and the Services Industry (AHTSI) Book Series

Maximiliano Korstanje
University of Palermo, Argentina

ISSN:2475-6547
EISSN:2475-6555

Mission

Globally, the hospitality, travel, tourism, and services industries generate a significant percentage of revenue and represent a large portion of the business world. Even in tough economic times, these industries thrive as individuals continue to spend on leisure and recreation activities as well as services.

The Advances in Hospitality, Tourism, and the Services Industry (AHTSI) book series offers diverse publications relating to the management, promotion, and profitability of the leisure, recreation, and services industries. Highlighting current research pertaining to various topics within the realm of hospitality, travel, tourism, and services management, the titles found within the AHTSI book series are pertinent to the research and professional needs of managers, business practitioners, researchers, and upper-level students studying in the field.

Coverage

- Tourism and the Environment
- Service Training
- International Tourism
- Cruise Marketing and Sales
- Sustainable Tourism
- Destination Marketing and Management
- Customer Service Issues
- Service Design
- Casino Management
- Leisure & Business Travel

IGI Global is currently accepting manuscripts for publication within this series. To submit a proposal for a volume in this series, please contact our Acquisition Editors at Acquisitions@igi-global.com or visit: http://www.igi-global.com/publish/.

The Advances in Hospitality, Tourism, and the Services Industry (AHTSI) Book Series (ISSN 2475-6547) is published by IGI Global, 701 E. Chocolate Avenue, Hershey, PA 17033-1240, USA, www.igi-global.com. This series is composed of titles available for purchase individually; each title is edited to be contextually exclusive from any other title within the series. For pricing and ordering information please visit http://www.igi-global.com/book-series/advances-hospitality-tourism-services-industry/121014. Postmaster: Send all address changes to above address.

Titles in this Series

For a list of additional titles in this series, please visit: http://www.igi-global.com/book-series/advances-hospitality-tourism-services-industry/121014

Prospects and Challenges of Community-Based Tourism and Changing Deographics
Ishmael Mensah (University of Cape Coast, Ghana) and Ewoenam Afenyo-Agbe (University of Cape Coast, Ghana)
Business Science Reference • © 2022 • 300pp • H/C (ISBN: 9781799873358) • US $195.00

Challenges and New Opportunities for Tourism in Inland Territories Ecocultural Resources and Sustainable Initiatives
Gonçalo Poeta Fernandes (CITUR, Polytechnic Institute of Guarda, Portugal)
Business Science Reference • © 2022 • 295pp • H/C (ISBN: 9781799873396) • US $195.00

Rebuilding and Restructuring the Tourism Industry Infusion of Happiness and Quality of Life
André Riani Costa Perinotto (Universidade Federal do Delta do Parnaíba, Brazil) Verônica Feder Mayer (Federal Fluminense University, Brazil) and Jakson Renner Rodrigues Soares (Universidade da Coruña, Spain & Universidade Estadual do Ceará, Brazil)
Business Science Reference • © 2021 • 330pp • H/C (ISBN: 9781799872399) • US $195.00

Handbook of Research on the Impacts and Implications of COVID-19 on the Tourism Industry
Mahmut Demir (Isparta University of Applied Sciences, Turkey) Ali Dalgıç (Isparta University of Applied Sciences, Turkey) and Fatma Doğanay Ergen (Isparta University of Applied Sciences, Turkey)
Business Science Reference • © 2021 • 906pp • H/C (ISBN: 9781799882312) • US $415.00

Socio-Economic Effects and Recovery Efforts for the Rental Industry Post-COVID-19 Strategies
Maximiliano Korstanje (University of Palermo, Argentina)
Business Science Reference • © 2021 • 281pp • H/C (ISBN: 9781799872870) • US $195.00

Risk, Crisis, and Disaster Management in Small and Medium-Sized Tourism Enterprises
Diego R. Toubes (University of Vigo, Spain) and Noelia Araújo-Vila (University of Vigo, Spain)
Business Science Reference • © 2021 • 308pp • H/C (ISBN: 9781799869962) • US $195.00

Resiliency Models and Addressing Future Risks for Family Firms in the Tourism Industry
Anita Zehrer (MCI The Entrepreneurial School, Austria) Gundula Glowka (MCI The Entrepreneurial School, Austria) Katrin Magdalena Schwaiger (MCI The Entrepreneurial School, Austria) and Victoria Ranacher-Lackner (MCI The Entrepreneurial School, Austria)
Business Science Reference • © 2021 • 361pp • H/C (ISBN: 9781799873525) • US $195.00

IGI Global
PUBLISHER of TIMELY KNOWLEDGE

701 East Chocolate Avenue, Hershey, PA 17033, USA
Tel: 717-533-8845 x100 • Fax: 717-533-8661
E-Mail: cust@igi-global.com • www.igi-global.com

Table of Contents

Detailed Table of Contents

The critical impact of the tourism sector on the development of local communities and economies has now been recognized, both theoretically and in practice, by local governments (OTAs) and other national or supranational policy makers. The OTAs, with their institutional role extending to the areas of shaping and promoting tourist destinations, are creating, in this direction, a network of synergies and actions with other stakeholders in the tourism sector. Therefore, the issue of collaborative marketing is major. At the same time, in addition to the strategy and management of the promotion and promotion of a tourist destination, there is a second level of vision of the action of OTA, that of the marketing facilitation strategy, which pursues the holistic coordinating the tourism sector of an area, ensuring the creation of synergies and the development of tourism development in the desired direction. For this reason, this chapter is going to examine the synergies created and how local authorities can contribute to this.

Conceptually drawing on network theory as its theoretical lens, this study examines two prime notions of network configuration of commercial expeditions. Exploring the role of both structural holes and network closure as indicators of team configuration for those venturing out in such extreme adventure, this study clarifies the impact of social structures, network closure, and structural holes in particular on performance outcomes in the context of expedition mountaineering. Presence and bridging of structural holes did turn out to be a significant predictor for the success or failure of an expedition. The findings show network closure to significantly influence the performance of mountaineering teams that make for a successful ascent. The capacity to span structural holes, commonly portrayed as serving as an eye-opener for options otherwise not found, does not appear to assist teams that make for successful ascents, however.

Digital transformation and artificial intelligence are considered among the most vital trending topics in the process of hospitality sector evolution. Many scholars found that digital transformation and artificial intelligence cause a massive shift in all aspects of the hospitality sector and digital technology application that impact the whole facet of internal and external stakeholders' lives. However, the adoption of digitalization and artificial intelligence is considered a strength. Corporate digital responsibility (CDR) is a strategy that enhances trust between the companies adopting digitalization and their primary stakeholders. Internal and external stakeholders' satisfaction develop contemporary social responsibility (CSR) challenges in the decision-making process in acquiring, analysing, implementing, and assessing for adopting digitalization in the hospitality sector. This chapter aims to give a literature review focusing on CDR and its relation analyses to hotel industry's internal stakeholders' satisfaction trough a Hungarian case study.

This chapter aims to fill the knowledge gap regarding the relationship between tourism and air quality, specifically in the Portuguese tourism industry, with a focus on tourist nationality. It examines whether this relationship differs according to tourist origin. This study uses an air pollutant, PM10, with a strong impact on human health that has been highly neglected in the literature. Despite the great use of CO2 in assessing the causal relationship between tourism and the environment, this is not the best indicator of air quality (AQ). This chapter presents results by applying vector autoregressive models (VAR) with monthly data for the period of 2007-2017, considering the nationality of tourists that visit Portugal. Results suggest that PM10 levels and tourism are negatively correlated (in the Pearson sense) with a link between them in the long run. This relationship is confirmed by the four methodologies tested. The negative relation in Pearson and cointegration results suggests that tourism can be affected by AQ in Portugal and may lead to better AQ.

Nasi Kandar is one of the most favourable delicacies for tourists when visiting Penang. However, the hygiene level in most of Nasi Kandar restaurants has slumped lately. Thus, the purpose of this study is to describe the relationship between visitor awareness towards revisit intention to Nasi Kandar restaurants. Three hundred eighty-four respondents from Penang Island were accessed using a simple random sampling. The results revealed that there is a positive relationship between visitor awareness and revisit intention. Price and service eventually give a strongly positive impact towards revisit intention. This result will help the local authorities to increase the quality and hygiene of Nasi Kandar in Penang, hence bring more tourists to visit this tourism destination.

The literature is highlighting that research on the internationalization of destinations is scarce. This chapter presents a validation of the governance factors which contribute to the internationalization of tourism destinations. A survey was carried out on Portuguese DMOs with responsibilities in the areas of tourism and local/regional development (N=147). The 'Questionnaire on Internationalization of Tourist Destinations' was submitted to content, reliability, and construct validations through exploratory and confirmatory second-order factor analysis. Governance factors influencing the internationalization of destinations were F1-Collaboration and F2-Networks. F1 associates items related to stakeholder involvement in destination decisions. F2 highlights the interaction between stakeholders, whose collective action is oriented towards active participation in decisions and leadership. The perception of Portuguese entities with different levels of influence on tourism development allows the systematization of results that meet some priority lines defined in the Tourism Strategy 2027.

Destination management organizations are functional structures that manage and market the tourist destinations operating to facilitate the cooperation among different stakeholders. A successful DMO supports tourism development, especially when tourism is an important economic driver for a destination. The idea of building different forms of DMOs and the concept of networking have guided tourism governance choices in different initiatives conducted in many countries. This chapter analyses the model of DMO adopted in Italy by Regione Toscana comparing it with the variable geometry approach by Beritelli et al. The study offers a critical reflection on the model of DMO, relevant from the perspectives of governance and management.

Tourism destinations can be considered as complex systems of interrelated and interdependent stakeholders. The complexity and limited power of influencing the number of stakeholders resulted in network approach to tourism destination governance. This approach is considered both theoretically and practically as a tool for strengthening its sustainable competitiveness, fostering innovation and knowledge sharing. Although the network analysis of tourism destinations has gained a significant attention in recent years, the complex understanding of its contribution to smart development is still missing. The aim of this chapter is to create a framework for smart approach in destination governance using the network science perspective. The chapter provides insights in using network analysis for strengthening the tourism destination governance. The chapter uses a case study methodology on two mature tourism destinations, providing an example of the use of network analysis for destination governance strengthening.

This study investigated the relationship among e-WOM, overall destination image, and European tourist behavioral intention to revisit and recommend Jordan as a tourism destination. The study examined the mediating effect of destination overall image on the relationship between e-WOM and tourist behavioral intention. A total of 339 questionnaires were collected from European tourists who visited Jordan in 2018 through a self-administered questionnaire. Structural equation modeling (SEM) was used. The result of SEM identified that e-WOM has a positive influence on the overall destination image and on the tourist behavioral intention. The overall image positively influenced the tourist behavioral intention. Overall destination image partially mediates between e-WOM and behavioral intention, and thereby, the positive impact of e-WOM on behavioral intention could be aggravated through the mediating effect of destination overall image. The findings have implications on the tourism industry, especially for key players in the Jordanian tourism board and travel companies.

This chapter presents the results of the exploratory study regarding the profile of local residents and visitors of the Historical Villages of Portugal Network (nHVP). This network integrates 12 villages located in low density territories in the center of Portugal. It is important to conduct this study in order to conceive strategies and policies that will contribute to reversing the desertification tendency and to retaining resources and surpluses that would otherwise leave the region. The authors concluded that the local residents present a medium/high level of satisfaction in terms of the impact of tourism in their villages and also with the association of their village with the nHVP brand. However, the local residents' participation in the activities and touristic events is low, which is in part due to the high age of most local residents. Regarding the visitors of the nHVP, they are generally satisfied or very satisfied with several points. The lowest level of satisfaction relates to the public transportation, the communication networks, and the medical care available in the territory.

The focus of this research is on the role of stakeholders and perceived destination competitiveness in ethnic enclave destination management. Ethnic enclave destination is one of the unique tourism products in terms of art, culture, heritage, and lifestyle. In Malaysia, Little India, Brickfields is one of the popular Indian enclave destinations significantly contributing to the development of tourism in Malaysia. The perceived destination competitiveness determines the destination attributes, and the destination management is ensured by the stakeholders. However, there are just a few pieces of research that look at the relationship between perceived destination competitiveness and destination management. As a result, this study used an integrated model of TDC to examine the mix of stakeholders and perceived destination competitiveness in Little India, Brickfields. This study will have a significant impact on ethnic enclave destination management by identifying stakeholders and their roles in boosting destination competitiveness at the destination of concern.

The European Capital of Culture (ECoC) title enables European citizens to learn about one another's culture, to admire the relics of their common past, and to experience the wealth of European values and the sense of belonging together. Winner cities and regions wish to sustain the effect of the title in order to maintain economic, social and cultural prosperity. Veszprém and Balaton region in Hungary won the title for 2023, which influences local and regional destination management. The aim of the research is to introduce the key role, factors, and pitfalls of local and regional stakeholders' network system and to highlight the challenges of sustainable cooperation.

The concept of "Legacy" might have many interpretations, going from tangible to intangible, and accordingly, it can include a series of items that can be bequeathed to different kind of "heirs." Either the giver or the community as main role players involved, the legacy is a bond for public administrations and destination managers. Legacy economics will be here connected to the innovative management of cultural destinations, where several and multi-goals-holders connect in order to provide resilience of villas, museums, and cultural assets together with increasing audiences and tourists. The collection of data about resilient creativity in Como (in the North of Italy) will show connections between legacy and management of public-private organizations for the last decade. A versatile and wide range of public and private interests together with national and international relevance will emerge. Community-type destinations will result as an original and innovative governance for resilience of culture and creativity connected with legacies and meta-management.

In today's modern world, creative expression is opening up new dimensions of business and new opportunities for economic development. One field of economic activities in which this is evident is tourism. Creativity in tourist destinations can be viewed in different ways, for example, through creative action (undertaken by destination management, residents, entrepreneurs, and tourists) and through creative spaces and creative events. Creativity plays a vital role in all elements involved in the creation of a destination's offering, regardless of which form of tourism is the focus of development efforts.

Given the growing role of self-actualisation of individuals in society and the displaying of social status, creativity has in the past 20 years begun to positively impact on economy activities taking place in tourist destinations. Creativity is especially important in developing cultural tourism in all its sub-types, where it is seen as a means of animating and adding value to cultural heritage locations.

Panoraia Poulaki, University of the Aegean, Greece
Antonios Kritikos, University of West Attica, Greece
Nikolaos Vasilakis, Ionian University, Greece
Marco Valeri, Niccolò Cusano University, Italy

Gastronomy as an important part of the culture, morals, customs, and traditions of the countries meets the needs of visitors and at the same time contributes to tourism development through the rich gastronomic heritage. Gastronomic tourism offers the possibility of increasing the income criteria of a population, increasing jobs, and developing the national economy. The purpose of this study is to highlight the role of female creativity in the development of a gastronomic destination with a case study of Naxos in the South Aegean in Greece. At the same time, it aims to highlight the great challenges and opportunities brought about by gastronomic tourism with the contribution of women's entrepreneurship. Women's creativity and entrepreneurship create new job opportunities. The dynamic presence and activity of the women of Naxos highlights the tourism destination of Naxos as a destination of high gastronomy. Secondary sources present gender discrimination in entrepreneurship, women's entrepreneurship in tourism, and the connection of tourism with gastronomic heritage.

Goretti Silva, Polytechnic Institute of Viana do Castelo, Portugal
Susana Rachão, Polytechnic Institute of Viana do Castelo, Portugal
Alexandra I. Correia, Polytechnic Institute of Viana do Castelo, Portugal
Luís André Soares, Polytechnic Institute of Viana do Castelo, Portugal

Increasing awareness of the relevance of natural resources for the practice of outdoor leisure and recreation purposes has contributed to the increasing attention upon the assessment and monitoring of the quality of sports spots. As such, evaluating the existing natural conditions and leisure-related infrastructures is crucial, helping public and private stakeholders in decision-making processes. This chapter aims at examining the conditions for surf tourism in the north of Portugal, enabling the monitoring and positioning as an international flagship destination for surf practitioners. This is done through a mixed-methods approach comprising the demand and supply perspectives and supported by an assessment tool developed to evaluate the conditions for the practice of surfing activities within a framework of sustainable tourism development. The relevance of assessing the potential and monitoring of the existing conditions for sport and outdoor activities, namely surf tourism, according to a defined set of attributes, is also discussed.

Preface

The book *New Governance and Management in Tourist Destinations* is the result of reflections involving researches of different nationalities. The book aims to provide a comprehensive collection of chapters including new insights for traditional paradigms, approaches, and methods, as well as more recent developments in research methodology in new approaches to governance and management of destination.

Management literature in the tourism industry deals with the problems and methods of managing a tourist destination. The establishment of an authority of governance is of fundamental importance for the competitiveness of a tourist destination. The governance of the destination can be conceived as a system of decisions and actions aimed at consolidating, and over time improving, the competitiveness of the destination, and defining clear and shared development goals among all parties in the tourism industry.

Establishing a governing authority for a destination is not always straightforward and depends on the specific features of each destination. In particular, there is no single person who can be chosen as the strategic coordinating authority of a tourist destination. The governing body has to record and appraise the tangible and intangible elements that make up the cultural identity of a destination and compare them with those of rival destinations, whether local, national or international. This is on the assumption that decisions are directly or indirectly fully agreed upon by the governing bodies of each company or each destination, and that all decisions are made with a view to improving the attractiveness and competitiveness of the destination.

Governance does not involve only shareholders, but also other people with a vested interest in the tourist destination (e.g., managers or service providers) and/or a direct interest in it - here we are referring to the 'stakeholders', bodies or organisations that have an important role in the area where the destination is located. Some of these factors (for example suppliers or tourist service industries) have an effect on the actual function of the destination, while others generate public interest towards the destination, such as state organisations or citizens' associations.

The setting up of a governing body is a necessary step in creating a combined tourist offer within a tourist destination, but in itself it is not enough. With this in mind, the governing body should be supported by a meta-management unit whose specific task is to guide the tourist companies in the destination in a direction that is coherent and strategic. Such a unit would support but not replace the governing body, and have the definite aim of furthering strategic decisions that are shared by all parties concerned, both public entities and private individuals, who interact with one another within the tourist destination.

As is the case for tourist companies, managing a tourism destination is not an easy task. The difficulties lie in how to best organise decision-making processes and how to allocate resource-controlling powers within the tourist destination. This can be the case in destinations that are arranged according to a strict hierarchy, with a well-defined managerial and procedural centre of gravity, as well as in destinations of a corporate or community type. Both of them have different ways of programming their tourist offer and consequently have very different governance issues.

In corporate-type destinations (e.g., theme parks, tourist villages, resorts, etc.), the tourist offer is designed by a company which, directly or indirectly, controls the tourist attraction resources of the destination through contractual arrangements. In this case the destination takes on the form of an actual business, whose governance problems are no different from any other tourist organisation. In community-type destinations, on the other hand, it is the territory and not the company that offers a series of natural and man-made attractions on the tourist market. In this case, governance is decidedly more complex, since the attraction resources do not belong to one single company, but to several independent businesses, each of whom is pursuing specific goals in terms of investment policies and profit generation.

Here local institutions are of crucial importance to the competitiveness of the tourist sector. They control the attraction resources, both natural and man-made, and can provide funding for the tourist offer as far as regards sustainability (Valeri et al., 2016).

The book adopts a multidisciplinary approach from management to local development, from entrepreneurship to sociology, from innovation to sustainability, providing illustration examples about destination governance and destination management. Its ultimate goal is to elaborate an historically-informed, future-focused research agenda that accounts for the needs and concerns of policy makers and practitioners in the tourism field.

STRUCTURE OF THIS BOOK

The book is structured in *two* parts. The first part focuses on *destination governance*. This part collects chapters that analyze the potential of decision process of a tourist destination planning the tourism offer and encouraging innovative and competitive forms of organization.

The second part focuses on *destination management***.** This part collects chapters that analyze management models that have increased the competitiveness of tourist destinations.

Chapters falling into this specific part of this book have as follows:

In Chapter 1 "Destination Governance: The Role of Local Authorities in Greek Tourism Marketing," the authors Dimitrios Belias, Ioannis Rossidis, Christos Papademetriou and Marco Valeri analyze the critical impact of the tourism sector on the development of local communities and economies. The OTAs, with their institutional role extending to the areas of shaping, promoting and promoting tourist destinations, are creating, in this direction, a network of synergies and actions with other stakeholders in the tourism industry. The issue of collaborative marketing is major. At the same time, in addition to the strategy and management of the promotion and promotion of a tourist destination, there is a second level of vision of the action of OTA, that of the marketing facilitation strategy, which pursues the holistic coordinating the tourism sector of an area, ensuring the creation of synergies and the development of tourism development in the desired direction. This chapter examines the synergies created and how local authorities can contribute on this.

In Chapter 2, "A Network-Based View on Extreme Expedition Team Composition in Alpine Tourism," the authors Rick Aalbers and Ramon Spildoorn explore the role of both structural holes and network closure as indicators of team configuration, on performance outcomes in the context of extreme expedition mountaineering. In a context of extreme sports tourism and drawing on a rich and longitudinal empirical dataset that of Himalayan mountaineering teams, they analyzed the network configuration of 144 climbing teams in relation to their historical network embeddedness in the broader Himalayan climbing community. Presence and bridging of structural holes did turn out to be a significant predictor for the success or failure of an expedition. As such their empirical study examines two prime notions of network configuration of extreme sports team. Exploring the role of both structural holes and network closure as indicators of team configuration. In doing so they clarify the impact of social structures present in extreme sports tourism as a direct lever for safety management in this exciting domain of alpine tourism.

In Chapter 3, "The Impact of Corporate Digital Responsibility (CDR) on Internal Stakeholders' Satisfaction in Hungarian Up-Scale Hotels," the authors Edit Kővári, Mohamad Saleh and Gyöngyi Steinbachné Hajmásy explore corporate digital responsibility (CDR) and its relation analyses to hotel industry's internal stakeholders' satisfaction trough a Hungarian case study. The results show that hospitality companies can be more efficient while adopting digitalization and artificial intelligence if they consider the economic, legal, ethical activities that would effectively attract, motivate, and retain their employees. The findings of the empirical research confirmed that there is a strong positive relationship between internal CDR and employee satisfaction.

In Chapter 4, "Cointegration and Causality Analysis of Portuguese Tourism and Air Quality: Differences Among International Tourism Markets," the author's Mara Madaleno, Margarita Robaina, Celeste Eusébio, Maria João Carneiro, Victor Rodrigues, Alexandra Monteiro, and Carla Gama aims to fill the knowledge gap regarding the relationship between tourism and air quality, specifically in the Portuguese tourism industry, with a focus on tourist nationality. It examines whether this relationship differs according to tourist origin. This study uses an air pollutant – PM10 - with a strong impact on human health that has been highly neglected in the literature. Results suggest that PM10 levels and tourism are negatively correlated (in the Pearson sense), with a link between them in the long run. This relationship is confirmed by the four methodologies tested. The negative relation in Pearson and Cointegration results suggests that tourism can be affected by AQ in Portugal and may lead to better AQ, possibly due to tourists' growing concern with the environment. Results point to differences among the nationality of visitors in Portugal.

In Chapter 5, "Georgetown as a Gastronomy Tourism Destination: Visitor Awareness Towards Revisit Intention of Nasi Kandar Restaurant," the authors Albattat Ahmad and Shufri Johari explored Penang the "Food Paradise" among tourists offering wide variety of tastes from various traditional cultural foods and ethnicities. Penang also the home of the best Mamak cuisine "Nasi Kandar" restaurants are highly recommended by Penang visitors due to their specialties and uniqueness. The results showed that most of the tourists choose nasi kandar restaurant as their food destination due to its taste. This is because it has a very rich aromatic and flavour with the influenced of Indian Muslim cuisine.

In Chapter 6, "Governance Factors That Influence the Internationalization of Tourism Destinations: The Perspective of Portuguese DMOs," the author/s Maria do Rosário Mira, Lisete Mónico, Zélia Breda, and Marco Valeri explore the 'destinations governance' dimension from the perspective of the Portuguese DMOs, with different levels of responsibilities and impacts on tourism development. The results show that the governance factors influencing the internationalization of destinations are 'collaboration' and 'networks'. The interaction and collaboration between stakeholders trigger one collective action is oriented towards active participation in the decisions and leadership of the destination project, with repercussions for the sharing of resources, the qualification of human resources, and residents' involvement in the project.

Chapter 7, "New Perspective of Networking in the DMO Model," by Vincenzo Asero and Sandro Billi analyses the governance model of DMOs and investigates the critical role of networking between the private and public sectors in the management of tourism systems, to explain the reason for many failures and weaknesses in practice. The research is based on the analysis of a selected case of DMO in

Tuscany that has been investigated considering some strategic factors, such as the relevant legislation supporting the different initiatives, the involvement of local stakeholders, the role played by the public bodies, the contribution of local communities and the creation of social capital.

In Chapter 8, "Complexity of Tourism Destination Governance: A Smart Network Approach," the authors Tomáš Gajdošík and Marco Valeri create a framework for smart approach in destination governance using the network science perspective. The findings show the contribution of real-time data exchange via destination management system for improvement of knowledge transfer and co-decisions among stakeholders, usefulness of social media for providing real-time information on tourists and the value of big data analytics for identification of strategic visitor flows.

Chapter 9, "Electronic Word-of-Mouth Effects on Middle East Destination Overall Image and Behavioral Intention: An Empirical Study in Jordan," by Ashraf M. Alfandi and Marco Valeri, focuses on tourists who visited Jordan and getting ready to go back to their countries, therefore, this study inspects tourists' perceptions of relevant and real destination attributes instead of general assessments which consider an important strategy to develop and recover the destination image.

In Chapter 10, "Profile of the Local Residents and Visitors of the Historical Villages of Portugal Network: An Exploratory Study," the authors Maria Manuela S. Natário, Carlos Santos, Ana I. Melo, Augusta Ferreira, Dalila Dias, Gonçalo Gomes, Graça Azevedo, Paula Rocha, Ricardo Biscaia, Rúben Duarte and Rui Pedro Marques present the results of an exploratory study about the profile and satisfaction of the residents and visitors of the 12 historical villages located in the central Portugal region. The results contribute to foster, conceive and elaborate strategies and policies that, on one hand, contribute to reverse the desertification trend of these territories while providing conditions for a sustained development and, on the other hand, contribute to cement their attractiveness and retain resources and surpluses that would otherwise leave the region.

In Chapter 11, "Role of Stakeholders and Perceived Destination Competitiveness Towards Ethnic Enclave Destination Management: An Overview of 'Little India' Brickfields, Malaysia," the authors Puvaneswaran Kunasekaran and Munieleswar Raju, explored the important role played by stakeholders such as Malaysian Government, Kuala Lumpur City Hall (Dewan Bandaraya Kuala Lumpur) (DBKL), local business traders and visitors in Little India, Brickfileds to maintain and enhance the destination as an 'Indian Ethnic Enclave' destination. In addition, the authors also discussed on the perceived destination competitiveness of this Little India, Brickfields. The authors clarified the elements of Endowed (inherited) Resources, Created Resources, Supporting Resources and Situational Conditions (extrinsic factors) which have been adapted from 'The Integrated Model of the Destination Competitiveness (IM Destination Competitiveness)' by Dwyer and Kim (2003). The importance and the relationship between the destination competitiveness on ethnic enclave destination have been highlighted in this chapter. Thus, the role of stakeholders and understanding the destination competitiveness are important on ethnic enclave destination to enhance the destination management for long run by attract more visitors and improve local communities (business traders and residents) lifestyle and socioeconomic well-being.

In Chapter 12, "Veszprém-Balaton 2023 European Capital of Culture Title for Sustainable Cooperation: The Key Role of Local and Regional Stakeholders' Network," the authors Edit Kővári, Katalin Lőrincz and Marco Valeri explore the key role, factors and pitfalls of local and regional stakeholders' network system and to highlight the challenges of sustainable cooperation of Veszprém, a city that will be the European Capital of Culture (ECoC) in 2023. The results show the challenges of sustainability, cooperation and networking of stakeholders ECoC cities, especially Veszprém-Balaton 2023. It provides an overview of the most important directions, content and the planned operating model of stakeholders' network system regarding how the ECoC programmes can be successful.

In Chapter 13, "Legacy Economics and Meta-Management of Creativity in Destinations: Legacy Economics for Destination Management," the authors Angela Besana and Maria Cristina Vannini explore the importance of tangible and intangible heritages, which can be a legacy for the attractivity of destinations. Legacy economics is then connected to the innovative management of destinations, where several public and private stakeholders provide resilience of cultural and creative assets together with the growth of communities' and tourists' advocacy and motivations for these assets. Como will be the focus of the analysis. A versatile and wide range of public, private, national and international interests will result for this destination from 2010 to 2017, as an original and innovative mixed governance and management for restitution of legacies. This chapter is to be considered a starting point for wider investigations on this topic and it is useful for managers of cultural foundations who want to go through bonds between legacy, community and tourism.

In Chapter 14, "Reaching the New Tourist Through Creativity: Sustainable Development Challenges in Croatian Coastal Towns," the authors Stipanović Christian, Rudan Elena and Zubović Vedran, all from University of Rijeka, Faculty of Tourism and Hospitality Management in Opatija, Croatia explore the creativity in developing the tourism offerings of Croatian coastal cities. The analysis of the current state of the creative offering of coastal Croatian cities and towns shows that creativity provides opportunities for the transformation of Croatia's coastal cities and towns from tourist destinations of marked seasonality whose tourism development hinges on natural preconditions (sea, sun, sand) into distinctive destinations with creative offerings. Creativity enables heritage and tradition to be interpreted in novel ways that are focused on the preservation of indigenous values and on sustainable development.

In Chapter 15, "The Contribution of Female Creativity to the Development of Gastronomic Tourism in Greece: The Case of the Island of Naxos in the South Aegean Region," the authors Panoraia Poulaki, Antonios Kritikos, Nikolaos Vasilakis and Marco Valeri analyze the role of female creativity in the development of a gastronomic destination with a case study of Naxos in the South Aegean in Greece. At the same time, the chapter aims to highlight the great challenges and opportunities brought about by gastronomic tourism with the contribution of women's entrepreneurship. Women's creativity and entrepreneurship create new job opportunities. The dynamic presence and activity of the women of Naxos highlights the tourism destination of Naxos as a destination of high gastronomy.

Chapter 16, "Assessing the Potential for Tourism Development: The North of Portugal as a Surf Destination," by Silvia Goretti, Susana Rachão, Alexandra I. Correia and André Luis aims at examining the conditions for surf tourism in the north of Portugal. This is done through a mixed-methods approach comprising the demand and supply perspectives and supported by an assessment tool developed to evaluate the conditions for the practice of surfing activities within a framework of sustainable tourism development. The relevance of assessing the potential and monitoring of the existing conditions for sport and outdoor activities, namely surf tourism according to a defined set of attributes, is also discussed.

In Chapter 17, "The Urban Sensescapes and Sensory Destination Branding," the authors Irma Imamović, António Azevedo and Bruno Sousa explore the multisensory features of tourist destinations. They suggest that places can be branded based on their unique sensory features, and therefore stimulate tourists' positive emotions and memory of a destination. The results advocate that urban areas possess different multisensory dimensions that can be introduced in marketing efforts to stimulate positive tourists' experience and to encourage local sustainable development. This approach depends on the actions of local government and tourism stakeholders who can work together and create experiencescapes and servicescapes to generate bettered strategies, convey meaningful sensory impressions, and stimulate positive tourists' reactions.

Section 1
Destination Governance

Chapter 1
Destination Governance:
The Role of Local Authorities in Greek Tourism Marketing

Dimitrios Belias
https://orcid.org/0000-0003-3409-6875
University of Thessaly, Greece

Ioannis Rossidis
University of the Peloponnese, Greece

Christos Papademetriou
Neapolis University, Pafos, Cyprus

Marco Valeri
https://orcid.org/0000-0002-9744-506X
Niccolò Cusano University, Italy

ABSTRACT

The critical impact of the tourism sector on the development of local communities and economies has now been recognized, both theoretically and in practice, by local governments (OTAs) and other national or supranational policy makers. The OTAs, with their institutional role extending to the areas of shaping and promoting tourist destinations, are creating, in this direction, a network of synergies and actions with other stakeholders in the tourism sector. Therefore, the issue of collaborative marketing is major. At the same time, in addition to the strategy and management of the promotion and promotion of a tourist destination, there is a second level of vision of the action of OTA, that of the marketing facilitation strategy, which pursues the holistic coordinating the tourism sector of an area, ensuring the creation of synergies and the development of tourism development in the desired direction. For this reason, this chapter is going to examine the synergies created and how local authorities can contribute to this.

DOI: 10.4018/978-1-6684-3889-3.ch001

INTRODUCTION

Tourism is an economic and social activity which affects and is affected from tourist destinations. This means that local governance has a crucial role on tourist development and it must be in cooperation with the tourist and local businesses. Local authorities are the ones who have a sound understanding of the needs of the local community but also they are able to coordinate the local stakeholders (Valeri, 2016; Rasoolimanesh et al., 2017; Valeri and Katsoni, 2021; Valeri, 2021; Trivelas et al., 2015; Belias et al., 2017; Belias et al., 2020).

It is important to mention that the existing literature often focuses on the role that tourist businesses have on the development of destinations though the role of local governance is often undermined, especially in Greece (Vassakis et al, 2019). Of course, during the past years local authorities in Greece have started taking an active role on tourist development. An example is Athens, where City of Athens has an active role on the establishment of Athens as a major tourist destination during the past years (Maguire, 2019). A significant contribution on this has the fact that since 2010 there has been a number of reforms on local governance – such as the Kalikratis reform – which gave to the cities and local governments a greater volume of independence in respect of tourist policy (Vassakis, et al, 2019; Trivellas et al, 2017). As mentioned above, local governments have a critical role on tourist development, something that has not been widely investigated (Maguire, 2019), while often is ignored the fact that the coordination between local governments and of the private sector can generate huge benefits for the tourist development and to bring wealth for local communities (Vassakis et al, 2019), while it is a factor which boosts entrepreneurship (Majewska, 2008; Valeri and Katsoni, 2021; Belias et al., 2020; Tsiotas et al., 2020). Nonetheless, there is the case of the lack of such research which will examine the contribution of local governance in tourist development, especially in Greece (Patikas, 2014).

From the above it is understood that the critical impact of the tourism sector on the development of local communities and economies has now been recognized, both theoretically and in practice, by local governments (OTAs) in Greece and other national or supranational policy makers. The OTAs, with their institutional role extending to the areas of shaping and promoting tourist destinations, are creating, in this direction, a network of synergies and actions with other stakeholders in the tourism sector. Therefore, the issue of collaborative marketing between the local authorities and the tourist industry is critical for the success of destination marketing. Therefore, this paper is going to fill-in the gap which exists by the existing literature, with emphasis on how the local governance merges with the private sector in order to promote a tourist destination.

This paper is a literature review. This mean that this research uses publications found on academic search engines such as EBSCO, Google Scholar and SCOPUS. Also, the paper will rely on official reports and other types of publications related with the topic. It is essential to note that the authors have used keywords such as "local governance + tourist development" so to identify the related papers and publications. The contribution of this paper lies on the fact that it is investigating a topic that has been hardly examined – in terms of how local governance contributes in Greece – and what are the best practices and policies adopted.

BACKGROUND

The aim of this chapter is to investigate the relationship between local governance and how it interacts with the local tourist sector. Indeed, there is evidence that can effective local governance and the local tourist industry have a strong interaction. Indeed, the synergies created between those two institutions can bring a number of benefits for the local communities, including the creation of jobs and an increase on income along with the launch of new ventures in the tourist industry (Majewska, 2008; Valeri and Katsoni, 2021; Belias et al., 2020; Tsiotas et al., 2020), including any startups (Dhir, 2019; Men et al, 2018) From the above it is understood that there is a gap in the existing literature which needs to be filled in.

MAIN FOCUS OF THE CHAPTER

This is a literature review. This means that the authors will rely on the existing literature and on reports. The literature would be taken from related databases such as SCOPUS and emerald. The authors have used key words, such as "local authorities & tourism marketing" in order to identify the related publications and to write down the paper. The paper indicates that an effective, efficient and sustainable tourism development policy with simultaneous care to address weaknesses and threats is critical to establishing a strong competitive position in the international tourism market. In order to implement it, it is necessary to have a more active partnership between the public and the private sector, with an emphasis on collaborative marketing practices. It is very important to notice how the recent changes on the way the local governance operates in Greece, is an indication that there is a need to have a better understanding on how the local government contributes in Greece' marketing practices and offers. This is a literature review which means that the data used is limited only to what other researchers have to say. The lack of empirical evidence constitutes are key limitation.

LOCAL GOVERNANCE AND TOURIST DEVELOPMENT

Conceptual Framework of Tourism Policy

As mentioned from Vassiliadis (2009, p.112), by the term tourism policy it is meant "the conscious promotion and shaping of tourism achieved through the influence of tourism-related data by public law organizations (state, local government bodies), institutions or private law associations (associations and unions) and by various interest groups (shareholders' associations) ", where as a conscious promotion is mentioned the process of setting goals by the related institutions. Tourism policies are classified into two categories: those of an operational nature (free market process) and those of a broader economic nature (planned process). Nowadays, tourism policy applies modern marketing methods and specific analyzes of the business environment and market that set goals for the implementation of more rational policies with a more effective distribution of means of achievement. Vassiliadis (2009) lists the main bodies of tourism policy at three levels: local, national and international (eg UNWTO, OECD).
Key areas of tourism policy are:

- Taking care to provide facilities during the tourist arrival, stay and movement.

- Creation of infrastructure.
- Development of transport and networks.
- Activation of the private initiative for tourism investments.
- View destination abroad.
- Provision of dynamic financing tools such as government grants, European financing programs, lending on flexible terms, etc.

The Institutional Role of Local Governments

In Greece the Law 3852/2010 (Government Gazette 87, vol. A,, also well known as KALLIKRATIS) has promoted the decentralization of local authorities and self-governance, replacing the previous organizational structure ("KAPODISTRIAS") of local governance. The first article of Law 3852 defines the 325 Municipalities and the 13 Regions that constitute the two levels (first and second) of local self-government and are fundamental institutions of local governance in Greece.

The Municipalities are governed by the Mayor and the Municipal Council, who are elected by universal suffrage for five years. They are divided into municipalities / departments that have their own councils with a more advisory role (decision-making only for minor / restrictive legislative issues). The Regions are governed by the Regional Governor and the Regional Council (elected by universal suffrage for a five-year term), divided into "Regional Units", each represented by its own Deputy Regional Governor, coming from the majority electoral combination (Albet, 2016).

The local governance in Greece, well known as OTA, occurs today with a limited level of self-autonomy and decision making. Of course, the Kallikratis reform has strengthen the local autonomy, but still it is very limited. In the secondary level, which refers on regions, they tend to have a greater level of autonomy and more resources, but still they are subject of control from the central government (Tsekos & Triantafyllopoulou, 2016).

The Role of Local Government in Tourism Development

The World Travel and Tourism Council (2018) argues that today the central and local government together with the local stakeholders can have an active role in tourist planning. For example, interventions from the local and national government can ensure the necessary funds and promotion, with the aim of increasing tourist flows and revenues. The related key stakeholders must develop a spirit of cooperation and have a common vision, a view to the balanced development of tourism, which will give to a tourist destination strong competitive advantage (Ritchie and Crouch, 2003, D'Angella and Go, 2009).

Nowadays, tourist management aims into creating an economy of experience based on the experiences provided on the visitor, on the local tourism consciousness. This fact makes self-evident the involvement of the local government in the tourism industry, as a regulator, exponent and balancer of the local will. As the tourism product is a special product, the production of which is based on all the structures of a destination, in order to have duration and maximum benefits for the local economy and society, organized treatment at the local level is required (Lopes et al, 2020).

In Greece, today, there is an autonomous Ministry of Tourism, which has the overall responsibility for tourism development with supervised bodies the EOT (Hellenic Tourist Organisation) and X.E.E. (Chamber of Greek hotels), while, significantly, private tourist organizations (S.E.T.E. (Chamber of Tourist businesses) contribute on tourism development. Overall it is widely accepted on many markets that the ministries and national tourist boards, perform critical tasks, such as approving large-scale investment projects, strategic planning, etc., but necessitated decentralized / their coordinated operation so that decision making is effective and immediate . Regarding Greece, two levels of tourism development have been instituted: the central government (through the Ministry and EOT), which holds a coordinating role, and the local government, which has been assigned a number of responsibilities, while SETE – through its affiliated which is Marketing Greece – has taken an important role in the digital promotion of Greek tourism but also on the self-regulation of the market along with XEE. Hence, there is evidence of the cooperation between the local and national government with the tourist authorities so to leverage Greek tourism. Actually, it should be mentioned that there is a high level of privatization on the decision making process since the tourist authorities and chambers have an important role along with the government (Katsios, 2019).

As far as local communities are concerned, the integration of resources into regional tourism development has, in addition to economic weight, political and social implications, such as:

- Weaker dependence of the destination on imported products and foreign investment funds, due to the use of local wealth-producing sources (reduction of cash outflows).
- Integration of tourism in the economic and social networks of the region, cultivating the "socially responsible" tourism education and utilizing the potential of the local community.
- Combating unemployment (especially young people and women) by supporting employment and training the workforce.
- Restraint of demographic shrinkage of rural areas
- Rising living standards of the local population.
- Stimulation of "social life" in remote, neglected and "aged" areas.
- Ensuring long-term business profitability, vitality and competitiveness.
- Existence of cross-sectoral synergies, with diffusion of benefits in the local community.
- Support and utilization of local products and skills (Patikas, 2014).

For the case of Greece, local authorities with their new strengthened role and the establishment of Tourism Committees on each Municipality as part of the Kallikratis reform, can contribute to the selection and adoption of a specific model of tourism development and, in general, to the formulation of tourism policy, intervening decisively in the field of both demand (promotion program, tourist traffic, tourist protection, price monitoring, etc.), as well as the offer (financial framework, tourism development planning, employment / education issues, etc.). At the same time, the O.T.A. as "transmitters" promote specific events, processes and "information products" (articles, news website, workshops, "open" electronic forum, etc.) to influence the recipients (interest groups, local community) for positive interaction and active their participation in tourism development / policy.

From an academic perspective, institutional and business interest is now oriented towards sustainable tourism development, which promotes the balanced correlation between ecoprotection, economic efficiency and social justice (Patikas, 2014). The search for policies, in this context, combined with the devaluation of the model of mass tourism and " homogenization ", as well as the wrong, hasty and unrestrained development, requires the adoption of strategies more "mild" and "integrated" in the "bearing capacity" of the host places, so that the possibility of meeting the needs of future generations is not endangered (Andriotis, 2008).

The sustainability requires harmonization natural and human environment, high quality tourism production, optimum use of fragile environmental resources, the local tradition and cultural heritage, promote unemployment combat operations, control of energy consumption (to promote renewable forms of) and encourage synergies / innovative activities between local productive industries to feed / expand the development of a destination and ensure long-term business vitality (Kumar et al., 2021; Iliopoulou-Georgudaki et al, 2016; Santos et al., 2021a; 2021b; Valeri et al., 2021, Vasilliadis et al., 2017).

The dominant directions of sustainable tourism development policies focus on the following fields (Kokkosis et al., 2011):

- specialized design, aimed at balancing the environment, society and economy.
- strengthening development feedback processes.
- formation of a special institutional framework for the promotion of local participation.
- taking measures to promote and protect the local structured and natural environment.
- use of alternative and special forms of tourism.

Local authorities, by creating appropriate conditions for cooperation and utilizing modern management tools, can implement sustainable tourism development programs by defining procedures, rules and the participation of all involved in the local tourism circuit, while adopting the measurement of sustainability indicators., in order to prevent and control the exceeding of the bearing capacity and the environmental degradation of an area. Using this approach, the maintenance of the attractiveness of a destination is checked, as well as the possibility of further tourist development (Iliopoulou-Georgudaki et al, 2016).

The institution of local self-government is connected, essentially and directly, with the sustainable tourism development. The responsibilities of the Local Authorities are including, as mentioned above, the preparation, participation and implementation of programs to promote alternative forms of tourism in their area of responsibility, as a counterpart to the general automation, uncontrolled growth and intensive exploitation which causes rapid and aggressive massification (Mylonopoulos and Parthenis, 2007). Alternative tourism is defined as environmentally sensitive, human-centered, controlled scale, allowing for positive social interaction and experiential experience (Andriotis, 2008).

It should here be noted that while the marketing of sustainable tourism development may form the visitor a particular perception of the destination, current practices still make it sustainable form of tourism impossible as a "rule", taking into account and The fact is that although more and more tourists are looking for sustainable tourism "options" for altruistic reasons, few are acting consistently / "sustainably" during their holidays (Hanna et al, 2018).

Cooperation Between the Private Sector and Local Government in the Promotion of Tourist Destinations (Collaborative Marketing)

According to Middleton et al . (2009) it is widely accepted from the tourist industry that the public-private partnerships with the involvement of all stakeholders in the local government (with a balanced structure and clear roles for all stakeholders) will contribute to optimal management of tourism development prospects, achieving higher returns and cost containment, so that the destination is in a competitive strategic position. Synergies are a necessary tool in the modern competitive environment, as it ensures better information on changes in demand, timely information on market trends, joint modernization and marketing actions, etc. (Mylonopoulos and Kontoudaki, 2011; Valeri, 2021b; Valeri and Fadlon, 2021).

For a tourism destination, a marketing program shall include the involvement of local community, private and state enterprises / institutions is an urgent need. The private sector is often more informed about modern influencing techniques and market demand information, more determined and flexible than the public sector, which relies on resources and supply. Participatory and consensual processes are a prerequisite for the sustainability and medium-term stability of tourism development (Sarantakou & Terkenli, 2019).

The competitive advantage of the destination and its operational efficiency can be distinguished only when all the participants work closely together in the analysis of options, the conclusion of agreements, the merging of financial resources and intellectual techniques. The perceived value of each tourism experience is the result of many combined factors in multiple environments, where the coordination and interrelationships of different value chains, whether from the private or public sector, have common characteristics and must perform appropriately (Pallis et al, 2019)

The objectives of public-private partnerships concern the upgrading of the tourism product, the mitigation of seasonality, the improvement of the quality of the offered services, the promotion of additional advantages, the innovative, with global repercussions, extroversion actions, the approach of developing new markets, thematic tourism, the emphasis on the interconnection of tourism, culture and the primary sector and the emphasis on research, innovation and smart specialization.

The role of the public sector is characterized as coordinating, developmental (decision-making for the use of tourism resources), executive (infrastructure creation), promotional (promotion), consulting (business and investment support), audit (quality assurance supervision). On the contrary, the role of the private sector is mainly managerial (sale), promotional with small-scale actions and, to a lesser extent, executive with the creation of private superstructure / infrastructure projects (Pallis et al, 2019).

Wang and Krakover (2008) support the existence of business relationships in collaborative destination marketing relationships with all stakeholders be characterized by a "hybrid" behavior, including competition and cooperation, referred to as coopetition (cooperation and competition).

This approach is influenced by the strategic thinking, the maturity of the approach to the destination marketing, the promotion campaign and the leadership of the DMO.

All the tourist circuit carriers to a destination (public and private) can participate in what we call cooperative marketing (Middleton et al., 2009), selecting types of relationship. The latter range from "loose" cooperatives to more formal and comprehensive ones, of good faith and mutual support, which, of course, in the pursuit of individual, but also collective, benefits compete in the provision of services, coming into contact with customers, and cooperating in promotion and promotional activities, "away" from them. Competition usually focuses on value-added strategies, while collaboration focuses on collective value-generating strategies. In the tourism sector, there are four main types of business relationships: interaction, coordination, cooperation and strategic networks (Valeri, 2016; Valeri and Baggio, 2020a; 2020b; 2020c; 2021; Baggio and Valeri, 2020). It should be noted that, although these forms follow a practice of low to high formatting, integration and structural complexity, they are indicative of the description of cooperative relationships between tourism businesses of a destination (Wang and Krakover, 2008).

A destination coexists with political (government and local) actors, national and European funding agencies, bureaucratic procedures, competing businesses, special interest groups, media and host communities with the management / governance of a destination at the crossroads of all of them. cooperation between them presupposes: the modeling of collaborative marketing, social inclusion, intergovernmental cooperation, collaborative innovation and the implementation of altruistic collaborative strategies (Pike and Page, 2014).

According to Wang (2008), public sector-private sector partnerships in the tourism market are interactive, as they take place through a process of exchange of ideas, knowledge, expertise and pooling of capital and human resources, leading to new opportunities, innovative solutions and a higher level of efficiency. The implementation, of course, of a destination-level cooperation strategy depends on coordination, communication, mutual compromise and building consensus on issues that require open, honest discussion, continuous communication, patience and successful conflict management.

Collaborative marketing is not characterized, as mentioned above, as a successive, static and linear process, but, in order to be characterized as smooth and successful, it must be cyclical and dynamic. Unfortunately, the diverse and fragmented nature of tourism products is an obstacle to the recognition, acceptance and adoption of collaborative practices, but vital to the tourism industry, especially at the destination level, where markets are interconnected and product / service coordination is quite complex. . Moreover, the modern need for tourism development compatible with the principles of sustainability, is an idea, with a short-term and long-term perspective of change, in joint decision-making regarding the management of tourist destinations.

In conclusion, the public-private partnership strategy is now seen as the only way to minimize external threats and maximize common opportunities for collective development, as opposed to the outdated individual approach to future challenges. Moreover, recognizing the value of trust, commitment and mutual understanding facilitates and promotes business creation and long-term vitality (Wang, 2008).

Friedman and Miles (as referred to D ' Angelo & Go, 2009) created a matrix "mapping" of the relationships of all involved in tourism development in accordance with relevant contractual forms and strategic actions, integrating four cells with different levels of participation, and two variables: the set of ideas and interests, compatible or incompatible, and the nature of the connections, which can be either occasional or fixed / necessary. Due to the need to achieve "reconciliation of goals" between stakeholders, the model integrates both the spatial and temporal dimensions.

Over time, the need for collaborative tourism marketing activities becomes increasingly urgent, with the private sector and DMOs realizing the mutual benefit of this collaboration in terms of achieving strategic goals and overcoming potential risks, while increasing the degree of their mutual trust (Albattat et al., 2020; Chemli et al., 2020; Toanoglou et al., 2021; Valeri, 2021b).

Destination Governance and Interaction Between Public and Private Sector's Marketing Initiatives

In tourist destinations there is often a complex governance due to competing stakeholders who participate in the production of tourism product and services. One of the key factors is the relationship that exists between governments and societies as well as locals who often work in the private sector and who are involved in the decisions that are made, is to manage tourist destinations with care, as tourism in particular in the current period due to the coronavirus, appears fragmented and has a special need for collaborations.

The governance that takes place in each region is considered as a key tool when there is cooperation with the private sector to overcome significant conflicts that arise in the field of tourism, also having an effective operation between stakeholders to address the complexity due to of the internal and external environment of organizations, which affects the way the tourist destination itself operates (Joppe, 2018).

This creates an interaction between structures and processes that influences the way responsibilities are distributed and exercised but also the way decisions are made as all participants, whether citizens or the private sector in general, have an opinion on how with which the tourist destination is managed. Although there are still many difficulties, a series of management and planning procedures have been gradually implemented, being necessary for the creation of a sustainable destination in which the negative effects are always identified when the tourist phenomenon affects the community in a negative way.

On the other hand, regarding tourism has several benefits for the local community which day by day are maximized and unemployment is strengthened, as well as the income that exists in the region (de Bruyn & Alonso, 2012).

Deguignet et al., (2014) add that there are also some areas that need special care due to the protection that must be applied to natural resources and ecosystems, resulting in the range of stakeholders increasing even more when it comes to taking a decision that will incorporate the management of a protected area, which means that the responsibilities are shared even more when it comes to the way according to which the tourist destination itself is managed.

In addition, cooperation between the private and public sector prevents the destruction of habitats as well as the protection against pollution and contamination of the ecosystem, something that protects the already difficult governance that exists in a protected area.

APPROACHES TO DESTINATION GOVERNANCE WHICH BOOST MARKETING OF THE DESTINATIONS

Regarding the way of governing a destination, various approaches have been developed according to which significant efforts have been made by using empirical formulation of new data for the management of destinations based on the Community model and the corporate model. Volgger, Pechlaner & Pichler, (2017) state that one of the key elements for governing a destination is the power of a coordination mechanism that completes the dimension of cooperation, while the analysis of social networks is important, which significantly reflects the true image of a tourist destination (Baggio and Valeri, 2020; Valeri and Baggio, 2020a; 2020b; 2020c; Valeri and Baggio, 2021). The cooperation that exists between the private and the public sector takes into account to a large extent the social roles and the trust that exists between these two sectors as well as their control and balanced participation, which means that the result of the cooperation pays off an improved picture in the governance of a destination.

The concept of destination management can be considered as a descriptive concept with a holistic perspective which requires collective action to stimulate a tourist destination and which must include management strategies as well as marketing and planning strategies. However, it is very important to make political decisions using structures and procedures that will ensure a balance between the private and the public interest. However, destination management does not always necessarily claim that a destination should be a business process for the market. Understanding destination management is a problem of choice between the organizations involved in its coordination and which should take into account to a large extent the financial performance of this destination but also how it should be managed by the business world of the private sector but also by the public administration itself.

Today, for the governance of a destination, a systematic review of the academic literature has been done, highlighting the basic dimensions of this structure and which shows that tourism has a great need for cooperation between the private and public sectors. Some of the elements that should govern this cooperation are responsibility, transparency, participation and balance in the decisions of the boards even if this tourism government is addressed to different industries and fields of research. There should also be effectiveness regarding the decisions and efficiency among the employees who participate in the specific governance by applying good practices as tourism follows the trends of political science (Islam et al., 2018). Other elements that need to be underlined are the structure of a strategic vision and leadership which always plays an important role in the implementation of strategic decisions in organizations. Thus the destination government records the interaction that exists between the two sectors which are represented in the results which are recorded in a tourist destination.

Given that in the current era of economic crisis, the tourism industry is a sensitive industry which has been much more affected by this crisis than other sectors of the economy, many tourist destinations have been hit hard. Therefore, today much more attention needs to be paid to the consequences that exist in the implementation of strategic management when it is to be implemented in the tourism industry. In the marketing and management of tourist destinations there are two basic steps that must be implemented after a crisis. The first element is the understanding of the crisis itself and the second element is the implementation of a strategy that will follow the crisis and that will allow the revitalization of the region. According to Mair, Ritchie, and Walters (2016), tackling tourism problems requires significant action to take appropriate strategies to enable all stakeholders to participate in strengthening a tourist destination. This shows that tourism today in terms of problem solving has gained a great deal of attention in the last decade (Cakar, 2018) due to the fact that it is a sector that has been deeply affected by both the coronavirus and previous pandemics as well as severe economic events (Farzanegan et al., 2020).

Another element that governs the cooperation between the private and public sector, is the innovation that must be developed in a destination while at the same time the phases of reduction of the final life cycles of a tourist product but also of the destination itself must be avoided regarding how public and private interests can be balanced and a long-term perspective be assigned to a destination.

In terms of research that has been done to date on the management of tourist destinations, one should not underestimate the fact that tourist destinations are affected by major events and therefore should always be based on the marketing strategy applied in a destination to develop its future growth trajectory. Thus, functionality should be created even in the abstract dimensions that tourism has and these should be transformed into essential concepts that play a role for the destination itself.

Related research indicates that the private and public sector can also cooperate on other areas related with tourism. For example, it is Maingi (2019) who mentions that the local authorities can cooperate with the private sector so to develop and implement training programs for tourist professionals aiming on the leverage of their skills. In some other research Yüzbaşıoğlu & Doğan (2021) are remarking the fact that local authorities need the contribution and support of the private sector so to develop special tourist activities such as marinas developed for destinations. The local authorities often do not posses the appropriate knowledge but they have the necessary resources and capabilities. Hence, the private sector contributes with its knowledge while the public sector with the appropriate resources so to maximise the end result.

CASES OF SYNERGIES BETWEEN LOCAL GOVERNMENT AND THE TOURIST SECTOR

The Case of the Cooperation Between Local Authorities and Airlines So to Establish Effective Distribution Channels

According to Papatheodorou et al., (2019), a typical case of cooperation is the case of airports with airlines as the aviation sector in recent years has undergone major changes to facilitate the creation of new transport packages that will cover all the needs of stakeholders. Greece is a country that has accepted the liberalization of transport since 1998, being one of the most important members of European airlines subject to the aviation market. Especially in 2017, when most countries were liberalized in the European Union in terms of their aviation, it was a milestone year as the European Union gave the third package of measures to support the Member States, which means that the market became accessible to all airline companies with a view to replacing national airlines with Community airlines by then. These changes that occurred in European aviation significantly affected tourism and created new partnerships in both Europe and Greece. Especially for Greece, as reported by Farmaki and Papatheodorou (2015), given that as a country suffered a severe economic crisis that affected the market, the liberalization of aviation strengthened international tourism and state revenues, creating both directly and indirectly new business opportunities but also investments in the tourism sector. Although the environment of the airlines connected to Greece is a free market, it is a market that performs and is sustainable with 14 regional airports that operate on a daily basis.

Thus, this liberalization contributed to the creation of a sustainable tourism which significantly enhanced the interest of all stakeholders. In fact, in terms of destination facilities, tourism management allows for discussions and agreements on strategic issues, especially when they involve destination management organizations that enable public-private partnerships (PPPs) to express and represent all stakeholders which are in fact reinforced by the legal framework governing this cooperation (Borenstein and Rose 2014).

In the case of airports and airlines, the good relationship that needs to exist between the transport company and the airport is a given as a bad cooperation can lead the airline to relocate to another airport and this can ultimately affect largely the entrepreneurship of local destinations. According to Nikolopoulos (2013) in this cooperation, the role of low cost companies such as Ryanair is important, which due to the fees imposed on air transport, and given its relatively low profit, has a strong bargaining power with airports and destinations she chooses even if the destination is not yet known. At this point, according to Ballis et al., (2018) at Kos, the island's authorities decide to negotiate with Ryan Air so to provide incentives including zero municipality taxes for the company's airplanes and other allowances in exchange that the airliner would promote Kos. Indeed, this partnership brought a huge benefit for the island. Ryan Air pushed the island's tourism from UK and helped Kos to increase by far its arrivals from the UK. This is an example how the local government can cooperated with the airliners and the local tourist community so to create policies which will benefit everyone involved with the tourist industry.

The Case of the Cooperation of the Public Sector with the Short-term Rental (Airbnb)

Another case of cooperation between the public and the private sector is the case of Airbnb which, given the impact it has had on the market over the last 10 years, is often cited by the research community as an innovation in the places where it operates (Guttentag & Smith, 2017). Airbnb, on the other hand, influences the local rental market, which means that a complex relationship in the rental market is created between the Airbnb rental process and the local landlords, which leads to the creation of a specialized legal framework that needs special attention. However, the research community relies heavily on Airbnb because on the one hand it is a platform that allows to generate a significant income and on the other hand it is a platform that is easy to use and enables high visitor attraction.

Other elements involved in this cooperation between the private and public sectors are the strengthening of employment and the creation of job opportunities, as well as the increased number of visits to secondary destinations that were not known until now and which have become accessible due to the possibility of renting through Airbnb, as a result of which they are chosen in many cases by tourists. However, it is a platform that needs specific policies to balance the problems that arise with the rest of the rental market, especially in urban centers (Ioannides et al., 2018). It seems that in addition to the fact of cooperation between the private and the public sector to create a new innovation in the form of rental housing, a form of balance must be found that will allow local markets to work together for the benefit of their visitors. and be able to create partnerships that overall will increase downtown traffic (Wachsmuth and Weisler, 2018). In this way, Airbnb will be able to be a business opportunity that will enable the best and highest quality apartment rental in areas that already have satisfactory traffic but also in areas that are now emerging as tourist destinations.

Destination Governance on the Post-COVID – 19 era

One such dimension that allows tourism at this time to develop after the coronavirus period, is the fact that there has been a cooperation between the private and public sectors in passenger transport in order for these transports to be sustainable and at the same time to protect the public in maximum possible degree. To do this, the private and public sectors are now cooperating in border control, the means of transport available that can be used to transport passengers, the restrictions that exist during a trip but also the rules of public health and such as applying masks and keeping distances to prevent transmission of the virus. These measures, which have resulted from the cooperation of the private and public sectors, allow the reopening of the borders for traveling and the possibility of the traveler to control the destinations that are available to travel, through the internet search and the of mobile applications that allow the visitor to search for all the necessary details using an interactive map created by private entities to know the individual measures that must be implemented in each country as a basic guide for visitors this country (Sheresheva, 2020).

Results – findings

As shown by the present study, cooperation between the private and public sector for the management of destinations is an important element for tourism development, especially when companies such as the case of Airbnb further expanded, as well as low-cost companies of air transport. It is important, however, to understand the needs of local businesses so that further action can be taken by the public sector as well as the needs of the local community itself in order to have further interest from those involved in local entrepreneurship.

As it turned out, destination governance is something that in Greece in the previous years was affected by the climate that prevailed due to the economic crisis, however, from 2017 onwards, actions have been taken by Europe so that even the low cost airlines support smaller and even more well known destinations. An important fact in this process is that many reforms, such as the reforms of Kallikrates, have facilitated many bureaucratic procedures and strengthened the existing tourism policy.

However, what needs to be said is the fact that the Greek government needs to pay more attention to tourism development and coordination between the private and public sector, in order to strengthen entrepreneurship but also to evolve in the modern era especially when it comes to the public sector. Thus, there should be such tourism policies which on the one hand are of a functional nature and on the other hand are planned so that by applying the appropriate marketing methods and strategies, they can enhance the business environment especially today that the coronavirus crisis has significantly affected tourism which in Greece, although it operated during the summer, it was significantly affected by several coronavirus cases. Therefore, there should be coordinated policies and such funds that will allow further cooperation with the private sector so that with the appropriate infrastructure and security that will be provided to the visitor, the tourism product would become competitive again.

FUTURE RESEARCH DIRECTIONS

It is a fact that to date, the public sector to some extent facilitates private entrepreneurship and the promotion of tourism. However, one problem that currently exists in the conduct of this cooperation is the fact that the public sector is characterized by bureaucracy. This bureaucracy often does not facilitate or significantly delays the cooperation that must take place in order to boost tourism, on the one hand, and to make the appropriate decisions in order to boost entrepreneurship. Thus, it is proposed to create smaller councils that will make decisions in the public sector in cooperation with the private sector, for initiatives that will strengthen entrepreneurship and tourism. It should be noted that there is already some advice to local authorities to facilitate the work of private companies, as shown by the cooperation between airports and air transport companies, however, this movement should be further extended to other areas of tourism in order for Tourists to visit Greece come from America and Europe but also from Asia, which is an emerging market for Greek tourism, so that eventually the Greek tourism product will be strengthened to a greater extent.

CONCLUSION

The paper idicates that an effective, efficient and sustainable tourism development policy with simultaneous care to address weaknesses and threats is critical to establishing a strong competitive position in the international tourism market. In order to implement it, it is necessary to have a more active partnership between the public and the private sector, with an emphasis on collaborative marketing practices. It is very important to notice how the recent changes on the way tha local governance operates in Greece, is an indication that there is a need to have a better understanding on how the local government contributes in Greece' marketing practices and offers. This is a literature review which means that the data used is limited only to what other researchers have to say. The lack of empirical evidence constitutes are key limitation.

REFERENCES

Albattat, A., Jamaludin, A., Zuraimi, N. S. M., & Valeri, M. (2020). Visit Intention and Destination Image in Post- Covid- 19 Crisis Recovery. *Current Issues in Tourism*. Advance online publication. doi :10.1080/13683500.2020.1842342

Albet, A. (2016). *Urban governance in southern Europe*. Routledge. doi:10.4324/9781315548852

Baggio, R., & Valeri, M. (2020). Network science and sustainable performance of family businesses in tourism. *Journal of Family Business Management*. doi:10.1108/JFBM-06-2020-0048

Ballis, A., Tsouka, D., Moschovou, T., & Kasselouris, G. (2018). The impact of airport development on tourism in the Greek islands of the South Aegean Sea. *Journal of Air Transport Studies*, *9*(2), 95–109. doi:10.38008/jats.v9i2.25

Belias, D., Vasiliadis, L., & Velissariou, E. (2020). Internal Marketing in Tourism: The Case of Human Resource Empowerment on Greek Hotels. *Cultural and Tourism Innovation in the Digital Era*, 559-573.

Belias, D., Vasiliadis, L., & Zaftis, E. (2020). The Impact of Cultural Routes on Traditional Settlements: The Case of Greece. In *Cultural and Tourism Innovation in the Digital Era* (pp. 123–134). Springer. doi:10.1007/978-3-030-36342-0_10

Belias, D., Velissariou, E., Kyriakou, D., Vasiliadis, L., Mantas, C., Sdrolias, L., Aspridis, G., & Kouste-lios, A. (2017). Customer Relationship Management and Social Media in Greek Tourism. In *International Conference on Strategic Innovative Marketing* (pp. 59-65). Springer.

Borenstein, S., & Rose, N. L. (2014). How airline markets work... or do they? Regulatory reform in the airline industry. In *Economic regulation and its reform: What have we learned?* (pp. 63–135). University of Chicago Press.

Cakar, K. (2018). Critical success factors for tourist destination governance in times of crisis: A case study of Antalya, Turkey. *Journal of Travel & Tourism Marketing*, *35*(6), 786–802. doi:10.1080/10548408.2017.1421495

Chemli, S., Toanoglou, M., & Valeri, M. (2020). The impact of Covid-19 media coverage on tourist's awareness for future traveling. *Current Issues in Tourism*, 1–8. Advance online publication. doi:10.1080/13683500.2020.1846502

d'Angella, F., & Go, F. M. (2009). Tale of two cities' collaborative tourism marketing: Towards a theory of destination stakeholder assessment. *Tourism Management*, *30*(3), 429–440. doi:10.1016/j.tourman.2008.07.012

de Bruyn, C., & Alonso, A. F. (2012). Tourism destination governance: Guidelines for implementation. In E. FayosSola, J. A. M. da Silva, & J. Jafari (Eds.), Knowledge management in tourism: Policy and governance applications (pp. 221–242). Bingley: Emerald Group.

Deguignet, M., Juffe-Bignoli, D., Harrison, J., MacSharry, B., Burgess, N., & Kingston, N. (2014). *2014 United Nations list of protected areas*. Retrieved from https://wdpa.s3.amazonaws.com/WPC2014/2014_UN_LIST_REPORT_EN.pdf

Farmaki, A., & Papatheodorou, A. (2015). Stakeholder perceptions of the role of low-cost carriers in insular tourism destinations: The case of Cyprus. *Tourism Planning & Development*, *12*(4), 412–432. doi:10.1080/21568316.2015.1013566

Farzanegan, M. R., Gholipour, H. F., Feizi, M., Nunkoo, R., & Andargoli, A. E. (2020). International Tourism and Outbreak of Coronavirus (COVID-19): A Cross-Country Analysis. *Journal of Travel Research*.

Guttentag, D. A., & Smith, S. L. (2017). Assessing Airbnb as a disruptive innovation relative to hotels: Substitution and comparative performance expectations. *International Journal of Hospitality Management*, *64*, 1–10. doi:10.1016/j.ijhm.2017.02.003

Hanna, P., Font, X., Scarles, C., Weeden, C., & Harrison, C. (2018). Tourist destination marketing: From sustainability myopia to memorable experiences. *Journal of Destination Marketing & Management*, *9*, 36–43. doi:10.1016/j.jdmm.2017.10.002

Iliopoulou-Georgudaki, J., Kalogeras, A., Konstantinopoulos, P., & Theodoropoulos, C. (2016). Sustainable tourism management and development of a Greek coastal municipality. *International Journal of Sustainable Development and World Ecology*, *23*(2), 143–153. doi:10.1080/13504509.2015.1102780

Ioannides, D., Röslmaier, M., & van der Zee, E. (2018). Airbnb as an instigator of 'tourism bubble' expansion in Utrecht's Lombok neighbourhood. *Tourism Geographies*, 1–19. doi:10.1080/14616688.2017.1409261

Islam, M. W., Ruhanen, L., & Ritchie, B. W. (2018). Adaptive co-management: A novel approach to tourism destination governance? *Journal of Hospitality and Tourism Management*, *37*, 97–106. doi:10.1016/j.jhtm.2017.10.009

Joppe, M. (2018). Tourism policy and governance: Quo vadis? *Tourism Management Perspectives*, *25*, 201–204. doi:10.1016/j.tmp.2017.11.011

Katsios, S. (2019). Tourism in 'Yellow Times': The De-formalisation of the Greek Economy and Its Impact on Tourism. In *Yellow Tourism* (pp. 209–225). Springer. doi:10.1007/978-3-319-94664-1_13

Lopes, V., Moreno Pires, S., & Costa, R. (2020). A strategy for a sustainable tourism development of the Greek Island of Chios. *Tourism: An International Interdisciplinary Journal*, *68*(3), 243–260. doi:10.37741/t.68.3.1

Maguire, K. (2019). Examining the power role of Local Authorities in planning for socio-economic event impacts. *Local Economy*, *34*(7), 657–679. doi:10.1177/0269094219889603

Maingi, S. W. (2019). *Sustainable tourism certification, local governance and management in dealing with overtourism in East Africa.* Worldwide Hospitality and Tourism Themes. Retrieved from https://www.emerald.com/insight/content/doi/10.1108/WHATT-06-2019-0034/full/html?casa_token=2HgIS_T2wVQAAAAA:hRTOgr1K-EapXp4i1ADRRImJ6dkZThwvUQgYHnLBWFUbs7NKRWa-NJw6gIZ-Abn8gZCdZVdVZqwrbQnFeGZGzMzgetMZsNziu_89rRSzoTQkci98TuTR-

Mair, J., Ritchie, B. W., & Walters, G. (2016). Towards a research agenda for post-disaster and post-crisis recovery strategies for tourist destinations: A narrative review. *Current Issues in Tourism, 19*(1), 1–26. doi:10.1080/13683500.2014.932758

Majewska, J. (2008). Local Governments' Involvement in the Development of Entrepreneurship in Emerging Tourism Destinations. *Zeszyty Naukowe/Akademia Ekonomiczna w Poznaniu*, (99), 127-145.

Middleton, V., Fyall, A., Morgan, M., & Ranchhod, A. (2009). *Marketing in travel and tourism.* Routledge Taylor and Francis Group.

Nikolopoulos, A. (2013). *Alone against all: The negotiation strategy.* Patakis Editions.

Pallis, C., Prasinos, A., & Pallis, P. (2017). The Relation Between Tourism, Local Development and Insularity: Case Study of Southern Aegean, Greece. *Journal of Tourism and Hospitality Management, 5*(6), 223–232.

Papatheodorou, A., Vlassi, E., Gaki, D., Papadopoulou-Kelidou, L., Efthymiou, M., Pappas, D., & Paraschi, P. (2019). The airline–airport–destination authority relationship: the case of Greece. In *Tourist Destination Management* (pp. 27–41). Springer. doi:10.1007/978-3-030-16981-7_3

Patikas, C. (2014), *Participatory destination management and creative tourism: From co-production of tourism products to co-creation of experiences. A reality and stakeholders' check for Thessaloniki, Greece* (Master's thesis). School of Economics, Business Administration & Legal Studies, International Hellenic University, Thessaloniki, Greece.

Pike, S., & Page, S. J. (2014). Destination Marketing Organizations and destination marketing: A narrative analysis of the literature. *Tourism Management, 41*, 202–227. doi:10.1016/j.tourman.2013.09.009

Rasoolimanesh, S. M., Ringle, C. M., Jaafar, M., & Ramayah, T. (2017). Urban vs. rural destinations: Residents' perceptions, community participation and support for tourism development. *Tourism Management, 60*, 147–158. doi:10.1016/j.tourman.2016.11.019

Ritchie, J. R. B., & Crouch, G. I. (2003). *The Competitive Destination: A Sustainable Tourism Perspective.* CABI Publishing. doi:10.1079/9780851996646.0000

Sarantakou, E., & Terkenli, T. S. (2019). Non-institutionalized forms of tourism accommodation and overtourism impacts on the landscape: The case of Santorini, Greece. *Tourism Planning & Development, 16*(4), 411–433. doi:10.1080/21568316.2019.1569122

Sheresheva, M. Y. (2020). Coronavirus and tourism. *Population and Economics, 4*(2), 72–76. doi:10.3897/popecon.4.e53574

Toanoglou, M., Chemli, S., & Valeri, M. (2021). The organizational impact of COVID-19 crisis on travel perceived risk across four continents. *Journal of Organizational Change Management*. Advance online publication. doi:10.1108/JOCM-12-2020-0369

Trivellas, P., Kakkos, N., Vasiliadis, L., & Belias, D. (2016). Residents' perceptions toward cultural, social and economic benefits and costs of tourism industry: An empirical survey. In *Tourism and Culture in the Age of Innovation* (pp. 171–181). Springer. doi:10.1007/978-3-319-27528-4_12

Trivellas, P., Kakkos, N., Vasiliadis, L., & Belias, D. (2017). Sustainability, social marketing and host attitudes about Tourism in the city of Larissa. In *Strategic Innovative Marketing* (pp. 89–95). Springer. doi:10.1007/978-3-319-33865-1_11

Tsekos, T. N., & Triantafyllopoulou, A. (2016). From municipal socialism to the sovereign debt crisis: Local services in Greece 1980–2015. In *Public and Social Services in Europe* (pp. 135–149). Palgrave Macmillan. doi:10.1057/978-1-137-57499-2_10

Tsiotas, D., Belias, D., & Niavis, S. (2020). Internal Marketing on the Tourism Sector: Examination of How the Staff Can Develop on Greek Hotels. *Strategic Innovative Marketing and Tourism*, 963-970.

Valeri, M. (2016). Networking and cooperation practices in the Italian tourism business. *Journal of Tourism, Heritage & Services Marketing*, *2*(1), 30–35. doi:10.5281/zenodo.376333

Valeri, M. (Ed.). (2021a). Tourism risk. Crisis and recovery management. Emerald Publishing.

Valeri, M. (2021b). *Organization Theories. Implications for the Strategic Management*. Springe.

Valeri, M., & Baggio, R. (2020a). Social network analysis: Organizational implications in tourism management. *The International Journal of Organizational Analysis*. Advance online publication. doi:10.1108/IJOA-12-2019-1971

Valeri, M., & Baggio, R. (2020b). Italian tourism intermediaries: A social network analysis exploration. *Current Issues in Tourism*, *24*(9), 1270–1283. doi:10.1080/13683500.2020.1777950

Valeri, M., & Baggio, R. (2020c). A critical reflection on the adoption of blockchain in tourism. *Journal Information Technology and Tourism*, *23*(2), 121–132. doi:10.100740558-020-00183-1

Valeri, M., & Baggio, R. (2021). Increasing the efficiency of knowledge transfer in an Italian tourism system: A network approach. *Current Issues in Tourism*, 1–16. Advance online publication. doi:10.1080/13683500.2021.1937960

Valeri, M., & Fadlon, L. (2018). Co-evolutionary prospects in tourism. *International Journal of Business and Management*, *13*(9), 95–105. doi:10.5539/ijbm.v13n9p95

Valeri, M., & Katsoni, V. (Eds.). (2021), Gender and tourism: challenges and entrepreneurial opportunities. Emerald Publishing. doi:10.1108/9781801173223

Vasiliadis, L., Trivellas, P., Belias, D., Meleas, J., Kyriakou, D., & Koustelios, A. (2016). Cultural tourism revisited: The case of Thessaly. In *Tourism and culture in the age of innovation* (pp. 69–78). Springer. doi:10.1007/978-3-319-27528-4_4

Vassakis, K., Petrakis, E., Kopanakis, I., Makridis, J., & Mastorakis, G. (2019). Location-based social network data for tourism destinations. In *Big data and innovation in tourism, travel, and hospitality* (pp. 105–114). Springer. doi:10.1007/978-981-13-6339-9_7

Volgger, M., Pechlaner, H., & Pichler, S. (2017). *The practice of destination governance: A comparative analysis of key dimensions and underlying concepts*. Academic Press.

Wachsmuth, D., & Weisler, A. (2018). Airbnb and the rent gap: Gentrification through the sharing economy. *Environment and Planning A. Economy and Space*, *50*(6), 1147–1170.

Wang, Y. (2008). Collaborative destination marketing: Understanding the dynamic process. *Journal of Travel Research*, *47*(2), 151–166. doi:10.1177/0047287508321194

Wang, Y., & Krakover, S. (2008). Destination marketing: Competition, cooperation or coopetition? *International Journal of Contemporary Hospitality Management*, *20*(2), 126–141. doi:10.1108/09596110810852122

World Travel & Tourism Council. (n.d.). *Travel & Tourism Economic Impact 2018-Greece*. Retrieved from https://www.wttc.org/-/media/files/reports/economic-impact-research/countries-2018/greece2018.pdf

Yüzbaşıoğlu, N., & Doğan, O. (2021). Strategic governance in the marina sector in the context of marine tourism cluster. *Almatourism-Journal of Tourism, Culture and Territorial Development*, *12*(23), 59–82.

Chapter 2
A Network-Based View on Extreme Expedition Team Composition in Alpine Tourism

Hendrik Leendert Aalbers
Radboud University, The Netherlands

Ramón Spildooren
Radboud Centre for Organization Restructuring, The Netherlands

ABSTRACT

Conceptually drawing on network theory as its theoretical lens, this study examines two prime notions of network configuration of commercial expeditions. Exploring the role of both structural holes and network closure as indicators of team configuration for those venturing out in such extreme adventure, this study clarifies the impact of social structures, network closure, and structural holes in particular on performance outcomes in the context of expedition mountaineering. Presence and bridging of structural holes did turn out to be a significant predictor for the success or failure of an expedition. The findings show network closure to significantly influence the performance of mountaineering teams that make for a successful ascent. The capacity to span structural holes, commonly portrayed as serving as an eye-opener for options otherwise not found, does not appear to assist teams that make for successful ascents, however.

INTRODUCTION

Tourist policy on mountaineering has turned towards regulating and restricting access to particular climbing routes, establishing policy to the operational side of touristic expedition mountaineering. Little is known however, in terms of academically infused policy advice, on how tourists can best pair up as they venture into extreme sport adventure, as part of a bustling, yet frequently overlooked branch of extreme sports tourism (Sharma, 2012; Neupane et al. 2021; Jones et al. 2021). Tourist governance however holds direct policy derived influence on the safety of alpine tourism, regulating and restricting access

DOI: 10.4018/978-1-6684-3889-3.ch002

to particular climbing routes, establishing policy to the operational side of touristicical activity (Lai et al 2016). This chapter hence takes on a network based view as it explores an aray of network indicators as input for such alpine tourism policy, revolving around the following research question: What is the role of both structural holes and network closure as indicators of team configuration, on performance outcomes in the context of alpine expedition mountaineering safety?

With eight of the world's highest mountains in the country, Nepal is a mountaineer's paradise that in the last decades has cranked out a considerable extreme sports tourism business. The ultimate glory of mountaineering is reaching the summit of Mt. Everest, but there are numerous technical challenges l on any of the other 326 climbing peaks in the the Nepal Himalaya. The touristic draw of one of the most formidable mountain ranges in the world has not only propelled commercial expedition initiatives that gear up, co-climb and advice the sporty tourists that attempt climb historical lines set out in the Annapurna range – scaled in 1950 for the first time by a French Expedition with Maurice Herzog as leader or at Everest – first summitted by Edmund Hillary and Tenzing Norgay in 1953 with a Bristish Expedition led by John Hunt.

Yet, despite the wide draw to scale the tallest mountains on the planet, not all of these touristical climbs – often truly extreme sports endeavors – are without risks. The term for this area of tourism is commercial expedition mountaineering. Tourist policy in general, and hence also in mountaineering has tuned towards regulating and restricting access to particular climbing routes, establishing policy to the operational side of touristicical activity (Lai et al 2016). Recently, work on govenance and management of tourist destination has been callig in particular to further insights into the safety aspects of sustainable tourism policy in particular (Ritchie and Jiang, 2019; Racuh et al. 20121), part of an ongoing stream of literature in tourism governance that has highligted safety as fundamental for ongoing touristic activty in the long run (Peters and Pikkemnaat, 2006; Hall, 2019). In this particular case of expedition mountaineering, such policy, in practice, revolves around advising climbers on how to responsibly climb, for instance translating into govwernmental or commercial agencies that can help the sporty tourist navigate these areas (Albattat et al. 2021; Toanoglou et al. 2021). Little is known however, in terms of academically infused policy advice on how tourists can best pair up as they consider teaming up with others, potentially steering tourists and tourist agency climbing specialists alike in their commercial climbing attempts in a safe and resposible manner (Volgger & Pechlaner, 2014).

Govenance and Management of Alpine Tourism: A Network Theory Lens

Inserting a network based view to the team configuration of commercial climbing expeditions this chapter clarifies the impact of social structures, network closure and structural holes in particular, on performance outcomes in the context of expedition mountaineering (Aalbers and Dolfsma, 2017; Aalbers, 2020). As such, this chapter revolves around the role of mountaineering team configuration as predictive of alpine expedition team succes. Drawing on network theory as its theoretical lens, and leveraging on a rich and longitudinal empirical dataset that of Himalayan mountaineering teams, we examine two prime notions for mountaineering team configuration choice. Exploring the role of both structural holes and network closure as indicators of team configuration, this chapter clarifies the impact of social structures, network closure and structural holes in particular, on performance outcomes in the context of expedition mountaineering, suggesting one more relevant than the other. We utilize a longitudinal empirical dataset that of Himalayan mountaineering teams, and examine the network configuration choices of 90 climbing teams in relation to their historical network embeddedness in the broader Himalayan climbing community.

The Himalayan Database Database (Salisbury & Hawley, 2004) this chapter draws on comprises data about all the mountaineering teams that went on an expedition in the Himalaya mountain range providing. The context offers a great setting to test the team network considerations undergirding high performance sporting teams in a context where low performance is certainly undersirable, with team error potentially being life threatening. Drawing on prior advancements from the field of organization network research, we postulate that it is the way that the team is organized that determines how well it can cope with these uncertainties (Bechky & Okhuysen, 2011). Specifically we postulate tand test the moderating effect of team network composition on team performance outcomes, in this case the succesful ascend and descend of some of the worlds most epic mountains.

Specifically, we analyzed the network configuration of 90 climbing teams in relation to their historical network embeddedness in the broader Himalayan climbing community. Conceptually drawing on network theory as its theoretical lens, we examine two prime notions of network configuration of commercial expeditions. Exploring the role of both structural holes and network closure as indicators of team configuration for those venturing out in such extreme adventure, this chapter hence clarifies the impact of social structures, network closure and structural holes in particular, on performance outcomes in the context of expedition mountaineering. Presence and bridging of structural holes did turn out to be a significant predictor for the success or failure of an expedition. Our findings show network closure to significantly influence the performance of mountaineering teams that make for a successful ascent. The capacity to span structural holes, commonly portrayed as serving as an eye-opener for options otherwise not found, does not appear to assist teams that make for a successful ascents, however. The focus on network closure ties in with a rich literature on network routines and the ensuing insights help to guide responsible extreme expedition mountaineering as part of an expanding alpine tourism industry (Neupane et al. 2021). Consolidating on these findings, this chapter closes with a number of managerial implications for those involved in the programing of future mountaineering expeditions and those faced with team compositioning choices as part of extreme sport management in more general terms.

Conceptual Framework and Hypotheses Development

Extending on extant sport science social network analysis techniques to evaluate overall team behavior (Clemente & Martins, 2017; Hambrick et al., 2018; Yan et al., 2018), this chapter defines the relations between individual actors – athlete mountaineers in this context - in terms of past task-related experiences. Within the context of mountaineering, this involves the joint participation of individual actors in previous expeditions. Zaheer and Soda (2009) have done similar analysis on the basis of past task-related experiences, and in the context of temporary teams. Ties that are repeated over time may become stronger, it nurtures and cements the relations (Aalbers, Koppius and Dolfsma, 2006; Aalbers and Dolfsma, 2015; Rueger et al. 2020). Structures typified by high levels of internal cohesion can generate norms, trust, obligations, and reciprocity (Zaheer & Soda, 2009). Prior experience with others has being related to improved team performance, also in a professional sports context, although too long of a shared past can be too much of a good thing - as empirical work on NBA players' team experience vividly points out (Berman et al., 2002). Within the specific context of mountaineering in the Himalaya, insights on dormant ties tell us a story of network embeddedness as driver for future team performance (Levin, Walter, & Murninghan, 2011). Dormant ties are relationships between two actors that have not communicated with each other for a longer period of time (Aalbers, 2012). Within the Himalayan mountaineering context, dormant ties are inevitable, as mountaineering is often a part-time hobby and actors only take part in an expedition every few years. Levin et al. (2011) demonstrate that dormant relationships can be a valuable source of social capital. They found that the reconnection with dormant relations can lead to the access of novel knowledge and insights, shared-perspective benefits, and preserved much of the strength it had before. Based on these arguments, it is relevant to define the relations between mountaineers based on past shared experience. Mountaineers reconnecting with others' they have shared previous experiences with can benefit from this relation. Exploring the role of both structural holes and network closure as indicators of expedition team configuration, this chapter zooms in on the impact of social structures, network closure and structural holes in particular, on performance outcomes in the context of expedition mountaineering.

Network Closure and Team Performance Outcomes

Network closure refers to a dense and interconnected set of relations between a network of individuals (Coleman, 1988). A network is closed when everyone is connected in such a fashion that no one can escape the notice of others (Burt, 2000). Network closure is beneficial to the speed of access to information and facilitates sanctions in case of opportunistic behavior at the individual level that make it less risky to trust others in the network (Coleman, 1988; Chemli et al., 2020; Valeri, 2021). As a prime indicator of network structure, network closure is believed to exert direct effect on performance outcomes (Aalbers and Dolfsma 2019). As actors in a highly closed network start to develop routines, stay away from opportunistic and selfish behavior, and create trust and shared meanings, these factors reinforce each other and "have a synergistic, positive effect on performance" (Soda et al., 2004, p. 896). Especially Coleman (1988) has contributed to this point of view, working from the notion that actors benefit more from a densely connected network as they are better able to rely on norms and sanctions against opportunism. The common norms that result from a dense network improve mutual understanding and thereby lower the chances of misunderstanding and misinterpretation of actions of individuals part of the network. This results in a reduction of mutually destructive competitive practices which harm the performance of the network (Zaheer & Bell, 2005). In a sports context, and here under condition of extreme sports in particular, routine based responses and their facilitating network structures at the team level, have been proposed as foundational to team performance (Warner al. 2012). Relatedly, and in a football performance context, Pena and Touchette (2012) note certain cliques within football teams to interact more frequently among themselves than others, suggesting the density of their relations to drive their team performance. Hence we pose the following:

Hypothesis 1: Network closure in a team's climbing network (being the accumulated joint network of all climbers within that team) has a positive effect on the success of the expedition, in terms of the safety of its ascent.

Structural Holes

Structural holes are bridged "when the focal actor (or "ego") is tied to others ("alters") who are not themselves connected" (Zaheer & Soda, 2009, p. 2), thereby connecting these otherwise disconnected individuals and possibly add value to the network in doing so. Research has also shown that personal benefits may arise when filling in this bridging position within the network (Balkundi, Kilduff, Barsness, & Michael, 2007). Contradictory outcomes have resulted, with one camp opting for a network *with* structural holes (e.g. Balkundi et al., 2007) and the other camp suggesting that network closure, rather than structural holes, lead to better performance outcomes for teams (e.g. Coleman, 1988).

Over time, the benefits of structural holes can be destroyed as they are typically short-lived because of the rate at which the information and opportunities decline in value (Burt, 2000). There is a reasonable chance that, even when other conditions remain ceteris paribus, other individuals in the network will gain access to the once-unique information (Soda et al., 2004). Burt (2002) also argues that bridged ties are more expensive to maintain than other ties, that they are more difficult to sustain, and, as a consequence, decay faster. But unlike the typically assumed empirical setting of companies and their competition, the performance outcomes of Himalayan expeditions are not dependent on the success or failure of others. Diffusion of unique information over the whole network of climbers, information that was once a result of a bridged structural hole, will rather lead to an increased total ascent rate of a peak as more and more climbers become familiar with the potential hazards that a peak might conceal. Therefore, the assumption that the benefits of structural holes reduce over time might not uphold perpetually.

In this empirical setting, theory suggests that structural holes lead to positive performance consequences. Higgs, Plewnia and Ploch (2005) found that teams rich in diversity are positively related to performance for complex tasks and negatively related for straightforward tasks. This is backed by Rodan and Galunic (2003), who further explain that especially the heterogeneity of knowledge amongst the team members help to overcome multifaceted and complex tasks. Hence, teams operating in the high pressure context of Himalayan mountaineering can benefit from the combined, diverse information and experience of the various team members, as it is one of the most dangerous and difficult sports on earth (Ortner, 1999). The whole team's performance may be enhanced due to the diverse ideas and skills that stem from previous expeditions that the members have been part of. As time passes, these benefits remain as the diminishing effect particularly found in a business context typified by competing forces (Soda, Usai, & Zaheer, 2004), does not uphold in a setting where another teams actions are not directly counterproductive for the success of someone else (Whelan et al., 2011). Based on the previous, it is expected that sports teams operating in a high-pressure context, such as in Himalayan mountaineering, that possess structural holes perform better than teams that do not. Teams that bridge these structural holes have (1) the advantage of accessing and combining diverse knowledge and resources which in turn leads to better team performance for the complex task they face; and (2) because expedition teams do not directly compete with each other in a fashion that they try to hamper one's success, the diminishing effect of structural holes does not uphold. This leads us to suggest the following:

Hypothesis 2a: The presence of structural holes in a team's climbing network (being the accumulated joint network of all climbers within that team) has a positive effect on the success of the expedition, in terms of the safety of its ascent..

Hypothesis 2b: The positive relation between structural holes in a team's climbing network and expedition success, in terms of the safety of its ascent and descent, is moderated by past internal network closure.

Empirical Setting and Methods

Empirical Setting

The Himalayan mountaineering context provides the empirical setting to answer the research question. This research makes use of expedition details gathered in the Himalaya from 1905 until 2016. The used database, the Himalayan Database (Salisbury & Hawley, 2004), is an extensive electronic compilation of Elizabeth Hawley's expedition notes that she recorded as a journalist while based in Kathmandu, Nepal. Elizabeth Hawley has been the undisputed chronicler of Himalayan expeditions for decades now and mountaineers wanting to gain worldwide recognition will first need her approval. Or as others say: "One of her biggest contributions is keeping the mountaineers honest" (Jolly, 2010).

The mountain range is without doubt the most impressive set of peaks on this planet. Spanning across five different countries, it contains over a thousand significant peaks, more than a hundred that are over 7,200 meters high, and eight of the world's 14 "eight-thousanders". The most well-known peak is the Mount Everest, the tallest mountain on this planet with an elevation of 8,848 meters. Apart from the tallest mountain, the Himalaya mountain range also encompasses some of the most deadly peaks in the world, with the Annapurna leading with a 38 per cent fatality ratio (Telegraph, 2013). These facts make the Himalaya mountain range the most popular destination for experienced climbers. Our dataset, referred to as 'The Himalayan Database' is especially useful as it contains details on almost every single expedition in the Himalaya, created by Salisbury and Hawley (2004), is used to test the hypotheses. The database is based on expedition archives of Elizabeth Hawley, who is a renowned journalist based in Nepal. The database is further extended by information gathered from books, alpine journals, and interviews conducted with Himalayan mountaineers. In total, the data covers all expeditions from 1905-2016 to more than 300 Nepalese peaks. UCINET (Borgatti, Everett, & Freeman, 2002) is used to compute the various network measures as a result of the aggregation of individual-level measures to the team level for all expeditions that took place in over the year 2011.

A few types of climbers were excluded from the analysis as they are not relevant or can bias the results, reducing the validity of the research. First, as the interest of this chapter focuses for a large part on structural holes, the expeditions part of the test need to consist of at least three members. In order for someone to bridge a structural hole, there needs to be a disconnection between two alters and one ego to bridge it. Secondly, climbers that only signed up to the expedition to climb up to the base camp of the peak were excluded. These climbers have no intention to reach the summit of their intended mountain, classifying them as a separate category under commercial trekking, shat we consider to stand separate from commercial expedition mountaineering. These individuals can be basically anyone, such as journalists or photographers, who want to see and feel what it's like to be at the base of a peak but don't want to subject themselves to the full adventure of reaching the summit.

Data Collection Process

The original data were gathered by Elizabeth Hawley, who has been based as a journalist in Kathmandu, Nepal, since 1960, monitoring all base camp activity and keeping log of ascents and descent across the Himalayan mountain range. Over 300 peak records, more than 4,500 expeditions records, over 35,000 member records and literature records exceeding the 6,000, the Himalayan Database is of considerable significance in the mountaineering community (The Himalayan Database - The Expedition Archives of Elizabeth Hawley, 2004).

Operationalization of the Variables

Team performance, as the *dependent variable* in this chapter, determines the performance of the expedition team operationalized as the ultimate success of ascent. This is a scale variable measuring the percentage of climbers that reached the summit relative to the total size of the team, in terms of the safety of its ascent. All reasons for failure or termination of the expedition are believed to be manifestations of coordination and teamwork failures. A total of 90 teams took on a peak within the scope of the above outlined dataset, 57.8% of these teams were successful in their ascent.

All *independent variables* and performance outcomes are measured at time *t*, whereas past independent variables are measured for the interval *t – 1* through *– n*, where *n* is ten years. Hence we incorporate up to 10 years of past mountaineering experience for each mountaineering team member.

Past internal network closure. Closure is here defined in terms of density of past ties among individuals that together form an expedition team. For the analysis, we have looked at past ties in the ten years prior to the expedition that is part of the analysis. A block density procedure is applied to the whole network of climbers, based on these past shared experiences that they have with each other. First, a square *7772 x 7772* matrix is computed. Then, a binary matrix sized *1709 x 90* containing information about which climber took part in which expedition during the year 2011 was computed. This vector is applied to the *7772 x 7772* matrix of all past relations. Then, the density of network ties among the current expedition teams is calculated on the basis of these past relationships.

Team size. The number of members of a team that participated in the actual attempt to reach the summit. This excludes base camp personnel. The expedition teams in this dataset averaged 6.20 climbers (s.d. = 3.73). The smallest team counted the minimum of 3 climbers per team whereas the largest team consisted of 26 climbers.

Control Variables

Network centrality. Previous research has already proven that centrality is related to performance (Tsai & Ghoshal, 1998). Also, when observing the effect between structural holes and performance it is important to include a degree centrality (Freeman, 1979) measure to ensure that the observed relationship is not spurious (Soda et al., 2004).

Average age. Age is controlled for as physical conditioning can reduce beyond a certain point, reducing the chances to reach the summit and increasing the chances for injury (Huey, Salisbury, Wang, & Mao, 2007). West (1984) found that younger people are better able to survive the hostile environment of high altitude climbing. This is the average age of the team members part of one team. The average age of the expedition teams was 43.39 years (s.d. = 7.98 years). The oldest team in the dataset averaged 67 years whereas the youngest team averaged 28 years.

Gender. Also, gender is controlled for as the proportion of women in an expedition as the gender effect can lead to different outcomes with respect to summiting success (Valeri and Katsoni, 2021). Although literature suggests similar physiological performance of men and women in high altitude environments (e.g. Schneider, Bernasch, Weymann, Holle, & Bärtsch, 2002), and Huey et al. (2007) demonstrated the similar odds of success, injury, and death for men and women, the inclusion of uncertainty in the analysis might shed new light on this topic.

Oxygen use. The use of oxygen increases the probability of successfully climbing a peak (Eguskitza & Huey, 2000). However, there is also a risk that less experienced climbers attempt peaks above their skill level, thereby endangering themselves (according to Krakauer (1997)). A total of 12.62% (s.d. = 27.39%) of the climbers used bottled oxygen with percentages ranging between 0% and 100% per team.

Support personnel. A higher ratio of Sherpas could possibly increase the probability of success. Teams had an average ratio of support personnel to climbers of 36.80% (s.d. = 42.33%) with a range that fell between 0 to 2.33 per climber. Appendix A further shows the distribution of the network- and control variables per level of uncertainty.

Mortality or injury. As injury and/or mortality can seriously jeopardize the chances of reaching the summit successfully we have also controlled for this effect. This was calculated as the percentage of climbers that suffered an injury or died relative to the total size of the team. On average, the mortality or injury rate per team was 6.7% (s.d. = 16.00%). In total, 14 expeditions suffered a death or serious injury.

Average experience of team. We have calculated the total average amount of times that the climbers part of an expedition have summited peaks in the Himalaya before. The average amount of past climbs of an expedition, being an average of the climbers part of that expedition, was 1.9 climbs (s.d. = 2.4). Some expeditions had zero experience as a whole in the Himalaya, whereas the most experienced expedition team averaged 12.29 past expeditions.

Data Analysis

Based on the data sample criteria, all relevant cases were filtered out from the Himalayan Database. The overall data set contains information of 7772 climbers that in one way or another can be linked to the 90 expeditions part of the analysis. The 90 expedition teams together total 558 climbers. For those 90 expeditions part of the analysis, the effect of past network structures on the current performance outcome was measured by looking at the relationships that each of the individual climbers have had in the period 2001-2010. This resulted in one 7772x7772 square matrix for that period. As part of our embeddedness measures, we do take into account the embeddedness of the climbers Figure 1 shows how you can look at one expedition team in different ways. Each expedition team is a network of its own, that then also becomes part of a larger network of climbers when you include the relationships they have external to the team. That expedition team is also part of the total network of 7772 climbers of those 90 teams in the total network of 7772 climbers.

The different levels of networks are visualized in this figure. In the top left, you see the relationships between climbers that together form an expedition team. Past network closure is relatively low, since few climbers within this expedition team share past experience. In the top right, the network of that same expedition team (in purple) can be seen with all of the relationships that these individuals have with other individuals – external to the expedition team (in blue). The team does score high on network centrality as it is well connected within the larger network. In the bottom, you see the complete network with all 7772 climbers and the relationships between them. Each expedition team is embedded within this larger network of climbers.

Results

A total of 90 expedition teams that have attempted to ascent the most thrilling peaks in the world in 2011 were tested on a variety of predictors for success. The whole model including all the 90 teams is tested on the basis of their success or failure. Table 1 present descriptive statistics including the means and standard deviations of all the variables that were included in the analysis.

Table 1. Descriptive statistics

	Mean	SD	(1)	(2)	(3)	(4)	(5)	(6)	(7)	(8)	(9)	(10)
(1) Average age	43.3995	7.98278	1									
(2) Female ratio	.1620	.22719	.043	1								
(3) Use of oxygen ratio	.1262	.27397	-.107	.190	1							
(4) Ratio of support personnel	.3680	.42330	.178	.109	.468**	1						
(5) Normalized past team centrality	.0019	.00274	.156	-.053	.099	.013	1					
(6) Past experience of climbers	1.9452	2.40341	.045	-.298**	.110	-.052	.462**	1				
(7) Mortality or injury rate	.0607	.16008	.002	-.092	.019	-.079	.634	.156	1			
(8) Team size	6.20	3.73324	.011	.035	.171	.033	.371**	-.141	-.110	1		
(9) Normalized past internal network closure	-.0001	1.00	.008	-.153	-.047	-.030	.029	.538**	.122	-.255*	1	
(10) Normalized past structural holes	.0001	1.0002	-.023	.184	-.136	.033	-.598**	-.700**	-.060	.075	-.398**	1
(11) Success of expedition	.3438	.34397	.033	-.031	-530**	.362**	.113	.082	-.033	.106	.184	-.089

* $p < .05$.
** $p < .01$.

Table 2 shows the results of the linear regression models with regard to the climbing team success. Model 2 consists of all the control variables. Of all the control variables, the use of oxygen within the expedition team was found to be significant ($B = .637, p < .01$). As more climbers within an expedition team make use of additional oxygen, their chances of successfully reaching the summit increases.

Model 3 reports past internal network closure to be a positive predictor of expedition success ($B = .109, p < .01$). Model 5, introducing the effect of structural holes, reports past internal network closure to remain statistically significant ($B = .113$, $p < .01$) while past structural holes is not significant. In Model 6, the interaction effect between past internal network closure and past structural holes is added. Within this model, again past internal network closure is significant ($B = .114$, $p < .01$) while past structural holes and the interaction effect between the two network variables is also not statistically significant. Across the models, the use of oxygen within the expedition team is significant. The results suggest that there is a positive relation between past internal network closure and the success of an expedition. As expedition teams are more interconnected based on past experience, the chances that they are successful in their expedition also increases. This leads us to accept hypothesis 1. Network closure in a team's climbing network has a positive effect on the success of the expedition. No evidence was found for the positive effect of structural holes in a team's climbing network on the success of the expedition, nor a positive relation between structural holes in a team's climbing network and expedition success is moderated by past internal network closure. This leads us to reject hypothesis 2a and 2b, suggesting the past network closure of a climbers assemble to be *the* network configuration variable of choice when considering team network levers influential on alpine expedition team success.

Table 2. Results of the Linear Regression Analysis for Success of Expedition

Linear Regression	Model 1	Model 2	Model 3	Model 4	Model 5	Model 6
Constant	.137 (.180)	.130 (.193)	.162 (.185)	.130 (.194)	.163 (.186)	.165 (.193)
Independent variables						
Past internal network closure			.109** (.038)		.113** (.039)	.114** (.039)
Past structural holes				-.001 (.051)	.027 (.050)	.025 (.066)
Interaction effects						
Past internal network closure * Past structural holes						.002 (.027)
Control variables						
Average age	.003 (.004)	.003 (.004)	.003 (.004)	.003 (.004)	.003 (.004)	.003 (.004)
Female ratio	-.230 (.149)	-.229 (.151)	-.254 (.145)	-.229 (.152)	-.258† (.146)	-.259 (.148)
Use of oxygen ratio	.640** (.140)	.637** (.143)	.681** (.138)	.673** (.144)	.687** (.139)	.688** (.141)
Ratio of support personnel	.100 (.088)	.101 (.089)	.086 (.085)	.101 (.089)	.083 (.086)	.082 (.90)
Past team centrality	1.461 (13.236)	.697 (15.221)	10.186 (14.937)	.539 (17.559)	15.109 (17.501)	14.737 (18.702)
Past experience of climbers	-.002 (.016)	-.001 (.017)	-.032 (.019)	-.002 (.020)	-.028 (.021)	-.028 (.022)
Mortality or injury rate	-.098 (.199)	-.095 (.201)	-.120 (.193)	-.095 (.203	-.131 (.195)	-.135 (.203)
Team size		.001 (.010)	.003 (.010)	.001 (.010)	.001 (.010)	.001 (.010)
F Ratio	5.596**	4.839**	5.622**	4.248**	5.046**	4.529**
R2	.266	.257	.319	.247	.313	.304
N	90	90	90	90	90	90

†p <.10.
* p <.05.
** p <.01.

DISCUSSION AND CONCLUSION

Discussion and Managerial Implications

Since the 1990s, the Himalayan mountaineering community has seen a rapid increase in commercial climbing teams. These climbing teams are often compromised of amateur climbers that pay guiding companies to lead them to the summit of a peak. The social dynamics of these teams is still ill understood however, and warrant future research to inform tourist policy that helps to mitigate the evident risks that come with poorly prepared expedition endeavors. Commercial teams are often formed of individuals that have no relevance to each other and clients select a guiding company primarily on aspects such as reputation, price, and success rates, making such tourist policy for this special category of extreme sports enthusiasts particularly warranted. Tfor instance, Jon Krakauer, a mountaineer who wrote the book *Into Thin Air* about the 1996 Mount Everest disaster further clarifies this difference: *I wasn't sure what to make of my fellow clients. In outlook and experience they were nothing like the hard-core climbers with whom I usually went into the mountains ... One climber's actions can affect the welfare of the entire team* (Krakauer, 1997, p. 44).

Scholars in alpine tourism and sports management research have for long been concerned with distinguishing between individual and group for an examination of group dynamics, and the safety of its team based ascent and descent. The strong focus on individual achievement, however, has notably concentrated in a dominant focus upon individual level factors such as athlete's cognition and perception in alpine mountaineering, while for long discounting how these factors agglomerate into a social system that is able to combine the attributes of individuals and their social relations (Lusher et al. 2010; Wasche et al. 2017). Nevertheless a plurality of network based studies come to the fore when reviewing the literature on sport team dynamics, with empirical work looking into the functioning of sort assemblies as diverse as soccer teams, volleyball-, handball- and basketball teams, explaining outcomes as diverse as overall team performance, leadership strength and even team innovative capability (Onody & De Castro 2004; Duch et al. 2010; Fransen et al. 2015). These team based dynamics are of particular interest too as they hold direct implications for alpine tourism safety policy, tuning in to calls to further establish policy to the operational side of touristicical activity (Lai et al 2016). Drawing on network theory as its theoretical lense, and extending beyond the current scoping around indoor sports, this chapter examines two prime notions of network configuration of extreme sports teams. It treats both notions as indicators of the team-based social infrastructure predicting team performance at the disposal of expedition management as they draw up their expedition teams: structural holes and network closure. As we examine these two network-based indicators of team configuration, this chapter makes an attempt to further clarify the impact of social structures, network closure and structural holes in particular, on performance outcomes in the context of expedition mountaineering. Where practitioners and researchers alike have sought to find the network configuration that is best for group effectiveness, little is known until date about such team configuration choices when it comes to the extreme sports, and the mounting of effective mountaineering teams in particular.

Drawing on a rich and longitudinal empirical dataset that of Himalayan mountaineering teams, we ventured into the team configurational choices in an extreme sports context where team composition takes substantial resources in preparation and training. Those teams that can strongly draw on past network closure, identify as the mountaineering teams that make for a successful ascent. The capacity to span structural holes, commonly portrayed as serving as an eye-opener for options otherwise not found, does not appear to assist teams that make for a successful ascents, however. The focus on network closure helps to guide team configurational choices in an expedition mountaineering context. Our findings provide guidance for where to scout for suitable team members as expedition management builds its team. The emphasis on the social embeddedness of climbers, next to the typical individual level attributes as past experience and climbing success rate, make for an interesting alternative managerial level to play around with. Our findings advise management that an informed choice on this matter holds direct economic effect for expedition management and her sponsorship.

LIMITATIONS AND DIRECTIONS FOR FUTURE RESEARCH

The conclusions should be moderated by several considerations regarding the execution of this research. First, the decay function of ties for the network concepts was not measured. It is plausible that network closure or structural holes start hurting or enhancing performance at different points in time. The Himalayan Database (2004) did not allow for such a variable to be included in the analysis. Future research should focus on an empirical setting in which the decay function of ties can be covered for. Alternatively, the decay of the structure as a whole might be an important consideration that is not unlikely. Second, as the Himalayan Database only covers peaks in the Himalaya mountain range accessible from Kathmandu, it is not possible to include a variable measuring the individual experience level regarding mountaineering. Outside the peaks covered in the database there are numerous peaks that are evenly challenging. It is impossible to come up with a standardized measure of individual levels experience within this empirical setting. Solely looking at the experience that mountaineers have within the Himalayan mountain range would lead to flawed results as it implicates nothing about the relative level of experience to any mountaineer that has earned his stripes in other mountain ranges. Future researchers should consider an empirical setting where individual levels of experience takes a prominent role to further infuse alpine tourism policy with hands-on indicators of alpine tourism safety.

REFERENCES

Aalbers, R. (2012). Organizing intra-organizational networks for innovation. PhD Theses in Economics and Business, University of Groningen.

Aalbers, R. (2020). Rewiring the Intrafirm Network under Downsizing: The Role of Tie Loss on Discretionary Tie Formation. *Long Range Planning*, *53*(3), 101858. Advance online publication. doi:10.1016/j.lrp.2018.11.002

Aalbers, R., & Dolfsma, W. (2017). Improving the value-of-input for ideation by management intervention: An intra-organizational network study. *Journal of Engineering and Technology Management*, *46*, 39–51. doi:10.1016/j.jengtecman.2017.10.006

Aalbers, R., & Dolfsma, W. (2019). Resilience of Information Flow during Restructuring: Characterizing Information Value being Exchanged and the Structure of a Network under Turmoil. *Journal of Business Research*, *100*, 299–310. doi:10.1016/j.jbusres.2019.03.003

Aalbers, R., Dolfsma, W., & Koppius, O. (2006). *On and off the beaten path: Transferring knowledge through formal and informal networksR Aalbers, O Koppius, W Dolfsma*. Circle Electronic Work. Paper Series.

Albattat, A., Jamaludin, A., Zuraimi, N. S. M., & Valeri, M. (2020). Visit Intention and Destination Image in Post- Covid- 19 Crisis Recovery. *Current Issues in Tourism*. Advance online publication. doi :10.1080/13683500.2020.1842342

Balkundi, P., Kilduff, M., Barsness, Z. I., & Michael, J. H. (2007). Demographic antecedents and performance consequences of structural holes in work teams. *Journal of Organizational Behavior*, *28*(2), 241–260. doi:10.1002/job.428

Bechky, B. A., & Okhuysen, G. A. (2011). Expecting the Unexpected? How SWAT Officers and Film Crews Handle Surprises. *Academy of Management Journal, 54*(2), 239–261. doi:10.5465/amj.2011.60263060

Beritelli, P. (2011). *Tourist destination governance through local elites: Looking beyond the stakeholder level* (Doctoral dissertation). Universität St. Gallen.

Berman, S. L., Down, J., & Hill, C. W. (2002). Tacit knowledge as a source of competitive advantage in the National Basketball Association. *Academy of Management Journal, 45*(1), 13–31.

Borgatti, S. P., Everett, M. G., & Freeman, L. C. (2002). *UCINET for Windows: Software for Social Network Analysis*. Harvard.

Burnette, J. L., Pollack, J. M., & Forsyth, D. R. (2011). Leadership in Extreme Contexts: A Groupthink Analysis of the May 1996 Mount Everest Disaster. *The Journal of Leadership Studies, 4*(4), 29–40. doi:10.1002/jls.20190

Burt, R. S. (2000). The network structure of social capital. *Research in Organizational Behavior, 22*, 345–423. doi:10.1016/S0191-3085(00)22009-1

Burt, R. S. (2002). Bridge decay. *Social Networks, 24*(4), 333–363. doi:10.1016/S0378-8733(02)00017-5

Chemli, S., Toanoglou, M., & Valeri, M. (2020). The impact of Covid-19 media coverage on tourist's awareness for future traveling. *Current Issues in Tourism*, 1–8. Advance online publication. doi:10.1080/13683500.2020.1846502

Clemente, F. M., & Martins, F. M. L. (2017). Network structure of UEFA Champions League teams: Association with classical notational variables and variance between different levels of success. *International Journal of Computer Science in Sport, 16*(1), 39–50. doi:10.1515/ijcss-2017-0003

Clemente, F. M., Martins, F. M. L., Kalamaras, D., & Mendes, R. S. (2015b). Network analysis in basketball: Inspecting the prominent players using centrality metrics. *Journal of Physical Education and Sport, 15*(2), 212.

Clemente, F. M., Martins, F. M. L., Wong, P. D., Kalamaras, D., & Mendes, R. S. (2015a). Midfielder as the prominent partici¬pant in the building attack: A network analysis of national teams in FIFA World Cup 2014. *International Journal of Performance Analysis in Sport, 15*(2), 704–722. doi:10.1080/24748668.2015.11868825

Coleman, J. S. (1988). Social Capital in the Creation of Human Capital. *The American Journal of Sociology, 94*, 95-120.

Duch, J., Waitzman, J. S., & Amaral, L. A. N. (2010). Quantifying the performance of individual players in a team activity. *PLoS One, 5*(6), e10937. doi:10.1371/journal.pone.0010937 PMID:20585387

Eguskitza, X., & Huey, R. B. (2000). Supplemental Oxygen and Mountaineering Deaths. *American Alpine Journal, 42*, 135–138.

Fewell, J. H., Armbruster, D., Ingraham, J., Petersen, A., & Waters, J. S. (2012). Basketball teams as strategic networks. *PLoS One, 7*(11), e47445. doi:10.1371/journal.pone.0047445 PMID:23139744

Fransen, K., Van Puyenbroeck, S., Loughead, T. M., Vanbeselaere, N., De Cuyper, B., Vande Broek, G., & Boen, F. (2015). Who takes the lead? Social network analysis as a pioneering tool to investigate shared leadership within sports teams. *Social Networks*, *43*, 28–38. doi:10.1016/j.socnet.2015.04.003

Freeman, L. C. (1979). Centrality in Social Networks. *Social Networks*, *1*(3), 215–239. doi:10.1016/0378-8733(78)90021-7

Hall, C. M. (2019). Constructing sustainable tourism development: The 2030 agenda and the managerial ecology of sustainable tourism. *Journal of Sustainable Tourism*, *27*(7), 1044–1060. doi:10.1080/09669582.2018.1560456

Hällgren, M. (2010). Groupthink in Temporary Organizations. *International Journal of Managing Projects in Business*, *3*(1), 94–110. doi:10.1108/17538371011014044

Hambrick, M. E., Svensson, P. G., & Kang, S. (2018). Using social network analysis to investigate interorganizational relationships and capacity building within a sport for development coalition. *Sport Management Review*. Advance online publication. doi:10.1016/j.smr.2018.12.002

Higgs, M., Plewnia, U., & Ploch, J. (2005). Influence of team composition and task complexity on team performance. *Team Performance Management*, *11*(7/8), 227–250. doi:10.1108/13527590510635134

Huey, R. B., Salisbury, R., Wang, J.-L., & Mao, M. (2007). Effects of age and gender on success and death of mountaineers on Mount Everest. *Biology Letters*, *3*(3), 498–500. doi:10.1098/rsbl.2007.0317 PMID:17698450

Jolly, J. (2010, August 28). *Elizabeth Hawley, unrivalled Himalayan record keeper*. Retrieved March 24, 2015, from BBC News: https://www.bbc.com/news/world-south-asia-10268549

Jones, T. E., Apollo, M., & Bui, H. T. (2021). Mountainous Protected Areas & Nature-Based Tourism in Asia. In *Nature-Based Tourism in Asia's Mountainous Protected Areas* (pp. 3–25). Springer. doi:10.1007/978-3-030-76833-1_1

Korte, F., & Lames, M. (2018). Characterizing different team sports using network analysis. *Current Issues in Sport Science, 3*(5). doi:10.15203/CISS_2018.005

Krakauer, J. (1997). *Into thin air: A personal account of the Mt. Everest disaster*. Villard.

Lai, P. H., Hsu, Y. C., & Wearing, S. (2016). A social representation approach to facilitating adaptive co-management in mountain destinations managed for conservation and recreation. *Journal of Sustainable Tourism*, *24*(2), 227–244. doi:10.1080/09669582.2015.1062018

Laws, E. (2011). *Tourist destination governance: Practice, theory and issues*. Cabi. doi:10.1079/9781845937942.0000

Levin, D. Z., Walter, J., & Murninghan, J. K. (2011). Dormant Ties: The Value of Reconnecting. *Organization Science*, *22*(4), 923–939. doi:10.1287/orsc.1100.0576

Lusher, D., Robins, G., & Kremer, P. (2010). The Application of Social Network Analysis to Team Sports. *Measurement in Physical Education and Exercise Science*, *14*(4), 211–224. doi:10.1080/1091367X.2010.495559

Neupane, P. C. (2021). Tourism Governance in the aftermath of COVID-19: A Case Study of Nepal. *The Gaze: Journal of Tourism and Hospitality*, *12*(1), 44–69. doi:10.3126/gaze.v12i1.35676

Ortner, S. B. (1999). *Life and Death on Mt. Everest: Sherpas and Himalayan Mountaineering*. Princeton University Press. doi:10.1515/9780691211770

Peters, M., & Pikkemaat, B. (2006). Crisis management in Alpine winter sports resorts—The 1999 avalanche disaster in Tyrol. *Journal of Travel & Tourism Marketing*, *19*(2-3), 9–20. doi:10.1300/J073v19n02_02

Rauch, S., Marzolo, M., Dal Cappello, T., Ströhle, M., Mair, P., Pietsch, U., ... Strapazzon, G. (2021). Severe traumatic brain injury and hypotension is a frequent and lethal combination in multiple trauma patients in mountain areas–an analysis of the prospective international Alpine Trauma Registry. *Scandinavian Journal of Trauma, Resuscitation and Emergency Medicine*, *29*(1), 1–10. doi:10.118613049-021-00879-1 PMID:33407690

Ritchie, B. W., & Jiang, Y. (2019). A review of research on tourism risk, crisis and disaster management: Launching the annals of tourism research curated collection on tourism risk, crisis and disaster management. *Annals of Tourism Research*, *79*, 102812. doi:10.1016/j.annals.2019.102812

Rodan, S., & Galunic, C. (2003). More than Network Structure: How Knowledge Heterogeinity Influences Managerial Performance and Innovativeness. *Strategic Management Journal.*

Rueger, J., Dolfsma, W., & Aalbers, R. (2020). Perception of Peer Advice in Online Health Communities: Access to Lay Expertise. *Social Science & Medicine*. Advance online publication. doi:10.1016/j.socscimed.2020.113117 PMID:33865092

Salisbury, R., & Hawley, E. (2004). *The Himalayan Database - The Expedition Archives of Elizabeth Hawley*. The American Alpine Club Press.

Sallent, O., Palau, R., & Guia, J. (2011). Exploring the legacy of sport events on sport tourism networks. *European Sport Management Quarterly*, *11*(4), 397–421. doi:10.1080/16184742.2011.599208

Schneider, M., Bernasch, D., Weymann, J., Holle, R., & Bärtsch, P. (2002). Acute mountain sickness: Influence of susceptibility, preexposure, and ascent rate. *Medicine and Science in Sports and Exercise*, *34*(12), 1886–1891. doi:10.1097/00005768-200212000-00005 PMID:12471292

Sharma, P. (2012). Tourism in Nepal 2030. Nepal, 2030, 73-92.

Soda, G., Usai, A., & Zaheer, A. (2004). Network Memory: The Influence of Past and Current Networks on Performance. *Academy of Management Journal*, *47*(6), 893–906.

Telegraph. (2013, July 24). *The world's most dangerous mountains*. Retrieved March 27, 2015, from Telegraph: https://www.telegraph.co.uk/travel/activityandadventure/10197830/The-worlds-most-dangerous-mountains.html?frame=2625884

Toanoglou, M., Chemli, S., & Valeri, M. (2021). The organizational impact of COVID-19 crisis on travel perceived risk across four continents. *Journal of Organizational Change Management*. Advance online publication. doi:10.1108/JOCM-12-2020-0369

Tsai, W., & Ghoshal, S. (1998). Social Capital and Value Creation: The Role of Intrafirm Networks. *Academy of Management Journal, 41*(4), 464–476.

Valeri, M. (2016). Networking and cooperation practices in the Italian tourism business. *Journal of Tourism, Heritage & Services Marketing*, *2*(1), 30–35. doi:10.5281/zenodo.376333

Valeri, M. (Ed.). (2021). Tourism risk. Crisis and recovery management. Emerald Publishing.

Valeri, M., & Katsoni, V. (Eds.). (2021). *Gender and tourism: challenges and entrepreneurial opportunities*. Emerald Publishing. doi:10.1108/9781801173223

Volgger, M., & Pechlaner, H. (2014). Requirements for destination management organizations in destination governance: Understanding DMO success. *Tourism Management*, *41*, 64–75. doi:10.1016/j.tourman.2013.09.001

Warner, S., Bowers, M. T., & Dixon, M. A. (2012). Team dynamics: A social network perspective. *Journal of Sport Management*, *26*(1), 53–66. doi:10.1123/jsm.26.1.53

Wäsche, H., Dickson, G., Woll, A., & Brandes, U. (2017). Social network analysis in sport research: An emerging paradigm. *European Journal of Sport Science*, *14*(2), 138–165. doi:10.1080/16138171.2017.1318198

West, J. B. (1984). Human Physiology at Extreme Altitudes on Mount Everest. *Science*, *223*(4638), 784–788. doi:10.1126cience.6364351 PMID:6364351

Whelan, E., Parise, S., De Valk, J., & Aalbers, H. L. (2011). Creating Employee Networks That Deliver Open Innovation. *MIT Sloan Management Review*, *53*(1), 37–44.

Yan, G., Watanabe, N. M., Shapiro, S. L., Naraine, M. L., & Hull, K. (2018). Unfolding the Twitter scene of the 2017 UEFA Champions League Final: Social media networks and power dynamics. *European Sport Management Quarterly*, 1–18.

Zaheer, A., & Bell, G. G. (2005). Benefiting from Network Position: Firm Capabilities, Structural Holes, and Performance. *Strategic Management Journal*, *26*(9), 809–825. doi:10.1002mj.482

Zaheer, A., & Soda, G. (2009). Network Evolution: The Origins of Structural Holes. *Administrative Science Quarterly*, *54*(1), 1–31. doi:10.2189/asqu.2009.54.1.1

Chapter 3
The Impact of Corporate Digital Responsibility (CDR) on Internal Stakeholders' Satisfaction in Hungarian Upscale Hotels

Edit Kővári
University of Pannonia, Hungary

Mohamad Saleh
University of Pannonia, Hungary

Gyöngyi Steinbachné Hajmásy
University of Pannonia, Hungary

ABSTRACT

Digital transformation and artificial intelligence are considered among the most vital trending topics in the process of hospitality sector evolution. Many scholars found that digital transformation and artificial intelligence cause a massive shift in all aspects of the hospitality sector and digital technology application that impact the whole facet of internal and external stakeholders' lives. However, the adoption of digitalization and artificial intelligence is considered a strength. Corporate digital responsibility (CDR) is a strategy that enhances trust between the companies adopting digitalization and their primary stakeholders. Internal and external stakeholders' satisfaction develop contemporary social responsibility (CSR) challenges in the decision-making process in acquiring, analysing, implementing, and assessing for adopting digitalization in the hospitality sector. This chapter aims to give a literature review focusing on CDR and its relation analyses to hotel industry's internal stakeholders' satisfaction trough a Hungarian case study.

DOI: 10.4018/978-1-6684-3889-3.ch003

INTRODUCTION

Digitalization and artificial intelligence affect nearly all markets and enterprises in all sectors (Chanias et al., 2019; Mattet et al., 2015). Digitalization and artificial intelligence and their consequences have become an extremely important competitive advantage. Many sectors, including tourism, admit the necessity of using digital technologies and artificial intelligence to make the work more efficient and effective (Buhalis et al., 2019; Camilleri, 2018). As a particular and energetic method of a successful hotel's daily activities, digitalization and artificial intelligence cannot be excluded (Lam, Law, 2019). Digitalization and artificial intelligence have become crucial for hotels in order to meet core market strategies. Such as enhancing service efficiency, cost saving, achieving a strategic advantage, information building, and profit maximalization (Camilleri; Xiang, 2018). However, the adoption of digitalization and artificial intelligence derives many necessary changes in hotels, such as value creation, financial aspects, identity, and corporate social responsibility (Matt et al., 2015). Based on Hinings et al. (2018), digitalization results from adopting various new technologies that create new roles, processes, procedures, and principles that would alter, replace, or supplement current laws. The impact of digitalization can be seen in hotels' operational procedures and their relationships with internal and external stakeholders and specially with their employees (Cobos et al., 2016; Reinartz et al., 2019). Moreover, adopting digitalization and artificial intelligence massively affect hotels' structure and function and also may eliminate some positions and roles (Hinings et al., 2018). The growing dependence on digitalization and artificial intelligence, in particular, has raised significant doubts about the future of many careers in the hospitality sector, especially in the hotel industry. Many scholars such as (Buhalis et al., 2019; Ivanov, 2019, Webster and Ivanon, 2020) anticipated that shifting to digitalization and the usage of artificial intelligence would result in a transition in employment from human to nonhuman workers. However, utilizing digitalization and artificial intelligence is a factor that increases employees' turnover, specifically in the hotel sector (Li et al., 2019). As a consequence, although some jobs may be impacted by digitalization, some will totally vanish (Makridakis, 2017; Navio-Marco et al., 2018), requiring workers to adapt to the evolving conditions, expertise, and career paths (Critten et al., 2017). With these massive changes in the hotels' operations, functions, systems, and activities. However, an effective CSR strategy should be implemented to enhance employees' satisfaction, as they are considered an essential component in the internal stakeholders' category.

Design/Methodology/Approach

We collect CDR-related research published in leading management and hospitality journals and applied content analysis method. These particular journals represent an appropriate overview of CDR research in academic hospitality literature, expressing digitalization and artificial intelligence usage. CDR is a relatively new line in hospitality literature regarding the adoption of digitalization and artificial intelligence. The paper apply a quantitative analyses of a questionnaire, measurs the relationship between internal CDR and employee satisfaction, as well as the internal CDR activities of hotels.

Results

The literature review findings indicate that hospitality companies can be more efficient while adopting digitalization and artificial intelligence if they consider the economic, legal, ethical activities that would effectively attract, motivate, and retain their employees. Surprisingly, only a few studies have examined CDR's impact on employees and the theme related to working life quality. An empirical study tested the questionnaire in the Hungarian hotel industry, in the upscale hotels of the Balaton region. The findings of the empirical research confirmed that there is a strong positive relationship between internal CDR and employee satisfaction.

Originality/Value

Most of the review papers of CDR research in hospitality journals focus on environmental issues and the business case for CDR, but those regarding internal stakeholder's satisfaction theories and research are few. This study contributes to the developing number of that kind of research.

Limits

The paper's limitation is that the methodology of the examined research and studies are very diverse to make global generalizations.

1.1. Corporate Digital Responsibility and Corporate Social Responsibility

Digitalization is an action in which changes occur in a certain organization by utilizing new technology to execute certain tasks more quickly and efficiently; these changes influence the degree of satisfaction of primary stakeholders, mainly employees (Loebbecke, Picot, 2015). As a result, numerous businesses have adopted the digital transition (Hess et al., 2016). Nevertheless, digital transformation is similar to any other type of transformation which has strength and threats for example digitalization can reduce cost, enhance effectiveness and efficiency of work (Herring, Roy, 2007). Furthermore, digital transformation can cause many negative impacts such as unemployment, decrease in the salary scale range, decrease in employees' productivity (Vial, 2019). As a result, digital transformation needs a strategy in order for the company to be responsible, and to eliminate or reduce the negative impacts of digitization (Vial, 2019). However, the strategy related to these responsibilities is called Corporate digital responsibility (Thorun, 2018). CDR is a strategy that enhances trust between the companies adopting digitalization and their primary stakeholders since this strategy protects the economic, legal, and ethical rights of the company's primary stakeholders (Koch, Windsperger, 2017).

Corporate Digital Responsibility is an augmentation of a company's obligations that considers the economic, legal, ethical, and philanthropic responsibilities toward the company's' primary and secondary stakeholders Identical to corporate social responsibility dimensions in Carroll's pyramid 1990 (Christina J. Herden, 2021). In this research we apply the term of CDR which roots in CSR though focuses in the digitalization instead of the social responsibility.

Figure 1. The CDR pyramid
Sources: (Christina J. Herden, 2021)

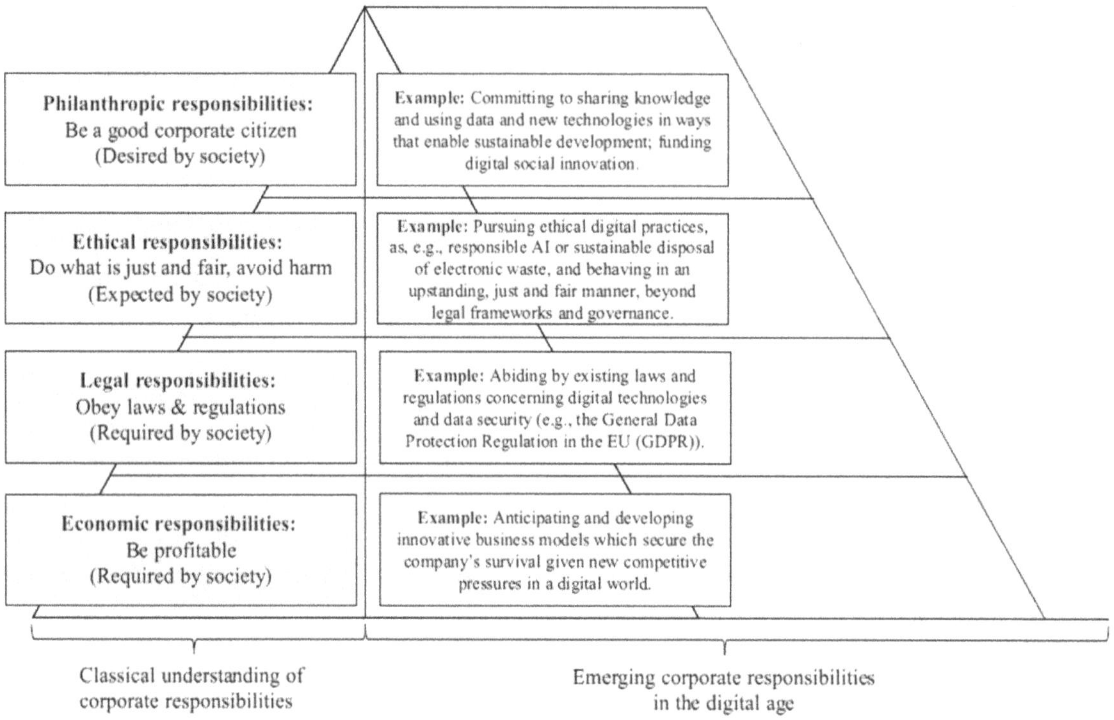

2. The Relationship Between Hotels Adopting Digitalization and Artificial Intelligence in Achieving Employees' Satisfaction Employee-Related CDR Activities

CDR is responsible for hotels' employees' engagement and satisfaction, especially in four and five-star hotels adopting digitalization and artificial intelligence in their daily operations (Lee, Park, 2013). Nowadays, hotels must have a clear strategy of handling social responsibility issues in the face of the problems emerging from the implementation and application of digitalization and artificial intelligence. Currently, digitalization and artificial intelligence impact should be taken into account in the philosophy of social responsibility, which is narrowly characterized as the principles and regulations that regulate decisions and actions in business activities (Moriarty, 2016). According to the concept of CDR, corporate digital responsibility is a collection of principles and basic guidelines that regulate a hotel's decisions related to internal and external digital issues (Floridi, 2010). The primary objective of corporate digital responsibility is to maintain a healthy working environment for the internal stakeholders and especially the employees and establish codes of conduct in which employees' rights are protected (Wirtz et al., 2018). Adopting digitization/ digitalization artificial intelligence can cause massive problems for employees if not well managed. Adopting digitalization and artificial intelligence requires not just brilliant concepts, money, machinery, professional workers, information security, but also requires responsibility toward the employees. The relationship between employee satisfaction and machines is based on different dimensions. However, formulating an adequate hotel corporate digital responsibility strategy requires considering the ethical, legal, and economic dimensions (Schwartz, Carroll, 2003).

2.1. The Ethical Dimension of CDR Toward Employees

As digital technologies and artificial intelligence are considered sensitive factors that influence organizational decision-making in many fields, especially in the labor field, hotels are required to formulate and implement moral standards and ethical codes to protect their labor force (Richter, Riemer, 2013; Valeri 2021a; 2021b; Valeri and Katsoni, 2021). However, embracing the principle that employees' rights should be regulated by the ethical dimension of CSR, then we should consider similar criteria of codes and conducts based on the ethical dimension of CDR to protect employees' ethical rights (Soltani, 2019). This point assumes that maintaining the ethical architecture of the application of new technologies and artificial intelligence is more than just a technical problem as it reaches employees' satisfaction. To control the advancement and application of new technologies and artificial intelligence that affect employees' satisfaction, hotels must create a robust, consistent collection of norms ingrained from their corporate culture (Vosoughi, Roy, Aral, 2018). This concept is known as corporate digital responsibility (CDR), which is characterized by a collection of standard variables that guide hotel activities in relation to employees' satisfaction (Brey, 2012).

2.2. The Economic Dimension of CDR Toward Employees

The adoption of digital technologies and artificial intelligence greatly influences the hotel's labor force on the economic side (Schwartz, Carroll, 2003). For this reason, every hotel is required to formulate and implement an economic responsibility strategy to proactively handle the economic consequences of adopting digitalization and artificial intelligence. Many articles related to this topic discussed the substitution of human jobs by robotics (Inman, Nikolova, 2017). However, digitalization and artificial intelligence changed traditional jobs, requiring different abilities and skills to undertake the new tasks, which necessitate retraining or replacing current employees with employees who already have certain capabilities. New jobs arose as a result of digitalization, while many other jobs become absolutely rendered and useless (Wirtz et al., 2018). Moreover, as a result, digitalization has significant consequences on the labor force. The volume and standard of labor required have been influenced by the adoption of digitalization and artificial intelligence. However, digitalization and artificial intelligence created new recruiting assessments and even new way of working and thinking (Schumann, 2014). For instance, how work is organized, the way of interaction between the employees and their managers, and salary scales.

2.3. The Legal Dimension of CDR Toward Employees

The legal dimension of CDR comprises legislative laws that direct the company in the procedure of choosing whom should be considered in the company CDR strategy. The General corporate digital responsibility legislative framework of the European Union is regarded as one of the most essential legislative frameworks for developing corporation-specific CDR rules (Council of the European Union, 2016). Although legal frameworks provide significant regulations to be followed, companies should build their own legal duties depending on the cultural characteristics of the country in which they operate (Acquist, Brandimarte, Loewenstein, 2015). Companies must focus mostly on their employees while developing or adopting their legal responsibilities of CDR since digitalization is considered as the most influential factor that affects employees' satisfaction (Yang, Rivers, 2009). Managing digitalization does, in fact, include certain strong guidelines that show already developed laws and regulations. Nevertheless, since CDR legal responsibilities are specific in each country, firms and especially international hotels face difficulties in adopting them literally (ISO, IEC, 2013). In this case international hotels should study the country laws as well as the employees' behaviors to adopt the best legal laws that satisfies their employees. Consequently, adopting digitalization and artificial intelligence into the daily operation of four- and five-star hotels has various advantages. Motivating employees and increasing their satisfaction are one of the most significant benefits of CDR.

3. THE IMPORTANCE OF EMPLOYEE SATISFACTION IN THE HOTEL INSDUSTRY

Employee satisfaction is a critical issue in the service-oriented organizations especially in the hotel industry and plays a crucial role in the success of the hotel firms (Matzler – Renzl, 2007). Locke (1976) defined employee satisfaction as a 'pleasurable or positive emotional state resulting from the appraisal of one's job or job experiences' (p. 1300). Bakacsi (2002) considered employee satisfaction as a work-related attitude. Grigoroudis – Siskos (2010) explained employee satisfaction as a pleasant, positive state resulting from the employee's work experience. At the same time employee satisfaction considered as a multi-dimensional construct. A number of studies highlighted the positive relationship between employee satisfaction, customer satisfaction and company performance. One of the most important conceptualizations is the 'Service-Profit Chain' (Heskett et al. 1994, Chiang et al. 2005, Abdullah et al. 2009, Chi – Gursoy 2009). The relationship between employee satisfaction and guest satisfaction is illustrated by the so-called 'Good Service Cycle' model *(Figure 1)* (in Rodríguez-Antón – Alonso-Almeida 2011)

Figure 2. Good Service Cycle model
Source: Rodríguez-Antón – Alonso-Almeida (2011)

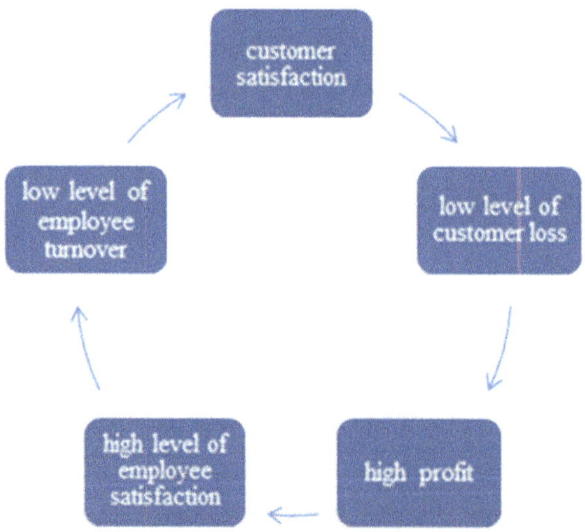

Obtaining this situation is significant in the service-oriented hotel industry (Matzler – Renzl 2007, Xiong et al. 2013, Lam et al. 2001). In the hotel sector, employees have direct connection with the guests, consequently they can largely contribute to customer satisfaction through appropriate service delivery (Rodríguez-Antón – Alonso-Almeida 2011, Williams – Buswell 2003). Consequently, the human element plays an extremely important role in the success of the service sector enterprises (Youn et al., 2018). A number of studies highlighted that when employees are more satisfied with their work, they are more productive and make fewer mistakes during their work (Rodríguez-Antón – Alonso-Almeida 2011). Authors such as Hallowell et al. (1996), Yeh (2013), Al Refaie (2015) and Yoon – Suh (2003) concluded that satisfied employees can provide a high-quality service.

3.1. The Relationship Between Employee Satisfaction and Internal CDR

Employees have a key role in the tourism sector, especially in the hotel industry. The CDR (corporate digital responsibility) concept, particularly internal CDR also focuses on the employees as the primary stakeholders of the organizations (Berkesné, 2018; Braun, 2015). Steinbachné (2020) explained internal CDR as activities, are directly related to the physical and psychological environment of employees and pay special attention to the health and well-being of employees, with an emphasis on employee training, equal opportunities and work-life balance. Besides applying different internal CDR activities provide the large scale of advantages for the hotels. Recruiting and retaining a skilled workforce is an essential element in the successful operation of the service-oriented companies. A number of studies have focused on the advantages of implementation the various CDR activities. CDR practices can enhance employees' loyalty (Kucukusta et al., 2013; Berezan et al., 2013), reduce turnover level, strengthen employee engagement and recruit skilled workforce (Levy – Park, 2011; Kim et al., 2017). Kim et al. (2017) examined CDR and its internal consequences among South Korean hotel workers and concluded that the different CSR initiatives has a positive impact on their quality of work, and their organizational commitment. Bohdanowicz et al. (2011) realized that companies applying CDR activities are more attractive to the current and potential workforce. Verčič – Ćorić (2018) also highlighted the positive relationship between CDR and employee performance and commitment. However, the above-mentioned researches used the widespread Carroll pyramid (economic, legal, ethical responsibility and philanthropy) to measure CDR and not focus on the relationship between internal CDR and employee satisfaction. To fill this gap Steinbachné (2020) conducted a comprehensive research to examine the employee satisfaction in the light of internal CSR activities with the involvement of four- and five-star hotels.

4. CASE STUDY – EXAMIINATINO OF EMPLOYEE SATISFACTION IN THE LIGHT OF INTERNAL CDR ACTIVITIES

The aim of the research was twofold. On the one hand the relationship between internal CDR and employee satisfaction was examined. On the other hand, the study also investigated to explore the link between the characteristics of the hotel, such as hotel ownership (a member of a hotel chain or an independent hotel) and the size of the hotel and the application of internal CDR activities.

As a first step of the research the employee-related CDR activities were identified after reviewing the CSR-related literature and the various international and domestic standards and guidelines based on the economic, ethical and legal dimensions. (UN Global Compact, ILO, OECD, GRI, EU 'Green Paper', CSR index, ISO26000, CSR EMAT) then adopting to the digitalization literature. These activities are illustrated in the abovementioned *Table 1*. Many scholars have found that adopting and deploying digitalization and artificial intelligence in the daily working process may impact the level of employees' satisfaction if serious actions were not taken. However, to assess employees' level of satisfaction at three- and four-star four- and five-star hotels, it was necessary to conduct a survey.

The survey included five variables related to CDR's economic dimension to measure the employees' level of satisfaction in these three- and four- and five-star hotels adopting digitalization and artificial intelligence located in Hungary, in the Balaton region. Following Budapest – the capital city of Hungary – the Balaton region is the second most popular destination in Hungary, visited by millions of domestic and foreign tourists every year. In terms of domestic tourism Lake Balaton is obviously the most popular destination.[1] The chosen variables were taken from different researches done in the same concern. Starting with the first variable related to the economic dimension, which is the wages and job security (Spiller, 2000), Many research findings have explained that the adoption and deployment of digitalization and artificial intelligence may impact the salary scale as well as the job security, another variable that should be treated is job creation (T. Raihan, R. AL Karim, 2017) since the deployment of digitalization and artificial intelligence causes a sharp decrease in job creation in which many human jobs are substituted by robotics. The third variable used to measure employees' level of satisfaction is the enhancement of employees' productivity (Anitha J.,2014; T. Raihan, R. Al Karim, 2017) since digitalization and artificial intelligence have been utilized in my productive tasks that were related to employees before adopting these technologies. Finally, the fourth variable which is motivation (T. Raihan, R. AL Karim, 2017), employees need to be motivated as much as possible since the decrease in work responsibilities resulted from adopting and deploying digitalization and artificial intelligence can cause a serious problem in the employees emotional and mental intelligence. In the same direction, the survey included 5 factors connected to CDR's ethical dimension to assess employees' satisfaction in the selected hotels located in Hungary. However, the CDR's ethical dimension variables were also taken from some articles related to the same topic. The first variable used in the survey to measure the level of employees' satisfaction is the fair treatment of employees concerning their salary (T. Raihan, R. AL Karim, 2017) due to the discrimination that may occur between the employees who have experience with new digitalized and automated machines and those who do not. The second variable is the working environment (United Nations Global Compact, 2000), where each hotel should provide a safe working environment for those employees who have direct contact with machines and new technologies. The third variable in this category is promoting health and safety programs (Burton, J.,2010; T. Raihan, R. AL Karim, 2017) as dealing with digitalized and automated machines may cause serious health problems. The fourth variable is learning and development (Spiller, 2000; Vostrá, H., Jindrová, A. and Dömeová, L., 2011) in which each employee should get trained on new technologies adopted by the hotel to increase employees' experience and productivity. The fifth and final variable is meaningful job (Spiller,2000; Rosso, Wrzesniewski, 2010) every employee should feel that his job is meaningful in order to stay motivated and engaged.

4.1. Variables Related to Employees' Satisfaction

As next step the factors influencing employee satisfaction were identified. The factors can be divided into two major groups, such as: work-related factors and organizational factors affecting employee satisfaction *(Table 1)*.

Table 1. Factors influencing employee satisfaction

1. Work-related factors
• Factors related to the nature of work
• Factors related to the amount of work
2. 2. Organizational factors affecting employee satisfaction
• Leadership style
• Career development and training opportunities
• Workplace communication
• Workplace safety
• Remuneration
• Relationship and teamwork with co-workers
• Working time arrangements
• Working conditions

Source: Steinbachné (2020 p. 87)

Work-related factors include factors related to the nature of work and the amount of work. Factors related to the nature of work embrace task variety and skill variety (Ineson, Benke & László, 2013). Job satisfaction depends on organizational factors as well. It is influenced by leadership style, involving the relationship with the supervisors and the management (Alegre, Mas-Machuca, & Berbegal-Mirabent, 2016; Byun, Karau, Dai & Lee, 2018; Matzler & Renzl, 2007). The other component of employee satisfaction is receiving regular feedback on the performance (Allen, Shore, & Griffeth, 2003; Alegre et al., 2016; Zhao, Ghiselli, Law & Ma, 2016). Career development and training opportunities are a very valuable tool for increasing job satisfaction (Ineson et al., 2013; Lundberg, Gudmundson & Andersson, 2009; Matzler & Renzl, 2007; Spinelli & Canavos, 2000; Yuen et al., 2018; Zopiatis, Constanti & Theocharous, 2014). Workplace communication can contribute significantly to employee satisfaction. Appropriate communication between employee-management, employee-supervisor and employee-coworker are very important (Alegre et al., 2016; Lundberg et al., 2009; Matzler & Renzl, 2007; Yuen et al., 2018). Job satisfaction also depends on job security and the fear of losing a job. Compensation is very valuable tool for employee satisfaction. It contains material and non-material remuneration (e.g. verbal praise) (Lundberg et al., 2009; Matzler & Renzl, 2007; Spinelli & Canavos, 2000; Yuen et al., 2018; Zopiatis et al., 2014). The relationship and teamwork with co-workers are also important sources of job satisfaction (Alegre et al., 2016; Lundberg et al., 2009; Matzler & Renzl, 2007; Yuen et al., 2018). As employees spend a major part of their lives at their workplace flexibility, work-life balance and working hours play a significant role (Alegre et al., 2016; Yuen et al., 2018). Job satisfaction also influenced by working conditions, involving the proper equipment of the workplace, and healthy and safe working conditions (Matzler & Renzl, 2007; Yuen et al., 2018; Zopiatis et al., 2014). Based on the determined factors a questionnaire was compiled to examine the relationship between internal CDR and employee satisfaction among the employees of four- and five-star hotels in the Balaton region. Kucukusta et al. (2013), McGehee (2009) and Strahilevitz - Myers (1998) highlighted that the implementation of different CDR activities is more important and even more common in upscale hotels. They believe that the application of CDR is essential for upscale hotels if they want to gain a long-term competitive advantage. For research purposes, four- and five-star categories include hotels that are Hotelstars Union (HSU) certified. In addition to the classification, the research also delimited the population geographically, as the Balaton region is the most popular destination in Hungary after the capital, so it is a significant study area in terms of tourism, and several hotels of higher categories can be found in the region (Steinbachné, 2020). The results confirmed that there is a strong positive relationship between employee satisfaction and internal CDR. The results of the correlation analysis clearly supported this statement (r^2=0.702; r=0.838). Analysis using the SEM model also demonstrated the positive relationship between the two variables (r=0.94). Consequently, it can be stated that the application of internal CDR is suitable for increasing employee satisfaction. In addition, it is worth to build into the daily operation and the corporate strategy of the hotel companies. However, examining the characteristics of hotels, it can be concluded that the application of internal CDR is not affected by the size and the ownership of the hotel. The independent sample t-test was performed to test the link between hotel ownership and the application of internal CDR activities. The results did not show a significant relationship (Sig.=0.777). However, the relationship between hotel capacity (size) and the application of internal CDR activities was examined using Pearson's correlation and the result did not show a significant relationship between the two factors (Sig.=0.853).

RESULTS

During the research the employee-related CDR activities and the factors influencing employee satisfaction were determined on the basis of the literature review were identified. With the help of these factors a questionnaire was compiled, which is suitable for measuring the relationship between internal CDR and employee satisfaction, as well as the internal CDR activities of hotels.

An empirical study tested the questionnaire in the Hungarian hotel industry, in the upscale hotels of the Balaton region. The findings of the empirical research confirmed that there is a strong positive relationship between internal CDR and employee satisfaction. However, examining the characteristics of hotels, it can be concluded that regardless of the size and ownership of the hotel, internal CDR can be implemented into the operation. Similar, science-based research has not been conducted yet. The previous researches approached this issue from a different perspective. During the case study the various international and domestic standards and guidelines (UN Global Compact, ILO, OECD, GRI, EU 'Green Paper', CSR index, ISO26000, CSR EMAT) as well as the literature related to internal CSR and CDR were reviewed to separate the different groups of employee-related CDR activities (ethical conduct, relationship between employees and employers, working conditions and social protection, social issues, occupational health and safety (in addition to legal requirements), human development and training and remuneration).

ORIGINALITY/VALUE

The aim of the current research was to examine the relationship between internal CDR and employee satisfaction among employees working in the upscale hotels of Balaton region (Hungary). Satisfied and motivated employees can deliver quality services, which can increase customer satisfaction and customer loyalty. It is significantly important in the hotel sector, where employees are critical elements of service quality. In spite of all hotel managers often do not put enough emphasis on CDR and employee satisfaction, but a similar hotel-specific research offers a great opportunity for any hotel to identify the strengths and weaknesses in the field. Besides the research and the case study can help hotels to apply different internal CDR activities effectively.

LIMITATIONS

Although the results of the current case study confirmed that there is positive relationship between internal CDR and employee satisfaction in the context of the Hungarian hotel industry, there are a number of limitations. First, the sample of the study was limited, because of the COVID-19. Consequently many workers in the tourism sector have been fired, others have changed careers. Second, the study investigated the hypothesized relationship by using a sample of upscale hotels in the Hungarian hotel industry. The generalizability of the findings to other accommodation categories and industries needs further investigation. Finally it should be noted that the current study explored limited hotel employees' work-related outcomes. Consequently, in the future other work-related outcomes should be investigated, such as organizational commitment and fluctuation.

REFERENCES

Abdullah, R., Karim, N. B. A., Patah, M. O. R. B. A., Zahari, H., Nair, G. K. S., & Jusoff, K. (2009). The linkage of employee satisfaction and loyalty in hotel industry in Klang Valley, Malaysia. *International Journal of Business and Management, 4*(10), 152–160. doi:10.5539/ijbm.v4n10p152

Acquisti, A., Brandimarte, L., & Loewenstein, G. (2015). Privacy and human behavior in the age of information. *Science, 347*(6221), 509–514. doi:10.1126cience.aaa1465 PMID:25635091

Al-Refaie, A. (2015). Effects of human resource management on hotel performance using structural equation modeling. *Computers in Human Behavior, 43*, 293–303. doi:10.1016/j.chb.2014.11.016

Alegre, I., Mas-Machuca, M., & Berbegal-Mirabent, J. (2016). Antecedents of employee job satisfaction: Do they matter? *Journal of Business Research, 69*(4), 1390–1395. doi:10.1016/j.jbusres.2015.10.113

Allen, D., Shore, L. M., & Griffeth, R. W. (2003). The role of perceived organizational support and supportive human resource practices in the turnover process. *Journal of Management, 29*(1), 99–118. doi:10.1177/014920630302900107

Anitha, J. (2014). Determinants of employee engagement and their impact on employee performance. *International Journal of Productivity and Performance Management, 63*(3), 308–323. doi:10.1108/IJPPM-01-2013-0008

Aziri. (2011). Job satisfaction: A literature review. *Management Research and Practice, 3*(4), 77-86.

Bagheri, N. (2013). *The relation between emotional intelligence and job satisfaction. European Journal.*

Bakacsi, G. (2002). Szervezeti magatartás és vezetés. Közgazdasági és Jogi Könyvkiadó.

Berezan, O., Raab, C., Yoo, M., & Love, C. (2013). Sustainable hotel practices and nationality: The impact on guest satisfaction and guest intention to return. *International Journal of Hospitality Management, 34*, 227–233. doi:10.1016/j.ijhm.2013.03.010

Berkesné, R. N. (2018). *CSR EMAT – A vállalatok társadalmi felelősségvállalásának kiválósági menedzsment értékelési eszköze* (PhD thesis).

Bohdanowicz, P., Zientara, P., & Novotna, E. (2011). International hotel chains and environmental protection: An analysis of Hilton's Wecare! programme (Europe, 2006-2008). *Journal of Sustainable Tourism, 19*(7), 797–816. doi:10.1080/09669582.2010.549566

Braun, R. (2015). *Vállalati társadalmi felelősségvállalás A vállalatok politikája*. Akadémiai Kiadó.

Brey, P. A. E. (2012). Anticipating ethical issue in emerging IT. *Ethics and Information Technology, 14*(4), 305–317. doi:10.100710676-012-9293-y

Buhalis, D., Harwood, T., Bogicevic, V., Viglia, G., Beldona, S., & Hofacker, C. (2019). Technological disruptions in services: Lessons from tourism and hospitality. *Journal of Service Management, 30*(4), 484–506. doi:10.1108/JOSM-12-2018-0398

Buhalis, D., Harwood, T., Bogicevic, V., Viglia, G., Beldona, S., & Hofacker, C. (2019). Technological disruptions in services: Lessons from tourism and hospitality. *Journal of Service Management, 30*(4), 484–506. doi:10.1108/JOSM-12-2018-0398

Burton, J. (2010). *WHO Healthy Workplace Framework and Model: Background and Supporting Literature and Practice*. World Health Organization

Byun, G., Karau, S. J., Dai, Y., & Lee, S. (2018). A three-level examination of the cascading effects of ethical leadership on employee outcomes: A moderated mediation analysis. *Journal of Business Research, 88*, 44–53. doi:10.1016/j.jbusres.2018.03.004

Camilleri, M. A. (2018). The promotion of responsible tourism management through digital media. *Tourism Planning & Development, 15*(6), 653–671. doi:10.1080/21568316.2017.1393772

Cascio, W. F. (2003). Responsible restructuring: Seeing employees as assets, not costs. *Ivey Business Journal, 68*, 1.

Cavico, F. J., Mujtaba, B. G., & Samuel, M. (2016). Code words and covert employment discrimination: Legal analysis and consequences for management. *International Journal of Organizational Leadership, 5*(3), 231–253. doi:10.33844/ijol.2016.60365

Chanias, S., Myers, M. D., & Hess, T. (2019). Digital transformation strategy making in pre-digital organizations: The case of a financial services provider. *The Journal of Strategic Information Systems, 28*(1), 17–33. doi:10.1016/j.jsis.2018.11.003

Chi, C., & Gursoy, D. (2009). Employee satisfaction, customer satisfaction, and financial performance: An empirical examination. *International Journal of Hospitality Management, 28*(2), 245–253. doi:10.1016/j.ijhm.2008.08.003

Chiang, C., Back, K.-J., & Canter, D. D. (2005). The Impact of Employee Training Satisfaction and Intention to Stay in the Hotel Industry. *Journal of Human Resources in Hospitality & Tourism*, *4*(2), 99–118. doi:10.1300/J171v04n02_06

Cobos, L. M., Mejia, C., Ozturk, A. B., & Wang, Y. (2016). A technology adoption and implementation process in an independent hotel chain. *International Journal of Hospitality Management*, *57*, 93–105. doi:10.1016/j.ijhm.2016.06.005

Council of the European Union. (2016). *General Data Protection Regulation*. Available at https://data.consilium.europa.eu/doc/document/ST-5419-2016-INIT/en/pdf

Crittenden, A. B., Crittenden, V. L., & Crittenden, W. F. (2017). Industry transformation via channel disruption. *Journal of Marketing Channels*, *24*(1–2), 13–26. doi:10.1080/1046669X.2017.1346974

European Commission. (2016). *Towards a European Pillar of Social Rights*. Author.

Floridi, L. (2010). Ethics after the information revolution. In L. Floridi (Ed.), *Cambridge Handbook of Information and Computer Ethics* (pp. 3–19). Cambridge University Press. doi:10.1017/CBO9780511845239.002

Frans, P. (2011). The Protection of Working Relationships in the Netherlands. In The Protection of Working Relationships: A Comparative Study. Leiden: Kluwer Law International.

Grigoroudis, E., & Siskos, Y. (2010). Customer Satisfaction Evaluation: Methods for Measuring and Implementing Service Quality. Springer.

Steinbachné Hajmásy, G. (2020). *A dolgozói elégedettség vizsgálata a belső CSR tevékenységek tükrében a Balaton régió négy- és ötcsillagos szállodáinak bevonásával* (PhD thesis).

Hallowell, R., Schlesinger, L. A., & Zornitsky, J. (1996). Internal Service Quality, Customer and Job Satisfaction: Linkages and Implications for Management. *Human Resource Planning*, *19*, 20–31.

Herring, H., & Roy, R. (2007). Technological innovation, energy efficient design and the rebound effect. *Technovation*, *27*(4), 194–203. doi:10.1016/j.technovation.2006.11.004

Heskett, J. L., Jones, T. O., Loveman, G. W., Sasser, W. E., & Schlesinger, L. A. (1994). Putting the service–profit chain to work. *Harvard Business Review*, *72*(2), 164–175.

Hess, T., Matt, C., & Wiesböck, F. (2016). Options for formulating a digital transformation strategy. *MISQE*, *105447*. Advance online publication. doi:10.7892/boris

Hinings, B., Gegenhuber, T., & Greenwood, R. (2018). Digital innovation and transformation: An institutional perspective. *Information and Organization*, *28*(1), 52–61. doi:10.1016/j.infoandorg.2018.02.004

Ineson, E., Benke, E., & László, J. (2013). Employee loyalty in Hungarian hotels. *International Journal of Hospitality Management*, *32*(1), 31–39. doi:10.1016/j.ijhm.2012.04.001

Inman, J. J., & Nikolova, H. (2017). Shopper-facing retail technology: A retailer adoption decision framework incorporating shopper attitudes and privacy concerns. *Journal of Retailing*, *93*(1), 7–28. doi:10.1016/j.jretai.2016.12.006

ISO/IEC. (2013). *Information technology – Security techniques – Information security management systems – Requirements*. American National Standards Institute.

Ivanov, S. (2019). Ultimate transformation: How will automation technologies disrupt the travel, tourism and hospitality industries? *Zeitschrift fur Tourismuswissenschaft, 11*(1), 25–43.

John, L. K., Acquisti, A., & Loewenstein, G. (2010). Strangers on a plane: Context-dependent willingness to divulge sensitive information. *The Journal of Consumer Research, 37*(5), 858–873.

Kim, H., Rhou, Y., Uysal, M., & Kwon, N. (2017). An examination of the links between corporate social responsibility (CSR) and its internal consequences. *International Journal of Hospitality Management, 61*, 26–34. doi:10.1016/j.ijhm.2016.10.011

Koch, T., & Windsperger, J. (2017). *Seeing through the network: competitive advantage in the digital economy*. *J Organ Des*. doi:10.118641469-017-0016-z

Kucukusta, D., Mak, A., & Chan, X. (2013). Corporate social responsibility practices in four and five-star hotels: Perspectives from Hong Kong visitors. *International Journal of Hospitality Management, 34*, 19–30. doi:10.1016/j.ijhm.2013.01.010

Lam, C., & Law, R. (2019). Readiness of upscale and luxury-branded hotels for digital transformation. *International Journal of Hospitality Management, 79*, 60–69. doi:10.1016/j.ijhm.2018.12.015

Lam, T., Zhang, H., & Baum, T. (2001). An investigation of employee's job satisfaction: The case of hotels in Hong Kong. *Tourism Management, 22*(2), 157–165. doi:10.1016/S0261-5177(00)00039-X

Lee, H. A., Denizci Guillet, B., & Law, R. (2013). An examination of the relationship between online travel agents and hotels: A case study of Choice Hotels International and Expedia.com. *Cornell Hospitality Quarterly, 54*(1), 95–107. doi:10.1177/1938965512454218

Levy, S., & Park, S.-Y. (2011). An analysis of CSR activities in the lodging industry. *Journal of Hospitality and Tourism Management, 18*(1), 147–154. doi:10.1375/jhtm.18.1.147

Locke, E. A. (1976). The nature and causes of job satisfaction. In M. D. Dunnette (Ed.), *Handbook of industrial and organizational psychology*. Rand McNally.

Loebbecke, C., & Picot, A. (2015). Reflections on societal and business model transformation arising from digitization and big data analytics: A research agenda. *The Journal of Strategic Information Systems, 24*(3), 149–157. doi:10.1016/j.jsis.2015.08.002

Lundberg, C., Gudmundson, A., & Andersson, T. D. (2009). Herzberg's Two-Factor Theory of work motivation tested empirically on seasonal workers in hospitality and tourism. *Tourism Management, 30*(6), 890–899. doi:10.1016/j.tourman.2008.12.003

Makridakis, S. (2017). The forthcoming Artificial Intelligence (AI) revolution: Its impact on society and firms. *Futures, 90*, 46–60. doi:10.1016/j.futures.2017.03.006

Malik, M. E. (2012). The Impact of Pay and Promotion on Job Satisfaction. *American Journal of Economics, 2*(4), 6–9. doi:10.5923/j.economics.20120001.02

Matt, C., Hess, T., & Benlian, A. (2015). Digital transformation strategies. *Business & Information Systems Engineering*, *57*(5), 339–343. doi:10.100712599-015-0401-5

Matzler, K., & Renzl, B. (2007). Assessing asymmetric effects in the formation of employee satisfaction. *Tourism Management*, *28*(4), 1093–1103. doi:10.1016/j.tourman.2006.07.009

McGehee, N. (2009). Corporate social responsibility within the US lodging industry: An exploratory study. *Journal of Hospitality & Tourism Research (Washington, D.C.)*, *33*(3), 417–437. doi:10.1177/1096348009338532

Moriarty, J. (2016). Business ethics. *Stanford Encyclopedia of Philosophy*. Available at https://plato.stanford.edu/entries/ethics-business/

Murali, V., Chaudhuri, S., & Jermaine, C. (2018). *Neural sketch learning for conditional program generation*. International Conference on Learning Representations (ICLR 2018), Vancouver, Canada.

Nav'ıo-Marco, J., Ruiz-G'omez, L. M., & Sevilla-Sevilla, C. (2018). Progress in information technology and tourism management: 30 years on and 20 years after the internet-Revisiting Buhalis & Law's landmark study about eTourism. *Tourism Management*, *69*, 460–470. doi:10.1016/j.tourman.2018.06.002

Porath, C. L. (2015). *Care: How to create an environment where employees and organizations thrive*. Academic Press.

Raihan, T., & Al Karim, R. (2017, April). CSR and employee job satisfaction: A case from MNCS Bangaldesh. *Global Journal of Human Resource Management*, *5*(3), 26–39.

Reagan, B. (2011). *Community Benefits and Accountable Development*. IRLE Job Creation and Local Communities Speaker Series.

Reinartz, W., Wiegand, N., & Imschloss, M. (2019). The impact of digital transformation on the retailing value Chain. *International Journal of Research in Marketing*, *36*(3), 350–366. doi:10.1016/j.ijresmar.2018.12.002

Richter, A., & Riemer, K. (2013). Malleable end-user software. *Business & Information Systems Engineering*, *5*(3), 195–197. doi:10.100712599-013-0260-x

Rodríguez-Antón, J., & Alonso-Almeida, M. M. (2011). Quality certification systems and their impact on employee satisfaction in services with high levels of customer contact. *Total Quality Management & Business Excellence*, *22*(2), 145–157. doi:10.1080/14783363.2010.529640

Rosso, B.D., Dekas, K. H., & Wrzesniewski, A. (2010). On the meaning of work: A theoretical integration and review. Research in Organizational Behavior, 30, 91-127. doi:10.1016/j.riob.2010.09.001

Schumann, J. H., Von Wangenheim, F., & Groene, N. (2014). Targeted online advertising Using reciprocity appeals to increase acceptance among users of free web services. *Journal of Marketing*, *78*(1), 59–75. doi:10.1509/jm.11.0316

Schwartz, M., & Carroll, A. B. (2003). Corporate social responsibility: A three domain approach. *Business Ethics Quarterly*, *13*(4), 503–530. doi:10.5840/beq200313435

Soltani, A. (2019). Abusability testing: Considering the ways your technology might be used for harm. Enigma 2019, Burlingame, CA.

Spiller, R. (2000). Ethical Business and Investment: AModel for Business and Society. *Journal of Business Ethics*, *27*(1/2), 149–160. doi:10.1023/A:1006445915026

Spinelli, M. A., & Canavos, G. C. (2000). Investigating the relationship between employees satisfaction and guest satisfaction. *The Cornell Hotel and Restaurant Administration Quarterly*, *41*(6), 29–33.

Strahilevitz, M., & Myers, J. G. (1998). Donations to charity as purchase incentives: How well they work may depend on what you are trying to sell. *The Journal of Consumer Research*, *24*(4), 434–446. doi:10.1086/209519

Thorun, C. (2018). Corporate Digital Responsibility: Unternehmerische Verantwortung in der digitalen Welt. In C. Gärtner & C. Heinrich (Eds.), *Fallstudien zur Digitalen Transformation* (pp. 173–191). Springer Gabler. doi:10.1007/978-3-658-18745-3_9

United Nation's Global Compact. (2000). United Nations.

Valeri, M. (Ed.). (2021a). Tourism risk. Crisis and recovery management. Emerald Publishing.

Valeri, M. (2021b). *Organization Theories. Implications for the Strategic Management*. Springer.

Valeri, M., & Katsoni, V. (Eds.). (2021). *Gender and tourism: challenges and entrepreneurial opportunities*. Emerald Publishing. doi:10.1108/9781801173223

Verčič, A. T., & Ćorić, D. S. (2018). The relationship between reputation, employer branding and corporate social responsibility. *Public Relations Review*, *44*, 442–452.

Vial, G. (2019). Understanding digital transformation: A review and a research agenda. *The Journal of Strategic Information Systems*, *28*(2), 118–144. doi:10.1016/j.jsis.2019.01.003

Vosoughi, S., Roy, D., & Aral, S. (2018). The spread of true and false news online. *Science*, *359*(6380), 1146–1151. doi:10.1126cience.aap9559 PMID:29590045

Vostrá, H., Jindrová, A., & Dömeová, L. (2011). The Position of the CR among the EU States Based on Selected Measures of the Lisbon Strategy. *Journal of Competitiveness*, *11*(3), 50–57.

Webster, C., & Ivanov, S. (2020). Robotics, artificial intelligence, and the evolving nature of work. In B. George & J. Paul (Eds.), *Digital Transformation in Business and Society Theory and Cases* (pp. 127–143). Palgrave-MacMillan. doi:10.1007/978-3-030-08277-2_8

Williams, C., & Buswell, J. (2003). Service Quality in Leisure and Tourism. Cromwell Press.

Wirtz, J., Patterson, P., Kunz, W., Gruber, T., Lu, V. N., Paluch, S., & Martins, A. (2018). Brave new world: Service robots in the frontline. *Journal of Service Management*, *29*(5), 907–931. doi:10.1108/JOSM-04-2018-0119

Wirtz, J., Patterson, P., Kunz, W., Gruber, T., Lu, V. N., Paluch, S., & Martins, A. (2018). Brave new world: Service robots in the frontline. *Journal of Service Management*, *29*(5), 907–931. doi:10.1108/JOSM-04-2018-0119

Wirtz, J., So, K. K. F., Mody, M., Chun, E. H., Liu, S., & Chun, H. (2019). Platforms in the peer-to-peer sharing economy. *Journal of Service Management*, *30*(4), 452–483. doi:10.1108/JOSM-11-2018-0369

Wirtz, J., So, K. K. F., Mody, M., Chun, E. H., Liu, S., & Chun, H. (2019). Platforms in the peer-to-peer sharing economy. *Journal of Service Management*, *30*(4), 452–483. doi:10.1108/JOSM-11-2018-0369

Xiang, Z. (2018). From digitization to the age of acceleration: On information technology and tourism. *Tourism Management Perspectives*, *25*, 147–150. doi:10.1016/j.tmp.2017.11.023

Xiong, L., King, C., & Piehler, R. (2013). "That's not my job": Exploring the employee perspective in the development of brand ambassadors. *International Journal of Hospitality Management*, *35*, 348–359. doi:10.1016/j.ijhm.2013.07.009

Yang, X., & Rivers, C. (2009). Antecedents of CSR practices in MNCs' subsidiaries: A stakeholder and institutional perspective. *Journal of Business Ethics*, *86*(2), 155–169. doi:10.100710551-009-0191-0

Yeh, C. M. (2013). Tourism involvement, work engagement and job satisfaction among frontline hotel employees. *Annals of Tourism Research*, *42*, 214–239. doi:10.1016/j.annals.2013.02.002

Yoon, M., & Suh, J. (2003). Organizational citizenship behavior and service quality as external effectiveness of contact employees. *Journal of Business Research*, *56*(8), 597–611. doi:10.1016/S0148-2963(01)00290-9

Youn, H., Lee, K., & Lee, S. (2018). Effects of corporate social responsibility on employees in the casino industry. *Tourism Management*, *68*, 328–335. doi:10.1016/j.tourman.2018.03.018

Yucel. (2012). Examining the Relationships among Job Satisfaction, Organizational Commitment, and Turnover Intention: An Empirical Study. *International Journal of Business and Management*, *7*(20), 44–58.

Yuen, K., Loh, H. S., Zhou, Q., & Wong, Y. D. (2018). Determinants of job satisfaction and performance of seafarers. *Transportation Research Part A, Policy and Practice*, *110*, 1–12. doi:10.1016/j.tra.2018.02.006

Zhao, X., Ghiselli, R., Law, R., & Ma, J. (2016). Motivating frontline employees: Role of job characteristics in work and life satisfaction. *Journal of Hospitality and Tourism Management*, *27*, 27–38. doi:10.1016/j.jhtm.2016.01.010

Zopiatis, A., Constanti, P., & Theocharous, A. L. (2014). Job involvement, commitment, satisfaction and turnover: Evidence from hotel employees in Cyprus. *Tourism Management*, *41*, 129–140. doi:10.1016/j.tourman.2013.09.013

ENDNOTE

[1] https://mtu.gov.hu/documents/prod/Kereskedelmi_szallashelyek_adatai_2018.pdf

Chapter 4
Cointegration and Causality Analysis of Portuguese Tourism and Air Quality:
Differences Among International Tourism Markets

Mara Madaleno
https://orcid.org/0000-0002-4905-2771
GOVCOPP, Universidade Aveiro, Portugal

Margarita Robaina
https://orcid.org/0000-0002-7919-5888
GOVCOPP, Universidade Aveiro, Portugal

Celeste Eusébio
https://orcid.org/0000-0002-2220-5483
GOVCOPP, Universidade Aveiro, Portugal

Maria João Carneiro
GOVCOPP, Universidade Aveiro, Portugal

Vitor Rodrigues
GOVCOPP, Universidade Aveiro, Portugal

Alexandra Monteiro
CESAM, Universidade Aveiro, Portugal

Carla Gama
CESAM, Universidade Aveiro, Portugal

ABSTRACT

This chapter aims to fill the knowledge gap regarding the relationship between tourism and air quality, specifically in the Portuguese tourism industry, with a focus on tourist nationality. It examines whether this relationship differs according to tourist origin. This study uses an air pollutant, PM10, with a strong impact on human health that has been highly neglected in the literature. Despite the great use of CO2 in assessing the causal relationship between tourism and the environment, this is not the best indicator of air quality (AQ). This chapter presents results by applying vector autoregressive models (VAR) with monthly data for the period of 2007-2017, considering the nationality of tourists that visit Portugal. Results suggest that PM10 levels and tourism are negatively correlated (in the Pearson sense) with a link between them in the long run. This relationship is confirmed by the four methodologies tested. The negative relation in Pearson and cointegration results suggests that tourism can be affected by AQ in Portugal and may lead to better AQ.

DOI: 10.4018/978-1-6684-3889-3.ch004

INTRODUCTION

There is a growing concern about the two-way relationship between tourism and air quality. If on one hand, air quality has a direct impact on certain touristic regions, affecting not only the tourism industry but also other economic activities, on the other hand, tourism has adverse impacts on the environment, namely through the emissions of greenhouse gases (GHG), waste production and natural resources consumption. Therefore, there is a rising awareness from the tourism industry to adapt and change in face of these two-way effects.

The relationship between tourism and environmental variables has been studied by different authors. Most of these studies use CO2 emissions, climate change, environmental pollution and air quality variables. Air pollution is a particular environmental problem nowadays, being responsible for many premature deaths. Almost all tourism activities have polluting impacts, but there is a consensus that the consumption of transport services, mainly road and air transport, have relevant impacts on air quality (WTO, 2003). On the other hand, air quality is a central factor in the selection of tourist destinations (Hu and Wall, 2005, Mihali, 2000, Zhang et al. 2015). First, because air pollution causes considerable health risks, as mental and emotional depression, and respiratory and cardiovascular diseases (Hadley et al., 2018, Szyszkowicz et al., 2016, Trasande and Thurston, 2005), and second, because air pollution considerably harms visibility, reducing the aesthetic satisfaction and pleasure obtained by tourists, affecting their willingness to travel (Mace et al., 2004, Poudyal et al., 2013). The seeking for fresh air can become a critical motivational stimulator for spatial mobility at the international level that may ultimately affect international tourism flows and outbound tourism demand (Wang et al., 2018).

Moreover, the tourist origin may affect this relationship, that is, the country of origin can influence, either the tourist's environmental behavior either the impact of the air quality in travel decisions (Xu and Dong, 2020; Valeri, 20216; Valeri and Baggio, 2020a; 2020b; 2020c; 2021; Baggio and Valeri, 2021). Tourist behavior is greatly shaped by social norms and values, which are closely contingent on geographical locations (Pizam and Milman, 1986). Moreover, the pulling effects of attributes involved in selecting among alternative tourism destinations correspond closely to the evolving environment tourists have in their place of residence (Santos et al., 2021a; 2021b). The influence of air quality on the tourism demand of people with high disposable income will be lower than that on people with low disposable income (Wang et al., 2018). Therefore, tourist origin is a relevant variable to be considered when studying the two-way relationship between tourism and air quality.

To investigate the link between these two variables - tourism and air quality – the present study intends to assess this relationship, specifically in the Portuguese tourism industry, examining whether this relationship differs according to tourist origin (nationality). PM10 is used as the air pollutant driver – since it is one with highest impact on human health, using monitoring monthly data collected for the period of 2007-2017.

The contribution to the literature is made in several aspects. First, this chapter studied the relationship between tourism demand and air quality using Vector Autoregressive (VAR) models, provided these allow for simultaneous relationships among variables and their lagged values. It is relevant to analyze this subject, given that no studies are known that, through cointegration and causality methods, examine the relationship between tourism flows and one of the most relevant air pollutants (PM10) in Portugal. As mentioned, the number of studies analyzing the relationship between tourism and PM10 is still very scarce (Eusébio et al., 2020). Second, this study measures the causality between these two variables by tourist nationality, exploring possible heterogeneities among different tourist groups. In this scope, the literature concerning tourist nationality is also very scarce, and it is not known any study for Portugal.

From the analysis we observe that PM10 levels and tourism are negatively correlated, with a link between them in the long run. The negative relation in Pearson and Cointegration results suggests that tourism and PM10 have a negative relationship, which may lead to a better AQ, possibly due to the growing tourists concerns with the environment. They also allowed us to conclude for differences among the nationality of visitors that do tourism trips in Portugal. Moreover, markets react with a delay showing evidence of past effects over the present ones, although Granger causality has not been very straight about the true relationship between AQ and tourist origin demand.

It is relevant to analyze the subject given that no studies are known that, through cointegration and causality methods, examine the relationship between tourism flows and one of the most relevant air pollutants (PM10) in Portugal. The number of studies analyzing the relationship between tourism and PM10 is still very scarce (Eusébio et al., 2020) and, as far as we know, no previous studies focuses on the nationality of the tourist demand. Nevertheless, the number of studies that assess the impact of tourism on AQ has been increasing (Russo et al., 2020). AQ can also influence tourism flows to several destinations, but the number of studies analyzing this effect is still very limited and has mostly focused on Asian countries (Eusébio et al., 2020). Therefore, a relationship between tourism and AQ is expected, which has led to an increase in the number of recently published studies examining this relationship (Zhou, 2019; Eusébio et al., 2020). However, most of these studies use carbon dioxide (CO_2) emissions as a proxy for AQ. Additionally, this relationship might change with the tourist origin and this chapter intends to verify if this is the case by answering the question: how does the relationship between tourism demand and AQ differs depending on the nationality of the tourist? A more in-depth analysis of the country of origin of tourists is extremely beneficial as supported by results.

Overall, according to the study results, it can be inferred that if the air quality in Portugal can be substantially improved, inbound tourist arrivals have the potential to rise. Policy directions are discussed and presented.

The remaining of the chapter is organized as follows. Section 2 presents a literature review and Section 3 describes the empirical model and data. Section 4 reports and discusses the estimation results based on the empirical model and associated implications. Finally, Section 5 concludes the paper and talks about directions for future studies.

2. LITERATURE REVIEW

2.1 Impact of Tourism on Air Quality

Several studies focus on the impact of tourism activities on air quality. The majority uses as an air quality proxy the variable CO_2 emissions (for instance, Azam et al. 2018, Ahmad et al. 2018, Wang and Wang, 2018), which is not directly linked to any atmospheric pollutant with human health effects associated and legislated with limit values. Only a few use particulate matter concentrations (PM10) - one of the most critical air pollutants in terms of registered concentration values and associated health impacts (as Saenz-de-Miera and Rosselló (2014) and Robaina et al. (2020)).

The general conclusion is that Tourism activities raise pollutants, and deteriorate air quality, as concluded for instance by Azam et al. (2018) for Malasia, Saenz-de-Miera and Rosselló (2014) for Mallorca,

Ahmad et al. (2018) for some Chinese regions (Ningxia, Qinghai, Gansu, and Shanxi), Paramati et al. (2017) for Eastern European Union (EU) countries and Robaina et al. (2020) for Switzerland, Cyprus, Great Britain, Austria and Italy.

Nevertheless, Azam et al. (2018) concluded that for Thailand and Singapore, the impact of tourism on environmental pollution is negative, and Paramati et al. (2017) concluded that CO_2 emissions in the Western EU decrease with the tourism growth.

2.2 Impact of Air Quality on Tourism

Despite the considerable literature regarding the impacts of tourism on air quality, there is a limited number of studies that assess the impact of the environment on tourism, specifically the impact of air quality on the tourist demand. Eusébio et al. (2020) reviewed the studies on this last impact, analyzing the different methods and approaches used, as well as the results obtained. The reviewed studies showed that tourists are more likely to avoid tourism destinations where the environment is severely polluted, especially if it concerns to air pollution.

The impact of air quality on tourism can be analyzed from two perspectives. First, air pollution causes great health risks to tourists, as concluded by Guindi et al. (2018) and Pant et al. (2018). Indeed Lu et al. (2015) and Trasande et al. (2005) stated that air pollution is directly correlated with the exacerbation of asthma, higher incidence of cardiovascular diseases, and increase mortality. Second, air pollution can reduce the quality of the visit to tourist attractions, in aesthetic features of sceneries and tourist experiences, or also causing traffic delays (Xu and Dong, 2020). This can damage the tourist perception of the destination, reducing future tourism demand (Poudyal et al., 2013; Becken et al., 2017).

The literature indicates that air quality can exert adverse impacts on tourists both physically and psychologically, so it is reasonable to argue that air quality is a critical influential factor in tourism development. Therefore, it is expected that air pollution would have a negative influence on the number of tourist arrivals (Xu and Dong, 2020).

2.3 Air Quality and Tourism Relationship Considering the Tourist Origin

Most of the studies that focus on the air quality and tourism relationship do not consider the origin country of the tourist, which can influence the tourist behavior due to, for instance, the economic level or living standards, and the environmental conditions at the origin country.

In fact, some studies concluded that the local air quality at the origin could affect its tourists' perception of the air quality when planning a tourism trip. In particular, it is verified that there is a positive correlation between local air pollution and the tourist awareness of and concern about air pollution (including sensory perceptions, symptoms, and risk perception) (Deguen et al., 2008).

Xu and Dong (2020) concluded that if the degree of air pollution in tourists' origin countries increases, on average, tourists probably care more about pollution and become less willing to visit China. The effect of pollution in tourist origin countries is explained by more awareness of and concern about air pollution by potential tourists if they live in more polluted countries. The impact of air pollution in destination regions was larger for tourists coming from more polluted areas (e.g. Asian countries), and visiting less polluted and more popular destinations, that is, inbound tourists coming from different origin countries and visiting different destination regions responded to air pollution dissimilarly. Other studies reached similar conclusions, as Atari et al. (2009) for Canada and Oglesby et al. (2000) for Switzerland.

Distinguishing the tourists between domestic and international, in China, Lui et al. (2019) concluded that domestic tourists are very sensitive to changes in PM2.5 concentration, whereas international tourists are less sensitive.

Considering the environmental impact of tourism, Prayag and Brittnacher (2014) studied the city of Nice, analyzing visitors from the two main generating markets of the destination: the UK and Germany. They concluded that eleven performance scores were attributed with high importance and only five scores were evaluated significantly differently by German and British visitors.

Another variable associated with tourist origin is the disposable income level. Wang et al. (2018) concluded that the local disposable income level negatively moderates the relationship between air quality and outbound tourism demand, that is, the influential power of air quality decreases when the local disposable income level is high. Indeed, faced with deteriorating air quality, people with high disposable income can invest in air purifiers and healthcare to scale down air-related health hazards. These people also consider travel as a means to maintain their social status because both travel and shopping are social status symbols (Kim et al., 2009; Park et al., 2010).

3. DATA AND METHODOLOGY

3.1 Study context

Portugal is an interesting case study for the importance of tourism on the national economy and for the impressive tourism growing rates of the last decade, a fact that worried academics, politicians, and the local population, concerning the phenomenon of over-tourism (especially in cities like Porto and Lisboa, and even so with no scientific proof) and the impact that this could have on the environment and, in particular, on air quality. Although the AQ situation in Portugal is significantly better when compared to other countries like China, the massive inbound tourism flows within the country are raising awareness among policymakers and environmentalists, turning clearer the need to infer about the current status of this relationship.

In 2019, there were 27.0 million tourists - an increase of 7.2% compared to 2018 - of which, 16.3 million were inbound (non-resident) tourists. With a total of 69.8 million overnight stays in 2019 (48.8 million overnight stays from inbounds and 21.1 million overnight stays from nationals), the main sending markets for Portugal were United Kingdom (9.4 million), Germany (5.9 million) and Spain (5.2 million) (see Figure *1* and Figure 2). It is relevant to notice that in the period 2009-2019, tourism (overnights) grew 92% and inbound tourism (non-resident overnights) grew 111%, while for residents the growth was 59%.

Figure 1. Overnights in Portugal, by Residents and non-Residents
Source: Own elaboration with data from INE (2008-2019). Estatísticas do Turismo. Lisboa: Instituto Nacional de Estatística.

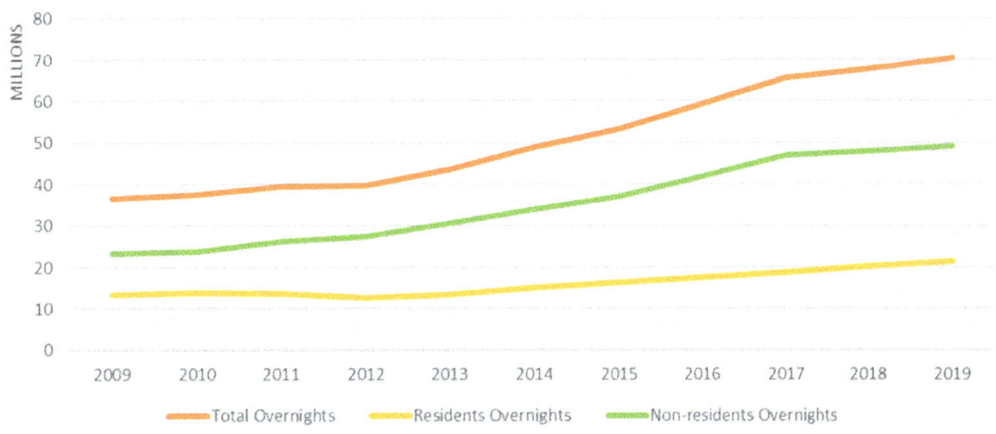

For the three main markets (UK, Germany, and Spain), the growth rates in this period were 65%, 77%, and 64%, respectively. Even more notable was the growth of the Chinese market (1186%), the USA (412%), and the Brazilian market (397%).

Figure 2. Overnights in Portugal by tourist origin
Source: Own elaboration with data from INE (2008-2019). Estatísticas do Turismo. Lisboa: Instituto Nacional de Estatística.

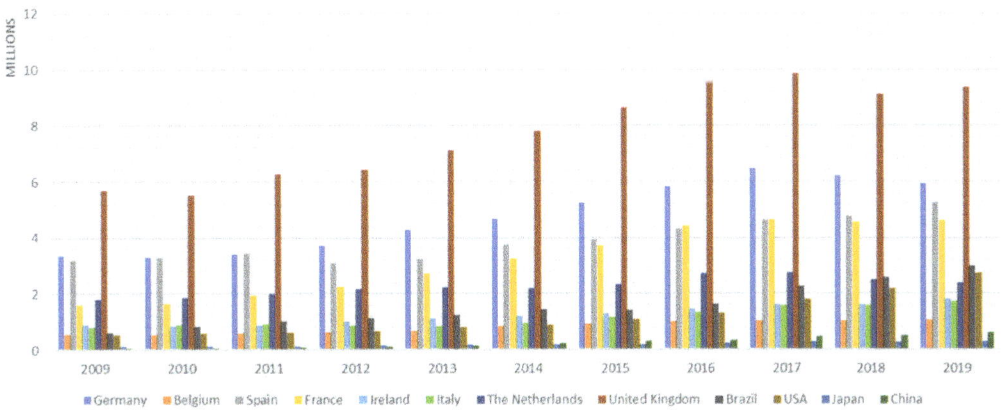

This growth was also reflected in revenues, with an increase of 8.1%, corresponding to 18.4 billion euros in 2019. Based on this indicator, the main foreign markets for Portugal were the United Kingdom (3.3 MM €), France (2.6 MM €), and Germany (2 MM €). The tourism sector is the largest export economic activity in the country, being responsible, in 2019, for 52.3% of exports of services and 19.7% of total exports, with tourist revenues registering a contribution of 8.7% in the Portuguese GDP (Turismo de Portugal, 2019). Additionally, in 2019, the tourism sector generated 336.8 thousand jobs (a 6.9% weight in the national economy) (data from 3 CAEs: accommodation, restaurants, and travel agencies).

The tourism expenditure in the percentage of GDP, was about 5.5% in 2010 and growing to 10.4% in 2019, exhibiting higher rates than in other European countries, with exception of Greece (see Figure *3*).

Figure 3. Tourism expenditure by total GDP
Source: Own elaboration with data from https://www.unwto.org/statistics/country-fact-sheets

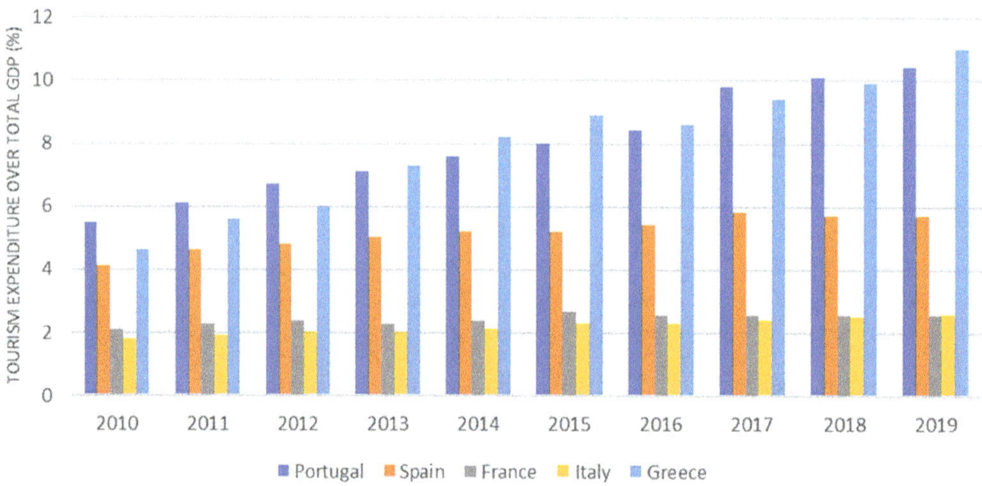

The weight of inbound tourism in the total population represented in Portugal 2.2% in 2009 and the percentage more than doubled to 4.8% in 2019, despite being in third place in the set of countries considered, behind Greece and Spain.

Figure 4. Inbound Tourism per capita
Source: Own elaboration with data from https://ec.europa.eu/eurostat/databrowser/view/TOUR_OCC_NINAT__custom_374745/default/table?lang=en and INE (2008-2019). Estatísticas do Turismo. Lisboa: Instituto Nacional de Estatística.

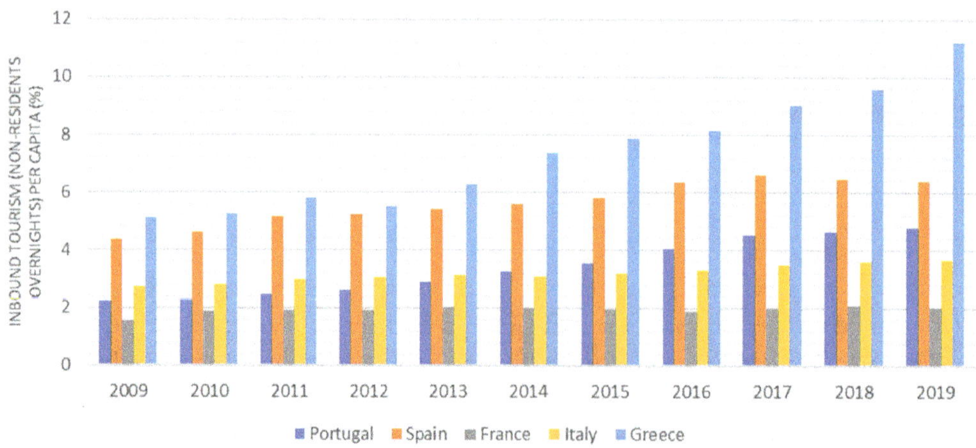

Although the pollutants concentration values have been decreasing in recent years, Figure 5 shows that regarding exposure of population to PM10 values, Portugal is below the European average, while for PM2.5 it is above.

Figure 5. Urban population exposure to air pollution by particulate matter
Source: Own elaboration with data from Eurostat.

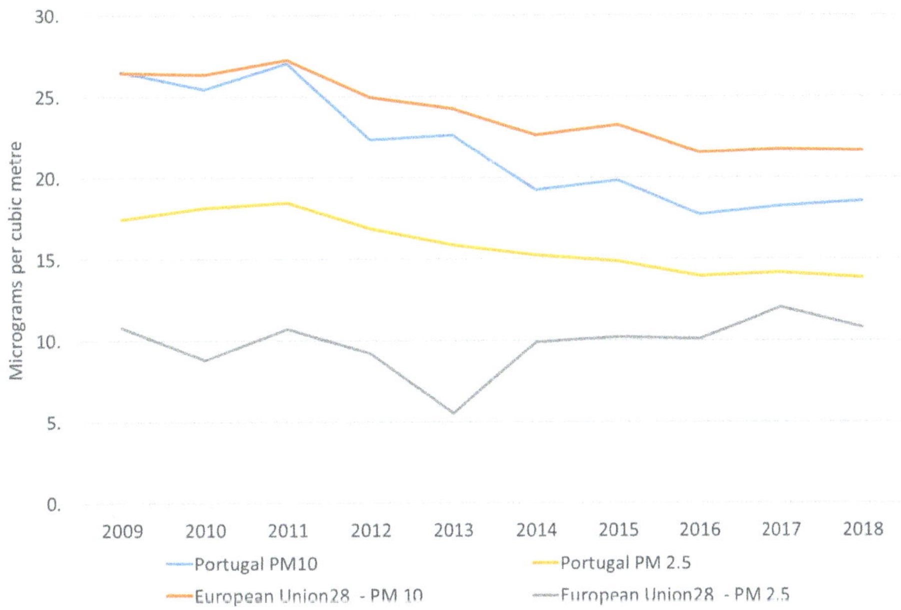

3.2 Data

This empirical research aims to estimate the relationship between inbound tourism and AQ (PM10 concentrations) in Portugal, using cointegration and causality tests, and exploring the differences in the direction of causality by each tourist origin country.

The proxy data selected for tourism is the total number of nights spent by tourists in accommodation establishments, by the origin of the tourist. Monthly figures from January 2007 to December 2017, taken from the National Statistics Institute (INE; Instituto Nacional de Estatística), were used for this study. The origin of the tourist countries considered were those with higher weight in total demand, namely, from Belgium (BEL), Brazil (BRA), China (CHI), France (FRA), Germany (GER), Ireland (IRE), Italy (ITA), Japan (JAP), The Netherlands (NET), Spain (SPA), The United Kingdom (UK), and the United States of America (USA). To facilitate the presentation of the results, these acronyms will be used to represent the origin of the tourist (being represented by the total of nights spent by the German (GER) tourist in Portuguese accommodation establishments, for example).

Data regarding PM10 observations was collected during the same period (January 2007 to December 2017) at the background monitoring stations belonging to the national AQ monitoring network. PM10 monthly mean concentrations were calculated from the hourly measurements. To maintain coherence with the available tourism data, PM10 concentrations were collected for Portugal.

3.3 Methodology

For estimates, the Eviews 10 software was used, for unit root tests, cointegration, and correlation analysis, as well as for all the previous necessary tests and the vector autoregressive models application. Since both PM10 and tourism data have seasonal variability, to obtain the deseasonalized data we applied the moving average technique (arithmetic means) when the number of observations is even. This method uses the concept of flattening the fluctuations of data by taking the means, measuring the trend by eliminating the changes or the variations using a centered moving average (Ahmed et al., 2010; Mansor et al., 2019). Additionally, as has often been used in similar studies, all the series are expressed in logarithms to facilitate the interpretation of coefficients (Ahmed et al., 2010; Mansor et al., 2019) and to adequately be able to compare the variables, since they were originally presented with different units.

The first step in the empirical analysis studies the stationary properties of the data by applying the commonly used unit root tests of Augmented Dickey-Fuller (ADF test) and the Kwiatkowski, Phillips, Schmidt, and Shin Test (KPSS test). All deseasonalized and log linearized series were revealed to be stationary. We also applied a common Vector Autoregressive (VAR) model with no restrictions added to be able to study the optimal number of lags to be included in each model by region, as well as the existence of cointegrating relationships. After this step, the model was adjusted considering the optimal number of lags and the number of cointegrated relationships when these existed. If the time series have a unit root, it makes sense that these variables have common dynamics that transform themselves into long-term relationships. The most appropriate methodology to estimate these long-term relationships is to investigate the presence of cointegration between the model variables and estimate error correction models. Regarding the VAR lag order selection tests, the Akaike Information Criterion (AIC) was considered. Various studies already indicate that the Akaike criterion is preferable, as in Tang et al. (2019), and the lag length criteria indicated 2 lags to be included. The most common method for cointegration testing is the method of Johansen as stated by Mansor et al. (2019). In the presence of cointegrating relationships, we need to adapt the model to be estimated from the VAR to the VEC (vector error correction model). In all pairwise estimations, the total number of cointegrated relationships identified was two. After all these model specification tests, and provided that for all tourist origin we have detected the presence of cointegrating relationships, model econometric estimations were made based on the VEC. The relationships between tourism demand and PM10 levels were studied through the estimation of VAR and VEC models to assess the existence of interdependent relationships and a long-term equilibrium relationship between tourism demand and PM10. This step was divided into three sections: (1) the estimation of VEC models; (2) the variance decomposition (VD) and (3) the Granger causality tests.

The VAR model is effective and flexible for the analysis of multivariate time series, being an extension of univariate models to dynamic ones. It is useful to describe the dynamic behavior of time series, and for forecasting, provided it allows estimations in a system of theory-based simultaneous equations. The model is specified as:

$$\begin{pmatrix} y_{1t} \\ y_{2t} \end{pmatrix} = \begin{pmatrix} c_1 \\ c_2 \end{pmatrix} + \begin{pmatrix} \beta_{11}^1 & \beta_{12}^1 \\ \beta_{21}^1 & \beta_{22}^1 \end{pmatrix} \begin{pmatrix} y_{1t-1} \\ y_{2t-1} \end{pmatrix} + \begin{pmatrix} \beta_{11}^2 & \beta_{12}^2 \\ \beta_{21}^2 & \beta_{22}^2 \end{pmatrix} \begin{pmatrix} y_{1t-2} \\ y_{2t-2} \end{pmatrix} + \begin{pmatrix} \varepsilon_{1t} \\ \varepsilon_{2t} \end{pmatrix} \tag{1}$$

where $cov\left(\varepsilon_{1t}, \varepsilon_{2s}\right) = \sigma_{12}$ for t = s; 0 otherwise.

Our VAR (p) model has p = 2 lags, providing the lag length estimation criteria implemented for each model specification. y_{1t} and y_{2t} are simultaneously dependent and independent variables, with y_1 in respect of tourist demand origin and y_2 of PM10. As we have one variable of PM10 and considered 12 different tourist origin country, we reached a total of 1 x 12 VEC model estimates. β_{it} refers to the coefficient estimate associated with variable y_i, where i is the variable and t the lagged value of the "explanatory" variable. ε_{it} refers to the error term. Next, the variance decomposition (VD) is a relevant analysis considering that it allows to calculate the chain reactions of a given shock. The VD method examines the effects of shocks on dependent variables. This technique determines how much of the estimated error of the variance, of any variable in the system, is explained by the "innovations" or shocks of each of the explanatory variables, given a series of time intervals (here from 1 to 12 months). Each VD coefficient is interpreted as the variance percentage of the error in predicting a variable due to a specific shock of another variable, over a given period. Finally, the Granger causality was estimated, to investigate the causality, and direction of that causality (unidirectional if just from one variable to the other, bidirectional if vice-versa), between two-time series variables. If we find Granger causality in our data, that does not mean there is a straight causal link, but rather that one variable Granger causes the other, or in other words, it is a "precedence", provided the tests allow us to study casual links between random variables. As well, impulse response functions are to be presented as a complement of the VD analysis.

4. EMPIRICAL RESULTS

Table 1 shows the Pearson correlation values between tourism (by tourist origin) and PM10 concentrations.

It is possible to observe negative and significant correlation values between PM10 and all tourist origin countries' demand. Therefore, higher concentrations of PM10 (bad AQ) are related to a lower number of total nights spent by tourists in accommodation establishments. Inversely, with higher tourism demand, independent of the tourist origin, there are lower values of PM10 and improved AQ. This could suggest that tourists are sensitive to AQ levels. Previous findings indicate that tourism has an important role in controlling CO_2 emissions in Western EU countries (Paramati et al., 2017) and that tourism reduces carbon emissions in high-income countries (Ren et al., 2019). These findings evidence a possible higher environmental concern of foreign tourists compared to domestic ones (Robaina et al., 2020).

Table 1. Pearson correlation coefficients and significance values

	BEL	BRA	CHI	FRA	GER	IRE	ITA	JAP	NET	SPA	UK	USA	PM10
BEL	1												
BRA	0.743***	1											
CHI	0.848***	0.924***	1										
FRA	0.878***	0.912***	0.974***	1									
GER	0.919***	0.778***	0.895***	0.905***	1								
IRE	0.868***	0.671***	0.767***	0.775***	0.877***	1							
ITA	0.828***	0.638***	0.718***	0.751***	0.836***	0.839***	1						
JAP	0.838***	0.783***	0.863***	0.858***	0.875***	0.817***	0.772***	1					
NET	0.835***	0.838***	0.872***	0.891***	0.880***	0.805***	0.759***	0.828***	1				
SPA	0.665***	0.533***	0.581***	0.590***	0.628***	0.657***	0.687***	0.590***	0.513***	1			
UK	0.785***	0.653***	0.718***	0.751***	0.826***	0.784***	0.841***	0.802***	0.810***	0.613***	1		
USA	0.894***	0.826***	0.913***	0.914***	0.954***	0.857***	0.873***	0.907***	0.879***	0.651***	0.867***	1	
PM10	-0.305***	-0.501***	-0.472***	-0.438***	-0.311***	-0.288***	-0.150*	-0.336***	-0.361***	-0.234***	-0.244***	-0.328***	1

Source: Own elaboration. Notes: *, **, *** statistically significant at 10%, 5% and 1%, respectively.

Coefficients in the cointegrating equation give the estimated long-run relationship among the variables. Results presented in Table 2 show that tourism demand (by tourist origin country) and PM10 have a long-run relationship, statistically significant at 1%. For all origin countries (tourist demand) this relationship is negative and therefore tourism and PM10 seem to have a negative long-run relationship. As previously identified, other studies also reveal a negative influence of tourism on air pollution (e.g. Azam et al., 2018 for Thailand and Singapore, and Paramati et al., 2017, for Western EU countries).

Table 2. Vector Error Correction estimation results

	BEL	PM10	BRA	PM10	CHI	PM10	FRA	PM10
CointEq1	-0.011	-0.120***	0.013	-0.129***	0.004	-0.073***	-0.014	-0.142***
Country T(-1)	-0.516***	0.302**	-0.497***	0.145	-0.612***	0.055	-0.406***	0.132
Country T(-2)	-0.149*	-0.216	-0.300***	0.140	-0.327***	0.117	-0.095	0.302
PM10(-1)	0.083	-0.254**	0.001	-0.153	-0.084	-0.151	0.045	-0.159
PM10(-2)	-0.058	-0.190**	0.018	-0.090	-0.039	-0.102	0.076**	-0.126
Adj. R-squared	0.262	0.395	0.206	0.372	0.274	0.381	0.168	0.386
F-statistic	10.076	17.705	7.650	16.132	10.683	16.751	6.178	17.103
Log likelihood	226.024	158.302	208.121	155.858	173.651	156.831	271.980	157.377
	GER	PM10	IRE	PM10	ITA	PM10	JAP	PM10
CointEq1	-0.016	-0.161***	0.000	-0.136***	-0.008	-0.068***	-0.018	-0.127***
Country T(-1)	-0.302***	0.104	-0.196**	0.052	-0.331***	0.101	-0.454***	0.103
Country T(-2)	-0.031	0.434	-0.127	0.060	-0.223**	-0.015	-0.307***	-0.064
PM10(-1)	0.054	-0.251**	0.123	-0.272**	0.026	-0.299***	0.078	-0.248**
PM10(-2)	0.031	-0.154*	-0.079	-0.158*	0.012	-0.180**	0.053	-0.158*
Adj. R-squared	0.070	0.360	0.104	0.344	0.092	0.334	0.187	0.354
F-statistic	2.930	15.420	3.983	14.451	3.579	13.842	6.877	15.037
Log likelihood	295.672	154.721	192.707	153.139	242.776	152.125	183.999	154.100
	NET	PM10	SPA	PM10	UK	PM10	USA	PM10
CointEq1	-0.051*	-0.332***	-0.009	-0.195***	0.000	-0.014***	-0.006	-0.073***
Country T(-1)	-0.366***	0.126	-0.919***	0.279***	0.035	-0.035	-0.349***	-0.095
Country T(-2)	-0.117	0.160	-0.405***	0.192**	0.011	-0.111	-0.021	0.186
PM10(-1)	0.064	-0.212*	-0.036	-0.269**	0.207	-0.309***	-0.037	-0.262**
PM10(-2)	0.009	-0.132	0.052	-0.151*	0.182*	-0.207***	0.011	-0.178**
Adj. R-squared	0.131	0.356	0.515	0.360	0.003	0.349	0.127	0.354
F-statistic	4.851	15.153	28.129	15.384	1.088	14.700	4.736	15.036
Log likelihood	272.743	154.288	156.047	154.662	129.503	153.549	261.367	154.098

Source: Own elaboration. Notes: *, **, *** statistically significant at 10%, 5% and 1%, respectively.

However, results are dissimilar between the effects of lagged tourist origin demand over-tourism demand and PM10 levels. Mostly no significant, the coefficient signs are positive suggesting that tourism demand increases PM10 levels, thus diminishing AQ levels. Moreover, previous levels of PM10 have a negative and most of the time significant impact, leading us to conclude that previous PM10 values tend to decrease current PM10 values. This may be justified by the learning process of tourists or even by the increased awareness of nationals and foreign tourists as to the importance of the maintenance of a safe environment and to the importance of increasing AQ.

Moving one step further, variance decomposition effects in Table 3 reveal that a shock of tourist origin country demand explains more of the variance decomposition of PM10 than does the opposite. Additionally, the effect is different in different countries of origin of the tourist, where the highest shocks come from Spain, Brazil, and The Netherlands, evidencing the importance of analyzing the relationship between tourism demand and AQ by assessing the country of origin of the tourist.

This analysis is further amplified by the impulse response functions presented in Figure 6, where the results of these three countries were presented solely to simplify the analysis.

Table 3. Variance decomposition effects

Period	BEL	PM10	BRA	PM10	CHI	PM10	FRA	PM10
Variance Decomposition of Country T:								
2	99.603	0.397	99.248	0.752	99.722	0.278	99.846	0.154
6	98.085	1.915	98.482	1.518	99.733	0.267	97.949	2.051
12	98.049	1.951	98.154	1.846	99.703	0.297	96.337	3.663
Variance Decomposition of PM10PT:								
2	1.013	98.987	4.480	95.520	0.023	99.977	5.157	94.843
6	5.920	94.080	6.302	93.698	1.026	98.974	5.972	94.028
12	8.980	91.020	9.148	90.852	2.485	97.515	7.341	92.659
Period	GER	PM10	IRE	PM10	ITA	PM10	JAP	PM10
Variance Decomposition of Country T:								
2	99.994	0.006	98.319	1.681	99.811	0.189	99.997	0.003
6	97.878	2.122	99.002	0.998	98.492	1.508	99.262	0.738
12	95.348	4.652	99.467	0.533	97.020	2.980	98.376	1.624
Variance Decomposition of PM10PT:								
2	0.112	99.888	2.621	97.379	0.038	99.962	0.510	99.490
6	1.511	98.489	7.312	92.688	0.425	99.575	4.778	95.222
12	3.848	96.153	16.379	83.621	1.349	98.651	9.710	90.290
Period	NET	PM10	SPA	PM10	UK	PM10	USA	PM10
Variance Decomposition of Country T:								
2	99.801	0.199	99.680	0.320	98.664	1.336	97.715	2.285
6	94.800	5.200	99.346	0.654	98.790	1.210	97.728	2.272
12	91.405	8.595	99.496	0.504	99.391	0.609	97.154	2.846
Variance Decomposition of PM10PT:								
2	0.632	99.368	0.828	99.172	2.916	97.084	0.947	99.053
6	4.428	95.572	5.905	94.095	7.416	92.584	1.604	98.396
12	12.492	87.508	16.316	83.684	8.704	91.296	2.583	97.417

Source: Own elaboration. Notes: *, **, *** statistically significant at 10%, 5% and 1%, respectively.

Figure 6 presents the results for the Spanish, the Brazilian, and the German tourist's demand. They suggest that the impact of a shock from PM10 on tourist demand initially decreases tourist demand in Portugal, being the shocks similar for each of these countries, but with higher intensity of some in detriment to others. Curiously, Chinese tourists demand shocks over PM10 are very low, maybe due to the highest difference among AQ levels in Portugal and China, even if they are among the few countries where higher PM10 levels decrease the tourist demand, as presented in Table 2. As stated by Xu and

Dong (2020), the effect of pollution in tourist origin countries is explained by more awareness of and concern about air pollution by potential tourists if they live in more polluted countries.

Figure 6. Impulse response functions (Spain, Brazil, and The Netherlands)

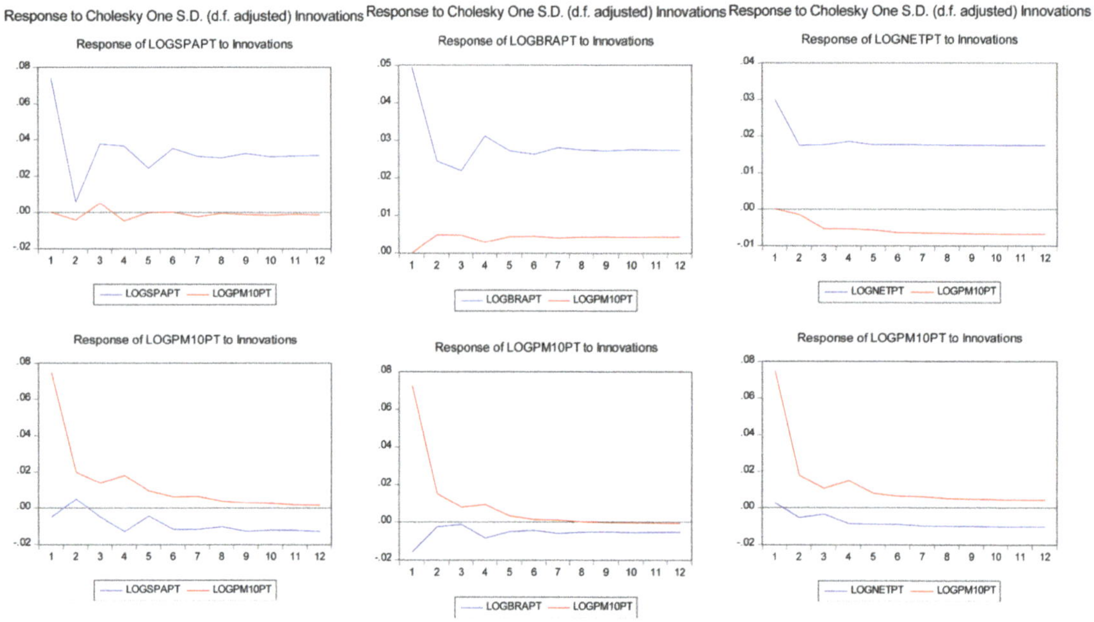

In terms of impulse response functions, results are very similar among countries, although not all presented here, due to space restrictions, where initial shocks from lower AQ decrease the demand from tourists for a destination such as Portugal. Results allow to confirm the negative pressure that tourism has on the environment and that the tourism demand is indeed influenced by AQ levels of the demanded tourism destination (at least, highlighted, for Portugal).

Finally, we present in Table 4 the Granger causality results. According to the Akaike Information Criteria (AIC) criteria, results show that PM10 Granger causes tourism demand from Belgium and Ireland, solely. There is evidence of bidirectional causality between tourism and PM10 solely for Belgium tourists. A bidirectional relationship between CO_2 emissions and tourism was found also in Paramati et al. (2017) through Granger causality tests. Tourism demand from Spanish tourists Granger causes unidirectionally PM10 levels in Portugal. This is somehow expected since we have seen previously that tourism from Spain is strong in Portugal, therefore, they increase the pressure on AQ and negative shocks are strictly related to tourism and AQ levels between the two countries.

Table 4. Granger Causality Results: VEC Granger Causality/Block Exogeneity Wald Tests

	Dep. Var.: BEL		Dep. Var.: BRA		Dep. Var.: CHI		Dep. Var.: FRA	
Excluded	Chi-sq	Prob.	Chi-sq	Prob.	Chi-sq	Prob.	Chi-sq	Prob.
PM10	7.415**	0.025	0.127	0.939	0.700	0.705	4.336	0.114
Dependent variable: PM10								
Country T	11.883***	0.003	1.637	0.441	1.504	0.472	2.113	0.348
	Dep. Var.: GER		Dep. Var.: IRE		Dep. Var.: ITA		Dep. Var.: JAP	
Excluded	Chi-sq	Prob.	Chi-sq	Prob.	Chi-sq	Prob.	Chi-sq	Prob.
PM10	2.274	0.321	8.544**	0.014	0.249	0.883	0.886	0.642
Dependent variable: PM10								
Country T	2.684	0.261	0.355	0.837	0.373	0.830	1.926	0.382
	Dep. Var.: NET		Dep. Var.: SPA		Dep. Var.: UK		Dep. Var.: USA	
Excluded	Chi-sq	Prob.	Chi-sq	Prob.	Chi-sq	Prob.	Chi-sq	Prob.
PM10	2.694	0.260	1.063	0.588	3.507	0.173	1.454	0.483
Dependent variable: PM10								
Country T	0.624	0.732	9.858***	0.007	2.425	0.297	1.610	0.447

Source: Own elaboration. Notes: *, **, *** statistically significant at 10%, 5% and 1%, respectively.

Different tourist origins reveal different behaviors concerning the relationship between pollutant concentrations and tourism demand. This can be justified by the attractiveness of the country and the number of tourists from different countries arriving, or the type of tourism which characterizes the region. A deeper analysis by regions in Portugal would allow us to better understand the relationship here evidenced between the origin of the tourist and AQ. Knowing the origin of the incoming tourism can explain some of these characteristics. Nevertheless, income per capita by origin should be included in this analysis since the geographical location and the language are the most indicative attractiveness factors that we observe in this relationship.

Paramati et al. (2017) also confirmed the impacts of tourism on CO_2 emissions in EU countries. The conclusions pointed to a significant difference between the two groups (Eastern and Western countries), concluding that the increase of tourism receipts in Western EU countries leads to a decrease in CO_2 emissions, improving environmental quality. However, the same tourism pattern in Eastern EU countries increases CO_2 emissions, which could indicate distinct management strategies with significant results in environmental quality. The same pattern is found in the analysis of different tourist origin countries. Finally, Ren et al. (2019) show that the increase of tourism decreases environmental pollution, playing an important role in raising environmental quality, but when the Granger causality test was performed, there was a non-significant short-run causal relationship between the variables.

5. CONCLUSION

This study examines the long-run causal relationship between AQ (using PM10 concentrations as a proxy) and tourism demand (using the number of overnights in tourism accommodations) in Portugal, considering also the origin country of the tourist (total number of tourists demand). We started by analyzing the Pearson Correlation, then applied pre-estimation specification tests, and finally estimating

Vector Autoregressive Models to obtain Granger Causality, Impulse Response Functions, and Variance Decomposition. Monthly data from between 2007 and 2017 were used.

Results allowed us to conclude from the Pearson Correlation results that a negative and significant correlation exists between tourism and PM10, independently of the country of origin of the tourist, evidencing that weaker AQ resulted in less tourists. Inversely, higher tourism demand is associated to lower levels of PM10 (better AQ). This could suggest that these tourist origin countries could have surpassed the threshold level of the environmental curve: tourists that visit them are sensitive to AQ levels or perhaps more tourism could improve environmental quality.

Cointegration results revealed that tourism and PM10 have a long-run relationship, valid for all tourist origins and that this relationship is negative, meaning that these two variables tend to evolve in opposite directions in the long run. Therefore, a growth in tourism levels can lead to better AQ (lower PM10 concentrations), and the growing environmental concern of tourists.

Higher dissimilarities arise when we look at the impact of shocks and magnitudes. In general, this relationship is confirmed by the methodologies used (with different interpretations, as they correspond to different tests). Moreover, the negative relation in Pearson and Cointegration results suggests that tourism can be affected by AQ in Portugal and may lead to better AQ, possibly due to tourists being more concerned with the environment. Tourism enterprises have also marketed themselves as, and attempt to be, environmentally friendly to satisfy a more aware and conscious tourist.

This chapter provides valuable theoretical and practical contributions to the world in general, not only to Portugal. For instance, in theory, some types of tourism and tourism development strategies that lead to an increase in tourism can contribute to an improvement in AQ. Nonetheless, the results highlight the importance of betting on more sustainable forms of tourism. Different results for different tourist origin countries can suggest different per capita income, geographic proximity, and language facilities (as evidenced by our literature review and the results obtained). Further studies in this field should include other macroeconomic variables, as the GDP of the tourist origin country, regional analysis as for tourism demand, cultural identity, energy consumption, transports used by tourists, kind of tourism practiced, among others, and all related to the tourist origin.

ACKNOWLEDGMENT

Financial Support from FEDER through Programa Operacional Fatores de Competitividade (COMPETE2020) and National Funds through FCT - MEC included in Project ARTUR – *O impacto da qualidade do AR na competitividade de destinos TURÍSTICOS* (POCI-01-0145-FEDER-029374). Thanks are also due to Thanks are due to FCT/MCTES for the financial support to CESAM (UIDP/50017/2020+UIDB/50017/2020) and GOVCOPP (Research Unit on Governance, Competitiveness and Public Policy (UIDB/04058/2020), through national funds. Thanks are due to the co-operative program for monitoring and evaluation of the long-range transmission of air pollutants in Europe (EMEP) for their effort in establishing and maintaining the air quality monitoring sites used in this investigation and the EBAS database for hosting the data.

REFERENCES

Ahmad, F., Draz, M. U., Su, L., Ozturk, I., & Rauf, A. (2018). Tourism and environmental pollution: Evidence from the one belt one road provinces of western China. *Sustainability*, *10*(10), 1–22. doi:10.3390u10103520

Ahmed, N. K., Atiya, A. F., Gayar, N. E., & El-Shishiny, H. (2010). An empirical comparison of machine learning models for time series forecasting. *Econometric Reviews*, *29*(5-6), 594–621. doi:10.1080/07474938.2010.481556

Atari, D. O., Luginaah, I. N., & Fung, K. (2009). The Relationship between Odour Annoyance Scores and Modelled Ambient Air Pollution in Sarnia, "Chemical Valley", Ontario. *International Journal of Environmental Research and Public Health*, 6. PMID:20054461

Azam, M., Alam, M. M., & Haroon Hafeez, M. (2018). Effect of tourism on environmental pollution: Further evidence from Malaysia, Singapore and Thailand. *Journal of Cleaner Production*, *190*, 330–338. doi:10.1016/j.jclepro.2018.04.168

Baggio, R., Valeri, M. (2020). Network science and sustainable performance of family businesses in tourism. *Journal of Family Business Management*. doi:10.1108/JFBM-06-2020-0048

Becken, S., Jin, X., Zhang, C., & Gao, J. (2017). Urban air pollution in China: Destination image and risk perceptions. *Journal of Sustainable Tourism*, *25*(1), 130–147. doi:10.1080/09669582.2016.1177067

Deguen, S., Pédrono, G., Ségala, C., & Mesbah, M. (2008). Association Between Pollution and Public Perception of Air Quality-SEQAP, a Risk Perception Study in France. *Epidemiology (Cambridge, Mass.)*, *19*, S216.

Eusébio, C., Carneiro, M.J., Madaleno, M., Robaina M., Rodrigues, V., Russo, M., Relvas, H., Gama, C. Lopes, M., Seixas, V., Borrego, C. & Monteiro, A. (2020). The impact of air quality on tourism: a systematic literature review. *Journal of Tourism Futures*. doi:10.1108/JTF-06-2019-0049

Guindi, M. N., Flaherty, G. T., & Byrne, M. (2018). Every breath you take: How does air pollution affect the international traveller? *Journal of Travel Medicine*, *25*(1), tay021. doi:10.1093/jtm/tay021 PMID:29718402

Hadley, M. B., Vedanthan, R., & Fuster, V. (2018). Air pollution and cardiovascular disease: A window of opportunity. *Nature Reviews. Cardiology*, *15*(4), 193–194. doi:10.1038/nrcardio.2017.207 PMID:29297510

Hu, W., & Wall, G. (2005). Environmental Management, Environmental Image and the Competitive Tourist Attraction. *Journal of Sustainable Tourism*, *13*(6), 617–635. doi:10.1080/09669580508668584

Kim, Y. G., Eves, A., & Scarles, C. (2009). Building a model of local food consumption on trips and holidays: A grounded theory approach. *International Journal of Hospitality Management*, *28*(3), 423–431. doi:10.1016/j.ijhm.2008.11.005

Lu, F., Xu, D., Cheng, Y., Dong, S., Guo, C., Jiang, X., & Zheng, X. (2015). Systematic review and meta-analysis of the adverse health effects of ambient PM2.5 and PM10 pollution in the Chinese population. *Environmental Research*, *136*, 196–204. doi:10.1016/j.envres.2014.06.029 PMID:25460637

Mace, B. L., Bell, P. A., & Loomis, R. J. (2004). Visibility and Natural Quiet in National Parks and Wilderness Areas: Psychological Considerations. *Environment and Behavior, 36*(1), 5–31. doi:10.1177/0013916503254747

Mansor, M. M., Green, D. A., & Metcalfe, A. V. (2019). Detecting Directionality in Time Series. *The American Statistician*, 1–16. doi:10.1080/00031305.2018.1545699

Mihali, C. T. (2000). Environmental management of a tourist destination: A factor of tourism competitiveness. *Tourism Management, 21*(1), 65–78. doi:10.1016/S0261-5177(99)00096-5

Oglesby, L., Künzli, N., Monn, C., Schindler, C., Ackermann-Liebrich, U., & Leuenberger, P. (2000). Validity of Annoyance Scores for Estimation of Long Term Air Pollution Exposure in Epidemiologic Studies: The Swiss Study on Air Pollution and Lung Diseases in Adults (SAPALDIA). *American Journal of Epidemiology, 152*(1), 75–83. doi:10.1093/aje/152.1.75 PMID:10901332

Pant, P., Huynh, W., & Peltier, R. E. (2018). Exposure to air pollutants in Vietnam: Assessing potential risk for tourists. *Journal of Environmental Sciences (China), 73*, 147–154. doi:10.1016/j.jes.2018.01.023 PMID:30290863

Paramati, S. R., Shahbaz, M., & Alam, M. S. (2017). Does tourism degrade environmental quality? A comparative study of eastern and western European Union. *Transport Res. D, 50*, 1–13. doi:10.1016/j.trd.2016.10.034

Parke, K. S., Reisinger, Y., & Noh, E. H. (2010). Luxury shopping in tourism. *International Journal of Tourism Research, 12*(2), 164–178. doi:10.1002/jtr.744

Pizam, A., & Milman, A. (1986). The Social Impacts of Tourism. *Tourism Recreation Research, 11*(1), 29–33. doi:10.1080/02508281.1986.11014414

Poudyal, N. C., Paudel, B., & Green, G. T. (2013). Estimating the Impact of Impaired Visibility on the Demand for Visits to National Parks. *Tourism Economics, 19*(2), 433–452. doi:10.5367/te.2013.0204

Prayag, G., & Brittnacher, A. (2014). Environmental Impacts of Tourism on a French Urban Coastal Destination: Perceptions of German and British Visitors. *Tourism Analysis, 19*(4), 461-475. doi:10.3727/108354214X14090817031116

Ren, T., Can, M., Paramati, S. R., Fang, J., & Wu, W. (2019). The impact of tourism quality on economic development and environment: Evidence from Mediterranean countries. *Sustainability, 11*(8), 2296. doi:10.3390u11082296

Robaina, M., Madaleno, M., Silva, S., Eusébio, C., Carneiro, M. J., Gama, C., Oliveira, K., Russo, M. A., & Monteiro, A. (2020). The relationship between tourism and air quality in five European countries. *Economic Analysis and Policy, 67*, 261–272. doi:10.1016/j.eap.2020.07.012

Russo, M. A., Relvas, H., Gama, C., Lopes, M., Borrego, C., Rodrigues, V., Robaina, M., Madaleno, M., Carneiro, M. J., Eusébio, C., & Monteiro, A. (2020). Estimating emissions from tourism activities. *Atmospheric Environment, 220*. doi:10.1016/j.atmosenv.2019.117048

Saenz-de Miera, O., & Rosselló, J. (2014). Modelling tourism impacts on air pollution: The case study of PM 10 in Mallorca. *Tourism Management, 40*, 273–281. doi:10.1016/j.tourman.2013.06.012

Santos, V., Ramos, P., Sousa, B., Almeida, N., & Valeri, M. (2021a). Factors influencing touristic consumer behavior. *Journal of Organizational Change Management*. Advance online publication. doi:10.1108/JOCM-02-2021-0032

Santos, V., Sousa, B., Ramos, P., & Valeri, M. (2021b). Emotions and Involvement in Tourism Settings. *Current Issues in Tourism*, 1–6. Advance online publication. doi:10.1080/13683500.2021.1932769

Szyszkowicz, M., Kousha, T., Kingsbury, M., & Colman, I. (2016). Air Pollution and Emergency Department Visits for Depression: A Multicity Case-Crossover Study. *Environmental Health Insights*, *10*, 155–161. doi:10.4137/EHI.S40493 PMID:27597809

Tang, J., Yuan, X., Ramos, V., & Sriboonchitta, S. (2019). Does air pollution decrease inbound tourist arrivals? The case of Beijing. *Asia Pacific Journal of Tourism Research*, *24*(6), 597–605. doi:10.1080/10941665.2019.1610004

Trasande, L., & Thurston, G. D. (2005). The role of air pollution in asthma and other pediatric morbidities. *The Journal of Allergy and Clinical Immunology*, *115*(4), 689–699. doi:10.1016/j.jaci.2005.01.056 PMID:15805986

Turismo de Portugal. (2019). https://travelbi.turismodeportugal.pt/pt-pt/Documents/Turismo%20em%20Portugal/turismo-em-numeros-2019.pdf

Valeri, M. (2016). Networking and cooperation practices in the Italian tourism business. *Journal of Tourism, Heritage & Services Marketing*, *2*(1), 30–35. doi:10.5281/zenodo.376333

Valeri, M., & Baggio, R. (2020a). Social network analysis: Organizational implications in tourism management. *The International Journal of Organizational Analysis*. Advance online publication. doi:10.1108/IJOA-12-2019-1971

Valeri, M., & Baggio, R. (2020b). Italian tourism intermediaries: A social network analysis exploration. *Current Issues in Tourism*, *24*(9), 1270–1283. doi:10.1080/13683500.2020.1777950

Valeri, M., & Baggio, R. (2020c). A critical reflection on the adoption of blockchain in tourism. *Journal Information Technology and Tourism*, *23*(2), 121–132. doi:10.100740558-020-00183-1

Valeri, M., & Baggio, R. (2021). Increasing the efficiency of knowledge transfer in an Italian tourism system: A network approach. *Current Issues in Tourism*, 1–16. Advance online publication. doi:10.1080/13683500.2021.1937960

Wang, L., Fang, B., & Law, R. (2018). Effect of air quality in the place of origin on outbound tourism demand: Disposable income as a moderator. *Tourism Management*, *68*, 152–161. doi:10.1016/j.tourman.2018.03.007

Wang, M. C., & Wang, C. S. (2018). Tourism, the environment, and energy policies. *Tourism Economics*, *24*(7), 821–838. doi:10.1177/1354816618781458

WTO. (2003). *Proceedings of the 1st International Conference on Climate Change and Tourism Djerba, Tunisia*, 9-11 April 2003. WTO.

Xu, B., & Dong, A. (2020). Evaluating the Impact of Air Pollution on China's Inbound Tourism: A Gravity Model Approach. *Sustainability*, *12*(4), 1456. doi:10.3390u12041456

Zhang, A., Zhong, L., Xu, Y., Wang, H., & Dang, L. (2015). Tourists' Perception of Haze Pollution and the Potential Impacts on Travel: Reshaping the Features of Tourism Seasonality in Beijing, China. *Sustainability*, *7*(3), 2397–2414. doi:10.3390u7032397

Chapter 5
Georgetown as a Gastronomy Tourism Destination:
Visitor Awareness Towards Revisit Intention of Nasi Kandar Restaurant

Albattat Ahmad
https://orcid.org/0000-0002-3127-4405
Management and Science University, Malaysia

Shufri Johari
Management and Science University, Malaysia

ABSTRACT

Nasi Kandar is one of the most favourable delicacies for tourists when visiting Penang. However, the hygiene level in most of Nasi Kandar restaurants has slumped lately. Thus, the purpose of this study is to describe the relationship between visitor awareness towards revisit intention to Nasi Kandar restaurants. Three hundred eighty-four respondents from Penang Island were accessed using a simple random sampling. The results revealed that there is a positive relationship between visitor awareness and revisit intention. Price and service eventually give a strongly positive impact towards revisit intention. This result will help the local authorities to increase the quality and hygiene of Nasi Kandar in Penang, hence bring more tourists to visit this tourism destination.

INTRODUCTION

Penang State is divided into two parts: Penang Island, which contains the capital city of Georgetown, and Seberang Perai, which contains the rest of the country. The majority of international and local tourists travel to Penang for gastronomic reasons, which is why the island has become so popular with tourists. When it comes to food, Penang is known as "Foodie Paradise" among tourists because it offers a wide variety of tastes from various traditional cultural foods and ethnicities. It is widely recognised as one of

DOI: 10.4018/978-1-6684-3889-3.ch005

the most famous street foods, and hawkers, vendors, and restaurants all serve it. Penang is also home to some of the best Mamak cuisine, which is known as "Nasi Kandar." Nasi kandar is a dish that consists of a variety of rice dishes served with various meat and vegetable curries. It was once used to transport kandar, or shoulder shafts, with crates of rice and curry suspended from either end of the shaft. Nasi kandar restaurants are highly recommended by Penang visitors due to their specialties and uniqueness. Most visitors will stop by at least once during their visit to complete their itinerary. Although there have been numerous complaints about the filthiness of restaurants, customers have continued to enjoy their visits to these establishments. Specifically, this is the main section of the researcher's key terms that are related to the factors that influence George Town as a gastronomy tourism destination, specifically on visitor awareness and their intention to return to the nasi kandar restaurant.

Nasi Kandar is one of the most popular delicacies among tourists who come to Penang for a vacation. However, the level of hygiene in the majority of nasi kandar restaurants has deteriorated in recent years. This study aims to describe the relationship between Visitor awareness and their intentions of returning to nasi kandar restaurant in order to achieve its objectives. A simple random sampling procedure was used to obtain 384 responses from residents of Penang Island. There is a positive relationship between Visitor awareness and the intention to return to the site. At the end of the day, price and service have a significant positive impact on the likelihood of returning. This outcome will assist the local authorities in improving the quality and hygiene of nasi kandar in Penang, thereby increasing the number of tourists who visit this tourist destination in Southeast Asia.

LITERATURE REVIEW

Governance and Management of Tourism Destinations

Tourism destinations are important units of analysis, albeit difficult to define (Baggio, Scott & Cooper, 2010), but they can be conceptualised as a cluster of interconnected stakeholders embedded in an underlying social network (Salpeteur, Calvet-Mir, Diaz-Reviriego & Reyes-García, 2017). A network of stakeholders interacts in order to meet the needs of visitors and to create the experience that travellers want to take away with them. Among the destination stakeholders are businesses such as hotels, attractions, tour companies, and other commercial service providers. Also included are government agencies and tourism offices, as well as representatives from the local community. Because of the complexity and dynamic nature of their interactions, they are particularly vulnerable to external shocks. Generally speaking, the fundamental premise of tourism destination management is that, through collaboration in planning and organisational activities, the effectiveness of these joint interactions can be improved to the benefit of all parties involved. Governance is a concept that refers to the relationships that exist between various stakeholders as well as the manner in which they communicate with one another. It is concerned with the process by which stakeholders determine, implement, and evaluate the rules governing their interaction with one another (Beritelli, Bieger & Laesser, 2007). As a result, it is possible that differences in tourism destination governance arrangements will result in differences in the effectiveness of joint stakeholder interactions and, as a result, improvements in destination competitiveness (Baggio, et. Al., 2010).

This chapter makes use of a network paradigm, which is based on the idea that the entire destination network, rather than just a single stakeholder, is a useful unit of analysis to be considered in this chapter.

Essentially, network governance is the coordination and protection of exchanges between a select, persistent, and structured set of autonomous firms that are engaged in the creation of products or services based on implicit and open-ended contracts, with the ability to adapt to environmental contingencies and coordinate and safeguard exchanges Valeri & Baggio, 2021). The structural properties of this network characterise the interactions of the group of stakeholders, and these properties appear to be useful in understanding destination governance and how it can be improved. A number of characteristics of such networks are discussed in this paper, and they are linked to issues of destination governance. With this paper, we will establish the key concepts involved in the study of network governance, show how they can be applied to illuminate aspects of existing destination governance networks, and offer some suggestions on how network simulations can improve information about how to improve the effectiveness of network governance arrangements (Scott, Baggio & Cooper, 2008).

When a destination is modelled as an entity subject to network governance, the governance system can be viewed as a tool that allows the destination to adapt to changes in the environment. However, we must acknowledge that the complexity of the dynamics of a destination system is a significant distinguishing feature of that system (Baggio & Sainaghi, 2011). In addition to Faulkner and Russell (1997) and McKercher (1999), who have all noted the complexity of how a destination change, they have all used complex systems approaches to analyse the destination, which are not particularly useful for simulation of destination dynamics. A lack of consensus among system components, as well as a lack of information about how they interact, is to blame for this. It is the nodes and their relationships in the network that serve as the building blocks of a model. A variety of pre-existing tools and heuristics are available to model the interaction of the nodes and relationships in the network. The remainder of this paper is devoted to an examination of this complex network approach as a means of informing theory and practise in tourism destination management and governance, as well as its limitations.

Hygiene

A report by The Star newspaper (2018) stated that one of famous nasi kandar restaurant in Penang was ordered to close its operation for two weeks after failing to comply with the basic hygiene and health restaurants' standards. The restaurant found to be dirty due to the presence of stools, dead carcasses, and dead cockroaches in the refrigerator at the premise.

Quality

Food quality is the most important thing that needs to be taken seriously for them to ensure tourists' revisit intention to come over to Georgetown in the future (Gaynor, 2019). To maintain the cleanliness and improve the food quality would require the knowledgeable food handlers that capable to avoid unwanted things like food poisoning which can cause death.

Price

The foods were sold with the reasonable price, and they serve with a big portion for one person (Fernandez, 2018). "Our prices are reasonable, and our portions are bigger to cater to an international customer base" (News Straits Times, 2017)

Services

Customer's perception about the restaurant, especially regarding the service dimension of food quality must be aware since it has a causal relationship to customers' satisfaction (Santos et al., 2021a; 2021b). High level of service quality is one of the features that can create extra customers' satisfaction towards a restaurant (Hanefors and Mossberg, 2015).

Research Framework

Figure 1. Research Framework

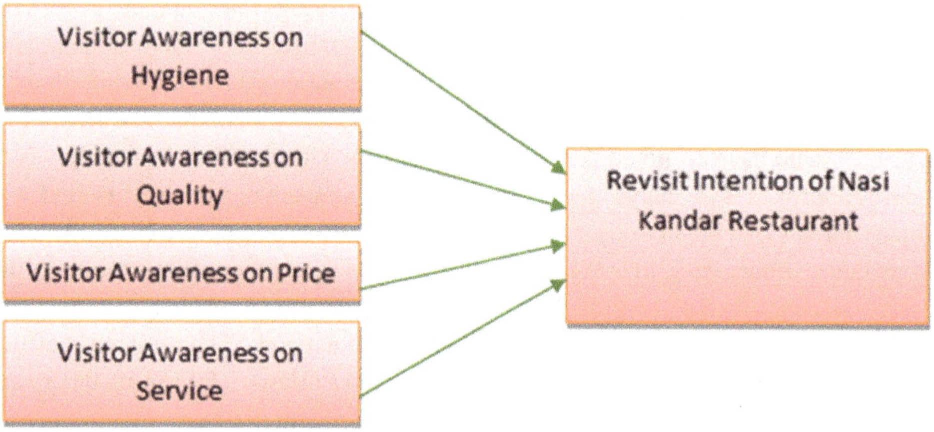

METHODOLOGY

Research design is an arrangement of the analyst's projects that considered as rule for the researchers to collect data and break down the research. Research design is characterised as a rule to confirm the data collection will be material and helpful. Quantitative research is the methodology utilised in this examination to obtain information and investigation throughout the exploration. Thus, this instrument centres around with measurable examination of numerical information gathered by means of the utilised of large-scale overview, and the used of techniques, such as questionnaire and online surveys. Therefore, it is through quantitative research for the researcher to be capable in determining and investigating the relationship between independent and dependent variables (Kılıç, M., & Kuzey, C. 2018; Valeri, 2016; Valeri and Baggio, 2020a; 2020b; 2020c; 2021; Baggio and Valeri, 2020). In quantitative approach, the researchers distribute questionnaires to answer some of the research questions. The specific target on Visitor who have spent their time in Penang basically would have information on the various type of food. Some people called Penang as Food Paradise as the state offers many types of food via the availability of trusted vendors, stall, restaurants, cafe, street and more. Therefore, 400 questionnaires were distributed but only 384 questionnaires were accepted as suggested by Krejcie and Morgan's table. The researchers' sampling criteria are among local and international tourists which indicates a simple random sampling method (Krejcie, R.V & Morgan, D.W., 1970) A suitable selection of sample criteria will be beneficial for the external and internal validity of the research.

RESULTS

Reliability Test

Scale measurement was utilised to recognise the dependability of the exploration brings by the study. The reason of having the reliability analysis was to identify how well the items in positive correlation affected with one another factors. As per Cronbach's Alpha dependable guideline by Zikmund (2003), the coefficient alpha worth 0.80-0.95 is viewed as a very good reliability. Meanwhile, for the coefficient alpha that worth 0.70-0.80 is viewed as acceptable good reliability, while for the value 0.60-0.70 is viewed as fair reliability, and the value which have below 0.60 is viewed as poor reliability. Table 1 indicated that the alpha coefficient for each factor of the independent variables is more than 0.7.

Table 1. Reliability Test

No	Variables	Cronbach's Alpha	Number of Items
1	Hygiene	0.793	11
2	Quality	0.783	10
3	Price	0.819	5
4	Service	0.814	6

Demographics

Table 2. Demographic Profile

Background factors	Categories	Frequency	Percentage %
Gender	Male Female	209 175	54.4 45.6
Age	18 – 20 21 – 25 26 – 30 31 – 35 36 – above	63 198 78 21 24	16.4 51.6 20.3 5.50 6.30
Nationality	Malaysian Non- Malaysian	342 42	89.1 10.9
Marital Status	Single Married Others	297 77 10	77.3 20.1 2.60
Educational Level	SPM STPM/Certificate Diploma/A Level Degree Master PHD	36 22 90 185 45 6	9.4 5.7 23.4 48.2 11.7 1.6
Purpose of the trip	Leisure Business Visit relatives/friends Education Others	284 38 25 37 4	74.0 8.9 6.5 9.6 1.0
Frequency visit Penang in a year	Once Twice Three and above	117 102 165	30.5 26.6 43.0

Referring the Table 2, the most respondents were Malaysian, which represent 89.1% or the frequency of 342. Meanwhile, non-Malaysian only hold a percentage of 10.9% or the frequency of 42. However, majority of the respondents were degree qualifiers, with the frequency of 185 or equal to 48.2%. This was followed by diploma holders with the frequency of 90, which represent 23.4% as its percentage. Next, it was followed by Master holders with the frequency of 45, which represent 11.7%. Besides, there are respondents who were SPM holders, which represent 9.4% or the frequency of 36. Next, respondents with STPM as their highest education level represent 5.7% with the frequency of 22, and lastly with the frequency of 6, it represents 1.6% from PhD holders. Most of the respondents who were coming for leisure represent 74.0% or the frequency of 284. This was followed by 9.6% or the frequency of 37 who are coming to Penang for education purpose. Next, about 8.9% or the frequency of 34 respondents came to the state for business matter. 6.5% or frequency of 25 were coming to visit their relatives of friends. Lastly, 1.0% or the frequency of 4 represent others. Next is regarding the respondents' frequency in visiting Penang in a year. As shown by the table, 43.0% with the frequency of 165 had visited the state three times and above. Next, 30.5% with the frequency of 117 had visited Penang for once per year. Lastly, 26.6% with frequency of 117 had visited Penang for twice in a year.

Foods are Safe to Consume

Table 3. Foods are Safe to Consume.

		Frequency	Percent	Cumulative Percent
Valid	Strongly Disagree	3	.8	.8
	Disagree	13	3.4	4.2
	Disagree/Agree	105	27.3	31.5
	Agree	209	54.4	85.9
	Strongly Agree	54	14.1	100.0
	Total	384	100.0	

According to Table 3, it is clearly shown that 209 frequencies of agreeing the foods as safe to be consumed by representing 54.4% as its percentage. Next, frequency of 105 or 27.3% is gathered from disagree/agree answers. Other than that, 54 frequency which is 14.1% strongly agree with the statement. Apart from that, 13 frequencies or 3.4% disagree, and lastly, 3 as the frequency which stated 0.8% as its percentage are strongly disagree with the statement related to the food safety in Penang. Acknowledgement Guidelines "Perfect, Safe and Healthy" (BeSS Guidelines) is an acknowledgement given to food premises to urge food handlers to keep up a spotless reason by giving protected and solid food decisions to clients (Mazlina & Rose, 2018).

The Foods Price is Same with the Menu

Table 4. Food Price and the Menu

		Frequency	Percent	Cumulative Percent
Valid	Strongly Disagree	1	.3	.3
	Disagree	15	3.9	4.2
	Disagree/ Agree	71	18.5	22.7
	Agree	182	47.4	70.1
	Strongly Agree	115	29.9	100.0
	Total	384	100.0	

According to Table 4, it is clearly shown that 182 frequency agree with the similarity of the food price while attending the restaurant and reviewing the menu, with the percentage of 47.4%. Next, frequency of 71 disagree/agree by representing 18.5% as its percentage value. Other than that, 115 frequency or 29.9% strongly agree with the statement. Apart from that, 15 frequency or 3.9% disagree, and lastly, 1 frequency which is 0.3% are strongly disagree. Based on Katrina and Abdul Ghani's (2012) article, the food price is important to be reviewed by the customers who dine in at "Mamak" restaurants.
The information on the menu is understandable

Table 5. Menu Information

		Frequency	Percent	Cumulative Percent
Valid	Strongly Disagree	2	.5	.5
	Disagree	12	3.1	3.6
	Disagree/Agree	68	17.7	21.4
	Agree	183	47.7	69.0
	Strongly Agree	119	31.0	100.0
	Total	384	100.0	

Table 6. The Right Decision to Visit Penang for Food Experience

		Frequency	Percent	Cumulative Percent
Valid	Strongly Disagree	2	.5	.5
	Disagree	15	3.9	4.4
	Disagree/Agree	76	19.8	24.2
	Agree	162	42.2	66.4
	Strongly Agree	129	33.6	100.0
	Total	384	100.0	

Table 7. Descriptive Analysis

Variables	Mean	Standard Deviation
Hygiene		
The foods are free from contamination.	3.70	0.957
The foods are safe to consume.	3.78	0.756
The foods are properly cook.	3.93	0.783
The staff uniforms are clean.	3.68	0.944
The staff maintain the hygienic appearance (i.e., hair and nails).	3.65	1.026
The staff are wearing an apron.	3.83	0.918
The restaurant is clean.	3.78	0.912
The restaurant environment is comfortable.	3.89	0.892
The restaurant's table and chair are clean.	3.89	0.823
The restaurant provides dustbin.	4.04	0.858
The restaurant is free from pest contamination (i.e., cockroaches, rats, and flies)	3.83	1.019
Quality		
The foods are served at the appropriate temperature	3.63	0.899
Serve hot foods while still hot.	4.03	0.815
The foods are served fresh.	4.04	0.858
The displayed foods are tempted.	3.85	0.918
The food-warmer area is well displayed with choices of foods.	4.02	0.606
There are no insects at the food warmer area.	4.04	0.858
The food smell good.	3.86	0.830
The food taste good.	4.22	0.776
The foods are rich in flavours.	4.04	0.858
The foods contain herbs and spices.	4.12	0.813
Price		
The foods price is reasonable.	3.70	0.957
The foods price is same with the menu.	4.03	0.815
The foods are good value for price.	3.68	0.944
The foods price influenced by the chosen of condiments.	3.85	0.918
The foods quality is worth the price.	3.78	0.756
Service		
The staff are welcoming the customers.	3.89	0.823
The staff are friendly and knowledgeable.	3.83	1.019
The information on the menu is understandable.	4.05	0.811
The service by the staffs is good.	3.70	0.957
The foods are served on time (min 15 minutes)	3.78	0.912
The staffs are always making sure that the order was served.	3.68	0.944
Revisit Intention		
I have made the right decision to visit Penang for food experience.	4.04	0.858
I am willing to eat Nasi Kandar in Penang in the future.	3.67	1.029
For me, consuming Nasi Kandar will be enjoyable.	3.78	0.915
For me, the cleanliness of the foods affects the foods quality.	3.83	1.013
If I choose to attend an event in Penang, I will be sure to come back for the food experience.	3.83	1.019

Continued on following page

Table 7. Continued

Variables	Mean	Standard Deviation
I would say positive things about food in Penang to others.	3.63	0.899
I would recommend the food in Penang to my friends and others.	4.05	0.811
Penang is suitable to be called as food paradise.	3.79	0.960
In overall, I am satisfied with the food experience in Penang.	3.95	0.811
I will revisit Penang for Nasi Kandar.	4.58	0.650

According to Table 5, it is clearly shown that 183 frequency agree with the menu information by representing 47.7% as its percentage. Next, frequency of 68 of disagree/agree which represent 17.7%. Other than that, 119 frequency which is 31.0% strongly agree. Apart from that, 12 frequency or 3.1% disagree, and lastly, 2 frequency which is 0.5% are strongly disagree. Based on Katrina and Abdul Ghani (2012), restaurant operators need to understand their customers by providing them with good information about the foods.

The Right Decision to Visit Penang for Food Experience

According to Table 6, it is clearly shown that the highest frequency is attained by the agree answers towards the statement of "I have made the right decision to visit Penang for food experience", where it shows the frequency of 162 that represents 42.2%. The strongly agree of frequency is 129 which represents 33.6% and 76 frequency of disagree/agree which represents 19.8%. The rest are 15 frequency of disagree answer which represent 3.9% as its percentage. Lastly, 2 frequency of strongly disagree represents 0.5%. Gani, Mahdzar, Abdullah, and Muhammad (2017) stated that the very clear attribute measured in local food is its pleasant taste among majority of foreign tourists. In addition, variety and specialty are another important points that intensify the quality of experience and overall satisfaction.

Correlation Analysis

Table 8. Correlation Analysis

Hypothesis	Remarks	Pearson Correlation	Strength
H1: There is a relationship in between Visitor awareness on quality and revisit intention.	Supported	.423	Moderate
H2: There is a relationship in between Visitor awareness on hygiene and revisit intention.	Supported	.398	Moderate
H3: There is a relationship in between Visitor awareness on price and revisit intention.	Supported	.518	High
H4: There is a relationship in between Visitor awareness on service and revisit intention.	Supported	.511	High

DISCUSSION AND CONCLUSION

Cleanliness is the persuasive element for customers to visit a restaurant without hesitant. Food handlers or restaurant owners need to be knowledgeable in food handling, to guarantee the food quality and its hygiene. Malaysians are turning out not only to be health aware, but progressively particular and alert with regards to the tidiness of the restaurants. Food could be part of tourists' experience and closely related with quality tourism for tourist in a way to search for new things and experience that could give a high value of satisfaction. Moreover, food quality is the quality attributes of food that is contributable to customers' satisfaction. Thus, to guarantee the food tourism is important among tourist, the food handlers, and the government. The food quality is the least important factor in comparison with other three variables (hygiene, service, and price) that influence tourists' revisit intention (Albattat et al., 2020; Chemli et al., 2021; Toanoglou et al., 2021). Most of the tourists choose nasi kandar restaurant as their food destination due to its taste. This is because it has a very rich aromatic and flavour with the influenced of Indian Muslim cuisine. Nasi kandar suits with the appetite of local and foreigner, so they do not really require any emphasisation in terms of its food taste and quality.

Food price is one of important elements for customers to dine in at "Mamak". Also, price can be considered as significant within the food choice decision making process. Customer today search for an entire unexpected nourishment experience. Customer may be more accentuation on the nourishment itself. Along these lines, restaurateurs can utilise this data to additionally improve their administrations in a way to increase upper hand just as led to rehash support. Therefore, the target to look at the connection between nourishment cost and client return to goal is accomplished. Customer would not return to the restaurant if the food were not helpful. Restaurant operators need to have common understanding of the customers by providing them with good information about the foods. Good service quality will create themselves as favourable towards customers behavioural intentions to have their dine at "Mamak" restaurants.

REFERENCES

Ahmad, A. T. (2015). Penang Museums, Culture and History. *Kajian Malaysia*, *33*, 153–174.

Al Mansoori, S. A. N. A., Ab Yazid, M. S., Khatibi, A., & Azam, S. F. (2018). Validating A Measurement Model for Strategic Management Planning and the Development of Healthcare Sector in Abu Dhabi (SEHA). *European Journal of Human Resource Management Studies*.

Al Mansoori, S. A. N. A., Ab Yazid, M. S., Khatibi, A., & Azam, S. F. (2018). Strategic Management Planning and the Development of Healthcare Sector in Abu Dhabi: Structural Equation Modeling (SEM) Approach. *European Journal of Social Sciences Studies*.

Albattat, A., Jamaludin, A., Zuraimi, N. S. M., & Valeri, M. (2020). Visit Intention and Destination Image in Post- Covid- 19 Crisis Recovery. *Current Issues in Tourism*. Advance online publication. doi :10.1080/13683500.2020.1842342

Albattat, A. R., & Amer, H. (2016). Backpackers Expectation and Satisfaction towards Budget Hotel: A Case Study in Penang. *Almatourism-Journal of Tourism, Culture and Territorial Development*, *7*(14), 168–180.

Anholt, S. (2016). *Places: Identity, Image and Reputation*. Springer.

Antón, C., Camarero, C., Laguna, M., & Buhalis, D. (2019). Impacts of Authenticity, degree of adaptation and cultural contrast on travellers' Memorable Gastronomy Experiences. *Journal of Hospitality Marketing & Management*, *28*(7), 1–22. doi:10.1080/19368623.2019.1564106

Atikahambar, Y., Zainal, A., Rahayu, M., & Mokhtar, R. (2019). Quality of Food and Tourists ' Satisfaction of Penang Delicacies in Predicting Tourists ' Revisit Intention Quality of Food and Tourists ' Satisfaction of Penang Delicacies in Predicting Tourists '. *Revisit Intention.*, *8*(12), 1606–1618.

Baggio, R., & Sainaghi, R. (2011). Complex and chaotic tourism systems: Towards a quantitative approach. *International Journal of Contemporary Hospitality Management*, *23*(6), 840–861. doi:10.1108/09596111111153501

Baggio, R., Scott, N., & Cooper, C. (2010). Improving tourism destination governance: A complexity science approach. *Tourism Review*, *65*(4), 51–60. doi:10.1108/16605371011093863

Baggio, R., & Valeri, M. (2020). Network science and sustainable performance of family businesses in tourism. *Journal of Family Business Management*. doi:10.1108/JFBM-06-2020-0048

Bakar, K. A., & Farinda, A. G. (2012). Consumers' Attitude Towards "Mamak" Food. *Malaysia 3rd International Conference on Business And Economic,* 1304–1316.

Barton, K. (2018). Guyana's Linden to Lethem Road: A Metaphor for Conservation and Development. *Focus On Geography*, *61*, 1–1. doi:10.21690/foge/2018.61.3p

Benson, M., & O'Reilly, K. (2018). Colonial Traces and Neoliberal Presents: Situating Malaysia and Panama. In Lifestyle Migration and Colonial Traces in Malaysia and Panama (pp. 41-80). Palgrave Macmillan.

Beritelli, P., Bieger, T., & Laesser, C. (2007). Destination governance: Using corporate governance theories as a foundation for effective destination management. *Journal of Travel Research*, *46*(1), 96–107. doi:10.1177/0047287507302385

Boniface, B., Cooper, R., & Cooper, C. (2016). *Worldwide Destinations: The geography of travel and tourism*. Routledge. doi:10.4324/9781315697666

Chahal, D. K. (2017). *The Determinants of Customer Satisfaction and Behavioral Intention among Restaurants in Klang Valley area of Malaysia : A Conceptual Study*. Academic Press.

Chemli, S., Toanoglou, M., & Valeri, M. (2020). The impact of Covid-19 media coverage on tourist's awareness for future traveling. *Current Issues in Tourism*, 1–8. Advance online publication. doi:10.1080/13683500.2020.1846502

De Silva, A. D. A., Khatibi, A., & Azam, S. F. (2018). Can Parental Involvement Mitigate—Swing Away from Science? Sri Lankan Perspectives. *Cogent Education*, *5*(1), 1467244. doi:10.1080/2331186X.2018.1467244

De Silva, A. D. A., Khatibi, A., & Azam, S. F. (2018). Do the Demographic Differences Manifest in Motivation to Learn Science and Impact on Science Performance? Evidence from Sri Lanka. *International Journal of Science and Mathematics Education, 16*(1), 47–67. doi:10.100710763-017-9846-y

Faida, R., Zabidi, A., & Jamaludin, M. (2018). The Validation of Halal Awareness and Halal Practice Instrument among Halal Restaurant Food Handlers. *Article in Asian Journal of Social Sciences & Humanities, 1*(3), 2600–9056. doi:10.26666/rmp.jssh.2018.3.1

Faulkner, B., & Russell, R. (1997). Chaos and complexity in tourism: In search of a new perspective. *Pacific Tourism Review, 1*(2), 93-102.

Foo, R., & Krishnapillai, G. (2018). Preserving the Intangible Living Heritage in the George Town World Heritage Site, Malaysia. *Journal of Heritage Tourism*, 1–13.

Gani, A. A., Mahdzar, M., Abdullah, N., & Muhammad, R. (2017). *Linking Image and Satisfaction of Food Tourism in Penang, Malaysia*. Academic Press.

Hammami, A. M., Stanton, J. L., Periklis, D., Baourakis, G., van Dijk, G., & Mamalis, S. (2018). Tourist's Behaviour Towards Local Cretan Food. *Sustainable Agriculture and Food Security*, 49-61.

Leong, Q., Ab. Karim, M., Othman, M., Mohd Adzahan, N., & Sridar, R. (2010). Relationships between Malaysian Food Image, Tourist Satisfaction and Behavioral Intention. *World Applied Sciences Journal, 10*(10), 164–171.

Ling, T.P., Noor, S. M., Mustafa, H., & Kiumarsi, S. (2019). *Mindfulness: Exploring Visitor and Communication Factors at Penang Heritage Sites*. Academic Press.

McKercher, B. (1999). A chaos approach to tourism. *Tourism Management, 20*(4), 425–434. doi:10.1016/S0261-5177(99)00008-4

Nor Azureen, R., Shahril, H., Ashraf Siddik Khan, A. R. S., Putri Dahlia, A. R., & Noor Saliza, S. (2016). The Influence of Food Quality on Customer Satisfaction in Fine Dining Restaurant. *Case in Penang, 2*(2), 45–50.

Omar, S. R., & Omar, S. N. (2018). Malaysian Heritage Food (MHF): A Review on its Unique Food Culture, Tradition and Present Lifestyle. *International Journal, 1*(3), 1-15.

Ozturk, A. B., Salehi-Esfahani, S., & Bilgihan, A. (2017). Social Media and Destination Marketing. In Advances in Social Media for Travel, Tourism and Hospitality (pp. 107-119). Routledge. doi:10.4324/9781315565736-9

Rahman, S. (2018). Emerging Built Heritage Commodification of Boutique Hotels In World Heritage Site: Evidence From George Town, Penang, Malaysia. *Planning Malaysia Journal, 16*(8).

Salpeteur, M., Calvet-Mir, L., Diaz-Reviriego, I., & Reyes-García, V. (2017). Networking the environment: Social network analysis in environmental management and local ecological knowledge studies. *Ecology and Society, 22*(1), art41. doi:10.5751/ES-08790-220141

Samaraweera, D., Hamid, J. A., Khatibi, A. A., Azam, S. F., & Dharmaratne, I. (2018). Measuring Burnout of Human Service Workers: Testing the Copenhagen Burnout Inventory in Sri Lankan Context. *European Journal of Education Studies*.

Santos, V., Ramos, P., Sousa, B., Almeida, N., & Valeri, M. (2021a). Factors influencing touristic consumer behavior. *Journal of Organizational Change Management*. Advance online publication. doi:10.1108/JOCM-02-2021-0032

Santos, V., Sousa, B., Ramos, P., & Valeri, M. (2021b). Emotions and Involvement in Tourism Settings. *Current Issues in Tourism*, 1–6. Advance online publication. doi:10.1080/13683500.2021.1932769

Scott, N., Baggio, R., & Cooper, C. (2008). *Network analysis and tourism*. Channel View Publications. doi:10.21832/9781845410896

Sulaiman, M. Z., & Wilson, R. (2019). Translating Tourism Landscapes: From Nature to Urban. In *Translation and Tourism* (pp. 89–127). Springer. doi:10.1007/978-981-13-6343-6_6

Toanoglou, M., Chemli, S., & Valeri, M. (2021). The organizational impact of COVID-19 crisis on travel perceived risk across four continents. *Journal of Organizational Change Management*. Advance online publication. doi:10.1108/JOCM-12-2020-0369

Valeri, M. (2016). Networking and cooperation practices in the Italian tourism business. *Journal of Tourism, Heritage & Services Marketing*, *2*(1), 30–35. doi:10.5281/zenodo.376333

Valeri, M., & Baggio, R. (2020a). Social network analysis: Organizational implications in tourism management. *The International Journal of Organizational Analysis*. Advance online publication. doi:10.1108/IJOA-12-2019-1971

Valeri, M., & Baggio, R. (2020b). Italian tourism intermediaries: A social network analysis exploration. *Current Issues in Tourism*, *24*(9), 1270–1283. doi:10.1080/13683500.2020.1777950

Valeri, M., & Baggio, R. (2020c). A critical reflection on the adoption of blockchain in tourism. *Journal Information Technology and Tourism*, *23*(2), 121–132. doi:10.100740558-020-00183-1

Valeri, M., & Baggio, R. (2021). Italian tourism intermediaries: A social network analysis exploration. *Current Issues in Tourism*, *24*(9), 1270–1283. doi:10.1080/13683500.2020.1777950

Valeri, M., & Baggio, R. (2021). Increasing the efficiency of knowledge transfer in an Italian tourism system: A network approach. *Current Issues in Tourism*, 1–16. Advance online publication. doi:10.1080/13683500.2021.1937960

Zainal, A., Nizan, A., & Nizam, M. (2010). *Malaysian Gastronomy Routes as a Tourist*. Academic Press.

Chapter 6

Governance Factors That Influence the Internationalization of Tourism Destinations:
The Perspective of Portuguese DMOs

Maria do Rosário Campos Mira
Coimbra Education School, Polytechnic Institute of Coimbra, Portugal

Lisete S. M. Mónico
Faculty of Psychology and Educational Sciences, University of Coimbra, Portugal

Zélia Maria Jesus Breda
https://orcid.org/0000-0002-5882-063X
Department of Economics, Management, Industrial Engineering, and Tourism, University of Aveiro, Portugal

Marco Valeri
https://orcid.org/0000-0002-9744-506X
Niccolò Cusano University, Italy

ABSTRACT

The literature is highlighting that research on the internationalization of destinations is scarce. This chapter presents a validation of the governance factors which contribute to the internationalization of tourism destinations. A survey was carried out on Portuguese DMOs with responsibilities in the areas of tourism and local/regional development (N=147). The 'Questionnaire on Internationalization of Tourist Destinations' was submitted to content, reliability, and construct validations through exploratory and confirmatory second-order factor analysis. Governance factors influencing the internationalization of destinations were F1-Collaboration and F2-Networks. F1 associates items related to stakeholder involvement in destination decisions. F2 highlights the interaction between stakeholders, whose collective action is oriented towards active participation in decisions and leadership. The perception of Portuguese entities with different levels of influence on tourism development allows the systematization of results that meet some priority lines defined in the Tourism Strategy 2027.

DOI: 10.4018/978-1-6684-3889-3.ch006

INTRODUCTION

This chapter aims to present the model explaining the governance of tourist destinations, with a view to their internationalization.

In addition, the relationships between the factors associated with this model are clarified and the steps inherent to the various validation studies of the instrument built for this purpose are explained, namely content, construct and reliability validation.

This study is part of a broader investigation, which proposes a model for the internationalization of destinations, in which the systemic perspective of tourism is strengthened, due to the relationship between the dimensions of territory, product, governance model and function of Destination Management Organizations (DMOs) (Mira, & Breda, 2019). The results indicate that each of the second order constructs (dimensions) significantly reflects the latent variable it sets out from ('internationalization of destinations'). In addition, the inter-correlation between the scales and the constituent factors indicates that the strongest associations are established between the 'territory' and 'product' dimensions, as well as between 'governance' and 'DMOs'. The connection between the territory's economy and the product's competitiveness is confirmed, as well as between DMOs and governance of the network destination, a link also defended at the theoretical level. Therefore, the DMOs and the network governance of destinations integrate the various elements of the tourism system, guarantee their coordination and the involvement of stakeholders in this process (Mira, & Breda, 2019; Valeri, 2016; Valeri and Baggio, 2020a; 2020b; 2020c; 2021; Baggio and Valeri, 2020). The perspective that the governance of the territory facilitates its organization as learning and innovation systems inspired by the uniqueness and identity of destinations is justified.

2. CONCEPTUAL FRAMEWORK

2.1 Governance

The governance models of the networked territory trigger intense collaborative, cooperation and negotiation relations necessary for the sharing and creation of resources, to respond to the challenges of competition. These relationships and the discursive activity linked to them generate a dynamic that is reflected in the design of new products and services, as well as in different practices that make their marketing and promotion more effective (Booyens, 2016; Boyens & Rogerson, 2016; Clavé & Wilson, 2017; Escah & Vaudor, 2014; Valeri, 2021).

The network management of innovation, knowledge management and marketing promotes joint work between entities from the public and private sectors and other interested parties, resulting in an increase in the power of communities and local governments to lead the destination's project (Badulescu, Hoffman, Badulescu, & Simut, 2016; Rovira, 2016; Sertakova, Koptseva, Kolesnik, Libakova, Luzan, & Sergeeva, 2016). This movement can have as multiplier effects the creation of new intersectoral and inter-destination networks, giving rise to the appearance of their own governance structures, shifting the emphasis of national policies and systems to regional economic structures (Blasco, Guia, & Prats, 2014a; Bohlin, Brandt, & Elbe, 2016). In this context, we are witnessing a new configuration of public management, which should start to contemplate the impact of market transactions and actors in decision-making on the most appropriate policies for the destination project, at the local level (Bohlin et al., 2016). These policies should not be static, their constant reconfiguration being fundamental to the development of destinations, in order to keep up with the frequent changes and changes in the markets (Brouder& Ioannides, 2014). In this context, the effectiveness of the governance of networked territories is characterized by applying a management model that relies on the horizontal coordination of stakeholders, instead of the usual models of vertical leadership, that is, promoting the sharing of power and decisions instead of hierarchical direction (Beritelli & Laesser, 2014; Hristov, 2015; Hristov & Zehrer, 2015; Farmaki, 2015; Lousã, 2013; Mintzeberg, 1979, 1983; Mumford, 2003; Spyriadis, Fletcher & Fyall, 2013). The focus is on the degree of autonomy of the actors and the processes in which they interact which, in turn, are inherent to the very motives, attributes or representations of the socio-cultural contexts from which they emerge (Scuttari,Volgger, & Pechlaner, 2016).

Makkonen and Weidenfeld (2016) add that the evolution of destination management towards governance models facilitates the creation and importing of knowledge, since these network structures enhance the inclusion of research projects in companies. Knowledge transfer mechanisms are one of the dimensions of governance that contribute most to the internationalization of destinations, since knowledge supports the innovation necessary to strengthen their position in the global market (Ferdinand & Williams, 2013; Nilsson, Eskilsson, & Ek, 2010; Weidenfeld, 2013). Bannò, Piscitello and Varum, (2015) say that the competitive capacity of destinations must go beyond the production and commercialization of tourism products. They also argue that concerted action between governments and different local entities is essential to include research and development (R&D) activities in companies.

In short, the dynamics of change and the proposals for improving tourist spaces create knowledge, leading local organizations and actors to assume their willingness to get involved in decisions of a political, economic and social nature that influence the evolutionary path of destinations (Farmaki, 2015; Soares, Baidal, & Gândar, 2015; Vermeulen, 2015; Volgger & Pechlaner, 2015).
Governance contributes to the internationalization of destinations when:

1. there is deregulation of direct government activity, resulting from the formation of local networks, made up of entities from the public and private sectors;
2. governments adapt to new ways of looking at the economy and politics;
3. the domains and activities that can be governed in a network, at the local level, are defined and clarified;
4. governance networks evaluate the effectiveness and results of organizations and destination management processes;
5. the networks are managed by DMOs that see themselves as local governance structures;
6. the sharing of knowledge and resources in a network triggers the emergence of true learning communities;
7. coordinating and governing tourism networks is based on fostering trust, negotiation and allocating a budget to the destination;
8. a set of rules and mechanisms for controlling the formulation of business policies, objectives and strategies are drawn up, involving all stakeholders.

2.1.1 Policies

The issue of the governance of tourist destinations raises some questions related to the role of local authorities in tourism development, as well as in managing and evaluating its effects (Badulescu et al.,

2016). Establishing and enforcing the laws and regulations necessary for tourism development and, at the same time, protecting natural and cultural resources, does not always facilitate working in partnership and effective cooperation between the public and private sectors. This process requires specially developed communication and business skills that help the State's policies translate into direct support for the destination project, as well as its products in general (Bernabé & Hernández, 2016).Traditionally, the role of government entities in the internationalization of tourist destinations has been to stimulate learning by creating structures and allocating resources that facilitate innovation (Booyens, 2016). However, for these authors, it is necessary to seek to go further, making these associative movements evolve into governance structures. It is the articulated interaction between governments and other actors, in terms of policy definition, planning and destination strategies that reinforces their competitive capacity, because this ensures that these policies and strategies mirror the territories' identity. Local economic and innovation policies are the engine, or the obstacle, of contemporary regional competitive capacity, because these create the legal, regulatory and financial conditions, at the territorial level, that stimulate innovation, knowledge and creativity (Booyens, 2016). Public decision-makers are the most influential actors in promoting innovation in the regions, because they create the measures and instruments that facilitate its operationalization (Makkonen & Rhode, 2016).

Strong political commitment to the economy and innovation also raises other problems, namely the need for sufficiently qualified human resources in the region (Escach &Vaudor, 2014). However, the greater the skills and knowledge of local actors, the greater their desire to participate and influence policies and strategies with an impact on their region. Applying territorial development models, based on politics and the knowledge economy, often triggers feelings of belonging and stakeholders' active participation in the destination project Booyens, 2016;Vodeb & Rudež, 2016;Weidenfeld, 2013). In other words, they facilitate the appearance of regional networked governance structures. However, Scuttari et al. (2016) attribute the difficulty of implementing territorial governance models to the political need to respond to short-term requests and interests, and mainly, to those posed by stakeholders with greater economic influence. In fact, transforming regions into territorial innovation systems requires a long-term strategy that integrates policies between different regions and provides measures that facilitate collaboration in terms of planning, marketing, access to markets and resources, allocation of resources and financing, among others (Blasco et al., 2014a, 2014b).

In short, there is still a widespread belief, especially in the political milieu, that foreign direct investment, by itself, is able to solve the competitiveness problems of destinations (Bannò, Piscitello, & Varum, 2015). Hence, the importance of assessing the effect of policies is of growing concern, leading to the call for attention to the need to redirect public programmes and incentives for innovation and knowledge creation. In the same connection, public policies must take into account regional differences and destinations' degree of maturity, since not all are at the same stage of development, not all have the same wealth of resources or support infrastructure, among other aspects (Bohlin, Brandt, & Elbe, 2016). The formulation of policies must take into account this territorial diversity and reward both foreign investors and entrepreneurs or local projects that truly contribute to innovating the provision, renewing and diversifying endogenous products and services (Sanz-Ibáñez & Clavé, 2016).

2.1.2 Networks

Regional networks can arise from the collective will to cooperate in a special way, creating critical mass and the ability to penetrate world markets (Escah & Vaudor, 2014). Regional innovation is affected by the patterns of collaboration between local actors and this shapes companies' evolutionary path (Makkonen & Rhode, 2016).

The integration of stakeholders at decisive moments of the destination project is essential to help discover and implement tourist attractions that are competitive in the global market (Badulescu et al., 2016; Blasco et al., 2014a). This territorial management model reinforces the process of knowledge transfer necessary for innovation, if it is conducted with the necessary flexibility for the organization of actors in a network (Brouder & Ioannides, 2014). Innovation also has effects on the internationalization of destinations, when collaboration between stakeholders emerges and this is assumed as a regional growth strategy (Booyens, 2016).

These movements, when reflected in the results desired individually by the organizational actors, reinforce adherence to this way of working, leading to the appearance of multiple associations that can go beyond the limits of their own destination itself. Inter-destination networks are one of the most effective strategies in regions' internationalization process (Farmaki, 2015; Ferdinand & Williams, 2013). However, the effectiveness of these collaborative movements depends on their management, transforming these initiatives, sometimes spontaneously, into configurations oriented to planning and results, although they can maintain an informal matrix. When this information and network sharing system is recognized by local actors as facilitating the aggregation of beneficial synergies for all interested parties, their adherence to the common project of the destination is more consistent (Sarasa, 2015).

One of the aspects that contributes most to stakeholders' participation in networking is related to recognition of its usefulness, in view of the interests of each one (Sanz-Ibáñez & Clavé, 2016). Business networks, the promotion of tourism products and services, innovation and knowledge transfer, and the sharing of resources, are the most effective in terms of local actors' adhesion to collaborative work (Booyens & Rogerson, 2015).

There are central and other peripheral actors, each with their own interests, sometimes competing, and not all with the same capacity to adapt to this new way of working (Brouder & Ioannides, 2014). This organizational model of regions challenges established standards, forcing leaders to clearly define the areas and levels in which the different stakeholders participate, as well as the type of contribution made by each (De Noni, Orsi, & Zanderighi, 2014). However, there is some consensus regarding the need to manage the identity symbolism of the destination at the collective level, because a shared image contributes to a long-term vision and strategic orientation of policies, planning and the decision-making process (Badulescuet al., 2016; Brouder & Ioannides, 2014; De Noni et al., 2014; Escah & Vaudor, 2014; Albattat et al., 2020; Chemli et al., 2020; Toanoglou et al., 2021).

Networks and cooperation between destinations have made it possible to achieve good results, especially in the field of marketing, training and qualification of human resources, organizing major events, sharing information and exchanging knowledge and experiences (Weidenfeld, 2013). If many tourism problems are associated with the fragmented action of investors, entrepreneurs, researchers and politicians, the organization of destinations in a network contributes to overcoming this shortcoming (Bernabé & Hernández, 2016). The success of this process depends on public entities' degree of commitment to regional governance systems, in which public actors become the mediators of negotiations between a wide range of stakeholders (Sanz-Ibáñez & Clavé; Sarasa, 2015; Vodeb & Rudež, 2016; Weidenfeld, 2013). Destination networks require different forms of organization from traditional ones. The coordination of the network must assume models of participatory governance, characterized by horizontal leadership styles, in order to bring about informal mechanisms that lead to the sharing of information and knowledge

(Volgger & Pechlaner, 2015). In this context, networks are models of territorial organization that help to increase the competitive capacity of actors, groups of actors, regions and countries.

Consolidating and understanding tourism as a system involves evaluating the results of networked governance systems, especially those resulting from partnerships between the public and private sectors, a role that should be attributed to DMOs (Booyens & Rogerson, 2015, 2016; Volgger & Pechlaner, 2015). Here, an attempt was made to highlight the link between the governance of network destinations and innovation, knowledge and competitiveness, which are fundamental ins consolidating the internationalization of territories.

Table 1. Sample characterization by DMO characteristics

		n	%
Organization	Municipal Council	102	69.4
	Development and / or promotion association	12	8.2
	Public entity for development and territorial planning	11	7.5
	Regional tourism entity	7	4.8
	Representative entity of the municipal association	6	4.1
	Sectorial Association	5	3.4
	Tourist Promotion Agency	2	1.4
	Missing values	2	1.4
	Total	**147**	**100**
Location (NUT II)	Center	47	32.0
	North	41	27.9
	Lisbon Metropolitan Area	17	11.6
	Alentejo	15	10.2
	Algarve	10	6.8
	Autonomous Region of Azores	8	5.4
	Autonomous Region of Madeira	7	4.8
	Missing values	2	1.4
	Total	**147**	**100**
Scope of influence	Local	104	70.7
	Regional	27	18.4
	National	10	6.8
	International	4	2.7
	Missing values	2	1.4
	Total	**147**	**100**

3. METHOD

3.1 Sample

The target population is composed of 407 public and private Portuguese non-profit organizations, with responsibilities in different areas of tourism, product competitiveness and local/regional development.

Are DMO 427 organizations and 43 are sector associations. Have a national scope of influence 30 organizations, 75 have a regional scope of influence, and 365 have a local level of influence. All these organizations, together with the Government, are responsible for the definition of priorities and actions in the development of the territory, of the products or in the management of the implementation of public policies concerning tourism as a whole. Tourism associations were included, as they play a relevant role in the product's competitiveness strategy.

Table 2. Characterization of participants (N = 147)

		n	%
Sex	Female	90	61.2
	Male	54	36.7
	Missing value	3	2.0
	Total	147	100
Age (years)	≤ 35 ≤ 49	93	63.3
	≤50 ≤64	31	21.1
	≤25 ≤34	19	12.9
	≤18 ≤24	1	0.7
	Missing value	3	2.0
	Total	147	100
Academic qualifications	Secondary Education (12th year)	8	5.4
	Bachelor	7	4.8
	Degree	67	45.6
	Postgraduate studies	23	15.6
	Master	36	24.5
	PhD	4	2.7
	Missing value	2	1.4
	Total	147	100
Professional category	Technician	101	68.7
	Director	14	9.5
	Assessor	7	4.8
	President	6	4.1
	Other	17	11.6
	Councilman	5	3.4
	Head of Division	5	3.4
	Head of Unit	3	2.0
	Executive	1	0.7
	Office clerk	1	0.7
	Graduate staff	1	0.7
	Assistant	1	0.7
	Missing value	2	1.4
	Total	147	100

The non-probabilistic and intentional sample composed of 147 organizations, being the majority Municipal Councils (69.4%). Seven NUTII location was considered, being the scope of influence from local (70.7%) to international (2.7%) (see Table 1).

The sample characteristics of the 147 subjects who filled out the questionnaire are shown in Table 2. Most respondents were aged from 35 to 49 years old (63.3%), and there were more females (61.2%) than males (42.5%). The majority is technician (68.7%) and aged from 35 to 49 years-old (63.3%). Most of them have higher education qualifications at the degree level (45.6%), followed by a master (24.5%) or a postgraduate degree (15.6%).

3.2 Materials

A survey was carried out using a self-administered questionnaire - the *Questionnaire on Internationalization of Touristm Destinations' (QITD)* - which aims to fill the lack of instruments to assess the internationalization of destinations.

Table 3. Factors and items proposed for the Governance (QITD)

Factors	Items	Authors
Evaluation	Destination results	Badulescu et al (2016); Bohlin et al. (2016); Bannò et al. (2015); Booyens & Rogerson (2015); Getz e Page (2016)
Collaboration	Stakeholder involvement in destination leadership	Farmaki (2015); Mintzeberg (1979, 1983); Vodeb e Rudež (2016)
	Stakeholder involvement in destination decisions	Badulescu et al. (2016); Bernabé & Hernández (2016); Bohlin et al. (2016); Booyens (2016); Brouder & Ioannides (2014); De Noni et al. (2014); Escah & Vaudor (2014); Farmaki (2015); Mintzeberg (1979, 1983); Sanz-Ibáñez & Clavé (2016); Sarasa (2015); Vodeb & Rudež (2016); Weidenfeld (2013)
	Involvement of residents in the destination project	Badulescu et al. (2016); Booyens e Rogerson (2015); Makkonen e Rhode (2016); Sarasa (2015); Soares et al. (2015); Vermeulen (2015)
	Stakeholder involvement in the qualification of human resources	Blasco et al (2014a); Weidenfeld (2013)
	Resource sharing among stakeholders	Booyens e Rogerson (2015); Soares et al. (2015); Vermeulen (2015)
	Cultural proximity between stakeholders	Blasco et al (2014a); Booyens & Rogerson (2015)
Planning	Destination project planning	Booyens (2016); Booyens & Rogerson (2016); Clavé e Wilson (2017); Escah & Vaudor (2014); Makkonen & Rhode (2016); Makkonen & Weidenfel (2016)
	Public participation in defining destination policies	Booyens e Rogerson (2015); Makkonen & Rohde (2016); Sertakova et al. (2016)
	Network planning of the tourist offer	Bernabé Hernández (2016); Blasco et al. (2014a); Blasco et al. (2014b); Escah e Vaudor (2014); Farmaki (2015)
	Destination infrastructure planning	Bernabé Hernández (2016); Blasco et al. (2014a); Blasco et al. (2014b); Escah & Vaudor (2014); Farmaki (2015)
	Policies for setting qualified human resources	Bernabé e Hernández (2016); Escach & Vaudor (2014); Sarasa (2015)
Networks	Destination promotion networks	Rovira (2016); Sertakova et al. (2016)
	Innovation networks	Makkonen e Rhode (2016); Sanz-Ibáñez & Clavé (2016); Scuttari et al. (2016); Sertakova et al. (2016)
	Knowledge networks	Sanz-Ibáñez e Clavé (2016); Sarasa (2015); Scuttari et al. (2016); Sertakova et al. (2016)
	International networks	Bohlin et al. (2016); Brouder & Ioannides (2014)
	Inter-destination networks	Blasco et al. (2014a); Bohlin et al. (2016); Farmaki (2015); Ferdinand (2013); Makkonen e Rhode (2016)
	Interorganizational networks	Blasco et al. (2014a); Bohlin et al. (2016); Clavé & Wilson (2017); Farmaki (2015); Volgger e Pechlaner (2015)
	Development networks	Booyens (2016); Escah e Vaudor (2014); Booyens & Rogerson (2015, 2016); Volgger & Pechlaner (2015)

The QITD is intended to be a measure for assessing the factors that contribute to the governance of tourism destinations. Based on the state of the art (see table 3), a set of items were composed to reflect the integrating factors (evaluation, collaboration, planning and networks

The QITD is composed of 19 items evaluated through a Likert Scale ranged from 1 (strongly disagree) to 7 (strongly agree), comprising evaluation (1 item), collaboration (6 items), planning (5 items), and networks (7 items) of the governance of tourism destinations (see Table 3).

3.3. Procedures

An online version of the questionnaire was built using Google Forms and was sent by e-mail to all Portuguese municipalities. The control of the responses was carried out monthly through the variable 'Municipality', being sent a reminder to the municipalities that had not yet responded, reinforcing the importance of their participation in the study. The questionnaire had the instruction that it should be filled in by representatives of the municipality with responsibilities in tourism. Information on the objectives of the study, completion instructions, the voluntary and anonymous nature of the participation, and the guarantee of data confidentiality were also included in the beginning of the questionnaire. Ethical issues were, thus, taken into account when doing data collection, namely respecting participants' anonymity and data confidentiality.

3.4. Data Analysis

The QITD was submitted to content, reliability and construct validations (Byrne, 2010; Costello & Osborne, 2005; Kline, 2011). Construct validation were carried out through Exploratory Factor Analysis (EFA) with Principal Component Analysis (PCA), performed with IBM® SPSS® Statistics (IBM Corp., 2013). Confirmatory Factor Analysis (CFA) was performed with Structural Equation Modeling, using with IBM® AMOS (v. 24.0, SPSS Inc, Chicago, IL; Arbuckle, 2013).

Missing values (1.08%) were replaced through series mean method. Outliers were analysed according to the Mahalanobis squared distance (Tabachnick & Fidell, 2013), with no relevant values being found. None of the variables presented Sk and Ku coefficients that could violate normal distribution assumption (Cohen, 1988), since $| Sk | < 2$ and $| Ku_{univariate} | < 5$.

The PCA assumptions were tested through the sample size (ratio of 7.1 subjects per item; Urbina, 2014), the normality and linearity of the variables, factoriability of R, and sample adequacy (Tabachnick & Fidell, 2013). Since we intend to retain as many independent factors as possible, we chose the VARIMAX rotation method with Kaiser's normalisation.

CFA followed the maximum likelihood estimation method (Jöreskog & Sörbom, 2004). Goodness of fit was analysed by the indices of NFI (Normed of fit index; good fit > .80; Schumacker & Lomax, 2012), SRMR (Standardized Root Mean Square Residual; appropriate fit < .08; Schumacker & Lomax, 2012), TLI (Tucker-Lewis Index - TLI; appropriate fit > .90; Kline 2011), CFI (Comparative fit index; good fit > .90; Hair, Anderson, Tatham, & Black, 2008), CMIN/DF (good fit < 2; Schumacker & Lomax, 2012), and RMSEA (Root Mean Square Error of Approximation; good fit < .05; Kline 2011). The fit of the model was improved by modification indices (MI; Urbina, 2014), leading to a correlation of the residual variability between variables with the highest MI. We followed Arbuckle's proposal (2013), which consists of analysing the MIs by their statistical significance ($\alpha < 0.05$).

Despite being used the same sample of 147 participants, with EFA we intend to examine the extraction of theoretically and statistically supported factors, whereas CFA aimed to analyze the fit of the obtained factorial model, as well as to test the second-order model.

Reliability was calculated by Cronbach's alpha (Nunnally & Bernstein, 1994). The score of .80 was taken as a good reliability indicator (Urbina, 2014). The composite reliability and the average variance extracted for each factor were evaluated as described in Schumacker and Lomax (2012).

4. RESULTS

4.1 Exploratory Factor Analysis

The requirements for a reliable interpretation of PCA were checked. The inter-correlation matrix and the identity matrix differed significantly [$X^2(36) = 1159.29$, $p < .001$ for the Bartlett's test], and the sampling was higher than the required value of .70 [Kaiser-Meyer-Olkin (KMO) = .857]. The anti-image matrix ranged between .938 and .807, being the highest score outside the matrix's diagonal of $r = -.658$.

Two criteria were used for factors' extraction: Eigenvalue over one and Scree plot. The combining of these two criteria lead to the extraction of two interpretable factors. Some items were excluded, since their communalities (h^2) and factorial loadings (s) were below .50. Other items were excluded based on high factorial loadings in more than one factor. Additionally, excluding these items leads to an improvement of the Cronbach's alpha.

Were excluded from the final solution items related with evaluation (item 'evaluation of the results of the destination'), planning (items 'planning of the destination project', 'public participation in the definition of destination policies', 'network planning of the tourist offer', 'policy for setting qualified human resources', 'destination infrastructure planning'), collaboration (item 'cultural proximity between stakeholders'), and networks (items 'inter-destination networks', 'inter-organizational networks', and 'development networks').

The final solution was responsible for 78.06% of the total variance, with factorial loadings higher than .45 in both dimensions (Tabachnick & Fidell, 2013). Factor 1 (F1) aggregates five items from the initial solution of the 'collaboration' dimension, so its original designation was maintained. This factor explains 41.83% of shared variance and associates constructs related to stakeholder involvement in destination decisions, namely: 'participation in the leadership of your project', 'resource sharing' and playing an 'active role in defining the qualification plan for human resources' (see Table 4). Despite residents belong to the stakeholders group they are included in this factor, highlighting the importance of their participation in the destinations' governance.

Factor 2 (F2) retained four items of the initially seven proposed for the 'networks' dimension, so its designation was also maintained, explaining 36.23% explains 41.83% of shared variance. This factor emphasizes the relationship between stakeholders, whose collective action is oriented towards active participation in the decisions and leadership of the destination, with repercussions on the sharing of resources, the qualification of human resources and the involvement of residents.

The scale showed a global high level of internal consistency ($\alpha = .928$), as well as for both factors (see Table 4).

Table 4. Descriptive statistics (M, SD), PCA, and Reliability of the QITD: Data from F1 and F2 after Varimax Rotation for the four countries

Itens	M	SD	F1	F2	h^2
GD11. Stakeholder involvement in destination decisions.	5.97	0.97	**.902**	.223	*.864*
GD13. Stakeholder involvement in the definition of the human resources qualification plan.	5.78	0.99	**.864**	.317	*.847*
GD10. Stakeholder involvement in leading the destination project.	6.01	0.98	**.820**	.271	*.745*
GD12. Involvement of residents in the destination project.	5.90	0.97	**.810**	.309	*.752*
GD14. Sharing resources among stakeholders.	5.88	1.00	**.725**	.377	*.667*
GD2. Innovation networks.	5.79	0.89	.296	**.876**	*.855*
GD3. Knowledge networks.	5.90	0.89	.255	**.873**	*.827*
GD4. International networks.	5.92	0.97	.341	**.797**	*.752*
GD1. Destination promotion networks.	6.28	0.89	.286	**.797**	*.717*
Eigenvalues			5.73	1.30	
Shared variance (%)			41.83	36.23	
Cronbach's α			.925	.909	

4.2 Confirmatory Factor Analysis

The CFA performed on the remaining nine items of the EFA showed good fit considering the NFI, CFI, and SRMR indices, a poor fit for the TLI, and a bad fit attending to the $\chi2$/df and RMSEA indices (see Table 5, Model 1).

Table 5. Fit indices for the QITD

Model	NFI	SRMR	TLI	CFI	χ^2/df	RMSEA	RMSEA 90% CI
1	.882	.066	.863	.901	5.40* (df = 26)	.174	.146-.202*
2	.952	.064	.951	.970	2.58* (df = 22)	.104	.071-.138*

* $p < .001$

A second-order CFA was performed with the QITD items (model 2), introducing a latent variable (Governance) which explains Collaboration and Networking factors. In this model error terms were correlated between the items belonging to the same factor, based on the higher MI (see Figure 1). This covariation might indicates non-random measurement errors, which can result from some semantic redundancy between items, sequential positioning in the scale, as well as the specific characteristics of the respondents (Aish & Jöreskog, 1990). Model 2 showed an excellent fit considering the NFI, CFI, and TLI indices, although remaining poor for the RMSEA. Even so, in general, model 2 can be considered acceptable (see Table 5). The graphical representation of the estimated factorial model is shown in Figure 1. All standardized regression coefficients were higher than .71.

The inter-correlation between GD10, GD11, GD13, and GD14 items, referring to the involvement of stakeholders in the destination's tourism project and specifying their form of participation in decisions and in the articulation between the components of the destination's governance, may explain the MI scores. GD1 and GD4 items refer to destination promotion networks, which includes international networks, may this redundance justifying the MI obtained.

Figure 1. Second order factor model of the QITD after correlate error terms based on modification indices: Standardized regression coefficients and explained variance

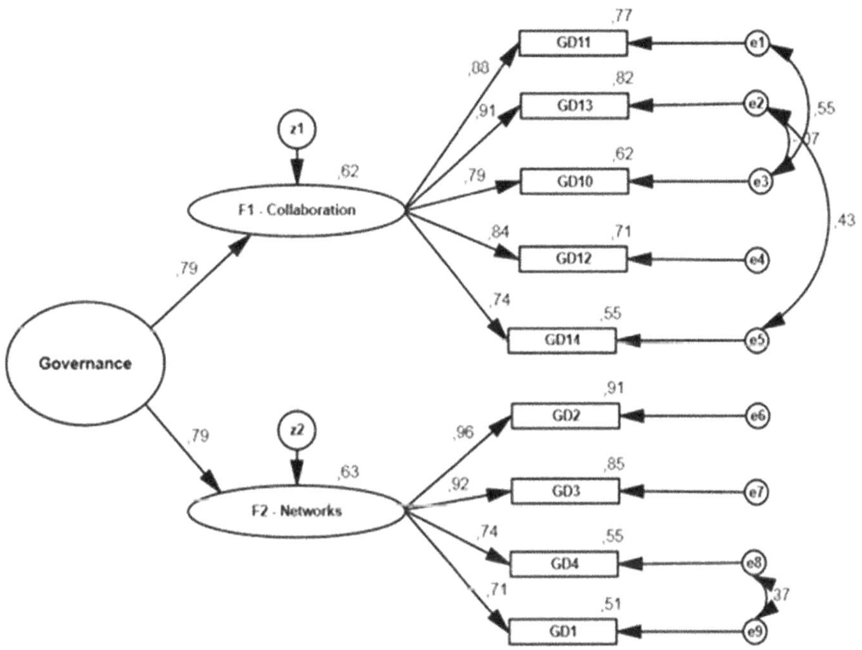

Table 6. Composite reliability (CR), average extracted variance (VE), internal consistency (α), descriptives, and determination coefficients (r^2) of the QITD

	CR	AVE	α	min	max	M	SD	Sk	Ku	$r^2_{(F1,F2)}$
Global scale	-	-	.928	2.78	7.00	5.94	0.76	-1.05	1.51	
F1- Collaboration	.918	.693	.925	2.60	7.00	5.91	0.86	-.85	.65	.40*
F2 - Networks	.904	.705	.909	3.00	7.00	5.97	0.81	-1.00	1.08	

* $p < .001$

The global scale and the two factors showed an excellent internal consistency (see Table 6), as well as the composite reliability and average variance extracted (AVE). The discriminant validity of the two factors is attested by the coefficient of determination ($r^2 = .40$), since the value is lower than the scores obtained in AVE. The mean score on the two factors is similar. The results are in line with the EFA, supporting 'collaboration' and 'networks' as the governance factors influencing the destinations' internationalization.

5. DISCUSSION

5.1 Governance and Internationalization of Destinations

The results validate what has already been defended at the theoretical level, that is, the governance model with effect on the internationalization of destinations must organize stakeholder networks and encourage their collaboration in development of the territory (Farmaki, 2015; Spyriadis et al.2013; Volgger & Pechlaner, 2015). This is the line of thought defended and supporting the proposed governance model for the internationalization of destinations.

The good convergent and discriminating validity of the two factors reveals that these, taken as a whole, contribute to forming the 'governance' construct. If these results clarify the domain of destination governance, translated into 'collaboration' and 'networks', the role of stakeholder collaboration in this system does not seem to be consensual.

5.1.1 Collaboration

Factor 1 of 'collaboration' reflects the items of 'stakeholder involvement in destination decisions', 'stakeholder involvement in defining the human resources qualification plan', 'stakeholder involvement in leading the destination project', 'resident involvement in the destination project' and 'resource sharing among stakeholders'. The theme of involvement stands out and the importance of managing the stakeholders that make up the tourism system (Badulescu et al., 2016; Booyens & Rogerson, 2015; Makkonen & Rhode, 2016; Makkonen & Weidenfeld, 2016; Vodeb & Rudež, 2016).

The association observed in the results between 'stakeholder involvement in the leadership of the destination' and their 'involvement in the destination decisions' is understood. In fact, leading implies making decisions; one without the other mirrors some incongruity (Clavé & Wilson, 2017; Farmaki, 2015; Mintzeberg, 1979, 1983; Vodeb & Rudež, 2016; Volgger & Pechlaner, 2015). In addition, the item that contributes most to explaining the latent variable is 'stakeholder involvement in the definition of the human resources qualification plan', which is associated with 'sharing of resources between stakeholders'. It seems that the respondents tend to answer that the resources to be shared between them may be human resources or else the way to qualify them. Hence the importance of qualifying human resources and stakeholders' active participation in this area (De Noni et al., 2014).

Despite the above, the results show that the qualification of human resources is slightly associated with leadership, albeit in a negative way. Looking only at the statistical interpretation of the data, these indicate that the more stakeholders collaborate in the qualification of human resources, the less they must participate in leadership of the destination or vice versa. Is it legitimate to question which governance model to follow and how to encourage stakeholder collaboration in this process? The need to involve stakeholders was not questioned; what domain they should be involved in seems to be unclear. The low association between leadership and the qualification plan for human resources seems to imply that definition of the training plan for tourism workers is not part of the attributions of the destination's strategic leadership, but is closer to technical functions. The negative association between these two concepts suggests that the qualification of human resources varies inversely with leadership. All the studies analyzed in the scope of this work, as well as the main theories on leadership (Lousã, 2013; Mumford, 2003) justify this interpretation of the data, because they argue that the more qualified the human resources are, the greater the autonomy they should be given, recognizing their ability to autonomously lead the processes they are responsible for (Farmaki, 2015; Mintzeberg, 1979, 1983). Therefore, vertical leaders should have less control. The extent to which these stakeholders should be involved in leadership and decision-making on the destination project is still somewhat diffuse. However, at the theoretical level, the need to share leadership and decision-making with stakeholders, taking into account destination governance, was previously established (Beritelli & Laesser, 2014; Hristov, 2015; Hristov & Zehrer, 2015; Spyriadis, Fletcher & Fyall, 2013).

It is concluded that collaboration in the governance model should contribute to: (i) stakeholder involvement;(ii) resource sharing;(iii) treating residents as an autonomous and distinct stakeholder group;(iv) joint definition of the typology of the qualifications of human resources, with a view to internationalization of the destination. In other words, the governance of destinations, so that they are able to internationalize, implies developing models of participatory and distributed leadership, which facilitate the organization of the destination in a network, integrating resources, structures and people who act and interact in tourism destinations.

5.1.2 Networks

Factor 2 of 'networks' implies the items of 'innovation networks', 'knowledge networks', 'international networks' and 'destination promotion networks'. Here, the need to create innovation and knowledge networks stands out (Makkonen & Rhode, 2016; Sanz-Ibáñez & Clavé, 2016; Sarasa, 2015; Scuttari et al., 2016; Sertakova et al., 2016). These should support the creation of international networks that enhance promotion of the destination. The association between international networks and promotion networks indicates the same objective, that is, to internationalize the destination. One of the functions of the governance model is also to promote the destination (Rovira, 2016; Sertakova et al., 2016).

These results explain the operating conditions and nature of the destination governance model, reinforcing the premise that the internationalization of destinations emerges from a collaborative system organized in a network. Internationalizing destinations depends on how stakeholders direct their interests to increasing innovation and knowledge, as well as connecting with international partners. This coordinated action by stakeholders becomes more consistent and effective when it incorporates the use of new technologies in its practices.

The high correlation of factor 1 'collaboration' with factor 2 'networks' is easily understood in light of what was explained earlier. Conducting a governance model in which leadership focuses on the requirements and results of innovation, instead of monitoring processes or people, because for this it relies on qualified and specialized human resources, facilitates organization of the network destination, since stakeholders know what it is and their role in this system has been recognized.

The results clarify the nature of the network collaboration that must support the governance model. Stakeholders must collaborate in decisions, in sharing resources, in the qualification of human resources and in leadership of the target project regarding: (i) application of technology in the development, commercialization and promotion of products;(ii) formulating strategies to attract tourists, develop endogenous products and assess the destination's potential;(iii) attracting financing and innovation structures to the destination that allow presenting innovative products;(iv) creation of tourism product companies inspired by the uniqueness of the territory.

6. CONCLUSION

The management of destinations, according to the principles of governance, implies analyzing the multiple processes of change that occur in regions, at different levels, because the introduction of changes in companies, resulting from innovation, causes changes within the economic and social structure of the tourism system. Change processes, oriented towards the creation of new paths, can sometimes be triggered by formal or informal mechanisms that encourage the collaboration of different actors in policies and in leading innovative actions. This evolution towards the knowledge economy has repercussions for organizations' development, with the effect, for the most part, of increasing their active participation in decision-making on the direction of the destination project. It is also these processes of change and innovation that increase destinations' competitive capacity, in the face of constant changes in the markets.

In order to make explicit this symbolic relationship of the various actors present, materialized by the effective transaction of products on a business basis, it is necessary to analyze the growing, multiple and simultaneous role of the various actors, which directs and affects the way in which power games develop, how they promote change and how they organize themselves in a networked system. This aspect becomes even more relevant if we consider that destinations, in their natural form, are regions where cooperation networks are formed in a somewhat fragmented way, so governance structures can be the strategy that unites these initiatives and integrates them at several levels.

Public entities have guided their action towards marketing and promoting destinations among potential markets. It is proposed that these entities redirect their attention to the consolidation of strategic networks, which ensure the transfer of external knowledge to inside destinations. In this way, local policies must focus on innovation and planning of the geographical area of the destination, focusing on skills, people and natural and cultural resources, aiming to present creative and differentiated tourist products.

The political focus is on the intrinsic assets and capabilities of a given territory, promoting creative solutions that stimulate growth and guarantee the region's competitiveness. This space planning has to take into account the existing human potential, as well as the ability to attract "talents" to the destination, requiring a continuous and systematic work of network organization of the different entities, from various sectors of activity. Public policies must be seen as a vector of development and economic diversification, integrating the political, scientific and technical dimensions. That is, the approximation of politics to the economy through knowledge.

The more effective local authorities are in defending the harmony between policies and innovation, based on the territory's identity, the easier it becomes to promote the destination as a whole and to attract markets with relevance for regional development. For these reasons, an interface between research and policy is crucial, because only knowledge can help define measures that increase local conditions able to harmonize relations between stakeholders; create new products and services and present a coherent, modern and differentiated offer; respond to tourists' demands, regarding the quality of the tourist experience, leading to the internationalization of destinations.

The coordination of stakeholder collaboration in the network is directly related to regions' tourism innovation capacity. However, management of these networks requires special leadership skills, without which the confidence necessary to build shared projects is not triggered. Trusting relationships between local actors are born from their feeling of belonging and their identification with the territory in which they live, as well as from the recognition that they intend to achieve similar results. It is the simultaneous existence of these feelings that strengthens interpersonal relationships in the long run. Sometimes, the existence of elements outside the network can help the formation of these affective bonds between its elements.

Networks explain the added value of a territory when the organized collective action of resources and opportunities is established. This combines the articulation between the interests of stakeholders, the sharing of knowledge between them and the incentive to innovate. Continuous negotiation of the different stakeholder interests, which are not always convergent, requires leadership skills, which promote empowerment and commitment, for which the production and dissemination of knowledge by the various actors seems essential. It is in this context that there must be structures that facilitate the transformation of individual initiatives into a competitive advantage for destinations, guiding territories towards offering a differentiated and authentic tourist experience.

From this perspective, the governance model of destinations is the facilitating element in constructing these learning cultures, mirroring a new way of working and cooperating, of establishing strategic alliances between organizations, supported by knowledge and innovation. There is an economy of the territory that must stimulate its competitive capacity; there is a governance model for the territory that should ensure stakeholders' involvement in the promotion and consolidation of differentiated innovation in the provision; there is a strategy for the territory that must guarantee the organization of the destination in a network based on knowledge.

ACKNOWLEDGMENT

We are grateful for the participation of Portuguese public decision-makers at the local level, who contributed with their experience and knowledge to the present research.

REFERENCES

Aires, L. (2011). *Paradigma qualitativo e práticas de investigação educacional.* UniversidadeAberta.

Aish, A., & Jöreskog, K. (1990). A panel model for political efficacy and responsiveness: An application of LISREL 7 with weighted least squares. *Quality & Quantity*, *19*, 716–723. doi:10.1007/BF00152013

Aish, A., & Jöreskog, K. (1990). A panel model for political efficacy and responsiveness: An application of LISREL 7 with weighted least squares. *Quality & Quantity*, *24*(4), 716–723. doi:10.1007/BF00152013

Albattat, A., Jamaludin, A., Zuraimi, N. S. M., & Valeri, M. (2020). Visit Intention and Destination Image in Post- Covid- 19 Crisis Recovery. *Current Issues in Tourism.* Advance online publication. doi :10.1080/13683500.2020.1842342

Arbuckle, J. L. (2013). *Amos 22 user's guide.* SPSS.

Badulescu, D., Hoffman, I., Badulescu, A., & Badulescu, R. (2016). Local authorities' involvement in fostering Hungarian-Romanian cross-border cooperation in tourism. *Lex Localis. Journal of Local Self-Government, 14*(3), 337–358. doi:10.4335/14.3.337-358(2016)

Baggio, R., & Valeri, M. (2020). Network science and sustainable performance of family businesses in tourism. *Journal of Family Business Management.* doi:10.1108/JFBM-06-2020-0048

Bagozzi, R., & Yi, Y. (1988). On the evaluation of structural equation models. *Journal of the Academy of Marketing Science, 16*(1), 74–94. doi:10.1007/BF02723327

Bannò, M., Piscitello, L., & Varum, C. (2015). Determinants of the internationalization of regions: The role and effectiveness of public policy measures. *Regional Studies, 49*(7), 1208–1222. doi:10.1080/00343404.2013.821570

Bentler, P. (1990). Quantitative methods in psychology: Comparative fit indices in structural models. *Psychological Bulletin, 107*(2), 238–246. doi:10.1037/0033-2909.107.2.238 PMID:2320703

Beritelli, P., & Laesser, C. (2014). Getting the cash-cow directors on board: An alternative view on financing DMOs. *Journal of Destination Marketing & Management, 2*(4), 213–220. doi:10.1016/j.jdmm.2013.10.001

Bernabé, C., & Hernández, M. (2016). Turismo en Albarracín y Comarca: Acción pública local y dinâmica reciente en clave de desarrollo turístico sostenible. *Anales de Geografía de la Universidad Complutense, 36*(1), 173–194. doi:10.5209/rev_AGUC.2016.v36.n1.52718

Blasco, D., Guia, J., & Prats, L. (2014a). Emergence of governance in cross-border destinations. *Annals of Tourism Research, 49*, 159–173. doi:10.1016/j.annals.2014.09.002

Blasco, D., Guia, J., & Prats, L. (2014b). Tourism destination zoning in mountain regions: A consumer-based approach. *Tourism Geographies, 16*(3), 512–528. doi:10.1080/14616688.2013.851267

Bohlin, M., Brandt, D., & Elbe, J. (2016). Tourism as a vehicle for regional development in peripheral areas – myth or reality? A longitudinal case study of Swedish regions. *European Planning Studies, 24*(10), 1788–1805. doi:10.1080/09654313.2016.1194807

Bollen, K. A. (1989). *Structural equations with latent variables.* Wiley. doi:10.1002/9781118619179

Booyens, I. (2016). Global–local trajectories for regional competitiveness: Tourism innovation in the Western Cape. *Local Economy, 31*(1–2), 142–157. doi:10.1177/0269094215618598

Booyens, I., & Rogerson, C. M. (2015). Creative tourism in Cape Town: An innovation perspective. *Urban Forum, 26*, 405–424. 10.100712132-015-9251-y

Booyens, I., & Rogerson, C. M. (2016). Unpacking the geography of tourism innovation in Western Cape Province, South Africa. *Bulletin of Geography. Socio-Economic Series, 31*(31), 19–36. doi:10.1515/bog-2016-0002

Brouder, P., & Ioannides, D. (2014). Urban tourism and evolutionary economic geography: Complexity and co-evolution in contested spaces. *Urban Forum, 25*, 419-430. 10.100712132-014-9239-z

Brown, T. A. (2015). *Confirmatory factor analysis for applied research* (2nd ed.). Guilford Press.

Byrne, B. M. (2010). Structural equation modeling with AMOS: Basic concepts, applications and programming (2nd ed.). London: Lawrence Erlbaum.

Chemli, S., Toanoglou, M., & Valeri, M. (2020). The impact of Covid-19 media coverage on tourist's awareness for future traveling. *Current Issues in Tourism*, 1–8. Advance online publication. doi:10.1080/13683500.2020.1846502

Clavé, S., & Wilson, J. (2017). The evolution of coastal tourism destinations: A path plasticity perspective on tourism urbanisation. *Journal of Sustainable Tourism*, *25*(1), 96–112. doi:10.1080/09669582.2016.1177063

Cohen, J. (1988). *Statistical power analysis for the behavioural sciences* (2nd ed.). Erlbaum.

Comissão Europeia. (2000). *Para um turismo urbano de qualidade: Gestão integrada da qualidade (GIQ) dos destinos turísticos urbanos*. Serviço das Publicações Oficiais das Comunidades Europeias.

Comissão Europeia. (2003). *A manual for evaluating the quality performance of tourist destinations and services*. Office for Official Publications of the European Communities.

Comissão Europeia. (2016). *The European tourism indicator system: ETIS toolkit for sustainable destination management*. Available at https://ec.europa.eu/growth/sectors/tourism/offer/sustainable/indicators/index_en.htm

ComissãoEuropeia. (2013). *Caixa de ferramentas do sistema europeu de indicadores de turismo para destinos sustentáveis*. Serviço de Publicações da União Europeia.

Costello, A. B., & Osborne, J. W. (2005). Best practices in exploratory factor analysis: Four recommendations for getting the most from your analysis. *Practical Assessment, Research & Evaluation*, *10*(7), 1–9.

Coutinho, C. P. (2015). Avaliação da qualidade da investigação qualitativa: Algumas considerações teóricas e recomendações práticas. In *F. Souza, D. Souza & A. Costa (Orgs.), Investigação qualitativa: Inovação, dilemas e desafios* (Vol. 2, pp. 101–121). Ludomedia.

De Noni, I., Orsi, L., & Zanderighi, L. (2014). Attributes of Milan influencing city brand attractiveness. *Journal of Destination Marketing & Management*, *3*(4), 218–226. doi:10.1016/j.jdmm.2014.06.001

Diamantopoulos, A., & Siguaw, J. A. (2000). *Introducing to Lisrel: A guide for the uninitiated*. Sage. doi:10.4135/9781849209359

Escach, N., & Vaudor, S. (2014). Réseaux de villes et processus de recomposition des niveaux: Le cas des villes baltiques. *Cybergeo: European Journal of Geography*, *679*, 1–24. doi:10.4000/cybergeo.26336

Farmaki, A. (2015). Regional network governance and sustainable tourism. *Tourism Geographies*, *17*(3), 385–407. doi:10.1080/14616688.2015.1036915

Ferdinand, N., & Williams, N. (2013). International festivals as experience production systems. *Tourism Management*, *34*, 202–210. doi:10.1016/j.tourman.2012.05.001

Fornell, C., & Larcker, D. F. (1981). Evaluating structural equation models with unobservable variables and measurement error. *JMR, Journal of Marketing Research*, *18*(1), 39–50. doi:10.1177/002224378101800104

Gorsuch, R. (1983). *Factor analysis*. Lawrence Erlbaum.

Hair, J. F., Anderson, R. E., Tatham, R. L., & Black, W. C. (2008). *Multivariate data analysis* (7th ed.). Pearson Prentice-Hall.

Hill, M., & Hill, A. (2008). *Investigação por questionário* (2nd ed.). EdiçõesSílabo.

Hristov, D. (2015). Investigating DMOs through the lens of social network analysis: Theoretical gaps, methodological challenges and practitioner perspectives. *Advances in Hospitality and Tourism Research, 3*(1), 18–39.

Hristov, D., & Zehrer, A. (2015). The destination paradigm continuum revisited: DMOs serving as leadership networks. *Tourism Review, 70*(2), 116–131. doi:10.1108/TR-08-2014-0050

Jöreskog, K. G., & Sörbom, D. (2004). *LISREL 8.7 for Windows* [Computer Software]. Scientific Software International, Inc.

Kline, R. (2011). Principles and practice of structural equation modeling (3rd ed.). New York: The Guilford Press.

Likert, R. A. (1932). A technique for the measurement of attitudes. *ArchivesofPsychology, 140*, 1–55.

Lima, L. P. (2000). Atitudes: Estrutura e mudança. In J. Vala & M. B. Monteiro (Eds.), *Psicologia social* (4th ed., pp. 187–225). Fundação Calouste Gulbenkian.

Lopes, C. (2005). *Documentos de apoio do módulo 6 (inquérito: amostragem e erros de amostragem, de não-resposta, de cobertura e de ajustamento) do curso de formação em métodos de investigação e análise de dados em ciências sociais*. Textos não publicados. Núcleo de Investigação e Intervenção em Psicologia Social da Faculdade de psicologia e Ciências da Educação da Universidade de Coimbra.

Lousã, E. (2013). *Liderança empreendedora e cultura de inovação em organizações de base tecnológica e análise comparativa entre setores de atividade*. Tese de Doutoramento, Universidade de Coimbra.

Makkonen, M., & Rohde, S. (2016). Cross-border regional innovation systems: Conceptual backgrounds, empirical evidence and policy implications. *European Planning Studies, 24*(9), 1623–1642. doi:10.1080/09654313.2016.1184626

Makkonen, T., & Weidenfeld, A. (2016). Knowledge-based urban development of cross-border twin cities. *InternationalJournalKnowledge-BasedDevelopment, 7*(4), 389–406. doi:10.1504/IJKBD.2016.080881

Marôco, J. (2010). *Análise estatística com utilização do SPSS* (3rd ed.). Edições Sílabo.

Marôco, J. (2011). *Análise de equações estruturais: Fundamentos teóricos, software& aplicações*. Pêro Pinheiro: Report Number.

Marôco, J., & Garcia-Marques, T. (2006). Qual a fiabilidade do alfa de Cronbach? Questões antigas e soluções modernas? *Laboratório de Psicologia, 4*(1), 65–90.

Mintzberg, H. (1979). *The structuring of organizations: a synthesis of the research*. Prentice-Hall.

Mintzberg, H. (1983). *Power in and around organizations*. Prentice-Hall.

Mira, M. R., & Breda, Z., (2019a). *Internacionalização de destinos turísticos: Uma abordagem sistémica.* BeauBassin, Mauritius: Novas Edições Acadêmicas, OmniScriptumPublishing.

Mira, M. R., & Breda, Z. (2019b). Internacionalização de destinos turísticos: O papel das DMO. In C. Estevão & C. Costa (Eds.), *Turismo, estratégia, competitividade e desenvolvimento regional: Uma visão estratégica no setor do turismo* (pp. 19–42). Novas Edições Acadêmicas, OmniScriptum Publishing.

Mumford, M. D. (2003). Where have we been, where are we going? Taking stock in creativity research. *Creativity Research Journal, 15*(2/3), 107–120. doi:10.1207/S15326934CRJ152&3_01

Nilsson, J., Eskilsson, L., & Ek, R. (2010). Creating cross-border destinations: Interreg Programmes and regionalisation in the Baltic Sea Area. *Scandinavian Journal of Hospitality and Tourism, 10*(2), 153–172. doi:10.1080/15022250903561978

Nunnally, J. C. (1978). *Psychometric theory*. McGraw-Hill.

Nunnally, J. C., & Bernstein, I. H. (1994). Psychometric theory (3rd ed.). New York: McGraw-Hill.

Rovira, N. (2016). ¿Ciudades en el mapa o en la guía turística? Venta de la ciudad y sentido del lugar. *Revista CIDOB d'AfersInternacionals, 113*(113), 89–105. doi:10.24241/rcai.2016.113.2.89

Sanz-Ibáñez, C., & Clavé, S. (2014). The evolution of destinations: Towards an evolutionary and relational economic geography approach. *Tourism Geographies, 16*(4), 563–579. doi:10.1080/14616688.2014.925965

Sarasa, J. (2015). Mito y estrategias para un destino romántico: La ciudad de Teruel. *Cuadernos de Turismo, 36*(36), 39–54. doi:10.6018/turismo.36.230871

Schumacker, R. E., & Lomax, R. G. (2012). *A beginner's guide to structural equation modeling*. Routledge Academic. doi:10.4324/9780203851319

Scuttari, A., Volgger, M., & Pechlaner, H. (2016). Transition management towards sustainable mobility in Alpine destinations: Realities and realpolitik in Italy's South Tyrol region. *Journal of Sustainable Tourism, 24*(3), 463–483. doi:10.1080/09669582.2015.1136634

Sertakova, E., Koptseva, N., Kolesnik, M., Libakova, N., Luzan, V., & Sergeeva, N. (2016). Brand-management of Siberian cities: Krasnoyarsk as a case study. *International Review of Management and Marketing, 6*(5), 185–191.

Soares, J., Baidal, J., & Gândar, J. (2015). La evolución de destinos turísticos litorales consolidados: Análisis comparado de balneario Camboriú (Brasil) y Benidorm (España). *Annales de Géographie, 35*(2), 143–166. doi:10.5209/rev_AGUC.2015.v35.n2.50118

Spyriadis, T., Fletcher, J., & Fyall, A. (2013). Destination management organisational structures. In C. Costa, E. Panyik, & D. Buhalis (Eds.), *Trends in European tourism planning and organization* (pp. 77–91). Channel View Publications. doi:10.21832/9781845414122-010

Tabachnick, B. G., & Fidell, L. S. (2013). *Using multivariate statistics* (6th ed.). Pearson Education.

Toanoglou, M., Chemli, S., & Valeri, M. (2021). The organizational impact of COVID-19 crisis on travel perceived risk across four continents. *Journal of Organizational Change Management*. Advance online publication. doi:10.1108/JOCM-12-2020-0369

UNWTO. (2007). *A practical guide to tourism destination management*. World Tourism Organization.

UNWTO. (2010). *Survey on destination governance: Evaluation report*. World Tourism Organization.

Urbina, S. (2014). *Essentials of psychological testing*. Wiley.

Valeri, M. (2016). Networking and cooperation practices in the Italian tourism business. *Journal of Tourism, Heritage & Services Marketing*, *2*(1), 30–35. doi:10.5281/zenodo.376333

Valeri, M. (2021). *Organization Theories. Implications for the Strategic Management*. Springer.

Valeri, M., & Baggio, R. (2020a). Social network analysis: Organizational implications in tourism management. *The International Journal of Organizational Analysis*. Advance online publication. doi:10.1108/IJOA-12-2019-1971

Valeri, M., & Baggio, R. (2020b). Italian tourism intermediaries: A social network analysis exploration. *Current Issues in Tourism*, *24*(9), 1270–1283. doi:10.1080/13683500.2020.1777950

Valeri, M., & Baggio, R. (2020c). A critical reflection on the adoption of blockchain in tourism. *Journal Information Technology and Tourism*, *23*(2), 121–132. doi:10.100740558-020-00183-1

Valeri, M., & Baggio, R. (2021). Increasing the efficiency of knowledge transfer in an Italian tourism system: A network approach. *Current Issues in Tourism*, 1–16. Advance online publication. doi:10.1080/13683500.2021.1937960

Vermeulen, R. (2015). Pursuing the peripheral path? A path-dependent analysis of the Frankfurt and Munich fairs. *European Planning Studies*, *23*(2), 332–348. doi:10.1080/09654313.2013.867318

Vodeb, K., & Rudež, H. (2016). Possibilities for inter-destination collaboration in tourism in the case of Opatija and Portorož: A managerial perspective. *RevijaZaSociologiju*, *46*(2), 205–227. doi:10.5613/rzs.46.2.4

Volgger, M., & Pechlaner, H. (2015). Governing networks in tourism: What have we achieved, what is still to be done and learned? *Tourism Review*, *70*(4), 298–312. doi:10.1108/TR-04-2015-0013

Weidenfeld, A. (2013). Tourism and cross border regional innovation systems. *Annals of Tourism Research*, *42*, 191–213. doi:10.1016/j.annals.2013.01.003

Chapter 7
New Perspective of Networking in the DMO Model

Vincenzo Asero
https://orcid.org/0000-0002-6933-5461
University of Catania, Italy

Sandro Billi
Side Note scc, Italy

ABSTRACT

Destination management organizations are functional structures that manage and market the tourist destinations operating to facilitate the cooperation among different stakeholders. A successful DMO supports tourism development, especially when tourism is an important economic driver for a destination. The idea of building different forms of DMOs and the concept of networking have guided tourism governance choices in different initiatives conducted in many countries. This chapter analyses the model of DMO adopted in Italy by Regione Toscana comparing it with the variable geometry approach by Beritelli et al. The study offers a critical reflection on the model of DMO, relevant from the perspectives of governance and management.

INTRODUCTION

In an era of globalized and highly dynamic relationships, networking plays a key role in the competitiveness, management and planning of tourism destinations (Hall, 2008; Valeri, 2016; Valeri and Baggio, 2020a; 2020b). Tourism is the sector with the highest number of inter-organizational relationships (Bickerdyke, 1996; Valeri, 2021) and the active involvement of diverse firms and public organizations play a crucial role in achieving competitive advantages in the tourism market (Pavlovich, 2003). Indeed, as Gibson et al. (2005) highlighted, networks provide a wide range of positive externalities stimulating partnerships and cooperation.

Many scholars underline how collaborative partnerships have increased in importance, becoming crucial in the coordination between tourism policies and their related actions. Some authors have em-

DOI: 10.4018/978-1-6684-3889-3.ch007

phasized that collective actions have a great impact on tourism development (e.g. Crouch & Ritchie, 1999; Dollinger, 1990). Similarly, Jamal and Jamrozy (2006) highlighted that various stakeholders take part in tourism destination management, although no one individually controls the decisional processes entirely. On this topic, Merinero-Rodríguez and Pulido-Fernández (2016) noted that relationships are the core element for understanding a tourism destination system.

The institution of a tourism network can occur due to initiatives promoted by private or public stakeholders, each with different preferences, strategies and contrasting interests. The literature has identified a large number of different reasons that can determine cooperation in a tourist region, but there are also some country-specific factors, which can stimulate cooperation. In some cases, these factors concern trade agreements and legal determinants arising from foreign and domestic regulations (Asero & Skonieczny, 2017).

A direct extension of enhancing tourism network in a destination is the creation of a Destination Management Organization (DMO). DMOs are functional structures that manage and market the destinations operating to facilitate dialogue among the private sector, public sector, and other stakeholders that may otherwise never collaborate. For this purpose, developing a DMO relies on identifying a destination vision through collaboration. A successful DMO supports tourism development, especially when tourism can be an important economic driver for a destination (Baggio and Valeri, 2020; Valeri and Baggio, 2021).

The concept of networking and the idea of building different forms of DMOs have guided tourism governance choices in different initiatives conducted in many countries. In Italy, for instance, it is important to mention some experiences such as the Local Tourist Systems, envisaged by the Italian Government with Law 135/2001; the Tourist Districts, introduced by the *Regione Siciliana* with Law 10/2005; the *Ambiti Turistici*, created by the *Regione Toscana* in 2018. In the case of Brazil, the Nautical and Cultural Tourism Network in the Baía de Todos os Santos, which was proposed by the *Secretaria de Turismo* of the state of Bahia in Brazil, deserves mention; the project to build up a local DMO in the Pelourinho neighborhood in Salvador, Bahia, Brazil.

Despite the importance of DMOs for the enhancement of the competitiveness of destinations, and the need for a high level of cohesion among different stakeholders to build up appropriate tourism networks, many initiatives have achieved limited results. One of the main reasons behind the failure of many projects has been the critical area of cooperation between private and public sectors in the construction of the partnerships. On the one hand, partnerships are not always naturally established, and in many cases have been fostered by national and local governments (Valeri and Katsoni, 2021). On the other, collaborative partnerships stem from voluntary agreements among actors that intend to share benefits from cooperation. Therefore, partnerships require joint actions that ensure interaction among autonomous stakeholders, while collaborations within the network should avoid the costs of resolving adversarial conflicts among the participants.

Although there is general agreement on the philosophy behind private-public sector cooperation and partnership, there are many different ways of developing the concept and putting it into practice (WTO, 2000). There is no single correct model or formula, because of the importance of local circumstances. Reflection is thus required on the role of the public sector in tourism activities, beyond its legitimate role as regulator and controller. The issue becomes critical if it is remembered that the principle of participation of public bodies in DMOs is being reviewed in some countries, such as the United States. The problem becomes even more pressing in the light of post-Covid-19, the fall in GDPs and the consequent decrease in tax revenues, which necessitates financial support and safety rules. At the same time, it is important to consider the extent to which the private sector is able and willing to assume part of the responsibility for the tourism development of a destination.

This chapter analyses the governance model of DMOs, and investigates the critical role of networking between the private and public sectors in the management of tourism systems, to explain the reason for many failures and weaknesses in practice. The research is based on the analysis of a selected case of DMO in Tuscany that has been investigated considering some strategic factors, such as the relevant legislation supporting the different initiatives, the involvement of local stakeholders, the role played by the public bodies, the contribution of local communities and the creation of social capital.

The chapter supports the argument that cooperative behaviour of organisations, enterprises and public institutions plays a key role in achieving competitive advantages for a destination. It is set out as follows: first, an extended literature review on tourism networks and the benefits of partnerships in DMO activities. The study then focuses on empirical experience, offering a critical reflection on the model of DMO, relevant from the perspectives of governance and management. In the last section, conclusions and implications are presented.

BACKGROUND: THE DMO MODEL

Currently, destination management is an increasingly important question, since competition tends more and more to be seen as existing among destinations instead of at the level of service providers, whereas competitiveness is a key factor influencing a destination's long-term existence. As Pearce (2014, p. 141) observed, the conceptualization of destination has practical implications for destination management since "the nature and the scope of the concepts used may determine the focus of management, who or what is managed, where management priorities lie, and where solutions might be found." However, while it is clear that the spatial structure of a tourist destination is central for policy makers, firms and destination managers to plan and implement coordinated actions of DMO, a number of researchers underline that destination is problematic to define as a concept (Framke, 2002; Saraniemi & Kylänen, 2011).

Buhalis (2000) defined a destination as a region where all the supply elements which a tourist would need are available. In a successful destination, service providers recognize the importance of partnerships, as they are actively co-operating in order to better satisfy the tourists' needs and provide them with a unique experience. However, in the scientific literature many authors emphasise that considering geographical and/or administrative boundaries as definers of a destination is not always appropriate, because tourists, while travelling, do not rely on administrative boundaries (Klepers & Rozite, 2010). Moreover, the boundaries of a destination can be considered as functional or administrative, fixed or fluid (Pearce & Schänzel, 2013). In this regard, Beritelli et al. (2014) affirmed that a destination is a space of 'variable geometry' defined by different supply networks activated by tourist movements. Along the same lines, Asero et al. (2020) found in their research that destinations are being shaped and re-shaped by tourists' mobility.

A destination is closely associated with the concept of network, which entails the different relationships and transactions that characterize a tourism system. Collaborative destination actions involve a number of public and private stakeholders working interactively on a common issue through a process of exchange of ideas and expertise, and pooling of financial and human resources (Vernon, Essex, Pinder, & Curry, 2005). Fruitful relationships are determined by some crucial factors, such as the perception that benefits will affect all partnership members, the recognition of interdependence, the coordination of skills, and the sharing of a strategic plan (Jamal & Getz, 1995).

Many scholars investigated the nature and formation of networks, strategic alliances, and partnerships in tourism (Gulati, 1999; Stokes, 2004; Ziakas & Costa, 2010). The models of cooperation have been considered from different perspectives, including territorial, static and dynamic aspects, impact on destination developments, evolution and transformation of partnerships (Czernek, 2013). In spite of such approaches, there is a debate in the literature on the advantages of formal networks vs informal ones. Strobl and Peters (2013) emphasized that informal relationships are important in forming alliances, while they contribute to the formation of formal partnerships. According to the authors, since formal rules are naturally associated with liability and safety, they should encourage cooperative actions within a tourism network. In the same vein, Vernon et al. (2005) showed that tourism cooperation involves various stakeholders working interactively on common problems or issues through a formal approach.

Although tourism networks are based on the connections among different actors, entrepreneurs play a central role in the evolution of a destination (Russell & Faulkner, 1999; Koh & Hatten, 2002). Similarly, Crouch and Ritchie (2010) noted that entrepreneurship contributes to its competitiveness in a number of ways. The type of activity they carry out depends greatly upon levels of entrepreneurship, in addition to the environmental conditions and the stage of the destination's development (Russell & Faulkner, 2004). This supports the idea that the more experienced network's actors are in tourism, the greater the benefits (Telfer, 2001). Thus, as Crouch and Ritchie (1999) argued, competitive advantage depends on tourism actors having common goals more than from the endowment of tourist resources at destination.

A logical extension of enhancing the competitiveness and the success of a tourism destination is establishing a DMO. DMOs are formal entities performing at different levels (national, regional, and local level) that operate along the entire tourism value chain. A DMO increases tourism through building private-public partnerships, developing marketing strategies, assuring the visitor experience at the destination, promoting the sustainability of the destinations and maximizing the benefits for the local communities and economy. According to Elbe et al. (2009), a main function of a DMO is that of being responsible for marketing a destination. From another perspective, Crouch and Ritchie (1999) highlighted that within the conceptualization of the organizational function of a DMO the 'M' emphasizes 'management', rather than simply 'marketing'.

Today, the role of the DMO is becoming far broader and requires guiding and coordinating destination development and management activities within the framework of a coherent strategy (UNWTO, 2007). Within this context, the World Tourism Organization (UNWTO, 2019) recently stated that DMOs are called to align their policies and actions with the 2030 Agenda for Sustainable Development and Sustainable Development Goals (SDGs), in order to maximize the contribution of tourism destination to all the 17 SDGs.

However, while the concept of destination management appears to be clear, in many cases it is difficult to put into practice. As the World Tourism Organization (UNWTO, 2019, p. 8) highlights "no 'one-fits-for-all' model can be applied, and each destination needs to adapt the nature and functions of its DMO to different variables". Then, although the characteristics, priorities and level of development may vary from one destination to another, every DMO requires proactive actions from tourism stakeholders increasingly focused on maximising the value of tourism for residents, businesses and visitors and to establish an effective market position, while ensuring the sustainability of the economic, cultural and social environment for a destination (OECD, 2020).

Principles, Organization and Advantages of the Variable Geometry Approach for the DMO

The Variable Geometry Approach (VGA) to the DMO has been deeply analysed by Berritelli et al. (2014). Development of the concept began with the perception that the destination is considered with a logic that depends on different points of view: territorial logic, on the perspectives of public sector and local population; experiential logic, of visitors; and finally business logic, of the tourism companies. It is difficult to put together the three logics and obtain a higher vision of the problems and needs in a destination, since each category of actors has a different point of view.

The idea of 'variable geometry' in the organization and management of a DMO comes from a spatial vision of the dislocation of the tourist in the reference areas. In the tourism areas, the visitors move between well-defined points of the territory; moreover, these movements can overlap and they can also change depending on seasons. Referring to some empirical research, we can assume that "tourist flows are not random, but patterned by the geographical and national background of tourists" (Bowden, 2003, p. 257). The points to which tourists move, where they spend most of their time and money, take on the role of *Strategic Business Areas* (SBA). Each SBA is characterized by specific segments, each with similar motivation and behaviour. Business areas that relate to different tourism products have different business portfolios. This means that every SBA needs a different strategy and a different approach to the market.

While some features of an area clearly show the above characteristics, in other cases they may be more difficult to perceive. In fact, while some tourism products have limited points of attraction, generating a concentration of visitors proportional to their power of desirability, other resources are more widespread, leading to a wider diffusion in the territory. The first situation is that of winter sports resorts. During the winter season, points of departure for uphill lifts concentrate the majority of visitors, around which are many direct and indirect business-related services: cafés and restaurants, ski equipment rentals, ski schools, shops specialized in winter equipment, souvenir shops, and even banks or car repair facilities. In the same area, during the summer season, we can perceive that the grouping of visitor changes. Places crowded in winter become less attractive, some activities close; we note other concentration points, for instance at the beginning of tracks or around sports courts or cycle paths. A mountain resort, therefore, respects both assumptions of the VGA: overlapping, and change depending on the season.

However, there is some difficulty applying the model in areas where resources and attractions are small scale and spread over the territory. This is the case, for instance, of cycle tourism or of wine tasting. The latter is usually proposed by many wineries scattered across an area, widely different in organization, size and capacity in attracting customers. Thus, while focusing on the core attraction of the destination, it is difficult to plot limited spaces for moving and staying. In the areas tied to wineries we do not find the generation of SBA, although the host company itself tends to diversify the offer by intercepting customer requests, proposing products that are not its core business. From another point of view, it must be recognized that spatial concentrations of tourists take place in these territories too, generating SBAs at points of aggregation that are not strictly linked to the main attraction: i.e., cultural events, urban centres, sports courts, etc.

In the Variable Geometry Approach, the DMO acquires new roles, goals and organizations working close with the SBA. The DMO does not operate indistinctly in a territory, but discusses and agrees with the SBA and other stakeholders, plans and projects focused on specific problems identified through research or proposed by the interlocutors. The DMO can give added value to the system if it can fulfil tasks in the productive process, tasks that are financed by the direct beneficiaries, directly or through collective sources.

In their study, Beritelli et al. (2014, p. 13) defined some principles for the development of the DMOs in relation to the VGA. The vision of DMOs passes from a stable, monolithic body, based on pre-defined and repetitive activities, to a subject that implements plans and projects together with the actors of the territory. The projects should be financed directly through private or public funds, and see the stakeholders closely involved in the financing aspect. The authors summarized the differences between the traditional view of the roles and tasks of the DMO and the new one, as in the following table:

Table 1. Traditional View and New Paradigm of Destination and DMO

	Traditional View	New Paradigm
With regard to the destination		
Basic Constructs	One area, with rather delimited boundaries, one territory, the area changes as the perspective (cruising altitude) of the organization or the institution changes (hierarchical perspective)	Multiple different, partly overlapping spaces, consisting of dynamic flows
Markets	Different markets, depending on the segmentation criteria, often aggregated	One market per given SBA, each one rather homogeneous in motivations, activities, and flows
Image, brand	One destination image, one destination brand as a rule, above many other product brands	Multiple images, multiple SBA brands, and a destination brand as the exception
Planning and research implications	One object of management/research with regard to competitiveness, sustainability, life cycle stage	Management/research at SBA level, SBAs with different competitive positions, and at different life cycle stages
Actor and institutional involvement	For all comprehensive, pertaining the whole territory/ area, everybody worries about everything	Public sector with regard to selected projects in the administrative boundaries and with neighbouring administrations, private and PPP pertaining to the relevant SBAs
With regard to the DMO		
Basic scope	Serve all the visitors, serve all the local suppliers	Create added value for visitors and suppliers in need of specific services
Activities, tasks	General, due to general finance sources (e.g., public sector, overnight taxes)	General tasks and specific tasks mandated by a single organization or groups of organizations and specific mandates and financial sources
Competencies	Stable, built around functions (e.g., public relations, promotion, ICT, social media); the more functions, the more influential the DMO, the better	Dynamic, built around the needs of the SBAs, process supportive
Organization and financial aspects	One big money pot, budget maximization, updating budget policy	Synergetic sum of differently financed mandates, fulfilling optimally the mandated task, activity- based costing
Adaptability	Low, incremental, depending on external factors such as pressure to act and/or new (public) financial funds	High, leaps due to adding or removing mandates
Performance measurement	Promotional input metrics, general frequencies in the destination, sales volume of capacities	Achievements according to the mandated task, quality of mid- to long-term relationships with partner organizations and institutions

Note: DMO = destination management organization; SBA = strategic business area; PPP = public–private partnership; ICT = information and communication technology.

The DMO in the New Tourism Legislation of the "Regione Toscana"

Starting from the description of the Variable Geometry Approach management for DMO, we will analyse the model adopted by the *Regione Toscana*, studying the organization then comparing it with the model.

In December 2016, the *Regione Toscana*, which has an administrative function for Tourism in Tuscany, adopted the Regional Law (L.R.) 86, "*Testo Unico del Sistema Turistico Regionale*" (Consolidated Text of the Regional Tourist System), modified less than two years later with the L.R. 24/2018. The aim of the law is to regulate the tourism organization system of the *Regione*, the accommodation facilities, tourism businesses and professional practices. The replacement of the previous LR 42/2000 was made necessary by the many amendments and changes that occurred in the legislation and in the market. Among others, specific purposes of the law are:

1. to define the tools of tourism policy, identifying the objectives for the enhancement and for the development of the Tuscan tourist system;
2. to promote and enhance, toward the local, the national and the foreign market, a unified image of the Tuscan tourism system and of each of its parts;
3. to foster agreements and collaboration with a plurality of subjects, including the State, the other regions, the local authorities, the Chambers of Commerce (CCIAA), and the Universities;
4. to define and implement policies for coordinated management of tourism resources, also to enhance the historical, monumental, naturalistic, cultural, rural and thermal heritage of the region; to favour the structural strengthening of the tourism offer system.
5. concerning the governance, the new law established that some activity should be managed by the municipalities in an associated form, by supra-municipal bodies called *Ambiti Turistici*.

Specifically, the *Ambiti* will be in charge of:

- reception and information;
- activation and management of an OTD – *Osservatorio Turistico di Destinazione,* a tourist monitoring centre.

The aim of the rules concerning the *Ambiti Turistici* was to complete the governance of tourism promotion, within territorial areas that will be defined by regional law; in addition to this model, there is also the possibility of associating municipalities with products defined as homogeneous.

To understand the task of the *Ambiti Turistici,* we can refer to the description of the activities foreseen within the spheres of reception and providing information. Specifically, the tourist information and reception services ensure information on the tourist attractions of the area included in the municipality of each *Ambito* concerning tourist products, landscape, cultural, historical, artistic and food and wine heritage. To reach their target, they can provide information and informative material about the local offer: on the availability of accommodation, on the offer of tourist services, on tours and excursions in the area. The reception and information system is called IAT.

The activity of the IAT can include the booking of tourism services and accommodation. The IAT can be managed directly by the *Ambito* or can be put on tender. For the activation of the IAT the *Regione Toscana* allocated 2,5 billion Euro for activities that, in addition to reception and information, include other functions that were not originally envisaged by law. In particular, the funding also covers:

- buildings for the governance of the destination.
- development of synergies and cooperation between public and private; subjects involved in government of the destination, including coordination of initiatives and activities.
- coordinated management of the tourist information and reception system in the territory, both in terms of tools and analysis of results.
- identification and sharing the peculiarities of the area and definition of strategies for the development of tourism products, including the organization of the offer.
- process of engagement with stakeholders.
- strengthening the network.
- management of information reception flows.
- elaboration of a Destination Management Plan, if opted for, that defines the strategic lines for the management of the tourist destination, through the identification of a common vision and objectives between the stakeholders, the identification of actions and organizations and related financial resources, both in terms of tools and analysis of results. The Destination Management Plan should also stimulate the birth of the DMO.

The Tuscan Vision and the Variable Geometry Approach

In the functional system of Tuscan tourism, the term DMO appears only in the call for the financing of the Start Up of the *Ambiti*. In this way the *Ambiti Turistici*, through the DMO, seem to take on activities that go beyond mere reception and provision of tourist information envisaged under the law, since they are also candidates for a strategic activity, to be conducted following a Destination Management Plan. The creation of DMOs follows a series of activities that include studies and engagement with the stakeholder, but without the vision of SBA, as above presented.

In this way, the theoretical lines illustrated in the study of Beritelli et al. (2014), with respect to the functioning and tasks of a DMO, seem to differ from the functions indicated - or rather not indicated - by the Tuscan legislation. A clear difference can be seen in a basic activity of DMOs, the provision of tourist information, which in the Tuscan vision will not be carried out directly by the manager of the destination, but will be outsourced.

Although the law establishing the *Ambiti Turistici* is very recent, it seems not to have assimilated the useful indications deriving from the VGA. In particular, the *Regione Toscana* has favoured a static and in many ways centralistic vision of the organization of the DMO.

In practice, six groups of functions have been identified for each area:

- Function 1: analysis, technical and design coordination.
    - coordination with individual administrations and other political bodies present in the area
    - study of trends of the destination
    - strategic analysis of supply and demand
    - relations with operators and with the population
    - drafting of the operational plan
    - identification of the thematic products of the area
    - organization, integration and coordination of the leisure, MICE and accessible offers
    - Fundraising
    - Projects related to resources, transport, mobility

- Coordination / organization of events
- Function 2: information service
  - Organization and management of tourist information *ad personam* and online
- Function 3: communication
  - Organization of printed communication material
  - Press office
  - Destination web portal and regional web portal
  - Social media management
  - Web marketing
- Function 4: Promotion and Marketing of Products and Services (DMC)
  - Creation of thematic products
  - Identification of markets, material preparation and promotion campaign
  - Organization of activities in the area in collaboration with *Toscana Promozione*
  - Participation in fairs and events on the basis of an operational plan defined with Toscana Promotion
  - Relationships with national and international buyers
  - Business development in the MICE segment
  - Sale of products
- Function 5: Consultancy
  - Consultancy in the creation of tourism promotion and distribution platforms and portals
  - Consultancy in identifying services necessary for particular market segments
  - Communication and marketing in line with destination activities
  - Support for the creation of promotional material
- Function 6: Destination Audit
  - Customer Satisfaction Studies
  - Quantitative monitoring of flows
  - Analysis and enhancement of human resources skills
  - Quality monitoring of the services offered

Concerning the governance, each *Ambito Turistico* has a decision-making body, the *Conferenza dei Sindaci*, constituted by the mayors of the municipalities that make up the area. Many of the activities have been outsourced, delivering the implementation to third parties, so reducing the contact with the territory and the opportunities to encourage the development of professionalism and indigenous initiatives.

As we can see in figure 1, the activity is cantered on a territorial vision, where institutional involvement is all comprehensive and everybody worries about everything, with a very low engagement of the private sector. The SBA related vision is not considered, and the same *Ambito* seems to be the depositary of the decision of which products to promote and develop, without a comparison with the requests and capacities of the territory. It thus appears to be a system that generates a decision from above, based on considerations that take little account of the productive system, the predisposition of local actors.

While the vision of 'variable geometry' provides for the construction of shared paths with groups of SBA, based on the study and the solution of problems, collective identification of necessary resources and professionalism, the functions of the *Ambiti* include the provision of consultancy, assuming the possession of specialized knowledge that can hardly be found in the market. Finally, in terms of function four - Promotion and Marketing of Products and Services (DMC). This is a sales activity, which requires great technical and relational knowledge, and instead of being developed by supporting the local actors, is delegated to a public body that carries it out by favouring the presence of external subjects, sometimes supported by public funding, in a curious form of competition.

Figure 1. Organization of Ambiti Turistici Regione Toscana

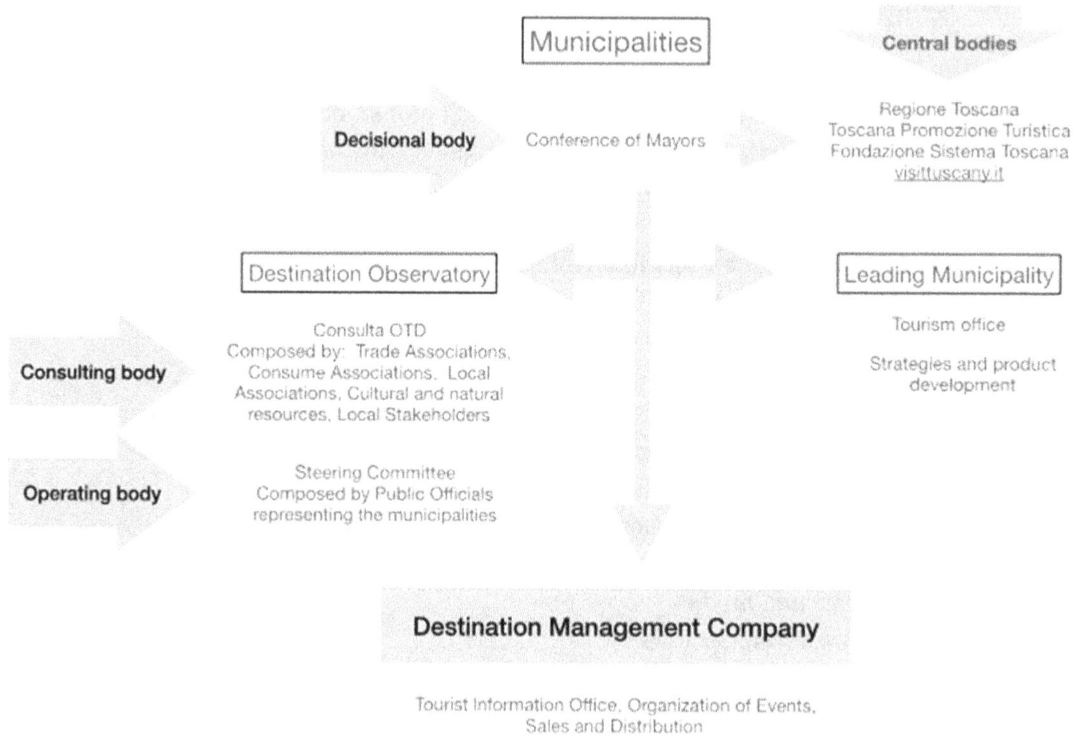

CONCLUSION

The increasing difficulty in achieving destination competitiveness necessitates DMOs taking a proactive interest in the stewardship of the destination's social, cultural, and environmental resources. However, as many scholars highlight, destination is a problematic concept, since it can be used in different ways and across a varying range of spatial scales.

Defining the boundaries of tourism destinations is recognised as an important issue in tourism, since space affects the way a tourism system works. The configuration of a tourism area, indeed, is crucial for policy makers, firms and destination managers to plan and implement coordinated actions of DMO. In this view, Bieger et al. (2009) noticed that the size of a destination influences the DMO functions, activities and budgeting.

The spatial dimension of a destination can be defined according to both cooperation among different locations and strategic partnerships among firms and public entities, rather than through administrative conditions. In this view, following Porter (1998), a destination can be considered as a tourism cluster, since it encompasses a geographical concentration of interconnected firms and institutions. By contrast, from another perspective, destinations are networks of social relations instead of areas with boundaries.

Destinations require the evaluation of the boundaries of the tourism places they contain, in order to increase market competition, and to meet globalizing tourism markets. Hence, on a practical level, the delimitation of a tourism destination is important for planning and managing tourism as well as defining marketing strategies. However, the geographical dimension of tourism destinations is often associated with blurred boundaries, since they can encompass either a single location or a network of different places (Dredge, 1999).

Based on these arguments, this chapter proposed the dynamic viewpoint of Beritelli et al. (2014) that considers a destination as a space of 'variable geometry', since it is the playground of different supply networks. The concept presents an alternative perspective of DMO closer to the complexity of managing tourism in dynamic systems, underlying the interdependencies and connections that determine the tourist supply in a destination.

In this framework, two issues appear critical: how to mobilize the various stakeholders, and how to develop cooperative behaviours and coordination in interaction among them. In this regard, as Asero and Skonieczy (2017) noted, while the heterogeneity of relationships provides several opportunities for a network, the manner in which the linkages are formed and maintained can pose some problems for management. Therefore, the structure and goal performances of a DMO can be affected by constraints, opportunistic behaviour and power imbalances.

As a consequence, while DMOs should not be benchmarked against each other, it is important to evaluate if the model adopted by a destination can help its firms to achieve and maintain competitive advantage and, at the same time, to enhance destination competitiveness. That has practical implications for informing investment and other resource allocation decisions by public and private sector tourism stakeholders.

In this context, this chapter highlighted that the Tuscan local tourism organization seems to be more oriented to a monolithic and static form with pre-established functions, often outsourced, under rigid administrative control and centralized strategies, rather than to a dynamic and flexible organization such as that proposed by the VGA, based on the interrelationships between tourism market needs, projects and destination attributes. A vision that does not favour the autonomy, growth and dynamism of tourism businesses, or support an effective market position, but proposes a rigid framework, based on strategies designed by the public sector that do not have direct relationships with the tourism markets.

REFERENCES

Asero, V., Gozzo, S., & Tomaselli, V. (2020). Shaping and re-shaping tourism areas. A network approach. In Destination Management and Marketing: Breakthroughs in Research and Practice (pp. 483-503), Hershey, PA: IGI Global.

Asero, V., & Skonieczny, S. (2017). Formal Networks among Tourism Enterprises: Peculiarities of the Italian Solution. *Eastern European Business and Economics Journal*, *3*(1), 48–66.

Baggio, R., Valeri, M. (2020). Network science and sustainable performance of family businesses in tourism. *Journal of Family Business Management.* doi:10.1108/JFBM-06-2020-0048

Beritelli, P., Bieger, T., & Laesser, C. (2014). The New Frontiers of Destination Management: Applying Variable Geometry as a Function-Based Approach. *Journal of Travel Research, 53*(4), 403–417. doi:10.1177/0047287513506298

Bickerdyke, I. (1996). Australia: The Evolving Structure and Strategies of Business Networks. In *Networks of Enterprises and Local Development: Competing and Co-operating in Local Productive Systems* (pp. 203–216). Organisation for Economic Co-operation and Development.

Bieger, T., Beritelli, P., & Laesser, C. (2009). Size matters! Increasing DMO effectiveness and extending tourism destination boundaries. *Tourism Review, 57*(3), 309–327.

Bowden, J. (2003). A Cross-national Analysis of International Tourist Flows in China. *Tourism Geographies, 5*(3), 257–279. doi:10.1080/14616680309711

Buhalis, D. (2000). Marketing the Competitive Destination of the Future. *Tourism Management, 21*(1), 97–116. doi:10.1016/S0261-5177(99)00095-3

Crouch, G. I., & Ritchie, J. R. B. (1995). *Destination competitiveness and the role of the tourism enterprise. In Conference of the International Management Development Association*. IMDA.

Crouch, G. I., & Ritchie, J. R. B. (1999). Tourism, competitiveness, and societal prosperity. *Journal of Business Research, 44*(3), 137–152. doi:10.1016/S0148-2963(97)00196-3

Crouch, G. I., & Ritchie, J. R. B. (2010). A model of destination competitiveness/sustainability: Brazilian perspectives. *Public Administration Review (RAP), 44*(5), 1049-1066.

Czernek, K. (2013). Determinants of cooperation in a tourist region. *Annals of Tourism Research, 40*(1), 83–104. doi:10.1016/j.annals.2012.09.003

Dollinger, M. J. (1990). The evolution of collective strategies in fragmented industries. *Academy of Management Review, 15*(2), 266–285. doi:10.2307/258157

Dredge, D. (1999). Destination place planning and design. *Annals of Tourism Research, 26*(4), 772–791. doi:10.1016/S0160-7383(99)00007-9

Elbe, J., Hallén, L., & Axelsson, B. (2009). The destination-management organisation and the integrative destination-marketing process. *International Journal of Tourism Research, 11*(3), 283–296. doi:10.1002/jtr.695

Framke, W. (2002). The Destination as a Concept: A Discussion of the Business-Related Perspective versus the Socio-Cultural Approach in Tourism Theory. *Scandinavian Journal of Hospitality and Tourism, 2*(2), 92–108. doi:10.1080/15022250216287

Gibson, L., Lynch, P. A., & Morrison, A. (2005). The local destination tourism network: Development issues. *Tourism and Hospitality Planning & Development, 2*(2), 87–99. doi:10.1080/14790530500171708

Gulati, R. (1998). Alliances and networks. *Strategic Management Journal, 19*(4), 293–317. doi:10.1002/(SICI)1097-0266(199804)19:4<293::AID-SMJ982>3.0.CO;2-M

Hall, C. M. (2008). *Tourism Planning. Policies, Processes and Relationship*. Pearson-Prentice Hall.

Jamal, T., & Jamrozy, U. (2006). Collaborative networks and partnerships for integrated destination management. In D. Buhalis & C. Costa, C. (Eds.), Tourism management dynamics: Trends, management and tools (pp. 164-172). Elsevier Butterworth-Heinemann. doi:10.1016/B978-0-7506-6378-6.50028-7

Jamal, T. B., & Getz, D. (1995). Collaboration theory and community tourism planning. *Annals of Tourism Research, 22*(1), 186–204. doi:10.1016/0160-7383(94)00067-3

Klepers, A., & Rozite, M. (2010). Administrative border as a challenge in spatial aspects of tourism competitiveness. In A. Clarke (Ed.), *Constructing Central Europe: Tourism Competitiveness* (pp. 23–33). University Press, University of Pannonia.

Koh, K. Y., & Hatten, T. S. (2002). The tourism entrepreneur. The overlooked player in tourism development studies. *International Journal of Hospitality & Tourism Administration, 3*(1), 21–48. doi:10.1300/J149v03n01_02

Merinero-Rodríguez, R., & Pulido-Fernández, J. I. (2016). Analysing relationships in tourism: A review. *Tourism Management, 54*, 122–135. doi:10.1016/j.tourman.2015.10.010

OECD. (2020). *Operationalisation of Destination Management Organisations in Romania. OECD Tourism Papers, No. 2020/01*. OECD Publishing. doi:10.1787/9074fc18-

Pavlovich, K. (2003). The evolution and transformation of a tourism destination network: The Waitomo Caves, New Zealand. *Tourism Management, 24*(2), 203–216. doi:10.1016/S0261-5177(02)00056-0

Pearce, D. G. (2014). Toward an Integrative Conceptual Framework of Destinations. *Journal of Travel Research, 53*(2), 141–153. doi:10.1177/0047287513491334

Pearce, D. G., & Schänzel, H. A. (2013). Destinations: Tourists' Perspectives from New Zealand. *International Journal of Tourism Research*. Advance online publication. doi:10.1002/jtr.1955

Porter, M. (1998). Clusters and the new economics of competition. *Harvard Business Review, 76*(6), 77–90. PMID:10187248

Russell, R., & Faulkner, B. (1999). Movers and shakers: Chaos makers in tourism development. *Tourism Management, 20*(4), 411–423. doi:10.1016/S0261-5177(99)00014-X

Russell, R., & Faulkner, B. (2004). Entrepreneurship, chaos and the tourism area lifecycle. *Annals of Tourism Research, 31*(3), 556–579. doi:10.1016/j.annals.2004.01.008

Saraniemi, S., & Kylänen, M. (2011). Problematizing the Concept of Tourism Destination: An Analysis of Different Theoretical Approaches. *Journal of Travel Research, 50*(2), 133–143. doi:10.1177/0047287510362775

Stokes, R. (2004). Relationships and networks for shaping events tourism: An Australian study. *Event Management, 10*(2/3), 145–158.

Strobl, A., & Peters, M. (2013). Entrepreneurial reputation in destination networks. *Annals of Tourism Research, 40*, 59–82. doi:10.1016/j.annals.2012.08.005

Telfer, D. J. (2001). Strategic alliances along the Niagara Wine Route. *Tourism Management, 22*(1), 21–30. doi:10.1016/S0261-5177(00)00033-9

Valeri, M. (2016). Networking and cooperation practices in the Italian tourism business. *Journal of Tourism. Heritage & Services Marketing*, *2*(1), 30–35. doi:10.5281/zenodo.376333

Valeri, M. (2021). *Organization Theories. Implications for the Strategic Management*. Springer.

Valeri, M., & Baggio, R. (2020a). Social network analysis: Organizational implications in tourism management. *The International Journal of Organizational Analysis*. Advance online publication. doi:10.1108/IJOA-12-2019-1971

Valeri, M., & Baggio, R. (2020b). Italian tourism intermediaries: A social network analysis exploration. *Current Issues in Tourism*, *24*(9), 1270–1283. doi:10.1080/13683500.2020.1777950

Valeri, M., & Baggio, R. (2021). Increasing the efficiency of knowledge transfer in an Italian tourism system: A network approach. *Current Issues in Tourism*, 1–16. Advance online publication. doi:10.1080/13683500.2021.1937960

Valeri, M., & Katsoni, V. (Eds.). (2021). *Gender and tourism: challenges and entrepreneurial opportunities*. Emerald Publishing. doi:10.1108/9781801173223

Vernon, J., Essex, S., Pinder, D., & Curry, K. (2005). Collaborative policymaking: Local sustainable projects. *Annals of Tourism Research*, *32*(2), 325–345. doi:10.1016/j.annals.2004.06.005

World Tourism Organization (UNWTO). (2000). *Public-Private Sector Cooperation-Enhancing Tourism Competitiveness*. World Tourism Organization.

World Tourism Organization (UNWTO). (2007). *A Practical Guide to Tourism Destination Management*. World Tourism Organization.

World Tourism Organization (UNWTO). (2019). *UNWTO Guidelines for Institutional Strengthening of Destination Management Organizations (DMOs) – Preparing DMOs for new challenges*. World Tourism Organization.

Ziakas, V., & Costa, C. A. (2010). Explicating inter-organizational linkages of a host community's events network. *International Journal of Event and Festival Management*, *1*(2), 132–147. doi:10.1108/17852951011056919

Chapter 8
Complexity of Tourism Destination Governance:
A Smart Network Approach

Tomáš Gajdošík
Faculty of Economics, Matej Bel University, Slovakia

Marco Valeri
https://orcid.org/0000-0002-9744-506X
Niccolò Cusano University, Italy

ABSTRACT

Tourism destinations can be considered as complex systems of interrelated and interdependent stakeholders. The complexity and limited power of influencing the number of stakeholders resulted in network approach to tourism destination governance. This approach is considered both theoretically and practically as a tool for strengthening its sustainable competitiveness, fostering innovation and knowledge sharing. Although the network analysis of tourism destinations has gained a significant attention in recent years, the complex understanding of its contribution to smart development is still missing. The aim of this chapter is to create a framework for smart approach in destination governance using the network science perspective. The chapter provides insights in using network analysis for strengthening the tourism destination governance. The chapter uses a case study methodology on two mature tourism destinations, providing an example of the use of network analysis for destination governance strengthening.

INTRODUCTION

Tourism destinations are complex adaptive systems of interrelated and independent stakeholders (Baggio, 2008; Pearce, 2014) where tourism demand meets the supply. Due to the changing conditions on the tourism market, the need for flexibility and dynamics in co-ordination of the tourism destination and lack of authority in management of destinations, destination management organizations have difficulties in performance of the tasks of the destination management (Pechlaner, Beritelli, Pichler, Peters, &

DOI: 10.4018/978-1-6684-3889-3.ch008

Scott, 2015). These difficulties had impact on the view of destinations and the concept of the governance applied to tourism started to be discussed. As Beritelli, Bieger and Laesser (2007) stress, the concept of destination governance consists of setting and developing the rules and mechanism for a policy, as well as business strategies, by involving all the institutions and individuals. Moreover, the complexity and limited power of influencing the number of stakeholders resulted in network approach to tourism destination governance (Valeri & Baggio, 2020b; 2021; van der Zee, Gerrets, & Vanneste, 2017). This approach is considered as a tool for strengthening the sustainable competitiveness of destinations (Van der Zee & Vanneste, 2015), fostering innovation (Gajdošík, Gajdošíková, Maráková, & Borseková, 2017) and knowledge sharing (Raisi, Baggio, Barratt-Pugh, & Willson, 2020).

The dynamic environment, in which DMOs now operate (Reinhold, Beritelli, & Grünig, 2019), pushes forward a new model of governance. Tourism literature stresses the crucial role of knowledge management and information and communication technologies (ICTs) in destination governance (Micera, Presenza, Splendiani, & Del Chiappa, 2013; Valeri and Baggio, 2020c). ICTs are seen as a source of knowledge generation and dissemination which can help DMOs to manage and reinforce their network value (Fortezza & Pencarelli, 2018; Fuchs, Höpken, & Lexhagen, 2014). Therefore applying the developments arising from the application of smart tourism concept (Sheehan, Vargas-Sánchez, Presenza, & Abbate, 2016) is welcomed. This concept creates opportunities for better data collection and analysis. Although, the smart ecosystem has been considered in a few studies dealing with network science in tourism destinations (see e.g. Baggio and Del Chiappa 2014; Raisi et al. 2018; Williams et al. 2017), the profound understanding of the potential of the smart initiative within this domain is still missing.

The Smart Approach in Destination Governance

Smart tourism describes the current stage of tourism development influenced by the evolution of information technologies. It is a step in the evolution of information technologies in tourism in which the physical and governance dimensions of tourism are entering the digital playing field and new levels of intelligence are achieved (Gretzel, 2011). Smart tourism is also connected with implementing new technology-based models of governance and planning (Lalicic & önder, 2018).

Although technological resources are in the structure of smart destinations, social capital and governance are the key factors for success of tourism destinations (dos Santos Júnior, Mendes Filho, Almeida García, & Simoes, 2017). Moreover, based on the smart city literature, governance is frequently mentioned as an instrumental layer of smart destinations (Gretzel, Ham, & Koo, 2018). In this sense, smart governance is defined as a sociotechnical approach, which aligns technological potential with novel forms of collaboration between local government and citizens with the aim of tackling urban issues based on the principles of sustainability (Tomor, Meijer, Michels, & Geertman, 2019).

Starting from the smart city concept that address the urban challenges, smart approach in destination governance includes structures and processes that are required for ensuring the commitment and effective coordination and integration of businesses, government and communities towards a holistic smart-oriented development plan for the tourism destination (Errichiello & Micera, 2015). Gretzel and Scarpino (2018) highlight that smart governance is also transparent, accountable, collaborative and participatory. Further, Gretzel et al. (2018) indicate, the role of smart destination is to lobby and partly sponsor the development of smart technology infrastructure and applications. This is done to curate and manage big data in a destination, in order to support the development and use of smart experiences by tourists and finally to link smart tourism with overall quality of life and sustainability in a destination.

Based on the above mentioned, it can be summarised that smart approach in destination governance includes processes and structures focused on decision-making based on available data by incorporating the network of all relevant stakeholders.

In order to access the complexity of destination governance, techniques and analysis methods of complex networks provide several opportunities (Baggio and Valeri, 2020; Valeri & Baggio, 2020a). The network science is focused on enumerating, mapping and analysing the patterns of connections between elements of a system (Baggio, 2017). The elements can include destination stakeholders (e.g. businesses, government agencies, residents) or tourists, while the connections can be based on cooperation, trust, knowledge sharing or mobility. Thus, network science is able to analyse the complexity of the tourism destination phenomenon (Baggio, 2020a; Gajdošík, Gajdošíková, Maráková, & Flagestad, 2017; Presenza & Cipollina, 2010) and provide useful knowledge for the theory and praxis. However, one of the main problems of the use of the network approach in the context of tourism destination is the data collection. Traditional methods of collecting data (e.g. surveys, archival records) provide past static data and their collection is many times time-consuming and misleading (e.g. high non-response rate in network surveys). Therefore, there is a challenge of shifting attention from a "big" to "smart" data, adding layers of information, facilitating real-time usage and appropriate dissemination of trends (Volo, 2020), The smart approach in network science can help to overcome this challenge.

The Contribution of Smart Approach to Destination Governance

Tourism destinations are facing a major "paradigm shift" due to the rapid development of information technologies (Buhalis & Law, 2008). Traditional DMOs' activities (e.g. marketing, branding) are nowadays in decline due to the development of ICT and more complex requirements of stakeholders and tourists (Reinhold et al., 2019). Todays' DMO should pay more attention to knowledge transfer (Raisi et al., 2020) and co-decision among stakeholders (Trunfio & Della Lucia, 2019). There is also a need to make information available to all stakeholders in a systematic and efficient way (Del Chiappa & Baggio, 2015).

Moreover, DMOs' knowledge about tourists relies on static and ad hoc data. Surveys and traditional statistics do not provide enough information about tourist behaviour, however big data analytics from web and social media, sensors, mobile and wearable devices helps DMOs to be more proactive and effective (Gajdošík, 2019a). Besides this, tourists seek more experience and want to be active in their creation (Xiang & Fesenmaier, 2017). Therefore, there is a pressure on DMOs to become co-creators of personalised experiences (Neuhofer, Buhalis, & Ladkin, 2014) based on real-time data (Buhalis & Sinatra, 2019).

Traditionally, destinations are defined by political borders (municipal, district), that causes several problems (Beritelli et al., 2007). Moreover, there has been a little attention to the concept of destination from tourist perspective (Pearce, 2014). Therefore the dynamic construct of a destination based on demand and supply intersection is needed (Beritelli, Bieger, & Laesser, 2013).

To overcome these challenges, it is essential that DMO is able to gain the knowledge form available data with the support of ICTs (Shaw & Williams, 2009). DMOs should provide an infrastructure designed to collect new information, disseminate, process and apply gained knowledge, while create opportunities for further learning and development of all stakeholders (Fortezza & Pencarelli, 2018). Therefore, there is a shift from traditional destination governance to smart destination governance based on knowledge perspective.

In order to show the contribution of smart destination governance to overcome the challenges of todays' DMOs the case study approach was selected. Case studies are used to go deeply into the analysed issue and to illustrate the specificity of the phenomenon (Eisenhardt & Graebner, 2007). Two destinations in the maturity stage of destination life cycle were selected in Slovakia – Liptov and High Tatras. These destinations were chosen based on their implemented projects focused on smart tourism development. Slovakia was chosen as a reference country where the use of Internet services and information technologies are at the average of the European Union countries (European Commission, 2019) and the country is recognised as a digital challenger (Novak et al., 2018). Therefore, the cases can be seen as typical examples of mature destinations in Europe, thus presenting the situation and possibilities for larger amount of destinations. The conceptual framework for smart approach in network science is presented based on smart data collection and analysis by the means of the social network analysis.

Knowledge Transfer and Co-decisions

Applying smart principles in a destination can create a knowledge-based destination, where knowledge transfer is an important element of destination competitiveness (Baggio & Cooper, 2010). In order to analyse knowledge transfer, several data collection methods have been used so far. Del Chiappa and Baggio (2015) used listings, management board compositions, catalogues of travel agencies, marketing leaflets and brochures, official corporate records together with in-depth interviews to identify the connections between stakeholders. Raisi et al. (2020) used online questionnaires to collect the data. However, these methods are time-consuming, retrospective and require active participation of stakeholders.

Following the recommendations of Buhalis and Amaranggana (2015), who state that there is a need for a technological platform on which information in a destination could be exchanged instantly and which will dynamically interconnect destination stakeholders, the data collection can be based on smart destination information system. In such a system, stakeholders are dynamically linked by information technology in order to collect and exchange real-time information about destination development. Therefore, to strengthen the knowledge transfer and to easily collect dynamic data, intelligent destination information system (Fig. 1) merging information from social media, web, mobile apps, reservation system and other external systems (e.g. sensors, cameras, GSM data, WiFi hotspots) can be a useful platform (Gajdošík, 2019b).

Based on the data collection from various sources, data processing is based on real-time processing and analysis. The data exchange layer use open data for stimulation of knowledge spread and provides real-time dashboards for destination stakeholders, which creates opportunities for co-decision.

In order to show the contribution of smart destination governance, network analysis of knowledge transfer among stakeholders in destination Liptov was performed. Two time periods were chosen. In 2014, the local DMO did not use any destination information system and information from meetings were rather occasional. Knowledge was spread mainly through formal and informal meetings and official website of the destination. In this period, the data were collected using information from websites and interviews among destination stakeholders.

In 2020, destination Liptov uses destination management system connected to mobile application and web. Fig. 2 indicates that after the implementation and active use of destination information system, the knowledge in destination is transferred more effectively and stakeholders cooperate in a greater extent.

From the quantitative point of view, topological measures of network analysis can reveal interesting information. Although network density is in both time periods lower than in a typical destination (Baggio, 2020b), the higher density in 2020 indicates that the connections among stakeholders are higher. Together with higher clustering coefficient, the knowledge transfer is more efficient and effective. Moreover, average path length was shorten leading to the greater global efficiency of the whole the network.

Figure 1. Architecture of intelligent destination information system

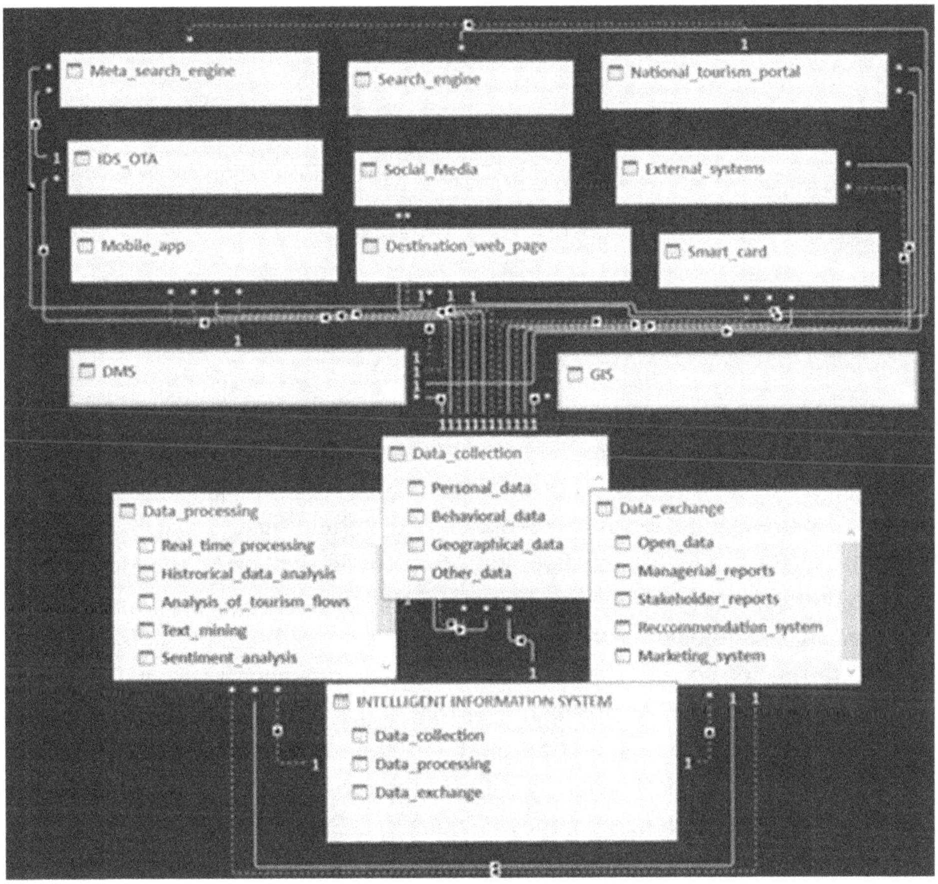

This kind of data collection and analysis presents the smart approach, as it uses complex and real-time data. From the network science perspective, it enables to examine more complex information. Further, it allows incorporating wider variety of information about stakeholders, For example, based on the size of data transfer, the amount of information exchanged can be measured. This overcomes the limitations of today's research (e.g. Baggio, Micera, & Del Chiappa, 2020), where the absorptive capacity, presenting the ability to acquire, retain and transfer the knowledge, has been so far arbitrary assigned. Moreover, it enables more precise modelling of a destination system, as a diffusion of knowledge can take into account more complex characteristics of stakeholders.

Figure 2. Network analysis of knowledge transfer in destination Liptov

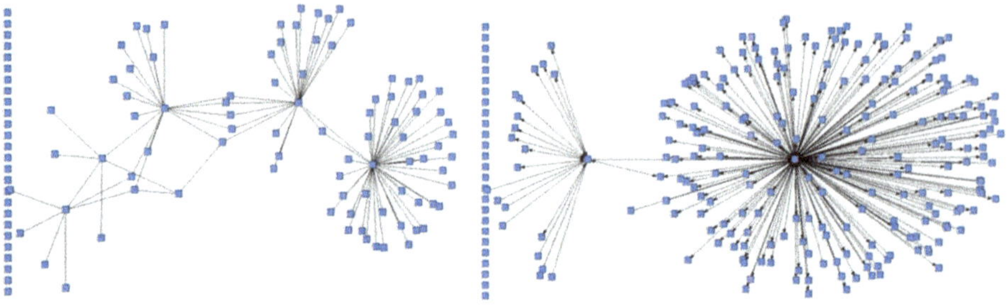

Quantitative characteristics/ year	2014	2020
density	0,002	0,006
average path length	3,4+/-1,5	2,1+/-0,4
global efficiency	0,296	0,466
clustering coefficient	0,072	0,335

Real-Time Information About Tourists' Behaviour Enabling Co-Creation

One of the greatest difficulties in analysing tourists' behaviour has been limited access to complete information, rather a collection of fragments has been combined to produce valuable information (Cousin, Chareyron, & Jacquot, 2017). So far, DMOs has relied mainly on accommodation statistics and knowingly and intentionally provided explicit tourists' feedback based on field surveys and e-reviews (Fuchs et al., 2014). With the rise of smart tourism initiative, some tourism destinations started to use sensors. These sensors embedded throughout destination environment include cameras, scanners, infrared ports, RFID, NFC and other sensing technologies capable of monitoring tourists' behaviour. Moreover, with the rise of web 2.0, other information sources started to be available to the majority of destinations. Nowadays, web and social media provide valuable information on tourist search and behaviour. Besides demographic and psychographic characteristics (age, gender, preferences), social media provide interesting textual and geotagged data. These new tools enable real-time monitoring of tourists and better understanding their needs during all phases of a tourist journey (Gajdošík, 2020).

Based on information on tourists and their behaviour, DMOs can co-create the destination product and build targeted marketing campaigns. Technology represents an opportunity to actively participate in destination activities and to take part in the construction of its own experience (Prebensen, Vittersø, & Dahl, 2013). There are several ways, how tourists share their data in order to co-create. Nowadays, the most important are enabling cookies on websites and location-based services in mobile apps or creating an account on a destination website. By using a mobile device and location-based services, tourists can exchange information in real time, be active in conversations and create opportunities to personalise. Moreover, DMOs can use behavioural and location-based targeting to ensure the marketing communication is delivered to the right customers. With the help of big data, the impact of marketing campaign, even as it is in progress, can be measured. Further, appropriate steps can be made to ensure the effectiveness of the campaign.

Nowadays, there are several tools available for DMOs to analyse data on tourists in real time. These tools also provide online dashboards for destination managers and include indicators such as number of mentions, source, location, sentiment, reach or influence score, which help to better understand tourists' behaviour. In order to show the importance of social media analytics and its monitoring to overcome the challenges of DMOs, the social media marketing campaign in High Tatras was analysed. Time series of influence score using hashtags #High Tatras and #Vysoke Tatry from Facebook, Instagram, Twitter, Web and Tripadvisor were exported using MediaToolkit (Figure 3).

As destination and its dynamics are seen as a complex systems, therefore the non-linear approach (Baggio & Sainaghi, 2016) was used, transforming time series into a network. There are several methods to create a network from time series. The most used are visibility graph methods and proximity networks. In destination context, the horizontal visibility graph algorithm is the most useful to map the time series.

While form the time series graph (Figure 3a) it is not evident, where the turning points in influence score are, the use of modularity of network can identify groups of densely connected nodes. Those nodes belonging to the same community represent periods with the same dynamics. The network analysis revealed five communities (Figure 3b). Based on these communities, turning points in time series were identified showing when the influence of the posts started to change. These changes were subsequently paired with posts within official marketing campaign of the DMO, showing which posts influenced these changes.

Figure 3. Identification of turning points in time series using modularity analysis of the network

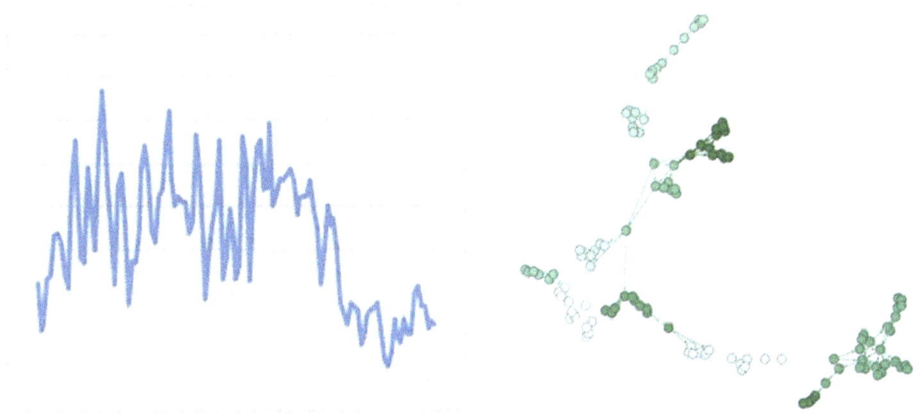

a) b)

Based on easily available data, destination managers can analyse the influence of their media campaigns, find out which and what kind of posts influence the most and thus make appropriate changes (e.g. content change, retargeting or retiming). The real time analytics allows changing strategies immediately, without losing valuable time, which is not able with traditional techniques of data collection. The smart approach to network science enables to use time series based on real-time data, which can supplement the official statistics and lead to timely identification of turning points in tourist behaviour. This enables faster and more accurate reaction of managers on demand changes in a destination.

Dynamic Construct of a Destination

The traditional administrative model of a destination is often an obstacle to the development of tourism in terms of limiting boundaries and emphasizing the supply side of a destination. Therefore, it is appropriate to enrich it with dynamic elements and factors influenced by visitor flows. This dynamic view on destination responds to the current problems of destination management organisations, where stakeholders question the ability of DMOs to attract tourists and influence their travel decision (Reinhold et al., 2019). Understanding the visitor flows enable destination managers to smartly create tourism experiences and include a demand-driven approach in the process (Cvelbar, Mayr, & Vavpotic, 2018).

Although visitor flows can be identified by surveys among destination stakeholders (Beritelli & Laesser, 2017; Hwang, Gretzel, & Fesenmaier, 2006; Shih, 2006), many times it is time consuming and difficult. In order to find strategic visitor flows, the use of big data is welcomed, which can include information from mobile positing, destination smart cards or geotagged information from social media. Specifically, passive mobile positioning can be viewed as efficient tool to strategically analyse tourist flows (Baggio & Scaglione, 2018). The sequence of data on tourist movement is stored in a database of a mobile operators and actualise each time person uses the mobile phone (call, messaging, mobile internet). DMOs thus have a clear picture of tourist mobility in and nearby the destination. This data can be further examined in order to find out the most visited attractions and thus determine tourism zones and boundaries of a destination.

In order to show the contribution of such a data, the network analysis of visitor flows in the destination High Tatras was performed. Firstly, the individual visitor flows were grouped to macro clusters using trajectory data mining and map-matching algorithms. Together, 105 macro clusters and 17 places were identified representing strategic visitor flows in the destination. Each macro cluster was rewritten into matrix format and then the overall matrix for visitor flows was constructed by merging all the trajectories and proceed by network analysis (Figure 4).

Figure 4. Network analysis of strategic visitor flows in the destination High Tatras

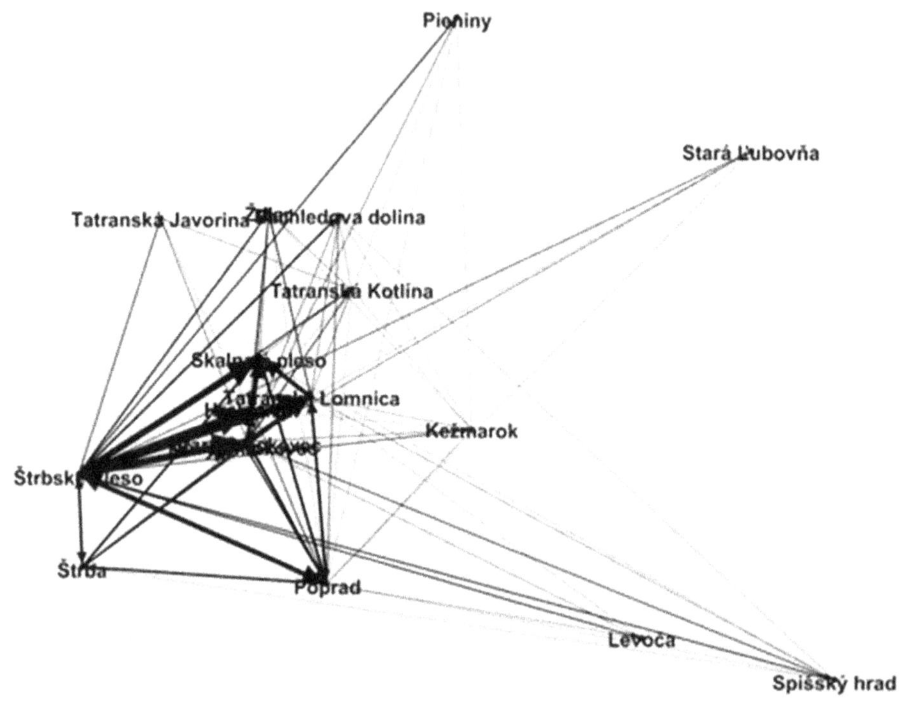

The network analysis can be further used to define the boundaries of a destination. The measure of connected components, where nodes are easily reachable, is useful in this sense. The strongly connected components indicate that there is a path between all pairs of places. Based on the list of strongly connected places the boundaries can assigned (Figure 5b).

Figure 5. Defining destination boundaries based on tourism flows

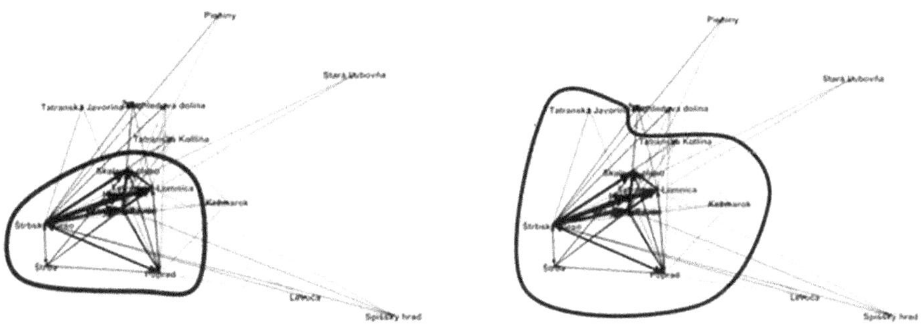

a) Traditional destination boundaries

b) New destination boundaries

The identification of visitor flows based on big data create more precise and dynamic outcomes. Network analysis is suitable for dealing with large-scale data, while traditional statistical techniques may suffer from large sample size biases (e.g. showing a large number of statistically significant results, even when they are not important). Therefore, the big data analytics within the smart approach brings more value if network science is used compared to traditional statistical techniques. Moreover, in a tourism destination context, it allows to incorporate the actual dynamics of tourist flows, which are the origins of destination development.

CONCLUDING REMARKS

This contribution aimed at creating a framework for smart approach in destination governance using the network science perspective. Smart destination governance pushes forward the decision-making based on available data and encompasses the network all relevant stakeholders. The findings from two mature tourism destinations in Slovakia show the contribution of real-time data exchange via destination management system in improvement of knowledge transfer and co-decisions among stakeholders. Further, the examination of time series obtained from social media provides useful real-time information on tourists and their behaviour. Finally, analysed passive mobile positioning can be viewed as the base line for identification of strategic visitor flows enabling to construct the destination from a demand side perspective. In this sense, data are seen as a "new oil" and smart governance as a new concept improving decision-making processes, stimulating co-creation and thus propelling destination competitiveness.

Consequently, the data, available thanks to governing the destination in a smart way, provide possibilities for enriching the network science. As every scientific approach is valid as far as the analysis is based on meaningful and precise data, the application of network science can profit from big data collection. Traditional network data sources (e.g. surveys, observation) have many biases. They rely on the willingness of subjects to cooperate in data collection and rely on the way questioned are formed, the manner in which the network boundaries are defined. Moreover, they provide static and past information. Therefore, they should be enriched by new sources creating possibilities of real-time analysis and incorporating wider variety of information from more stakeholders. Comparing to other statistical techniques, large-scale data are suitable for network analysis, as it do not suffer from large sample biases. The smart approach in network science pushes forward the use of dynamic data in order to be more proactive, rather than reactive. The further research should focus on analysing more in-depth the behaviour of tourists and the role of DMO in the destination network. Such an outcomes will enrich the concept of smart destination governance.

ACKNOWLEDGMENT

The research was supported by the research project VEGA 1/0237/20 Tourism 4.0: Smart and sustainable tourism development in a competitive environment

REFERENCES

Asero, V., Gozzo, S., & Tomaselli, V. (2016). Building Tourism Networks through Tourist Mobility. *Journal of Travel Research*, *55*(6), 751–763. doi:10.1177/0047287515569777

Baggio, R. (2008). Symptoms of complexity in a tourism system. *Tourism Analysis*, *13*(1), 1–20. doi:10.3727/108354208784548797

Baggio, R. (2017). Network science and tourism – the state of the art. *Tourism Review*, *72*(1), 120–131. doi:10.1108/TR-01-2017-0008

Baggio, R. (2020a). The science of complexity in the tourism domain: A perspective article. *Tourism Review*, *75*(1), 16–19. doi:10.1108/TR-04-2019-0115

Baggio, R. (2020b). Tourism destinations: A universality conjecture based on network science. *Annals of Tourism Research*, *82*(April), 102929. doi:10.1016/j.annals.2020.102929

Baggio, R., & Cooper, C. (2010). Knowledge transfer in a tourism destination: The effects of a network structure. *Service Industries Journal*, *30*(10), 1757–1771. doi:10.1080/02642060903580649

Baggio, R., & Del Chiappa, G. (2014). Real and virtual relationships in tourism digital ecosystems. *Information Technology & Tourism*, *14*(1), 3–19. doi:10.100740558-013-0001-5

Baggio, R., Micera, R., & Del Chiappa, G. (2020). Smart tourism destinations: A critical reflection. *Journal of Hospitality and Tourism Technology*, *11*(3), 407–423. doi:10.1108/JHTT-01-2019-0011

Baggio, R., & Sainaghi, R. (2016). Mapping time series into networks as a tool to assess the complex dynamics of tourism systems. *Tourism Management*, *54*, 23–33. doi:10.1016/j.tourman.2015.10.008

Baggio, R., & Scaglione, M. (2018). Strategic visitor flows and destination management organization. *Information Technology & Tourism, 18*(1–4), 29–42. doi:10.100740558-017-0096-1

Baggio, R., & Valeri, M. (2020). Network science and sustainable performance of family businesses in tourism. *Journal of Family Business Management.* doi:10.1108/JFBM-06-2020-0048

Beritelli, P., & Laesser, C. (2017). The Dynamics of Destinations and Tourism Development. In D. R. Fesenmaier & Z. Xiang (Eds.), *Design Science in Tourism, Tourism on the VergeVerge* (pp. 195–214). doi:10.1007/978-3-319-42773-7_13

Beritelli, P., Bieger, T., & Laesser, C. (2007). Destination Governance: Using Corporate Governance Theories as a Foundation for Effective Destination Management. *Journal of Travel Research, 46*(1), 96–107. doi:10.1177/0047287507302385

Beritelli, P., Bieger, T., & Laesser, C. (2013). The New Frontiers of Destination Management: Applying Variable Geometry as a Function-Based Approach. *Journal of Travel Research, 53*(4), 403–417. doi:10.1177/0047287513506298

Buhalis, D., & Amaranggana, A. (2015). Smart Tourism Destinations Enhancing Tourism Experience Through Personalisation of Services. In Information and Communication Technologies in Tourism 2015 (pp. 377–389). doi:10.1007/978-3-319-14343-9_28

Buhalis, D., & Law, R. (2008). Progress in information technology and tourism management: 20 years on and 10 years after the Internet—The state of eTourism research. *Tourism Management, 29*(4), 609–623. doi:10.1016/j.tourman.2008.01.005

Buhalis, D., & Sinatra, Y. (2019). Real-time co-creation and nowness service: Lessons from tourism and hospitality. *Journal of Travel & Tourism Marketing, 36*(5), 563–582. doi:10.1080/10548408.2019.1592059

Cousin, S., Chareyron, G., & Jacquot, S. (2017). Big data and tourism. In L. Lowry (Ed.), *The SAGE international encyclopedia of travel and tourism* (pp. 151–155). SAGE Publications.

Cvelbar, L. K., Mayr, M., & Vavpotic, D. (2018). Geographical mapping of visitor flow in tourism: A user-generated content approach. *Tourism Economics, 24*(6), 701–719. doi:10.1177/1354816618776749

Del Chiappa, G., & Baggio, R. (2015). Knowledge transfer in smart tourism destinations: Analyzing the effects of a network structure. *Journal of Destination Marketing & Management, 4*(3), 143–144. doi:10.1016/j.jdmm.2015.02.001

dos Santos Júnior, A., Mendes Filho, L., Almeida García, F., & Simoes, J. M. M. (2017). Smart Tourism Destinations: a study based on the view of the stakeholders. *Revista Turismo Em Análise -RTA, 28*(3), 358–379.

Eisenhardt, K., & Graebner, M. (2007). Theory Building From Cases: Opportunities And Challenges. *Academy of Management Journal, 50*(1), 25–32. doi:10.5465/amj.2007.24160888

Errichiello, L., & Micera, R. (2015). Smart Tourism Destination Governance. In J. Spender, G. Schiuma, & V. Albino (Eds.), *FKAD 2015 International Forum on Knowledge Asset Dynamics. Culture, Innovation and Entrepreneurship: connecting the knowledge dots* (pp. 2179–2191). Bari: IKAM - Institution of Knowledge Asset Management.

European Commission. (2019). *Digital economy and society index 2019. Country report Slovakia*. Retrieved from https://ec.europa.eu/digital-single-market/en/news/digital-economy-and-society-index-desi-2019

Fortezza, F., & Pencarelli, T. (2018). A comprehensive picture of the social media challenge for DMOs. *Anatolia*, *29*(3), 456–467. doi:10.1080/13032917.2018.1439759

Fuchs, M., Höpken, W., & Lexhagen, M. (2014). Big data analytics for knowledge generation in tourism destinations – A case from Sweden. *Journal of Destination Marketing & Management*, *3*(4), 198–209. doi:10.1016/j.jdmm.2014.08.002

Gajdošík, T. (2019a). Big Data Analytics in Smart Tourism Destinations. A New Tool for Destination Management Organizations? In V. Katsoni & M. Segarra-Oña (Eds.), *Smart Tourism as a Driver for Culture and Sustainability* (pp. 15–33). doi:10.1007/978-3-030-03910-3_2

Gajdošík, T. (2019b). Towards a Conceptual Model of Intelligent Information System for Smart Tourism Destinations. In R. Silhavy (Ed.), *Software Engineering and Algorithms in Intelligent Systems* (pp. 66–74). doi:10.1007/978-3-319-91186-1_8

Gajdošík, T. (2020). Smart tourists as a profiling market segment: Implications for DMOs. *Tourism Economics*, *26*(6), 1042–1062. doi:10.1177/1354816619844368

Gajdošík, T., Gajdošíková, Z., Maráková, V., & Borseková, K. (2017). Innovations and networking fostering tourist destination development in Slovakia. *Quaestiones Geographicae*, *36*(4), 103–115. Advance online publication. doi:10.1515/quageo-2017-0039

Gajdošík, T., Gajdošíková, Z., Maráková, V., & Flagestad, A. (2017). Destination structure revisited in view of the community and corporate model. *Tourism Management Perspectives*, *24*, 54–63. Advance online publication. doi:10.1016/j.tmp.2017.07.009

Gretzel, U. (2011). Intelligent systems in tourism: A Social Science Perspective. *Annals of Tourism Research*, *38*(3), 757–779. doi:10.1016/j.annals.2011.04.014

Gretzel, U., Ham, J., & Koo, C. (2018). Creating the City Destination of the Future: The Case of Smart Seoul. In Y. Wang, A. Shakeela, A. Kwek, & C. Khoo-Lattimore (Eds.), *Managing Asian Destinations* (pp. 199–214). Perspectives on Asian Tourism. doi:10.1007/978-981-10-8426-3_12

Gretzel, U., & Scarpino-Johns, M. (2018). Destination resilience and smart tourism destinations. *Tourism Review International*, *22*(3), 263–276. doi:10.3727/154427218X15369305779065

Hwang, Y. H., Gretzel, U., & Fesenmaier, D. R. (2006). Multicity trip patterns. Tourists to the United States. *Annals of Tourism Research*, *33*(4), 1057–1078. doi:10.1016/j.annals.2006.04.004

Lalicic, L., & Önder, I. (2018). Residents' involvement in urban tourism planning: Opportunities from a smart city perspective. *Sustainability (Switzerland)*, *10*(6), 1852. Advance online publication. doi:10.3390u10061852

Micera, R., Presenza, A., Splendiani, S., & Del Chiappa, G. (2013). SMART Destinations: new strategies to manage tourism industry. In G. Schiuma, J. C. Spender, & A. Pulic (Eds.), *IFKAD 2013: 8th International Forum on Knowledge Asset Dynamics. Smart Growth: Organizations, Cities and Communities* (pp. 1405–1422). Zagreb: IKAM - Institute of Knowledge Asset Management.

Neuhofer, B., Buhalis, D., & Ladkin, A. (2014). A Typology of Technology-Enhanced Tourism Experiences. *International Journal of Tourism Research*, *16*(4), 340–350. doi:10.1002/jtr.1958

Novak, J., Purta, M., Marciniak, T., Ignatowicz, K., Rozenbaum, K., & Yarwood, K. (2018). *The rise of digital challengers. How digitization can become the next growth engine for Central and Eastern Europe*. Academic Press.

Pearce, D. G. (2014). Toward an Integrative Conceptual Framework of Destinations. *Journal of Travel Research*, *53*(2), 141–153. doi:10.1177/0047287513491334

Pechlaner, H., Beritelli, P., Pichler, S., Peters, M., & Scott, N. (2015). *Contemporary Destination Governance: A Case Study Approach*. Emerald Group Publishing. doi:10.1108/S2042-144320156

Prebensen, N. K., Vittersø, J., & Dahl, T. I. (2013). Value Co-creation significance of tourist resources. *Annals of Tourism Research*, *42*, 240–261. doi:10.1016/j.annals.2013.01.012

Presenza, A., & Cipollina, M. (2010). Analysing tourism stakeholders networks. *Tourism Review*, *65*(4), 17–30. doi:10.1108/16605371011093845

Raisi, H., Baggio, R., Barratt-Pugh, L., & Willson, G. (2018). Hyperlink Network Analysis of a Tourism Destination. *Journal of Travel Research*, *57*(5), 671–686. doi:10.1177/0047287517708256

Raisi, H., Baggio, R., Barratt-Pugh, L., & Willson, G. (2020). A network perspective of knowledge transfer in tourism. *Annals of Tourism Research*, *80*(March), 102817. doi:10.1016/j.annals.2019.102817

Reinhold, S., Beritelli, P., & Grünig, R. (2019). A business model typology for destination management organizations. *Tourism Review*, *74*(6), 1135–1152. doi:10.1108/TR-03-2017-0065

Shaw, G., & Williams, A. (2009). Knowledge transfer and management in tourism organisations: An emerging research agenda. *Tourism Management*, *30*(3), 325–335. doi:10.1016/j.tourman.2008.02.023

Sheehan, L., Vargas-Sánchez, A., Presenza, A., & Abbate, T. (2016). The Use of Intelligence in Tourism Destination Management: An Emerging Role for DMOs. *International Journal of Tourism Research*, *18*(6), 549–557. doi:10.1002/jtr.2072

Shih, H. Y. (2006). Network characteristics of drive tourism destinations: An application of network analysis in tourism. *Tourism Management*, *27*(5), 1029–1039. doi:10.1016/j.tourman.2005.08.002

Tomor, Z., Meijer, A., Michels, A., & Geertman, S. (2019). Smart Governance For Sustainable Cities: Findings from a Systematic Literature Review. *Journal of Urban Technology*, *26*(4), 3–27. doi:10.1080/10630732.2019.1651178

Trunfio, M., & Della Lucia, M. (2019). Engaging Destination Stakeholders in the Digital Era: The Best Practice of Italian Regional DMOs. *Journal of Hospitality & Tourism Research (Washington, D.C.)*, *43*(3), 349–373. doi:10.1177/1096348018807293

Valeri, M., & Baggio, R. (2020a). Italian tourism intermediaries: A social network analysis exploration. *Current Issues in Tourism*, *0*(0), 1–14. doi:10.1080/13683500.2020.1777950

Valeri, M., & Baggio, R. (2020b). Social network analysis: Organizational implications in tourism management. *The International Journal of Organizational Analysis*. Advance online publication. doi:10.1108/IJOA-12-2019-1971

Valeri, M., & Baggio, R. (2020c). A critical reflection on the adoption of blockchain in tourism. *Journal Information Technology and Tourism*, *23*(2), 121–132. doi:10.100740558-020-00183-1

Valeri, M., & Baggio, R. (2021). Increasing the efficiency of knowledge transfer in an Italian tourism system: A network approach. *Current Issues in Tourism*, 1–16. Advance online publication. doi:10.1080/13683500.2021.1937960

van der Zee, E., Gerrets, A. M., & Vanneste, D. (2017). Complexity in the governance of tourism networks: Balancing between external pressure and internal expectations. *Journal of Destination Marketing & Management*, *6*(4), 296–308. doi:10.1016/j.jdmm.2017.07.003

Van der Zee, E., & Vanneste, D. (2015). Tourism networks unravelled; a review of the literature on networks in tourism management studies. *Tourism Management Perspectives*, *15*, 46–56. doi:10.1016/j.tmp.2015.03.006

Volo, S. (2020). Tourism statistics, indicators and big data: A perspective article. *Tourism Review*, *75*(1), 304–309. doi:10.1108/TR-06-2019-0262

Williams, N. L., Inversini, A., Ferdinand, N., & Buhalis, D. (2017). Destination eWOM: A macro and meso network approach? *Annals of Tourism Research*, *64*, 87–101. doi:10.1016/j.annals.2017.02.007

Xiang, Z., & Fesenmaier, D. R. (2017). Big Data Analytics, Tourism Design and Smart Tourism. In F. Zheng (Ed.), *Analytics in Smart Tourism Design, Concepts and Methods*. doi:10.1007/978-3-319-44263-1_17

Section 2
Destination Management

Chapter 9
Electronic Word of Mouth Effects on Middle East Destination Overall Image and Behavioral Intention:
An Empirical Study in Jordan

Ashraf Mohammad Alfandi
Irbid National University, Jordan

Valeri Marco
https://orcid.org/0000-0002-9744-506X
Niccolò Cusano University, Italy

ABSTRACT

This study investigated the relationship among e-WOM, overall destination image, and European tourist behavioral intention to revisit and recommend Jordan as a tourism destination. The study examined the mediating effect of destination overall image on the relationship between e-WOM and tourist behavioral intention. A total of 339 questionnaires were collected from European tourists who visited Jordan in 2018 through a self-administered questionnaire. Structural equation modeling (SEM) was used. The result of SEM identified that e-WOM has a positive influence on the overall destination image and on the tourist behavioral intention. The overall image positively influenced the tourist behavioral intention. Overall destination image partially mediates between e-WOM and behavioral intention, and thereby, the positive impact of e-WOM on behavioral intention could be aggravated through the mediating effect of destination overall image. The findings have implications on the tourism industry, especially for key players in the Jordanian tourism board and travel companies.

DOI: 10.4018/978-1-6684-3889-3.ch009

1. INTRODUCTION

Since we have begun to witness new tourism destinations that shape the way we choose our preferred tourism attractions, we began encountering new challenges and difficulties, particularly in the Middle East and specifically a country surrounded by an abundance of conflicts and disputes, such as Jordan. From a marketing point of view, the destination must hold a strong and positive image that ensures its popularity with tourists, as well as creating an accommodating atmosphere that positively influences both tourists' re-purchase and intention to recommend the destination to other tourists (El-Said and Aziz, 2019). However, this image may be distorted by negative Word-of-Mouth (WOM) especially by the new electronic WOM (e-WOM) that potential tourists started to rely more and more on it in searching for information about any specific destination (El-Said and Aziz, 2019). Since the tourism industry is service-oriented which is intangible in its nature, potential tourists depend on reliable communication sources of information such as WOM to reduce risks associated with their purchase decision (Gruen et al., 2006). As such, the present study attempts to empirically examine the relationships between e-WOM and tourists' overall image of Jordan towards tourists' behavioral intention. The current study argued that it is necessary to examine the relationship between these factors which will assist a destination with future marketing campaigns designed to increase market share by correct the negative perception and reinforce the positive perception. However, few studies -to the best of the researcher's knowledge- have examined the relationship between e-WOM, destination image, and intention to travel especially in the case of European travelers to Jordan. In addition, studying tourist's behaviour intentions whether by their desire to revisit or their willingness to recommend the destination and separate positive WOM is very essential (Ren and Hong, 2017), as it can help to forecast whether the target customers will become long-term customers and bring more profits to the enterprises by building up an attractive destination image and expand their marketing effort to maximize their use of resources (Su and Fan, 2011). The study focus on tourists who visited Jordan and getting ready to go back to their countries, therefore, this study inspects tourists' perceptions of relevant and real destination attributes instead of general assessments which consider an important strategy to develop and recover the destination image.

2. TOURISM IN JORDAN

In Jordan, the tourism industry is considered as a paramount economic sector that contributed a considerable 7,632.8 million USD to its GDP in 2017 and was ranked as number 69 out of 185 countries worldwide by the World Travel and Tourism Council (WTTC) as of 2017. According to the same report, the Jordanian Ministry of Tourism and Antiquities (JMTA) is hoping for the total contribution of travel and tourism to rise by 23.5% of GDP in 2028 to reach 13,563.6 USD Million WTTC, 2018). To achieve this, the tourists must perceive Jordan as a safe, secure, attractive and worthy detestation since, in deciding to travel to a certain place; tourists must consider a multidimensional culture that involves a wide range of different factors (Farajat et al., 2017).

Jordan as a tourism destination offers a wide range of cultural and natural touristic attractions, varying from the virtually untouched landscape of the protected area of Wadi Rum, to the ancient rosy-red city and designated UNESCO World Heritage Site of Petra, and the mineral lined shores of the Dead Sea (Al-Oun and Al-Homoud, 2008; Liu et al., 2016). However, Jordan has recently been struggling with surrounding conflicts in neighboring countries and is no stranger to disturbances. In particular, this region has experienced many extended conflicts over time such as the Arab Spring that started in 2010, the civil war in Syria that is still running to date, the on-going conflict in Iraq, the rise of the so-called Islamic State of Iraq and Syria (ISIS) and the conflict between Palestine and Israel. Of course, the western media wouldn't leave all of these events without conveying, and therefore, the region image where Jordan is located has been harmfully affected and the tourist perception will be so negative even though Jordan is still considered as a safe country. For example, a study by Ahmed and Kadir (2013) confirmed that information sources especially the mass media negatively impact the destination image particularly after events such as political instability. Nevertheless, Jordan is seen as one of the countries that have been affected indirectly by terrorism threats and therefore has experienced many declines in the number of tourists and visitors (Farajat et al., 2017).

3. LITERATURE REVIEW

3.1. E-WOM and Tourism Destination Image

WOM is described as a real-time conversation between consumers regarding a particular product or service (Sen and Lerman, 2007). However, the appearance of the internet has dramatically revolutionized many different fields in our everyday life. Through using social networking platforms and consumer reviews sites, the internet gives the chance to the consumers to share information about their experiences and to spread advice regarding their experience with products and services (Brown et al., 2007; Valeri, 2016; Valeri and Baggio, 2020a; 2020b; 2020c; 2021; Baggio and Valeri, 2020). The internet has created a communication atmosphere (web-based technology) that enables information flow along consumer-to-consumer, consumer-to-business and business-to-consumer channels (Litvin et al., 2008; Litvin et al., 2003), which resulted in extending the concept of WOM to become online WOM or the so-called e-WOM (Abubakar and Ilkan, 2016).

The technologies that could be used to separate positive or negative WOM (consumption-related advice) through the internet could be a website, twitter, Facebook, Instagram, and platform with web-based (Hennig-Thurau et al., 2004). Ladhar and Michaud (Ladhar and Michaud, 2015) describe the e-WOM as a communication toll about particular goods and services directed at customers through internet-based technologies. It is a statement, opinion, or perception about product or service made by potential, actual, or former customers, which could be negative or positive and made available to a multitude of people and institutions via the Internet (See-To and Ho, 2014). In sum, e-WOM is a previous consumer written memo on the web (Abubakar and Ilka, 2016).

Potential tourists prefer to use the WOM over the information provided by travel companies. They look for information about the brand through credible sources, especially when they perceive high psychological, performance, or social risks with their purchase (Wangenheim and Bayón, 2004). It is proposed in the literature that potential consumers perceived the WOM as more reliable and attested (Abubakar and Ilkan, 2016), easier (Ladhar and Michaud, 2015) more influential, (Smith et al., 2005), and more into details source of information (Jalilvand and Samiei, 2012).

Jalilvand et al., 2013 assures that consumers trust other consumers more than companies especially when they had previous experience with the brand. The theory of message source confirms the previous discussions as it suggested that a high source of information credibility makes the consumer highly persuaded by the message (Eagly and Chaiken, 1993). Due to the intangibility and inseparability of tourism services, WOM becomes a strong predictor for tourism destination image and potential tourists' behavioural intention (Casaló et al., 2015). It is argued that potential tourists nowadays rely more on e-WOM to reduce uncertainty, ambiguity, and risks associated with purchase decisions (Casaló et al, 2015; Abubakar and Ilkan, 2016; Albattat et al., 2020; Chemli et al., 2020a; 2020b; Santos et al., 2021a; 2021b; Toanoglou et al., 2021; Valeri, 2021). As the literature supports the significant impact of WOM, it is assumed that e-WOM to be even more effective than WOM in the off-line world due to its greater accessibility and high reach (Chatterjee, 2001). This could be due to the fact that potential tourists searching for information and recommendations simultaneously about certain destination in order to learn about it and reduce uncertainty (Bickart and Schindler, 2001; Navqi et al., 2018), at the same time, e-WOM provides information about the tourism destination as well as serves as a recommendation which make the e-WOM a significant information source that meet their information needs (Park and Njite, 2010). Several previous studies have provided evidence on the role of e-WOM in creating the tourism destination image. For instance, a study by Jalilvand and Samiei (2012) showed that e-WOM positively influences destination image in Teheran, Iran. Prayogo et al. (2016) find that e-WOM has a significant impact on the destination image of Malioboro Street, in Yogyakarta. A study by Ladhar and Michaud (2015) revealed that positive e-WOM about a specific hotel creates a strong brand image which results in a significantly higher booking intention for it. Jalilvand and Samiei (2012) research indicate that e-WOM is one of the most effective elements influencing brand image.

This evidence highlights that in the context of tourism destination selection, e-WOM as an informal source of information has been thought of as a key factor in tourism destination image formulation and selection choice. As such, against these findings the current study proposes the following hypothesis:

H1: *E-WOM has a positive and significant impact on tourist's overall destination image.*

3.2. E-WOM and Behavioural Intention

Generally, the intention is a subjective judgment about how we will behave in the future (Blackwell et al., 2001). The future intention could be categorized into four categories; these are referrals, price sensitivity, repurchase, and complaining behaviour (Zeithaml and Bitner, 1996). An earlier study by Fishbein and Ajzen (1975) described the behavioural intention as the function of evaluative beliefs, normative beliefs, and situational factors that can be anticipated at the time of the vacation plan or commitment. In the tourism literature, tourist's behavioural intention is the tourist's planned future action (Barlas et al., 2010) and can be described as an intention to return and willingness to recommend the destination to others (Lončarić et al., 2012; Castro et al., 2007).

Some of the earlier studies argued in favour of looking at the tourist's behavioural intention as their intention to revisit suggesting that revisit intention is the main element of behavioural intention. For instance, Thiumsak and Ruangkanjanases (2016) argued that the intention to revisit is one of the most significant consequences of tourist's participation. However, since tourism is a service-oriented industry that is intangible in its nature, and the tourists can't evaluate the service before buying it, the tourist's behavioural intentions will differ accordingly (Litvin et al., 2008; Lončarić et al., 2012). It can't be said that the repurchase intention is the only predictor for behavioural intention. That is because of many potential tourists seeking novelty and therefore their intention to recommend the destination to others and separate a positive WOM reflect good indicators for their behavioural intention (Phillips et al., 2013; Jani and Nguni, 2016). Previous studies consider WOM as a significant information source in making purchase decisions (Confente and Russo, 2015), an important marketing tool (Bao and Chang, 2014), essential when the service is complex (Zeithaml, 1988), and a critical information source especially when the service has a high perceived risk (Litvin et al., 2008). As such, the present study argued in favour of conceptualizes tourist's behavioural intention by looking at both factors which are; revisit intention and the willingness to recommend the destination.

Potential tourists may use e-WOM as it considers an easy way of searching for information about any specific destination (Ladhar and Michaud, 2015). The tourist's intention to behave is the probability of what tourists feel for a certain time so it can build a subjective perception that affects their attitude and behaviour toward the destination (Whang et al., 2016). Some refreshers argued that satisfied tourists expected to recommend the brand to their network which created an intention to revisit (Liu et al., 2016; Litvin et al., 2008). Without a doubt, alongside several other factors, credible information that potential tourists obtained about the destination will influence their decision to recommend and/or travel to the destination (Chen et al., 2014). Jalilvand and Samiei (2012) confirmed the direct and significant impact of e-WOM as an information source on the tourist's behavioural intention. A study by Ladhar and Michaud (2015) showed that information perceived from e-WOM plays an influential role in the tourist's decision-making process in selected specific hotels. Jalilvand et al., 2013 provides evidence on the significant effects of e-WOM on the tourist behavioural intention to revisit Iran. Furthermore, the intention to revisit certain places in Turkey is also influenced by e-WOM (Abubakar and Ilkan, 2016). One hypothesis was derived from this argumentation:

H2: *Electronic word of mouth has a positive and significant impact on tourist's behavioral intention.*

3.3. Destination Image and Behavioural Intention

It's argued that destination image is a combination of the product, behaviour and attitude, and the environment (Milman and Pizam, 1995). The earlier work of Gunn (1972) identified that potential tourist regarding the absorption of information such as information from the news and movies to shape an organic image of the tourism destination which will motivation those to start gathering more information about the destination. After this stage of collecting information about the destination, the induced image will be formulated. The induced image of the destination is the image formulated in the potential tourist mined from receiving the tourism destination marketers advertising or any other activities to promote the destination. Fakeye and Crompton (1991) added the complex image which is formulated after the actual visit and direct experience with tourism destination. In this context, the present study focuses on the complex image (overall image) as the population of the current study is the European tourists who already visited Jordan. In this regard, the destination image is operationalized as an individual's mental representation of knowledge (beliefs), feelings and overall perception of a particular destination (Fakeye and Crompton, 1991). In other words, it is the overall evaluative construct measuring tourists' holistic impression of Jordan as a tourism destination (Echtner and Ritchie, 1993). Previous studies showed that a complex image (overall image) serves as a strong proxy for capturing destination image (Prayag, 2009; Prayag et al., 2013). Destination image is commonly accepted as an important aspect of successful tourism management (Molina et al., 2013). To be effective, destination marketers try to build strong and positive destination image that influence consumer perceptions of the destination. Brand image, in general, is defined by Keller (1993, p. 3), as an association or perception consumers made based on their memory towards a product. The significance of the perceived brand image is that brand image enables the potential customer to recognize a product, evaluate the quality, lower purchase risks, and obtain certain experience and satisfaction out of product differentiation (Anastasiei and Dospinescu,

2018; Bigne et al., 2001). In addition, brand image is often used by consumers as an extrinsic cue to make a purchase decision (Kim et al., 2003; Pereira et al., 2018). A brand with a positive image has the ability to eliminate risks that the potential customers will face when they make their decision (Molina et al., 2013). In terms of destination image, Milman and Pizam (1995) defined the destination image as the visual or mental impression of a place or a product experienced by the general public. The tourists are affected by the destination image that influences directly their decision-making process and their behavioural intention (Castro et al., 2007; Xu et al., 2018).

Earlier studies found a direct relationship between destination image and tourist behavioural intention. For instance Bigne et al. (2001) found that destination image is a direct antecedence of tourists to visit and revisit intention and their intention to recommend the destination. Another study by Prayag (2003) found that destination image has a direct impact on the tourist's behavioural intention. A study by Park and Njite (2010) showed that different destination image attributes significantly affect tourist behavioural intention. Chen and Tsai (2007) proposed a more integrated tourist behaviour model by including destination image and perceived value into the ''quality–satisfaction–behavioural intentions'' paradigm. The study found that the destination image has both direct and indirect effects on behavioural intentions. Amore recent study, Tavitiyaman and Qu (2013) studied the impact of destination image on the behavioural intention of travellers to Thailand. Their study found that destination image indirectly influences the behavioural intention through the tourists' overall satisfaction. On the base of the above, it could be concluded that besides the direct connection of destination image on the behavioral intention, it is believed that the overall destination image plays an important mediating role in this relationship (Litvin et al., 2008). Positive online reviews about the destination can create an appositive image which attracts tourists to the destination which confirmed the mediating impact of destination on the behavioral intention as supported by several previous studies (Jalilvand et al., 2013; Liu et al., 2016; Jalilvand and Samiei, 2012). Given the extant literature, the following two hypotheses are proposed:

H3: *Destination overall image has a positive and significant impact on the tourist's behavioral intention.*

H4: *Destination overall image has mediation role for the relationship between e-WOM and tourist's behavioural intention.*

4. METHODOLOGY

4.1 Sampling

The present study intended to examine the impact of e-WOM and overall tourism destination image on European tourist's behaviour intentions directly after their visit to Jordan in 2019 (June 2019-August 2019). The study used a purposive sampling method since the accurate size of the population cannot be ascertained. A self-administered survey was conducted as a data collection technique on European tourists. However, Veal (2006) suggested as a minimum of 550, and the sample size of 10,000 populations equals 370 sample units and 500,000 and above equals 384. Hence, a total of 400 questionnaires were distributed among European tourists taking into consideration the incomplete, erroneous and not returned survey, out of which 339 were usable for analysis resulting in a response rate of 85%. Trained researchers assistances approached the European tourists while they are waiting to leave Jordan through its major Queen Alia International Airport (QAIA) based on face-to-face, and asked them whether they were Jordanian residents or not. In case they were not Jordanian residents and they are visiting Jordan for tourism activities and over 18 years, they considered as potential respondents for the current study (Mill and Marrison, 2002). As the study sample was the European tourists, the questionnaire was ready in four languages; these are original English version, German, French, and Russian translated by professional translators and retranslated back to English to assure the accuracy of meaning.

4.2 Instruments

A self-administered survey was conducted as a data collection technique on European tourists. Before conducting the study, the pilot test on 20 postgraduate tourism management students to assess how well the scale captured the construct it was supposed to measure was conducted. Items that loaded less than 0.40 were deleted as suggested by (Hair et al., 2006). Furthermore, scale validity and reliability were examined. Items that their composite reliabilities estimate below the recommended threshold of 0.70 were deleted (Fornell and Larcker, 1981). After examining the significance and magnitude of factor loading as well as the average variance extracted, the scale was convergent valid as the average variance of all items extracted was above 0.50, and significant ($p<0.001$) with t values exceeding the critical value of 3.29 as recommended by (Fornell and Larcker, 1981). Regarding the discriminant validity, the result showed also that the scale is discriminant valid. As a result, the final version of the questionnaire was comprised of 12 items structured into four parts. The first part presents the respondents socio-demographic variables such as nationality, age, gender, income, and education via a categorical scale (Valeri and Katsoni, 2021). In the line with the previous literature review, the second part consisted of 6 items related to e-WOM adopted from (Jalilvand and Samiei, 2012) who adopted it from previous work of Bambauer-Sachse and Mangold's (2011) and modified it for tourism context. The scales for e-WOM were divided as (1) strongly disagree, (2) disagree, (3) neutral, (4) agree, (5) strongly agree in the line of five points Likert scale. The third part of the questionnaire contained 5 items regarding the tourist's behavioural intentions adopted from earlier work of [65,66]. The scales for tourist's behavioural intentions were divided as (1) strongly disagree, (2) disagree, (3) neutral, (4) agree, (5) strongly agree in the line of five points Likert scale. The last part contained a single question regarding the overall destination image. The last section of the questionnaire is a single item anchored on a 5-points scale rating from (1- unfavorable) to (5- favorable) used to measure the overall image of Jordan as a tourism destination. Respondents asked to answer the following question "How would you describe your impression (the overall image) of Jordan as a tourism destination". This method of measuring the overall image of the tourism destination is largely used in previous studies (Bigne et al., 2021; Baloglu and McCleary, 1999; Beerli and Martin, 2004a). These earlier studies argued in favour of using this method as the destination image has been described as an overall impression greater than the sum of its parts (Fakeye and Crompton, 1991; Beerli and Martin, 2004b; Prayag et al., 2013) rather than using former approach which is sum of all attributes (Echtner and Ritchie, 1993) that omitted some of the relevant destination attributes (Castro et al., 2007; Prayag, 2009). Regarding using a single item in the SEM analysis, Petrescu (2013) in a theoretical and practical overview confirmed that using a single item can be successful and can be employed in statistical procedures such as SEM.

5. FINDINGS

5.1 Demographic profile of the respondents

The table below shows the demographic profile of the respondents. Out of the 339 participants, the majority of respondents were female representing 57.2%. The majority of the age group was respondents age between 26 to 40 years consists of 44.9%. The majority of the respondents were with bachelor's degrees representing 43.9%. In terms of monthly income, 46.1% of the respondents failed in the category of 1500 - 3000 euros per month. The top three Europeans nationality visiting Jordan were Russia (14%), UK (13%), and Germany (10%). Finally, about 70% of the respondents using the group packed tour as their travel arrangement method.

Table 1. Respondent's demographic profile.

%	N.	Age	%	N.	Nationality
8.2	28	18- 25	4.4%	15	Italy
44.9	152	26 to 40	5.0%	17	Portugal
36.6	124	41 to 60	13%	44	UK
10.3	35	61 and above	10%	34	Germany
%	N	Gender	2.0%	7	Belgium
42.8	145	Male	7.4%	25	France
57.2	194	Female	14%	47	Russia
%	**N**	**Qualification**	2.3%	8	Denmark
18.6	63	Below bachelor	5.9%	20	Sweden
43.9	149	Bachelor	7.4%	25	Switzerland
33.6	114	Master	0. 6%	2	Hungary
3.9	13	Above master	0.9%	3	Czech rep.
%	**N**	**Income level**	2.0%	7	Norway
21.8	74	Less than 1500 euro	3.5%	12	Netherlands
46.1	156	1500 - $3000 euro	5.0%	17	Ukraine
29.8	101	3000 - 4500 euro	2.6%	9	Poland
2.3	8	Above 4500 euro	2.9%	10	Spain
%	**N**	**Travel method**	0.9%	3	Finland
66.9	227	Group package tour	0.6%	2	Croatia
22.1	75	Individually arranged	1.4%	5	Austria
11.0	37	Other	3.5%	12	Turkey
			4.4%	15	Other European Countries

5.2 Descriptive statistics

In the following section, the results of the descriptive statistical test of the study variables will be discussed. In terms of the average value of the sample towards the e-WOM, the arithmetic means ranges from (3.9) in their least limit for the paragraph " I often consult other tourists' online travel reviews to help choose an attractive destination" to (4.4) in their highest limit for the paragraph " To make sure I choose the right destination, I often read other tourists' online travel reviews" on a Likert 5-point scale, with 5 being (strongly agree). The general arithmetic mean was very high and reached (4.3). The European tourists frequently gather information from tourists' online travel reviews and strongly believe in using this communication tool as it makes them more confident about their decision and reduces the risk associated with their choice.

Regarding the overall image of Jordan as a tourism destination, the respondents were asked to answer the single question of "How would you describe your impression (the overall image) of Jordan as a tourism destination" that anchored on 5-points scale rating from (1- unfavorable) to (5- very favorable). The result shows that European tourists have a favorable overall image of Jordan as a tourism destination with general arithmetic mean reaching (3.6). This describes a good value for the tourism destination but not very high. This could be due to the unsatisfied tourists about the high prices in Jordan especially when they feel they did not get value for their money.

In terms of the European tourist's intention to revisit Jordan in the future as well as their willingness to recommend the destination to others (tourist's behavioural intention), the general arithmetic means reached (4.0). This high mean value shows that respondents will plan to revisit to Jordan in the future as well as encourage and recommend Jordan as a tourism destination to others. The highest limit for the behavioural intention paragraph was for the "I will encourage friends and relatives to visit Jordan" with arithmetic mean reaching (4.4).

6. RESULTS

The indicator loadings, critical ratios, Cronbach's alpha and AVE of each variable are presented in the table (2) below. It is a roll of thumb that an indicator value below (0.4.) will be eliminated. As indicated by the table (2), all values were above 0.4., as such, the testing phase has been fulfilled. In terms of the Cronbach's alpha values of the variables included, they were higher than 0.70 and were ranged from (0.795) overall destination image to (0.875) for e-WOM. The Average Variance Extracted (AVE) as measurements of composite validity was used. The AVE of all factors is higher than (0.5), ranging from (0.661) for e-WOM to (0.794) for overall destination image. Hence, the validity of all variables was confirmed.

Table 2. Value Factor Loading, Average Variance Extracted (AVE), Composite Reliability (CR), and Cronbach Alpha (CA) from Each Latent Variable.

Variable	Indicator Loadings	CR	CA	AVE
E-WOM (6 items)				
I often consult other tourists' online travel reviews to help choose an attractive destination.	0.546	5.422	0.875	0.661
When I travel to a destination, tourists' online travel reviews make me confident in traveling to the destination	0.665	6.755		
I frequently gather information from tourists' online travel reviews before I travel to a certain destination	0.787	6.542		
I often read other tourists' online travel reviews to know what destinations make good impressions on others.	0.655	7.413		
If I don't read tourists' online travel reviews when I travel to a destination, I worry about my decision.	0.877	5.455		
To make sure I choose the right destination, I often read other tourists' online travel reviews.	0.568	5.785		
Overall destination image (one item)				
your impression of Jordan as tourism destination	0.669	7.552	0.795	0.794
Travel intentions (5 items)				
I have a strong intention to revisit Jordan in the future	0.558	5.456	0.844	0.777
I will visit this destination in the next 2 years	0.767	6.422		
I will recommend Jordan to other people	0.547	5.774		
I will encourage friends and relatives to visit Jordan	0.554	6.112		
I would say positive things about Jordan to other people	0.644	5.488		

In order to obtain an understanding of the relationship between study variables, correlation analyses were conducted on the data of the survey based on the independent variable of overall tourism destination image and e-WOM against the dependent variables of tourist's behavioural intention. Correlation is significant when the value is less than (0.05) which means the interactions between variables could be analyzed. A majority of the correlation values of the variables showed correlations coefficients with values below 0.68 and in the expected direction (Raykov and Marcoulides, 2006). Table (3) below shows the inter-factor correlation analysis between each construct included in the study. The result shows that all variables are significantly correlated. The correlation between the e-WOM and overall image was the highest with (0.665, $p \leq 0.01$). The correlation between the e-WOM and behavioural intention came next with (0.555 $p \leq 0.01$). Finally, the correlation between the overall image and behavioural intention was the lowest with (0.555 $p \leq 0.01$). Based on these findings, preliminary support for the hypothesized relationships could be concluded.

Table 3. Inter-factor Correlations of constructs.

Variables	E-WOM	Overall Image	Behavioural Intention
E-WOM	-		
Overall Image	0.665 *	-	
Behavioural Intention	0.598 *	0.555 *	-

After ensuring that all study variables are significantly correlated, the model well fit should be examined. To do that, and to confirm the hypotheses, the Structural Equation Modelling (SEM) was employed as it can simultaneously estimate each hypothesized path while incorporating corrections for measurement error.

The results that came from the maximum likelihood estimation showed that the three-factor model improved significantly and fits well the data. Particularly, the comparative fit index (CFI), and the root mean square error of approximation (RMSEA) were used to assessing the model fit. The RMSEA shows a good fit if the value equal or less than (0.05) and adequate fit if equal or less than (0.10). The result demonstrated that the model adequate fit as the value was (.084). In terms of the CFI, the model fit was confirmed since the value was greater than (0.90), particularly (0.978). In addition, the results showed that the x2 was (68.654), d.f was (82), the normed x2 was (2.254), GFI was (0.932), and the RMR was (0.39).

Table 4. Structural equation modelling (Hypothesized models testing)

Hypothesis	IV	DV	β	S-error	t-statistic	*p*
H1	e-WOM	Overall image	0.786	.086	6.445	significant
H2	e-WOM	Behavioural intention	0.614	.135	5.254	significant
H3	Overall image	Behavioural intention	0.774	.147	5.988	significant

Based on the results provided in the above table (4), the first three proposed hypotheses (H1, H2, and H3) were supported and confirmed. More specifically, e-WOM found to have a significant effect on the tourists' overall image and tourist's behavioural intention. This suggests that e-WOM is an important antecedent for both European tourists' behavioural intention to revisit Jordan and their overall image of Jordan as a tourism destination (H1 and H2: supported). In addition, the tourism destination's overall image was significantly influencing the tourists' behavioural intention. This indicates that e-WOM is an important antecedent for European tourists' behavioural intention to revisit (H3: supported). In fact, the empirical results from the structural model showed that the influence of e-WOM on destination overall image (β=0.786, t= 6.445, $p < 0.05$) was greater than its influence on the tourists' behavioural intention (β=0.614, t=5.254, $p < 0.05$).

In order to test the last hypothesis (H4), Hair et al. (2006, pp. 866–867) suggested four-step procedures for the mediating effect of overall destination image on the relationship between e-WOM and tourist behavioural intention. These are, (a) whether e-WOM significantly impact behavioural intention; (b) whether e-WOM significantly impact overall destination image; (c) whether overall destination image significantly impacts behavioural intention; and (d) whether the impact of e-WOM on behavioural intention reduces statistically when overall destination image is included in the model to predict tourist's behavioural intention. The first three steps are fulfilled as we found significant effects of e-WOM on overall destination image and tourists behavioural intention, likewise, overall destination image predict the tourist's behavioural intention. In the last step, the direct effect of e-WOM on the tourist's behavioural intention without including the overall destination image in the model was tested. The results demonstrated a significant effect of e-WOM on tourist's behavioural intention, however, the impact has shown a greater effect estimated as many as (0.804) at CR (8.87). This implies that e-WOM impact on tourist's behavioural intention partially mediated by overall destination image, as such, the last hypothesis (H4) was supported.

7. DISCUSSIONS

The present study examined the effect of e-WOM on the overall image of Jordan as a tourism destination and the European tourist's behavioral intention. In addition, the study examined the impact of destination image on the behavioral intention as well as the mediating effect of destination image between e-WOM and behavioral intention. According to findings, e-WOM among tourists has a significant, positive, and direct impact on tourists' behavioral intention (Abubakar and Ilkan, 2016; Jalilvand and Samiei, 2012; Ladhar and Michaud, 2015), as well as on the destination overall image (Litvin et al., 2008; Litvin et al., 2008). This confirms that tourism destination image and e-WOM are strong indicators of tourist's intention to revisit the destination as well as recommend the destination to others. Additionally, the findings validate the impact of destination image on the behavioral intention (Whang et al., 2016; Jalilvand et al., 2013; Abubakar and Ilkan, 2016), and the mediating effect of destination overall image on the relationship between e-WOM and tourist's behavioral intention. These findings offer important implications for tourism managers. Basically, as we found that e-WOM plays an important role in increasing tourists' travel intentions and creating a favorable destination image, destination marketers should acquaint themselves with the phenomenon of virtual worlds that offer an inspiring perspective for enhancing the quality tourists experience through collaborating with tourists. Destination marketers should understand that more and more potential tourists go online looking for information about the destination that they planning to visit, which influences their decision-making process. At the same time, an increasing number of previous tourists evaluate the destination that they experienced using online websites. The traditional WOM is limited to friends and family members, however, the new WOM that technology communication offers (e-WOM), made it possible for anybody in any place in the world to evaluate his/her experience either positively or negatively. As such, e-WOM offers a way to obtain a significant competitive advantage especially in the service industry such as the tourism industry where its products mostly are intangible. In this case, the potential tourists consider e-WOM as an information source that is trustworthy and experienced. Hence, it suggested the previous tourist reviews and comments should analyze by measuring the positive and negative words that reviewers have used, which provides insights on how to meet tourist's expectations, desires and needs (Marine-Roig et a., 2017). Understanding what mad previous tourists unsatisfied could help destination marketers to improve their products and services to meet the tourist's expatiations. In the case of satisfied tourists, destination marketers should encourage them to express their good experiences online.

Furthermore, the study suggests that the more the current tourists have a favourable image of the destination that they visiting, the more they are willing to revisit it in the future as well as recommend it to others such as their friends and family. Tourists rely more on the destination image when they are deciding a tourist destination; in addition, they have a big concern about the quality of the services they will get during their visit, for that reason, tourism services providers in Jordan should double their effort regarding offering high-quality tourists' experience to ensure that tourists will gain valuable experiences during their stay.

8. CONCLUSION AND LIMITATIONS

The study, however, had few limitations that should be noted. First, the study is limited to the European tourists who visited Jordan. Thus, the sample could not be treated as representative of all international tourists. As such, the study could be replicated with other market sources such as Arab or Asian tourists. Secondly, the sample was limited to the tourists who visited Jordan and preparing to leave Jordan from the international airport. Another study could focus on potential tourists rather than current or previous tourists. This could provide more insights into their behavioral intentions as they never experience the destination before. Thirdly, a future study could be conducted on what type of information the tourists searching for online and how this information used by the tourists to make their travel decision. Fourthly, the destination image could be measured using multiple variables rather than a single item that this study used.

REFERENCES

Abubakar, A. M., & Ilkan, M. (2016). Impact of online WOM on destination trust and intention to travel: A medical tourism perspective. *Journal of Destination Marketing & Management*, *5*(3), 192–201. doi:10.1016/j.jdmm.2015.12.005

Ahmed, T., & Abdul-Kadir, A. (2013). Impact of information sources on the decision makes process of travel to the Egyptian tourist destination after January 25. *Tourism and Hospitality Management*, *61*, 395–423.

Al-Oun, S., & Al-Homoud, M. (2008). The potential for developing community-based tourism among the Bedouins in the Badia of Jordan. *Journal of Heritage Tourism*, *3*(1), 36–54. doi:10.1080/1743873X.2008.9701249

Albattat, A., Jamaludin, A., Zuraimi, N. S. M., & Valeri, M. (2020). Visit Intention and Destination Image in Post- Covid- 19 Crisis Recovery. *Current Issues in Tourism*, *24*(17), 2392–2397. doi:10.1080/13683500.2020.1842342

Anastasiei, B., & Dospinescu, N. (2018). A model of the relationships between the Big Five personality traits and the motivations to deliver word-of-mouth online. *Psihologija (Beograd)*, *4*(2), 1–12. doi:10.2298/PSI161114006A

Baggio, R., & Valeri, M. (2020). Network science and sustainable performance of family businesses in tourism. *Journal of Family Business Management*. doi:10.1108/JFBM-06-2020-0048

Baloglu, S., & McCleary, K. (1999). A Model of Destination Image Formation. *Annals of Tourism Research*, *26*(4), 868–897. doi:10.1016/S0160-7383(99)00030-4

Bambauer-Sachse, S., & Mangold, S. (2011). Brand equity dilution through negative online word-of-mouth communication. *Journal of Retailing and Consumer Services*, *18*(1), 38–45. doi:10.1016/j.jretconser.2010.09.003

Bao, T., & Chang, T. (2014). Finding disseminators via electronic word of mouth message for effective marketing communications. *Decision Support Systems*, *67*, 21–29. doi:10.1016/j.dss.2014.07.006

Barlas, A., Mantis, K., & Koustelios, A. (2010). Achieving positive word-of-mouth communication: The role of perceived service quality in the context of Greek ski centres. *World Leisure Journal*, *52*(4), 290–297. doi:10.1080/04419057.2010.9674654

Beerli, A., & Martin, J. D. (2004a). Tourists' characteristics and the perceived image of tourist destinations: A quantitative analysis—a case study of Lanzarote, Spain. *Tourism Management*, *25*(5), 623–636. doi:10.1016/j.tourman.2003.06.004

Beerli, A., & Martin, J. D. (2004b). Factors influencing destination image. *Annals of Tourism Research*, *31*, 657–681. https://www.sciencedirect.com/science/article/abs/pii/S0160738304000349

Bickart, B., & Schindler, R. M. (2001). Internet forums as influential sources of consumer information. *Journal of Interactive Marketing*, *15*(3), 31–40. doi:10.1002/dir.1014

Bigne, J. E., Sanchez, M. I., & Sanchez, J. (2001). Tourism image, evaluation variables and after purchase behaviour: Inter-relationship. *Tourism Management*, *22*(6), 607–617. doi:10.1016/S0261-5177(01)00035-8

Blackwell, R., Miniard, P., & Engel, F. (2001). *Consumer Behavior*. Harcourt College.

Brown, J., Broderick, A. J., & Lee, N. (2007). Word of mouth communication within online communities: Conceptualizing the online social network. *Journal of Interactive Marketing*, *21*(3), 2–220. doi:10.1002/dir.20082

Casaló, F., Flavián, C., Guinalíu, M., & Ekinci, Y. (2015). Do online hotel rating schemes influence booking behaviors? *International Journal of Hospitality Management*, *1*, 28–36. doi:10.1016/j.ijhm.2015.05.005

Castro, C., Armario, M., & Ruiz, M. (2007). The influence of market heterogeneity on the relationship between a destination's image and tourists' future behavior. *Tourism Management*, *28*(1), 175–187. doi:10.1016/j.tourman.2005.11.013

Chatterjee, P. (2001). Online reviews: Do consumers use them? *Advances in Consumer Research. Association for Consumer Research (U.S.)*, *28*, 129–133.

Chemli, S., Toanoglou, M., & Valeri, M. (2020). The impact of Covid-19 media coverage on tourist's awareness for future traveling. *Current Issues in Tourism*, 1–8. Advance online publication. doi:10.1080/13683500.2020.1846502

Chen, C., & Tsai, D. C. (2007). How destination image and evaluative factors affect behavioural intentions? *Tourism Management*, *28*(4), 1115–1122. doi:10.1016/j.tourman.2006.07.007

Chen, Y. C., Shang, R. A., & Li, M. J. (2014). The effects of perceived relevance of travel blogs' content on the behavioral intention to visit a tourist destination. *Computers in Human Behavior*, *30*, 787–799. doi:10.1016/j.chb.2013.05.019

Confente, I., & Russo, I. (2015). Why do People Talk? A Comparison Between Offline WOM and Online WOM. In *ECSM2015-Proceedings of the 2nd European Conference on Social Media: ECSM 2015* (p. 81). Academic Conferences Limited.

Eagly, A., & Chaiken, S. (1993). *The psychology of attitudes*. Harcourt Brace Jovanovich College Publishers.

Echtner, C., & Ritchie, J. (1993). The measurement of destination image: An empirical assessment. *Journal of Travel Research*, *31*(4), 3–13. doi:10.1177/004728759303100402

El-Said, O., & Aziz, H. (2019). Egypt's Competitiveness: Empirical Examination of the Relationship Between Destination Attributes, Tourist Satisfaction, and Behavioral Intentions Among the Hotel Guests. *Tourism and Hospitality Management*, *25*(1), 25. doi:10.20867/thm.25.1.2

Fakeye, P. C., & Crompton, J. L. (1991). Image differences between prospective, first-time, and repeat visitors to the lower Rio Grande valley. *Journal of Travel Research*, *30*(2), 10–15. doi:10.1177/004728759103000202

Farajat, S. A., Liu, B., Pennington-Gray, L. (2017). Addressing travel writer's role as risk brokers: the case of Jordan. *Journal of Policy Research in Tourism Leisure and Events*, *9*, 23-39. doi:.2016.1231194 doi:10.1080/19407963

Fishbein, M., & Ajzen, I. (1975). *Beliefs, Attitude, and intention behavior*. Addison-Wesley Publishing Company.

Fornell, C., & Larcker, D. (1981). Evaluating structural equation models with unobservable variables and measurement error. *JMR, Journal of Marketing Research*, *48*(1), 39–50. doi:10.1177/002224378101800104

Gruen, T. W., Osmonbekov, T., & Czaplewski, A. J. (2006). EWOM: The impact of customer-to-customer online know how exchange on customer value and loyalty. *Journal of Business Research*, *59*(4), 449–456. doi:10.1016/j.jbusres.2005.10.004

Gunn, C. A. (1972). *Vacation scape: Designing Tourist Regions*. Taylor & Francis.

Hair, J. F., Black, W. C., Babin, B. J., Anderson, R. E., & Tatham, R. L. (2006). Multivariate Data Analysis (6th ed.). Prentice Hall.

Hennig-Thurau, T., Gwinner, K., Walsh, G., & Gremler, D. (2004). Electronic word-of-mouth via consumer-opinion platforms: What motivates consumers to articulate themselves on the Internet? *Journal of Interactive Marketing*, *18*(1), 38–52. doi:10.1002/dir.10073

Huang, S., & Hus, K. (2003). Determinants of User Intention toward IT Instruction: An Examination of Internal and External Factors. Knowledge Management and E-Learning. *International Journal (Toronto, Ont.)*, 1.

Jalilvand, M., & Samiei, N. (2012). The effect of electronic word of mouth on brand image and purchase intention.: An empirical study in the automobile industry in Iran. *Marketing Intelligence & Planning*, *30*(4), 460–476. doi:10.1108/02634501211231946

Jalilvand, M. R., Ebrahimi, A., & Samiei, N. (2013). Electronic Word of Mouth Effects on Tourists' Attitudes Toward Islamic Destinations and Travel Intention: An Empirical Study in Iran. *Procedia: Social and Behavioral Sciences*, *81*, 484–489. doi:10.1016/j.sbspro.2013.06.465

Jani, D., & Nguni, W. (2016). Pre-trip vs. post-trip destination image variations: A case of inbound tourists to Tanzania. *Original Scientific Paper*, *64*, 27–40.

Keller, K. L. (1993). Conceptualizing, Measuring, and Managing Customer-Based Brand Equity. *Journal of Marketing*, *57*(1), 1–22. doi:10.1177/002224299305700101

Kim, H., & Richardson, L. (2003). Motion picture impacts on destination images. *Annals of Tourism Research*, *30*(1), 216–237. doi:10.1016/S0160-7383(02)00062-2

Kim, S., & Jun, J. (2016). The impact of event advertising on attitudes and visit intentions. *Journal of Hospitality and Tourism Management*, *29*, 1–8. doi:10.1016/j.jhtm.2016.04.002

Ladhar, R., & Michaud, M. (2015). EWOM effects on hotel booking intentions, attitudes, trust, and website perceptions. *International Journal of Hospitality Management*, *46*, 36–45. doi:10.1016/j.ijhm.2015.01.010

Litvin, S. W., Blose, J. E., & Laird, S. T. (2003). Tourists' use of restaurant webpages: Is the internet a critical marketing tool? *Journal of Vacation Marketing*, *11*(2), 155–161. doi:10.1177/1356766705052572

Litvin, S. W., Goldsmith, R. E., & Pan, B. (2008). Electronic word-of mouth in hospitality and tourism management. *Tourism Management*, *29*(3), 458–468. doi:10.1016/j.tourman.2007.05.011

Litvin, S. W., Goldsmith, R. E., & Pan, B. (2008). Electronic word-of-mouth in hospitality and tourism management. *Tourism Management*, *29*(3), 458–468. doi:10.1016/j.tourman.2007.05.011

Liu, B., Schroeder, A., Pennington-Gray, L., & Farajat, A. (2016). Source market perceptions: How risky is Jordan to travel to? *Journal of Destination Marketing & Management*, *5*(4), 294–304. doi:10.1016/j.jdmm.2016.08.005

Lončarić, D., Prodan, M., & Ribarić, I. (2012). The influence of a visitor's perceptions of a museum's website design on behavioural intentions. *Tourism and Hospitality Management*, *18*, 297–312. https://hrcak.srce.hr/ojs/index.php/ekonomski-vjesnik/article/view/3969

Marine-Roig, E., Martin-Fuentes, E., & Daries-Ramon, N. (2017). User-generated social media events in tourism. *Sustainability*, *9*, 2250.

Mill, R. C., & Marrison, A. M. (2002). *The tourism system: An introductory text* (4th ed.). Kemdall/Hunt.

Milman, A., & Pizam, A. (1995). The role of awareness and familiarity with a destination: The central Florida case. *Journal of Travel Research*, *33*(3), 21–27. doi:10.1177/004728759503300304

Molina, R., Fri`as-Jamilena, D-M., & Castaneda-Garc`ia, A. (2013). The moderating role of past experience in the formation of a tourist destination image and in tourists' behavioral intention. *Current Issues in Tourism*, *16*, 107-127. https://www.tandfonline.com/doi/abs/10.1080/13683500.2012.665045

Naqvi, M. H. A., Jiang, Y., Naqvi, M. H., Miao, M., Liang, C., & Mehmood, S. (2018). The Effect of Cultural Heritage Tourism on Tourist Word of Mouth: The Case of Lok Versa Festival, Pakistan. *Sustainability*, *10*(7), 2391. doi:10.3390u10072391

Park, Y., & Njite, D. (2010). Relationship between destination image and tourists' future behavior: Observations from Jeju Island, Korea. *Asia Pacific Journal of Tourism Research*, *15*(1), 1–20. doi:10.1080/10941660903510024

Pereira, L., Limberger, P., Silva, L., & Pereira, M. (2018). An Empirical Investigation of Destination Branding: The Case of the City of Rio de Janeiro, Brazil. *Sustainability*, 11.

Petrescu, M. (2013). Marketing research using single-item indicators in structural equation models. *Journal of Marketing Analytics*, *1*(2), 99–117. doi:10.1057/jma.2013.7

Phillips, J., Wolfe, K., Hodur, N., & Leistritz, L. (2013). Tourist word of mouth and revisit intentions to rural tourism destinations: A case of North Dakota, USA. *International Journal of Tourism Research, 15*(1), 93–104. doi:10.1002/jtr.879

Pike, S., Bianchi, C., Keer, G., & Patti, C. (2010). Consumer-based brand equity for Australia as a long-haul tourism destination in an emerging market. *International Marketing Review, 27*(4), 434–449. doi:10.1108/02651331011058590

Prayag, G. (2009). Tourists' Evaluation of Destination Image, Satisfaction and Future Behavioral Intentions—The Case of Mauritius. *Journal of Travel & Tourism Marketing, 26*(8), 836–853. doi:10.1080/10548400903358729

Prayag, G., Hosany, S., & Odeh, K. (2013). The Role of Tourists' Emotional Experiences and Satisfaction in Understanding Behavioural Intentions. *Journal of Destination Marketing & Management, 2*(2), 118–127. doi:10.1016/j.jdmm.2013.05.001

Prayogo, R., Ketaren, F., & Hati, R. (2016). Electronic Word Of Mouth, Destination Image, And Satisfaction Toward Visit Intention: An Empirical Study In Malioboro Street, Yogyakarta. *Advances in Social Science, Education and Humanities Research*, 81.

Raykov, T., & Marcoulides, G. A. (2006). *A first course in structural equation modeling* (2nd ed.). Lawrence Erlbaum Associates Publishers.

Ren, G., & Hong, T. (2017). Investigating Online Destination Images Using a Topic-Based Sentiment Analysis Approach. *Sustainability, 9*(10), 9. doi:10.3390u9101765

Santos, V., Ramos, P., Sousa, B., Almeida, N., & Valeri, M. (2021a). Factors influencing touristic consumer behavior. *Journal of Organizational Change Management*. Advance online publication. doi:10.1108/JOCM-02-2021-0032

Santos, V., Sousa, B., Ramos, P., & Valeri, M. (2021b). Emotions and Involvement in Tourism Settings. *Current Issues in Tourism*, 1–6. Advance online publication. doi:10.1080/13683500.2021.1932769

See-To, W. K., & Ho, K. (2014). Value co-creation and purchase intention in social network sites: The role of electronic Word-of-Mouth and trust – A theoretical analysis. *Computers in Human Behavior, 31*, 182–189. doi:10.1016/j.chb.2013.10.013

Sen, S., & Lerman, D. (2007). Why are you telling me this? An examination into negative consumer reviews on the Web. *Journal of Interactive Marketing, 12*(4), 76–94. doi:10.1002/dir.20090

Smith, D., Menon, S., & Sivakumar, K. (2005). Online Peer and Editorial Recommendations, Trust, and Choice in Virtual Markets. *Journal of Interactive Marketing, 19*(3), 15–37. doi:10.1002/dir.20041

Su, L., & Fan, X. A. (2011). *Study on the Relationships between Service Quality, Satisfaction, Trust and Loyalty among Rural Tourism*. Paper presented at 8th International Conference on Service Systems and Service Management, Tianjin. https://ieeexplore.ieee.org/document/5959418

Tavitiyaman, P., & Qu, H. (2013). Destination Image and Behavior Intention of Travelers to Thailand: The Moderating Effect of Perceived Risk. *Journal of Travel and Tourism Management, 3*(3), 169–185. doi:10.1080/10548408.2013.774911

Thiumsak, T., & Ruangkanjanases, A. (2016). Factors influencing international visitors to revisit Bangkok, Thailand. *Journal of Economics, Business and Management*, *4*, 220-230. http://www.joebm.com/index.php?m=content&c=index&a=show&catid=57&id=709

Toanoglou, M., Chemli, S., & Valeri, M. (2021). The organizational impact of COVID-19 crisis on travel perceived risk across four continents. *Journal of Organizational Change Management*. Advance online publication. doi:10.1108/JOCM-12-2020-0369

Travel and tourism economic impact. (2018). *World – WTTC*. https://www.wttc.org/-/media/files/reports/economic-impact-research/regions-2018/world2018. pdf

Valeri, M. (2016). Networking and cooperation practices in the Italian tourism business. *Journal of Tourism, Heritage & Services Marketing*, *2*(1), 30–35. doi:10.5281/zenodo.376333

Valeri, M. (Ed.). (2021). Tourism risk. Crisis and recovery management. Emerald Publishing.

Valeri, M., & Baggio, R. (2020a). Social network analysis: Organizational implications in tourism management. *The International Journal of Organizational Analysis*, *29*(2), 342–353. doi:10.1108/IJOA-12-2019-1971

Valeri, M., & Baggio, R. (2020b). Italian tourism intermediaries: A social network analysis exploration. *Current Issues in Tourism*, *24*(9), 1270–1283. doi:10.1080/13683500.2020.1777950

Valeri, M., & Baggio, R. (2020c). A critical reflection on the adoption of blockchain in tourism. *Journal Information Technology and Tourism*, *23*(2), 121–132. doi:10.100740558-020-00183-1

Valeri, M., & Baggio, R. (2021). Increasing the efficiency of knowledge transfer in an Italian tourism system: A network approach. *Current Issues in Tourism*, 1–16. Advance online publication. doi:10.1080/13683500.2021.1937960

Valeri, M., & Katsoni, V. (Eds.). (2021). *Gender and tourism: challenges and entrepreneurial opportunities*. Emerald Publishing. doi:10.1108/9781801173223

Veal, A. J. (2006). *Research methods for leisure and tourism: A practical guide* (3rd ed.). Financial Times - Prentice Hall/Pearson Education.

Wangenheim, F., & Bayón, T. (2004). The effect of word of mouth on services switching: Measurement and moderating variables. *European Journal of Marketing*, *38*, 1173–1185. doi:10.1108/03090560410548924

Whang, H., Yong, S., & Ko, E. (2016). Pop culture, destination images, and visit intentions: Theory and research on travel motivations of Chinese and Russian tourists. *Journal of Business Research*, *69*(2), 631–641. doi:10.1016/j.jbusres.2015.06.020

Xu, F., Lin, X., Li, S., & Niu, W. (2018). Is Southern Xinjiang Really Unsafe? *Sustainability*, *10*(12), 4639. Advance online publication. doi:10.3390u10124639

Zeithaml, A. (1988). Consumer Perceptions of Price, Quality, and Value: A Means-End Model and Synthesis of Evidence. *Journal of Marketing*, *52*(3), 2–22. doi:10.1177/002224298805200302

Zeithaml, A., & Bitner, J. (1996). *Services marketing*. McGraw-Hill.

Chapter 10
Profile of the Local Residents and Visitors of the Historical Villages of a Portugal Network:
An Exploratory Study

Maria Manuela Santos Natário
https://orcid.org/0000-0002-5983-8399
Polytechnic of Guarda, Portugal

Carlos Santos
https://orcid.org/0000-0001-9728-8538
University of Aveiro, Portugal

Ana I. Melo
University of Aveiro, Portugal

Augusta Ferreira
https://orcid.org/0000-0003-1659-3309
University of Aveiro, Portugal

Dalila Dias
Historical Villages of Portugal, Portugal

Gonçalo Gomes
Turismo Centro de Portugal, Portugal

Graça Azevedo
https://orcid.org/0000-0002-6346-4035
University of Aveiro, Portugal

Paula Rocha
University of Aveiro, Portugal

Ricardo Biscaia
University of Aveiro, Portugal

Ruben Duarte
University of Aveiro, Portugal

Rui Pedro Marques
https://orcid.org/0000-0001-8449-0526
University of Aveiro, Portugal

ABSTRACT

This chapter presents the results of the exploratory study regarding the profile of local residents and visitors of the Historical Villages of Portugal Network (nHVP). This network integrates 12 villages located in low density territories in the center of Portugal. It is important to conduct this study in order to conceive strategies and policies that will contribute to reversing the desertification tendency and to retaining resources and surpluses that would otherwise leave the region. The authors concluded that the local residents present a medium/high level of satisfaction in terms of the impact of tourism in their villages and also with the association of their village with the nHVP brand. However, the local residents'

DOI: 10.4018/978-1-6684-3889-3.ch010

participation in the activities and touristic events is low, which is in part due to the high age of most local residents. Regarding the visitors of the nHVP, they are generally satisfied or very satisfied with several points. The lowest level of satisfaction relates to the public transportation, the communication networks, and the medical care available in the territory.

INTRODUCTION

The Historical Villages of Portugal Network (nHVP) includes 12 historical villages located in Portugal's central region (i.e., Almeida, Belmonte, Castelo Mendo, Castelo Novo, Castelo Rodrigo, Idanha-a-Velha, Linhares da Beira, Mêda, Monsanto, Piódão, Sortelha, and Trancoso), part of three municipal communities (i.e., Beiras and Serra da Estrela, Beira Baixa, and the Coimbra region).

The nHVP villages are located in sparsely populated territory, but have great potential. From the perspective of touristic attractiveness, they are distinguished by having interesting and diversified cultural, historical, and built heritage as well as by an assembly of endogenous products from diversified landscapes of a thousand colours and by mineral water springs. They are also distinguished by the different river valleys, from the Águeda, Côa, Douro, Erges, Mondego, and Tejo rivers to the Zêzere, and by the mountainous landscapes of Açor, Gardunha, Marofa, and Penha Garcia.

Each village has its own specific characteristics; some have more Jewish heritage, others are located closer to the border, and others are equipped with thermal spa treatments. Yet they all have vast, diversified, and rich built heritage in common, ranging from castles, forts, and churches to other monuments. These villages have a history dating back to the Middle Ages. However, they have lost much of their population, mainly young people and working age population, resulting in a sparsely populated region.

The nHVP was created to enhance and promote the region's realities in a joint and integrated way and to raise the potential for a joint common identity based on history, culture, and heritage as valued elements of these territories. The promotion of the nHVP is assumed to be a factor in the promotion and development of these territories, contributing to enhancing their attractiveness and retaining resources and surpluses that tend to leave the region.

In turn, tourism is renowned as one of the key sectors in the development of all countries and an important source of income and employment; it also plays a vital role in the creation of wealth while having implications at the social and environmental level. In addition, it fulfils a larger role in image promotion and international perception while also influencing internal policies. It is assumed to be a factor of competitiveness for countries and regions (Dupeyras & Maccallum, 2013).

According to Petrevska and Gerasimova (2012), tourism and regional development are closely linked. In fact, in many regions, the tourism industry is one of the greatest sources of economic growth and job creation. Thus, the regional development of tourism can trigger general economic growth by creating a new dynamic. In low-density territories, tourism is a strategic factor for the development of these regions; a producer of different social, cultural, and economic effects; and a generator of competitive factors (Ramos & Fernandes, 2016).

In the case of the nHVP, tourism assumes an anchoring characteristic in its sustainability and development. However, the presence of tourism and visitors in these villages affects the lives of the recipient communities, so the study of their satisfaction with the impact of this economic activity is important.

The Historical Villages of Portugal must provide attractiveness, whether for the residents or those visiting the area. In this context, this study about the region's residents and visitors is important to identify their profiles and help conceptualize and create strategies and policies that. On one hand, these will contribute to invert the desertification tendency while simultaneously providing conditions for sustained development and, on the other hand, will contribute to enhancing the villages' attractiveness and retaining resources and surpluses that would otherwise leave the region.

In terms of what has already been shown, the purpose of this chapter is to present the results of an exploratory study about the profile and satisfaction of the residents and visitors to the nHVP. This study is part of a broader project aligned with the project "PLowDeR – Framework for the Analysis of the Economic and Social Impact of Tourist Activities on Low Density Territories: The Case of Portuguese Historical Villages" (Melo, et al., 2020).

To materialize the objective of this chapter, it was divided into five sections after the initial introductory part. Section 1 presents the theoretical framework about tourism and the development of the low-density territories. Section 2 presents the methodology. Sections 3 and 4 present the profiles of the nHVP residents and visitors, and in Section 5 we discuss the results.

1. THEORETICAL FRAMEWORK

According Liu (2016), a very important issue in local and regional development that has emerged it is destination governance. The governance practices are necessary for increasing the competitiveness and sustainability of many destination (Gill & Williams, 2011). *The relationship between tourist destinations and the environment demonstrates that sustainability is relevant to competitiveness of destinations. The win-win link between sustainability and competitiveness is based on the belief that businesses that pursue environmental and economic and social performance improvements may benefit from these activities.* (Valeri, 2015:204)

Within the context of destination governance is becoming important the participation of multi-stakeholders and the establishment of wide-ranging relationships at various levels and actors (Zahra, 2011; Çakar & Uzut, 2020) in a dense network to create goods and services at tourist destinations (Çakar & Uzut, 2020).

Existing studies about tourism in low-density territories (Ramos & Fernandes, 2016), which are often considered peripheral (Bohlin, Brandt, & Elbe, 2016) and rural areas (Almeida, 2017; Carson, 2018; Salvatore, Chiodo, & Fantini, 2018) fit into the territorial approaches to endogenous and bottom-up development that search for factors which were able to generate differentiating and competitive territorial dynamics to promote the development of these territories.

Some studies consider tourism to be an important driver of socio-economic change (Salvatore et al., 2018) that is increasingly viewed as being able to rejuvenate regions facing economic difficulties and contribute to the reduction of disparities among regions (Bohlin et al., 2016). According to Petrevska and Gerasimova (2012), tourism has emerged as a major factor for regional economic development. Regardless of its nature, tourism has major economic and social impacts at the regional and local levels in areas where tourism activities take place. Almeida (2017), for example, analysed the critical factors of success in terms of tourism development in rural areas.

Low-density territories (LDT) hold a plethora of resources that are loved by post-Fordist travellers and are structured as being opposed to mass tourism. Tourism practices in these territories tend to favour local culture and traditional activities over an urban way of life (Umbelino, 1998). In this sense, tourism allows for the creation of various tourism products based upon resource endowment – a set of material or immaterial elements, activities, and places (Ramos & Fernandes, 2016).

According to Ramos and Fernandes (2016), development of low-density territories through tourism must be based on the preservation and encouragement of cultures, beliefs, and traditional values, deepening identities while enabling territories to take part of the global order. Respect for the culture and way of life of both the community and region is important, thereby making residents' participation necessary to achieve sustainable development through tourism (Vieira, 2014).

Vieira (2014) noted that, regarding the economic impact of tourism, residents' perception is mainly positive. However, there are negative aspects in terms of the cultural, social, and environmental views, so tourism must readjust its strategy, taking into consideration the rights, desires, and quality of life of the local communities. It must also provide the opportunity for broad sectors of the community to contribute to the decisions, planning, and administration.

Visitors' satisfaction with the visited destination – namely, the global experience (attractions found there, commodity and service supplies) – is also important for the promotion of tourism and is part of the satisfaction dimension developed by Melo et al. (2020). Dupeyras and Maccallum (2013) underscored that visitors' satisfaction tends to measure attractiveness, making it essential for the territories' current and future competitiveness. Visitors' satisfaction emerges as an important qualitative indicator of the demand side. The challenge is to collect data based on solid strategic methods and allow for comparison over time.

2. METHODOLOGY

The goal of this chapter is to present the results of the exploratory study about the profile and satisfaction of nHVP residents and visitors. The study was conducted in collaboration with the project "PLowDeR – Framework for the Analysis of the Economic and Social Impact of Tourist Activities on Low Density Territories: The Case of Portuguese Historical Villages" (Melo, et al., 2020).

To develop this project, we made several surveys according to the different profiles of the social actors of the nHVP. Specifically, we surveyed the residents and the visitors. The first survey was made in person (on paper or via the ILDA platform[1]); the second was left in the tourist office of each historical village.

In order to obtain information about the reality of these territories, we had previously gathered global data about the number of residents and visitors from an initial visit to the tourist offices. Table 1 summarizes this information, to date.

According to this information we made some methodological decisions in order to make data collection effective and sustainable:

i) *Residents' survey*: The data collection work in the smaller villages would last a maximum of two days, while in larger villages it would last a maximum of four days. This data collection phase took place between September 11 and October 31, 2018. To ensure higher levels of participation with the survey, we carried out some preparatory work to establish trust toward the researcher within the community. To this end, we sent out advance notice of the dates of the visit to each village through the several parish councils, parishes, and tourist offices. In addition, on each day we were in a village to administer the survey, we sought out a privileged informant to facilitate the interac-

Table 1. Global data of residents and visitors of the Historical Villages of Portugal

Villages	Residents	Visitors
Almeida	500	January to August 2018: 63 342
Belmonte	3500	-
Castelo Mendo	50	-
Castelo Novo	90-100	In intermediate months:1150. Triple in high season
Castelo Rodrigo	55-60	January to August 2018: 51 253
Idanha-a-Velha	-	-
Linhares da Beira	49	Annual visitors: 30 000
Marialva	30	About 1000 people/month
Monsanto	700/800	-
Piódão	62	August 2018: 2528
Sortelha	1	August: 100/day
Trancoso	-	January to August 2018: 15 708

Source: (Santos, et al., 2019)

tions with the remaining members of the community and to establish ideal levels of trust. In this way we were able to reach a total of 465 individuals (about 10% of the 12 villages' population).

ii) *Visitors' survey*: Data were collected at the tourist offices during the second half of 2018, which allowed us to ascertain that some variation occurred between the high and low seasons. The responses totalled a sample of 258 individuals.

In the first phase, the analysis and identification of the profile of the nHVP residents were based on the measurement of the most significant differences between the profiles in terms of the dependent variables (i.e., level of satisfaction with the impact of tourism in their village; level of satisfaction with being associated with the nHVP brand; pride in belonging to the community and culture of the nHVP; willingness to recommend the visit of the nHVP; and collaboration in the promotion of activities and touristic events in the village).

In a second phase, the analysis focused on the relationship between the dependent variables and the independent variables (i.e., gender, education level, motor difficulties, and number of persons in the household). We also tried to verify which of the nHVP villages presented a higher level of satisfaction in the previously mentioned categories. In both cases, the tests used were based on an ANOVA analysis.

Regarding the visitors, we initially tried to analyse their level of satisfaction regarding a number of variables (see Table 2) and later tried to relate the dependent variables identified in Table 2 to the level of visitors' satisfaction based on the independent variables (i.e., age, gender, education level, and nationality). To this end, variables were crossed and an ANOVA analysis was used.

Table 2. Variables used for the study of visitors

Level of satisfaction with the visit/experience in the AHP.
Level of satisfaction with security in the territory of the AHP.
Level of satisfaction in relation to the quality of the catering establishments.
Level of satisfaction in relation to the quality of the drinks and similar establishments,
Level of satisfaction in relation to the quality of the accommodation.
Level of satisfaction in relation to the quality of sports activities practised.
Level of satisfaction in relation to the quality of the cultural activities carried out.
Level of satisfaction in relation to the cleaning/hygiene of public spaces.
Level of satisfaction with the information provided (transport, routes, monuments, cultural agenda, etc.).
Level of satisfaction in relation to local/traditional commerce.
Level of satisfaction in relation to tourist entertainment companies.
Level of satisfaction in relation to places of interest (e.g. Interpretive infrastructures, interpretive centres, museum spaces, viewpoints, among others).
Level of satisfaction in relation to access routes.
Level of satisfaction in relation to means of public transport.
Level of satisfaction in relation to routes and pedestrian routes (associated information, signs, road status).
Level of satisfaction in relation to communication networks (mobile, wireless).
Level of satisfaction in relation to the medical care existing in the tourist territory (health technicians, mobile units and proximity to a medical post).

For the analysis of the collected data, the responses to the questionnaires were entered into Excel 2016® and SPSS 25®.

3. ANALYSIS OF RESULTS

3.1. Residents

3.1.1. General Characterization of the Resident Sample

Graph 1 summarizes the sample of the resident population (questionnaire respondents in each HVP); Belmonte stands out as the HVP with the most responses of the respondent residents (117), followed by Almeida (78) and Trancoso (68). It should be remembered that these are village and county seat HVPs. In terms of administration, Castelo Rodrigo and Castelo Novo are in fact the villages with the highest number of respondent residents, 31 and 29 respectively.
Table 3 indicates the age of the residents who replied to the survey.

The analysis in Table 3 shows that the sample population living in the HVP is on average 61 years and 75 years of age. By age group (Graph 2), Idanha-a-Velha is the nHVP village with the oldest population, with 76% of residents over 65 years of age, followed soon by Marialva (71%) and Castelo Mendo (62%). Excluding Sortelha, a village with only one working-age population, the villages with the largest number of young people are Castelo Rodrigo, Castelo Novo, Trancoso and Almeida. If we ignore the blue spot, over the age of 65, we understand that the villages with the most active population are Castelo Novo, Castelo Rodrigo, Trancoso and Piódão.

Figure 1. Number of residents per village

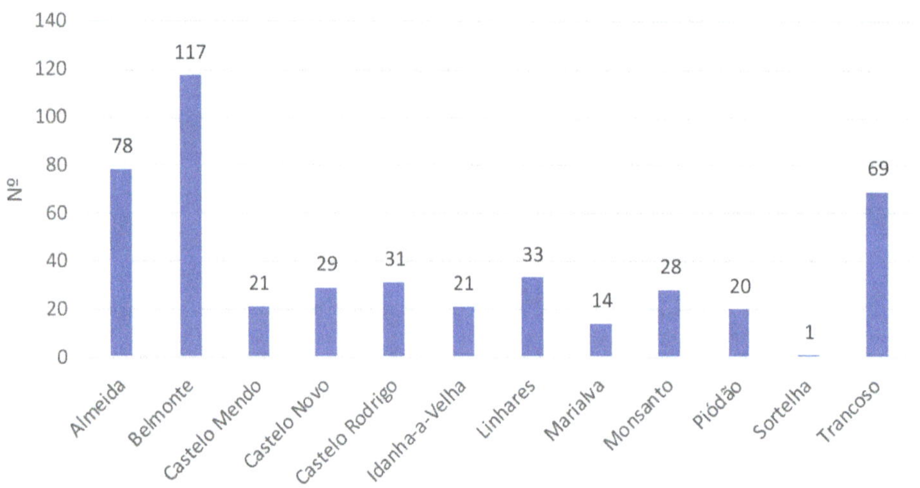

Table 3. Descriptive statistics on the age of residents

	Descriptive statistics						Percentiles		
	Interval	**Minimum**	**Maximum**	**Average**	**Deviation error**	**Moda**	**25**	**50**	**75**
Age of residents	83	17	100	61.14	19.287	75	48	64	77

Figure 2. Age group of residents in nHVP

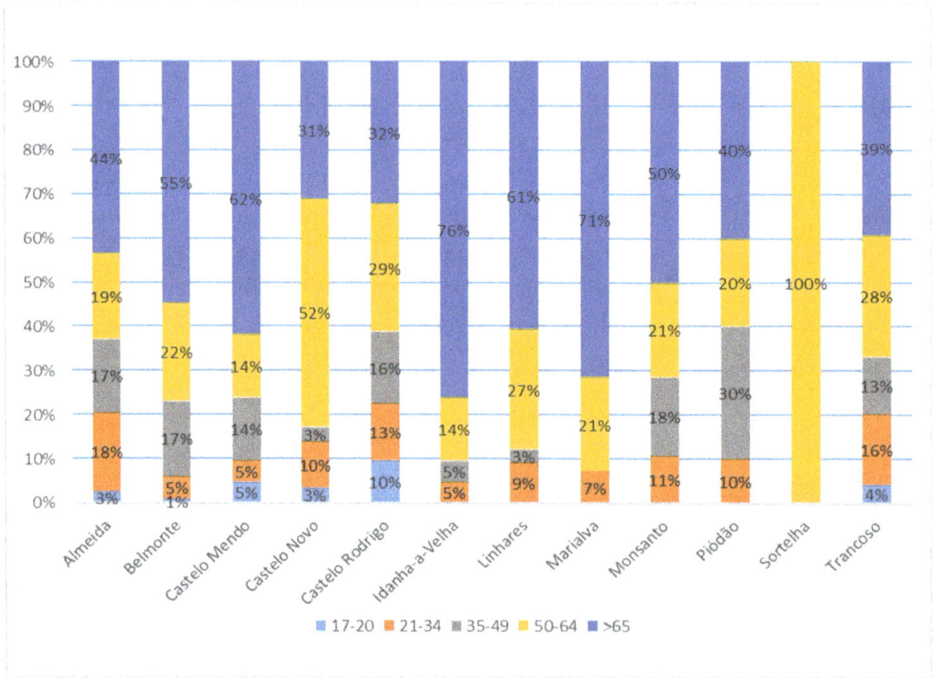

Figure 3. Sex of residents in nHVP

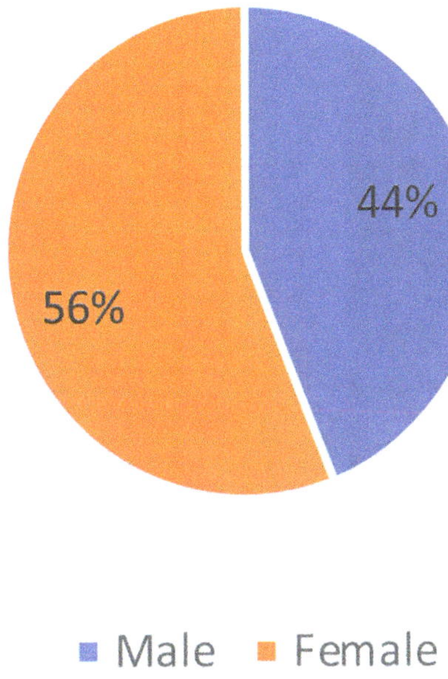

Figure 4. Sex of residents by HVP

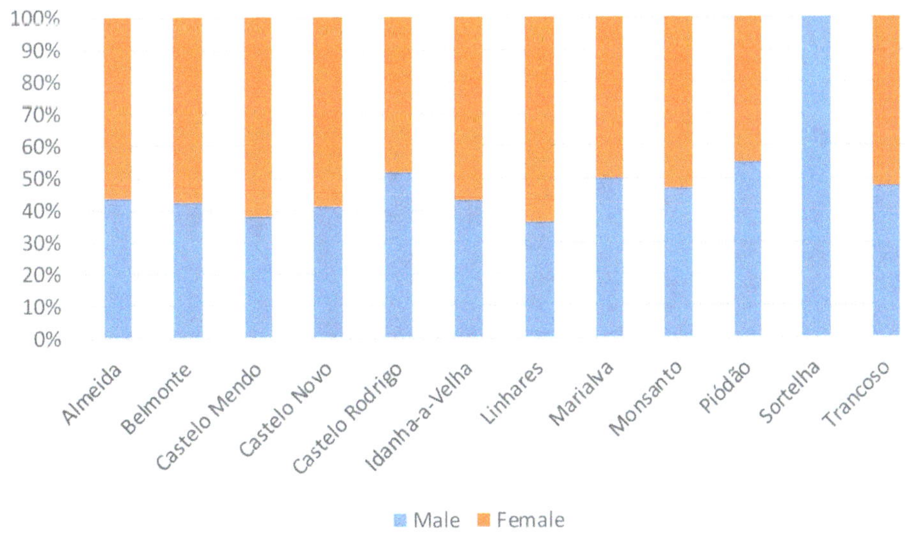

As regards the gender of the respondents, there is a predominance of women in all villages, with the exception of Sortelha, which is justified by the fact that there is only one inhabitant, male (Graphs 3 and 4).

As far as the level of education is concerned, it is the case across the nHVP that the majority of the population of the sample has basic education – 1st cycle (Graph 5), with the exception of Castelo Rodrigo, where the largest representation has basic education – 3rd cycle (26%).

Figure 5. Education level of residents in nHVP

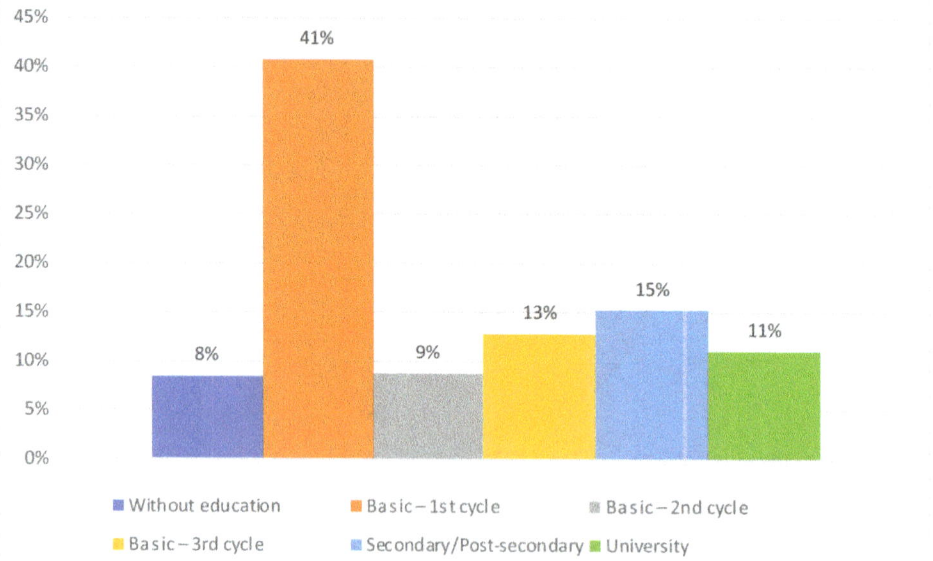

In most villages in the nHVP, there is more residual representation at the extreme level: without schooling and higher education, as shown in Graph 6.

Figure 6. Education level of residents by HVP

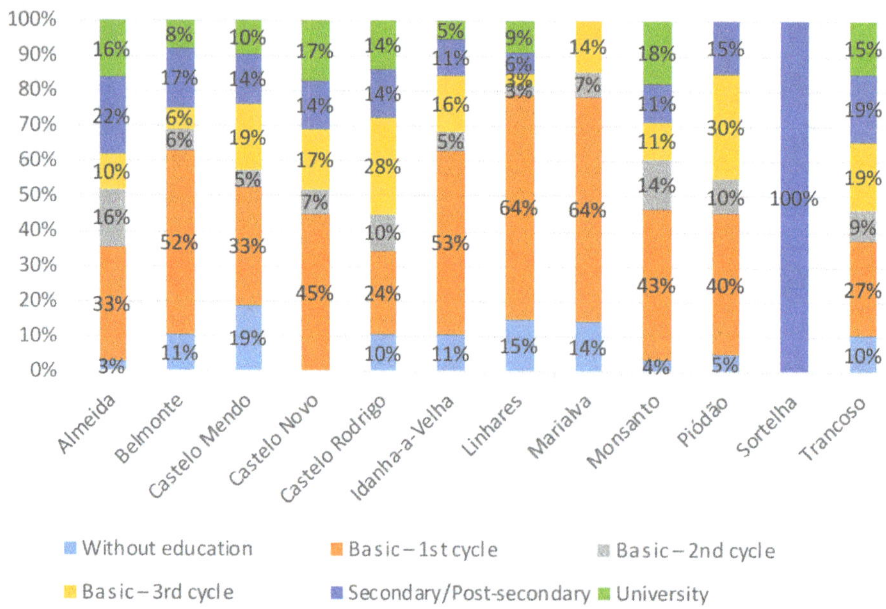

Figure 7. Naturality (country) of residents in the nHVP

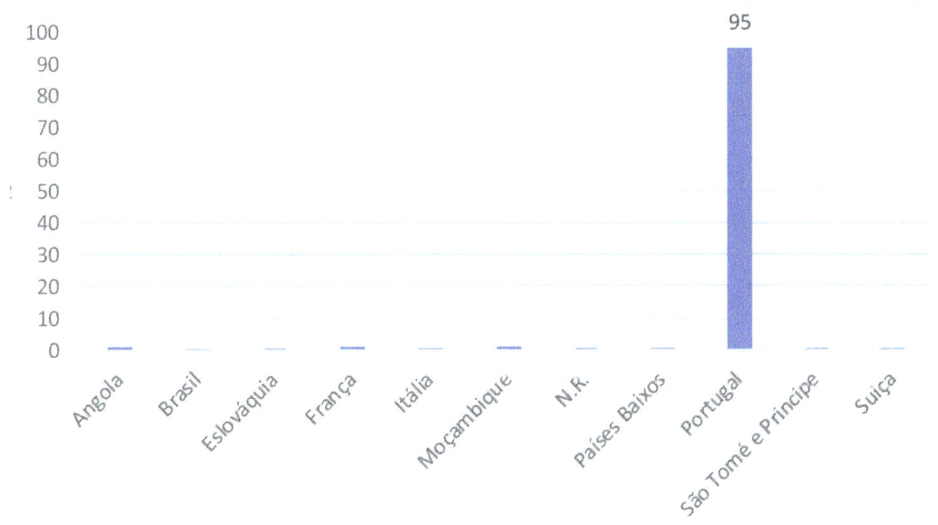

Figure 8. Place of birth (municipality) of residents

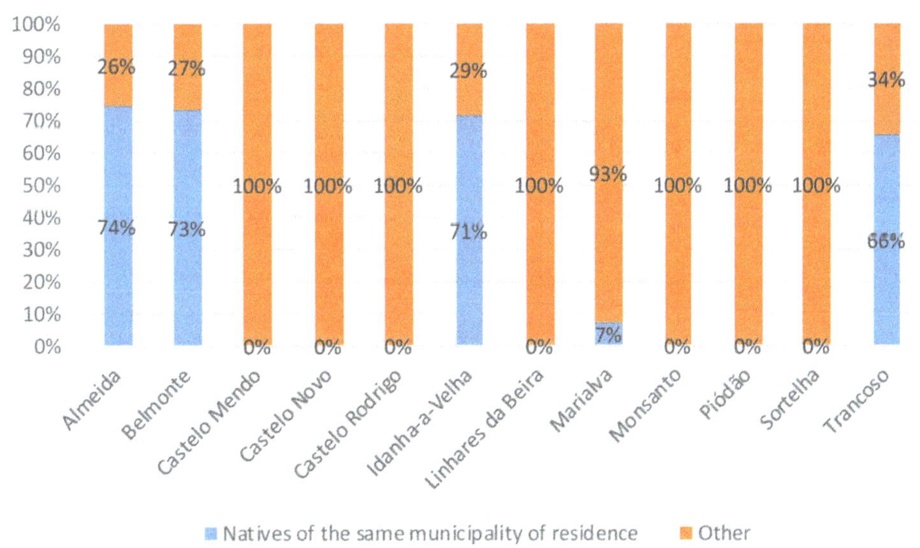

Analysing Graph 7, it can be seen that there are no significant levels of mobility when it comes to fixing the population in foreign HVPs.

Graph 8 also shows that the inhabitants of the HVP are, for the most part, natives of the same municipality of residence, with it only being visible in the villages of Almeida, Belmonte, Castelo Novo, Castelo Mendo, Castelo Rodrigo and Linhares that some of the population, albeit mainly residual, are from neighbouring counties.

Figure 9. Number of elements that make up the household in the nHVP

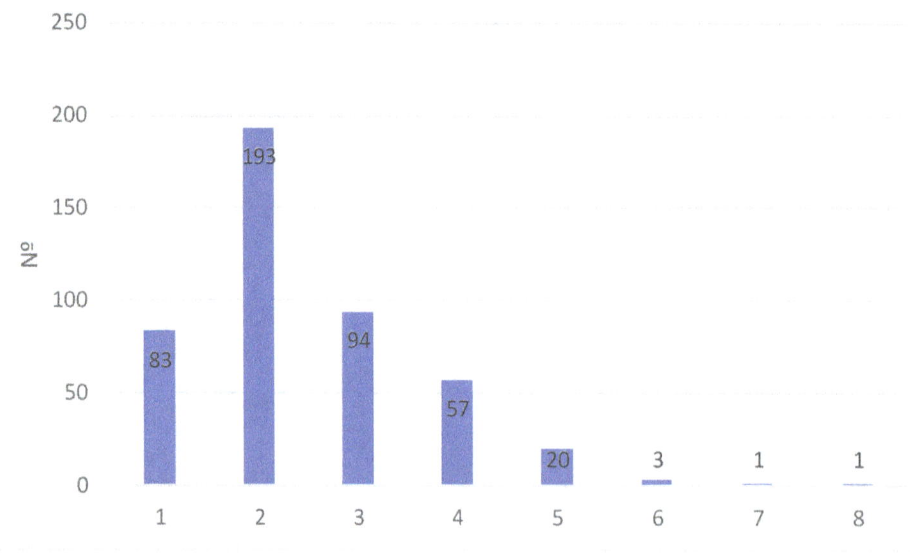

Figure 10. Type of resident housing in the nHVP

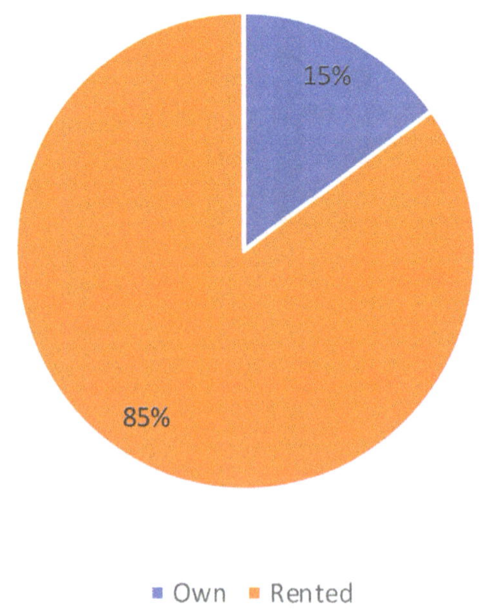

National households are composed by an average of 2.6 people per household (Censos, 2011). In HVP, households are composed by an average of 2.4 people and are divided as follows: in nHVP, 193 individuals are part of two-person households (Graph 9), equivalent to 43 per cent, followed by one in three (21 per cent) and 18 per cent of the population living alone.

Analysing Graph 10, it can be seen that most of the inhabited houses in the HVP belong to the residents themselves.

Analysis of Graph 11 shows that there are more cases of residents living in rented houses in Almeida, Belmonte and Trancoso.

Figure 11. Type of housing of residents by HVP

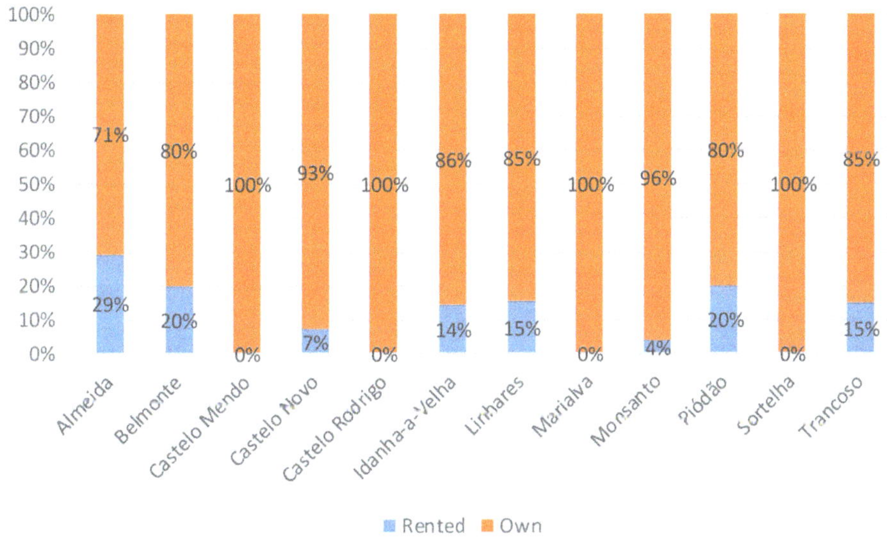

3.1.2. Profile and Satisfaction of Residents

In order to identify the profile of the residents of the network of historic villages of Portugal (nHVP), in the first phase, we sought to assess whether there were significant differences between the residents of the villages of nHVP with regard to the dependent variables: the level of satisfaction with the impact of tourism in their village; level of satisfaction with the association with the HVP brand; pride in belonging to the HVP community and culture; willingness to recommend other people to visit the HVP and collaboration in boosting tourist activities and events in the village. In the second phase, an attempt was made to assess the relationship between the variables stated above and the independent variables: sex, educational level, motor difficulties and the number of members that make up the household. We also tried to check which of the villages in the nHVP had the highest level of satisfaction in the categories mentioned above. The tests used in both cases were based on analysis of the ANOVA table.

As regards the differences in the profile of residents between the villages that make up the nHVP, the results can be seen in Table 4.

At a level of 1%, it can be seen that there are significant differences in the level of satisfaction of the villagers with regard to the impact of tourism in their village and the association with the HVP brand. At a significant level of 10%, it appears that there are differences in the level of collaboration between tourist events. With regard to the pride of belonging to the community and the fact that they would recommend a visit to the village, there are no significant differences between the HVP.

Table 4. Satisfaction with pride, visit recommendation and residents' collaboration

Dependent variables	Sum of squares	df	Mean square	Z	Sig.
The level of satisfaction with the impact of tourism on your village (e.g. access, cost of living, service offer, etc.)	57.408	11	5.219	5.987	0.000***
The level of satisfaction with the association with the brand of the historic villages of Portugal	33.717	11	3.065	3.259	0.000***
Pride of belonging to the HVP community and culture	0.859	11	0.078	1.513	0.124
Willingness to recommend other people to visit the HVP	0.443	11	0.040	1.465	0.142
Collaboration to promote tourism activities and events in the village	4.057	11	0.369	1.684	0.074*

Note: *** level of statistical significance of 1%; ** level of statistical significance of 5%, * level of statistical significance of 10%.

Since there are differences in the level of satisfaction of residents, Table 5 shows the average satisfaction of residents in each of the villages. As far as the level of satisfaction with the impact of tourism on their village is concerned, it appears that the villages in which the residents are most satisfied are Castelo Rodrigo and Linhares da Beira, and the villages in which the residents are least satisfied are Trancoso and Sortelha. As for the level of satisfaction of the residents with the HVP brand, Castelo Rodrigo, Belmonte and Linhares da Beira are the villages in which the residents have the highest level of satisfaction. Sortelha is a village where the residents are less satisfied with the association with the brand HVP. As for participation in activities and tourist events in the village, this is low, though higher in Castelo Rodrigo and Castelo Mendo.

Table 5. Average level of satisfaction of residents, by HVP

	Level of satisfaction with the impact of tourism on your village (e.g. access, living costs, service offer, etc.) (average)	Level of satisfaction with the association with the brand of the historic villages of Portugal (average)	Collaboration on the promotion of tourism activities and events in the village (percentage Yes)
Almeida	3.48	3.69	35.1%
Belmonte	3.98	4.04	24.6%
Castelo Mendo	3.62	3.57	50.0%
Castelo Novo	3.44	3.38	48.3%
Castelo Rodrigo	4.23	4.13	51.6%
Idanha-a-Velha	3.81	3.71	28.6%
Linhares da Beira	4.21	4.00	27.3%
Marialva	3.71	3.64	21.4%
Monsanto	3.75	3.43	37.0%
Piódão	4.00	3.90	45.0%
Sortelha	3.00	1.00	0.0%
Trancoso	3.12	3.57	29.0%
Total	3.72	3.78	33.3%

There is a relationship between the level of satisfaction of residents and age, sex, educational level, motor difficulties and the number of elements that make up the household.

The results of the ANOVA test on differences between the age of residents and the dependent variables under analysis can be found in Table 6.

Table 6. Residents' level of satisfaction with some dependent variables: by age

Dependent variables	Sum of squares	df	Mean square	Z	Sig.
The level of satisfaction with the impact of tourism on your village (e.g. access, cost of living, service offer, etc.)	80.774	77	1.049	1.089	0.300
The level of satisfaction with the association with the brand of the historic villages of Portugal	73.443	77	0.954	0.953	0.592
Pride of belonging to the HVP community and culture	4.391	77	0.057	1.109	0.264
Willingness to recommend other people to visit the HVP	2.275	77	0.030	1.076	0.324
Collaboration to promote tourism activities and events in the village	27.625	77	0.359	1.838	0.000 ***

Note: *** level of statistical significance of 1%; ** level of statistical significance of 5%, * level of statistical significance of 10%.

It can be concluded that, given the age of the residents, there are only statistically significant differences at the level of 1% regarding the collaboration of residents in the promotion of activities and tourist events in the village.

As far as collaboration in tourism events is concerned, there are significant differences between age groups of nHVP residents. It is the youngest residents (average age 55 years) who claim to be involved in these events, as opposed to those who, on average, are 64 years of age who say that they do not cooperate in promoting village activities and tourist events.

The results presented in Table 7 show that there are no differences between the gender of the population and the dependent variables in question.

Table 7. Residents' level of satisfaction with some variables depending on sex

Dependent variables	Sum of squares	df	Mean square	Z	Sig.
The level of satisfaction with the impact of tourism on your village (e.g. access, cost of living, service offer, etc.)	0.044	1	0.044	0.045	0.833
The level of satisfaction with the association with the brand of the historic villages of Portugal	2.586	1	2.586	2.614	0.107
Pride of belonging to the HVP community and culture	0.000	1	0.000	0.001	0.980
Willingness to recommend other people to visit the HVP	0.005	1	0.005	0.196	0.658
Collaboration to promote tourism activities and events in the village	0.145	1	0.145	0.650	0.421

Note: *** level of statistical significance of 1%; ** level of statistical significance of 5%, * level of statistical significance of 10%.

With regard to the level of education, the results presented in Table 8 allow us to conclude that there are only significant differences (the level of importance of 1%) in terms of the dependent variables: recommendation for a visit to the village and collaboration in the promotion of activities and tourist events in the village.

Table 8. Residents' level of satisfaction with some dependent variables, by level of education

Dependent variables	Sum of squares	df	Mean square	Z	Sig.
The level of satisfaction with the impact of tourism on your village (e.g. access, cost of living, service offer, etc.)	10.084	6	1.681	1.737	0.111
The level of satisfaction with the association with the brand of the historic villages of Portugal	5.459	6	0.910	0.916	0.483
Pride of belonging to the HVP community and culture	0.323	6	0.054	1.031	0.404
Willingness to recommend other people to visit the HVP	0.818	6	0.136	5.168	0.000***
Collaboration to promote tourism activities and events in the village	9.742	6	1.624	7.959	0.000***

Note: *** level of statistical significance of 1%; ** level of statistical significance of 5%, * level of statistical significance of 10%.

Table 9 shows the percentage of residents, by level of education, who replied that they recommended other people to visit the villages of the nHVP and that they cooperated in promoting activities and tourist events in the village.

Table 9. Recommendation of residents to visit the village and collaborate on events, by level of education

Education	Willingness to recommend other people to visit the HVP	Collaboration to promote tourism activities and events in the village
Without education	83.8%	5.4%
Basic – 1st cycle	97.9%	25.3%
Basic – 2nd cycle	100.0%	38.5%
Basic – 3rd cycle	100.0%	47.5%
Secondary/Post-secondary	95.7%	44.9%
University	100.0%	54.9%

From the analysis in Table 9, it can be seen that residents with basic education (2nd and 3rd cycles) and higher education are the ones who most recommend visiting their HVP, reaching 100%, and those who least recommend it are those who have no education at all. With regard to cooperation in the promotion of tourist activities and events in the village, there is a tendency for residents not to participate. Those who participate most are those with a higher education level of 54.9% and those who participate least are those who do not have a higher education level of 5.4%.

With regard to motor difficulties for residents, the results presented in Table 10 show that there are only significant differences (the statistical significance level of 1%) in relation to the dependent variable of collaboration in the promotion of activities and tourist events in the village. See Chart 12 for a more detailed analysis. Taking into account the robust geography of many of the HVPs, as demonstrated by the research team when visiting the villages, people with motor difficulties do not usually work together to promote activities and tourist events.

Table 10. Level of satisfaction of residents regarding some dependent variables, due to motor difficulties

Dependent variables	Sum of squares	df	Mean square	Z	Sig.
The level of satisfaction with the impact of tourism on your village (e.g. access, cost of living, service offer, etc.)	1.254	2	0.627	0.641	0.527
The level of satisfaction with the association with the brand of the historic villages of Portugal	1.040	2	0.520	0.523	0.593
Pride of belonging to the HVP community and culture	0.074	2	0.037	0.707	0.494
Willingness to recommend other people to visit the HVP	0.096	2	0.048	1.737	0.177
Collaboration to promote tourism activities and events in the village	3.216	2	1.608	7.424	0.001***

Note: *** level of statistical significance of 1%; ** level of statistical significance of 5%, * level of statistical significance of 10%.

Finally, with regard to the number of members that make up resident households, the results presented in Table 11 allow us to conclude that there are only significant differences in terms of collaboration in boosting tourism activities and events in the village (the statistical significance level of 1%).

Figure 12. Residents with motor difficulties who collaborate in promoting activities and events

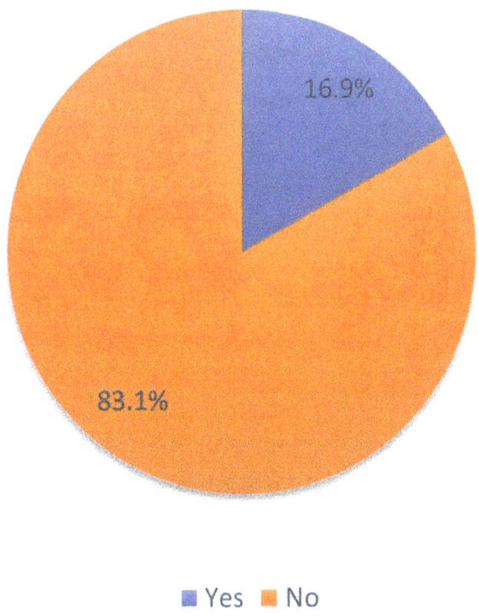

Table 11. Residents' level of satisfaction with dependent variables, by household size

Dependent variables	Sum of squares	df	Mean square	Z	Sig.
The level of satisfaction with the impact of tourism on your village (e.g. access, cost of living, service offer. etc.)	11.512	7	1.645	1.709	0.105
The level of satisfaction with the association with the brand of the historic villages of Portugal	6.542	7	0.935	0.964	0.457
Pride of belonging to the HVP community and culture	0.438	7	0.063	1.226	0.287
Willingness to recommend other people to visit the HVP	0.052	7	0.007	0.306	0.951
Collaboration to promote tourism activities and events in the village	6.654	7	0.951	4.531	0.000***

Note: *** level of statistical significance of 1%; ** level of statistical significance of 5%, * level of statistical significance of 10%.

Thus, analysis of Table 12 shows that, on average, residents with a larger number of household members are the ones who collaborate the most in promoting activities and tourist events in the village.

Table 12. Participation in boosting tourism activities and events by household size

Collaboration to promote tourism activities and events in the village	Number of members that make up the household (average)
Yes	2.72
No	2.33

3.2 Visitors

3.2.1 General Characterization of Sample Visitors

The sample consists of a total of 258 visitors, as indicated above, with Graph 13 showing the breakdown by village. It appears that the most visited village during the period under analysis was Almeida, followed by Sortelha and Castelo Rodrigo, and the least visited was Castelo Mendo.

The average age of visitors who replied to the survey is 41 years (Table 13). In addition, Graph 14 shows that it is those in the the age group between the ages of 36 and 45 years that most visit nHVP (23 per cent).

nHVP is slightly more visited by women, corresponding to 52% of visitors (Graph 15).

Due to the features of nHVP, which are poorly available to people with limited mobility, it is easy to understand that 94% of visitors do not have any motor problems (see Graph 16).

With regard to the level of education, it is the people with the highest level of education who are most likely to visit the nHVP, with a focus on tourists with higher education (33%) (Figure 17).

Graph 18 reveals the nationality of nHVP visitors. They are mostly Portuguese (65.%), followed by the English (13%) and the Spanish (11%).

Figure 13. Number of visitors per HVP

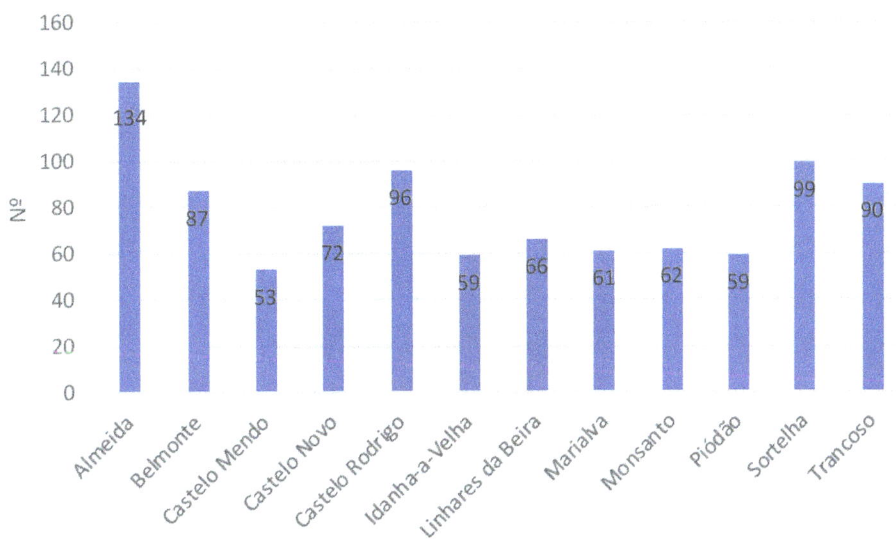

Table 13. Descriptive statistics on the age of visitors

	Descriptive statistics						Percentiles		
	Interval	Minimum	Maximum	Average	Deviation error	Moda	25	50	75
Visitor Age	75	0	75	41,07	14,49	40	30	40	51

Figure 14. Age range of visitors (%)

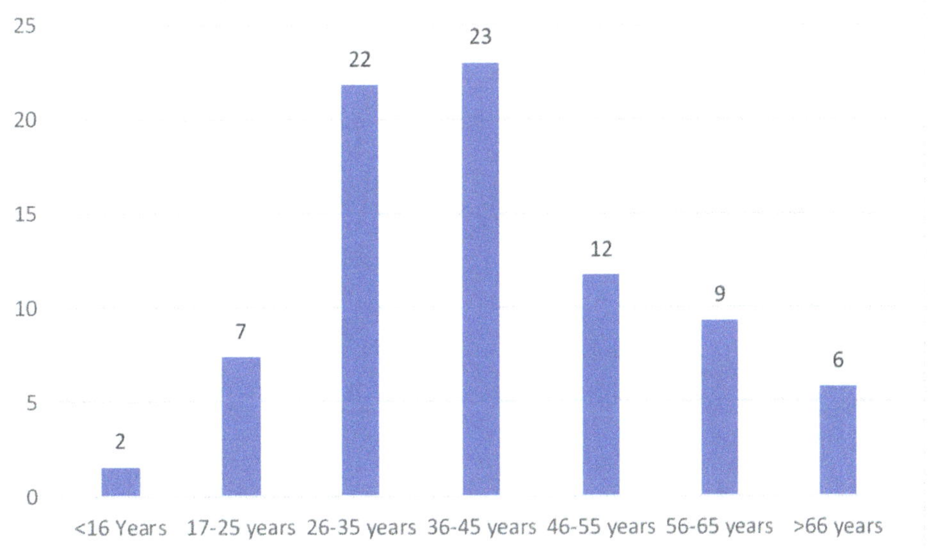

Figure 15. Percentage of visitors by sex (%)

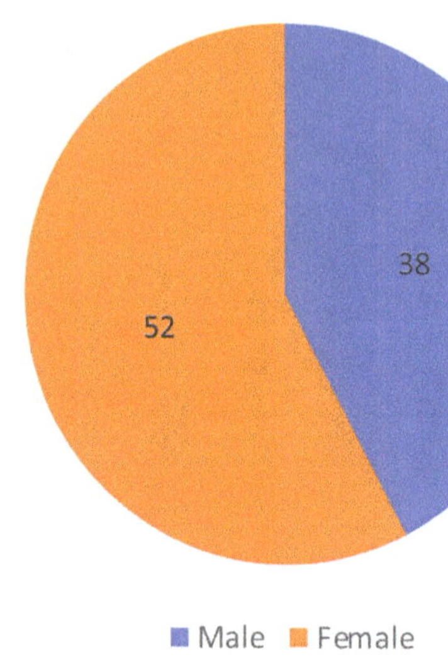

Figure 16. Percentage of visitors who have motor disabilities (%)

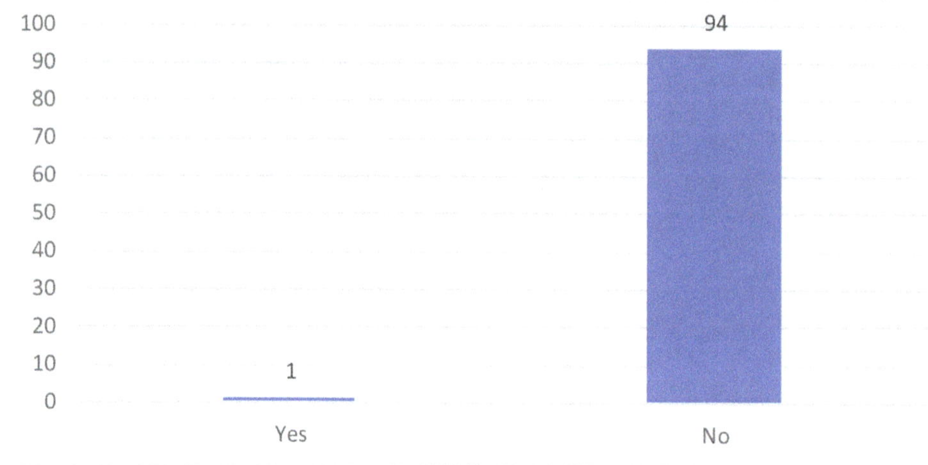

Figure 17. Education level of visitors to the nHVP (%)

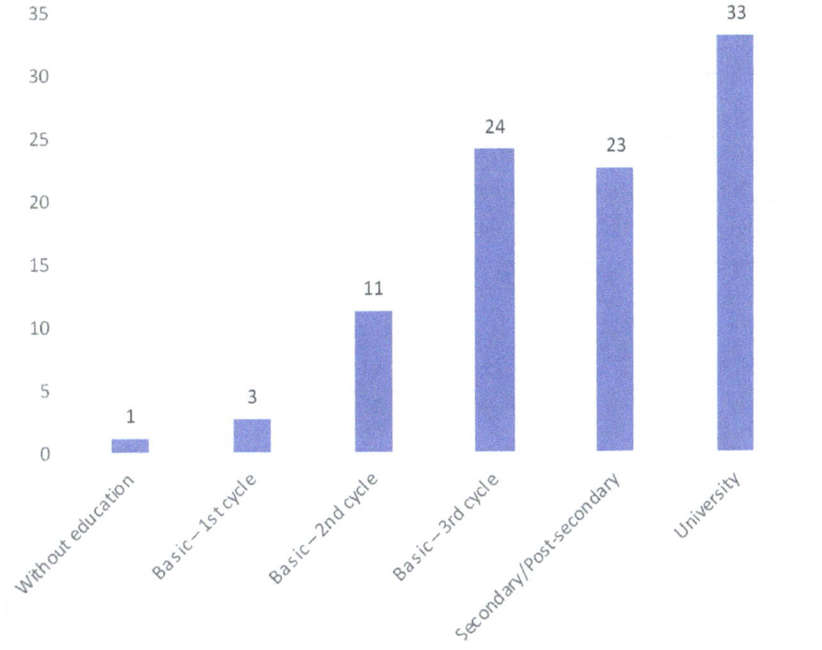

Figure 18. Nationality of visitors (%)

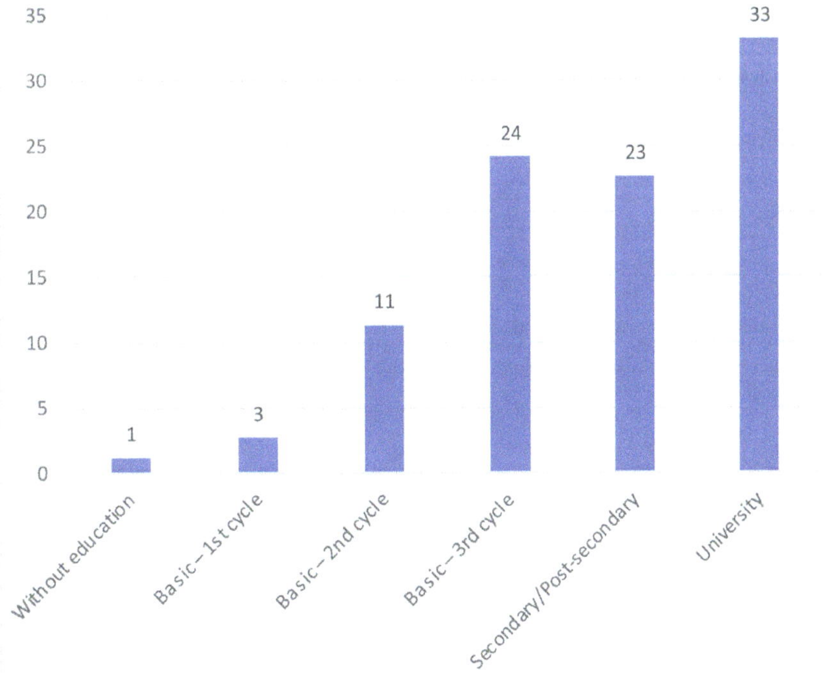

3.2.2. Profile and Visitor Satisfaction

As regards the level of satisfaction of visitors to the HVP, it can be seen (Table 14) that, in general, visitors have a level of satisfaction of more than 3.2 points (on the Likert scale ranging between 1-Very dissatisfied and 5-Very satisfied). The highest level of satisfaction of the HVP tourists is linked to the visit/experience in general in the HVP, followed by the protection felt in the HVP. Visitors are less pleased in relation to the public means of transport available, in relation to the current communication networks (mobile network, wireless) and in relation to the medical care available in the tourist area (health technicians, mobile units and proximity to medical offices).

Table 14. Level of satisfaction of nHVP visitors

	Average	Deviation error
Level of satisfaction with the visit/experience in the HVP	4.54	0.589
Level of satisfaction with security in the territory of HVP	4.43	0.695
Level of satisfaction with the price of catering establishments	4.31	0.688
Level of satisfaction in relation to the quality of beverage and similar establishments	4.21	0.725
Level of satisfaction in relation to the quality of the accommodation	4.28	0.698
Level of satisfaction in relation to the quality of sports activities practised	3.88	0.816
Level of satisfaction in relation to the quality of cultural activities carried out	3.90	0.690
Level of satisfaction regarding cleanliness/hygiene of public spaces	4.04	0.734
Level of satisfaction with the information provided (transport, routes, monuments, cultural agenda, etc.)	3.91	0.707
Level of satisfaction in relation to local/traditional commerce	3.92	0.697
Level of satisfaction in relation to tourist entertainment companies	3.74	0.781
Level of satisfaction in relation to places of interest (e.g. Interpretive infrastructures, interpretive centres, museum spaces, viewpoints, among others)	4.00	0.729
Level of satisfaction in relation to access routes	3.74	0.745
Level of satisfaction with public transport	3.20	0.760
Level of satisfaction in relation to routes and pedestrian routes (associated information, signs, road status, ...)	3.55	0.839
Level of satisfaction in relation to communication networks (mobile, wireless)	3.28	1.017
Level of satisfaction in relation to medical care in the tourist territory (health technicians, mobile units and proximity to a medical centre)	3.28	0.887

In the visitor profile study, we tried to connect the dependent variables regarding the degree of satisfaction of nHVP visitors to the independent variables: age, sex, educational level and nationality.

From the findings reported in Table 15, it can be shown that there are variations of statistical significance in terms of visitor satisfaction, taking into account their age, in relation to routes and walking routes (5 per cent statistical significance level), communication networks (1 per cent statistical significance level) and medical treatment in the tourist area (statistical significance level of 10%). As for the remaining satisfaction variables, there are no age differences to be reported.

Table 15. Level of satisfaction of nHVP visitors: differences in means by age

Dependent variables	Sum of squares	df	Mean square	Z	Sig.
Level of satisfaction with the visit/experience in the HVP	14.015	43	0.326	1.399	0.088
Level of satisfaction with security in the territory of HVP	20.700	43	0.481	1.288	0.153
Level of satisfaction with the price of catering establishments	19.542	44	0.444	1.176	0.254
Level of satisfaction in relation to the quality of beverage and similar establishments	20.418	39	0.524	1.014	0.466
Level of satisfaction in relation to the quality of the accommodation	19.677	40	0.492	1.019	0.462
Level of satisfaction in relation to the quality of sports activities practised	19.275	35	0.551	0.806	0.753
Level of satisfaction in relation to the quality of cultural activities carried out	19.989	36	0.555	1.236	0.221
Level of satisfaction regarding cleanliness/hygiene of public spaces	22.253	42	0.530	1.004	0.481
Level of satisfaction with the information provided (transport, routes, monuments, cultural agenda, etc.)	20.359	38	0.536	1.112	0.337
Level of satisfaction in relation to local/traditional commerce	13.770	39	0.353	0.638	0.939
Level of satisfaction in relation to tourist entertainment companies	18.950	35	0.541	0.832	0.718
Level of satisfaction in relation to places of interest (e.g. Interpretive infrastructures, interpretive centres, museum spaces, viewpoints, among others)	22.693	40	0.567	1.101	0.348
Level of satisfaction in relation to access routes	25.133	43	0.584	1.048	0.418
Level of satisfaction with public transport	23.322	33	0.707	1.404	0.135
Level of satisfaction in relation to routes and pedestrian routes (associated information, signs, road status, …)	36.762	40	0.919	1.537	0.053**
Level of satisfaction in relation to communication networks (mobile, wireless)	61.982	40	1.550	1.940	0.006***
Level of satisfaction in relation to medical care in the tourist territory (health technicians, mobile units and proximity to a medical centre)	33.361	33	1.011	1.491	0.087*

Note: *** level of statistical significance of 1%; ** level of statistical significance of 5%, * level of statistical significance of 10%.

In view of the disparities set out above, Table 15 shows the degree of satisfaction of residents in relation to the roads, communication networks and medical care that exist in the tourist area. The study reveals that visitors who are less happy with the roads and the pedestrian roads and the contact networks are the least young, with an average age of 52 years and 51 years, respectively. Those with an average age of 48 years and 46 years respectively are very happy. As far as medical treatment in the tourist region is concerned (health workers, mobile units and proximity to medical centres), it is the youngest who are the most disappointed.

The different levels of visitor satisfaction are not related to the gender variable (Table 17).

Table 16. Level of satisfaction of visitors regarding nHVP, by age

	Average Ages / Routes	Average Ages / Wireless	Average Ages / Health Care
Very unsatisfied	42.50	51.27	20.00
Dissatisfied	51.67	49.30	35.94
Neither satisfied nor dissatisfied	39.20	36.38	36.53
Pleased	41.50	43.92	42.93
Very satisfied	48.40	45.88	48.22
Total	41.93	42.57	39.16

Table 17. Level of visitor satisfaction with some dependent variables, by sex

Dependent variables	Sum of squares	df	Mean square	Z	Sig.
Level of satisfaction in relation to the quality of sports activities practised	0.024	1	0.024	0.070	0.792
Level of satisfaction in relation to the quality of cultural activities carried out	0.785	1	0.785	1.619	0.205
Level of satisfaction regarding cleanliness/hygiene of public spaces	0.050	1	0.050	0.104	0.747
Level of satisfaction with the information provided (transport, routes, monuments, cultural agenda, etc.)	0.033	1	0.033	0.064	0.801
Level of satisfaction in relation to local/traditional commerce	0.000	1	0.000	0.000	0.982
Level of satisfaction in relation to tourist entertainment companies	0.754	1	0.754	1.188	0.278
Level of satisfaction in relation to places of interest (e.g. Interpretive infrastructures, interpretive centres, museum spaces, viewpoints, among others)	0.120	1	0.120	0.246	0.621
Level of satisfaction in relation to access routes	0.007	1	0.007	0.012	0.912
Level of satisfaction with public transport	0.131	1	0.131	0.261	0.611
Level of satisfaction in relation to routes and pedestrian routes (associated information, signs, road status, …)	0.001	1	0.001	0.002	0.962
Level of satisfaction in relation to communication networks (mobile, wireless)	0.011	1	0.011	0.017	0.896
Level of satisfaction in relation to medical care in the tourist territory (health technicians, mobile units and proximity to a medical centre)	0.001	1	0.001	0.001	0.974
Level of satisfaction in relation to the quality of sports activities practised	0.137	1	0.137	0.240	0.625
Level of satisfaction in relation to the quality of cultural activities carried out	0.221	1	0.221	0.375	0.542
Level of satisfaction regarding cleanliness/hygiene of public spaces	0.231	1	0.231	0.325	0.570
Level of satisfaction with the information provided (transport, routes, monuments, cultural agenda, etc.)	0.009	1	0.009	0.008	0.928
Level of satisfaction in relation to local/traditional commerce	0.531	1	0.531	0.666	0.416

Table 18. Visitor satisfaction with some dependent variables, by level of education

Dependent variables	Sum of squares	df	Mean square	Z	Sig.
Level of satisfaction with the visit / experience in the HVP	0.798	5	0.160	0.450	0.812
Level of satisfaction with security in the territory of HVP	2.406	6	0.401	0.825	0.552
Level of satisfaction with the price of catering establishments	0.686	6	0.114	0.234	0.965
Level of satisfaction in relation to the quality of beverage and similar establishments	1.570	5	0.314	0.587	0.710
Level of satisfaction in relation to the quality of the accommodation	1.984	5	0.397	0.809	0.546
Level of satisfaction in relation to the quality of sports activities practised	1.558	6	0.260	0.375	0.893
Level of satisfaction in relation to the quality of cultural activities carried out	1.845	5	0.369	0.767	0.576
Level of satisfaction regarding cleanliness/hygiene of public spaces	0.779	6	0.130	0.233	0.965
Level of satisfaction with the information provided (transport, routes, monuments, cultural agenda, etc.)	2.301	5	0.460	0.917	0.473
Level of satisfaction in relation to local/traditional commerce	0.812	5	0.162	0.326	0.897
Level of satisfaction in relation to tourist entertainment companies	1.325	5	0.265	0.421	0.833
Level of satisfaction in relation to places of interest (e.g. Interpretive infrastructures, interpretive centres, museum spaces, viewpoints, among others)	2.746	6	0.458	0.856	0.530
Level of satisfaction in relation to access routes	3.327	6	0.555	0.999	0.429
Level of satisfaction with public transport	3.063	5	0.613	1.064	0.386
Level of satisfaction in relation to routes and pedestrian routes (associated information, signs, road status, …)	8.538	6	1.423	2.139	0.054**
Level of satisfaction in relation to communication networks (mobile, wireless)	15.555	6	2.592	2.713	0.017**
Level of satisfaction in relation to medical care in the tourist territory (health technicians, mobile units and proximity to a medical centre)	3.297	5	0.659	0.831	0.531

Note: *** level of statistical significance of 1%; ** level of statistical significance of 5%, * level of statistical significance of 10%.

As regards the level of schooling, as can be seen in Table 18, there are discrepancies of statistical significance at the level of 5% in terms of the level of satisfaction with the roads and walking routes and the level of satisfaction with the contact networks.

It can be seen from the study in Table 19 that those with the 2nd cycle in basic education and higher education are the most comfortable in these questions.

With regard to motor difficulties, there are statistically important variations in satisfaction with the consistency of beverage establishments and the like (10 per cent significance level) (see Table 20).

As far as motor problems are concerned, tourists who do not have any motor problems are the most unhappy with the consistency of drinks and similar establishments.

As for nationality, as can be seen in Table 21, there are substantial variations in the quality of beverage establishments and the like (5 per cent statistical significance level), in relation to the information received

Table 19. Level of satisfaction of nHVP visitors, by level of education (average)

Education	Level of satisfaction in relation to routes and pedestrian routes (associated information, signs, road status, ...)	Level of satisfaction in relation to communication networks (mobile, wireless)
Without education	3.00	3.00
Basic – 1st cycle	2.00	100
Basic – 2nd cycle	4.00	3.00
Basic – 3rd cycle	3.20	2.86
Secondary / Post-secondary	3.59	3.33
University	3.72	3.55
Total	3.57	3.31

Table 20. Visitor satisfaction with some dependent variables, taking into account motor difficulties

Dependent variables	Sum of squares	df	Mean square	Z	Sig.
Level of satisfaction with the visit / experience in the HVP	0.406	2	0.203	0.580	0.561
Level of satisfaction with security in the territory of HVP	1.697	2	0.848	1.777	0.173
Level of satisfaction with the price of catering establishments	1.977	2	0.988	2.122	0.124
Level of satisfaction in relation to the quality of beverage and similar establishments	2.581	2	1.290	2.510	0.085*
Level of satisfaction in relation to the quality of the accommodation	1.684	2	0.842	1.751	0.178
Level of satisfaction in relation to the quality of sports activities practised	0.799	2	0.399	0.595	0.554
Level of satisfaction in relation to the quality of cultural activities carried out	0.042	2	0.021	0.043	0.958
Level of satisfaction regarding cleanliness/hygiene of public spaces	0.859	2	0.430	0.796	0.453
Level of satisfaction with the information provided (transport, routes, monuments, cultural agenda, etc.)	0.555	2	0.278	0.551	0.578
Level of satisfaction in relation to local/traditional commerce	0.020	2	0.010	0.021	0.980
Level of satisfaction in relation to tourist entertainment companies	0.184	2	0.092	0.148	0.862
Level of satisfaction in relation to places of interest (e.g. Interpretive infrastructures, interpretive centres, museum spaces, viewpoints, among others)	0.508	2	0.254	0.474	0.624
Level of satisfaction in relation to access routes	0.308	2	0.154	0.274	0.760
Level of satisfaction with public transport	0.722	2	0.361	0.620	0.541
Level of satisfaction in relation to routes and pedestrian routes (associated information, signs, road status, ...)	0.344	2	0.172	0.241	0.786
Level of satisfaction in relation to communication networks (mobile, wireless)	0.278	2	0.139	0.132	0.876
Level of satisfaction in relation to medical care in the tourist territory (health technicians, mobile units and proximity to a medical centre)	1.325	2	0.662	0.840	0.435

Note: *** level of statistical significance of 1%; ** level of statistical significance of 5%, * level of statistical significance of 10%.

(5 per cent statistical significance level), in relation to access routes (1 per cent statistical significance level) and in relation to communications (5 per cent statistical significance level).

As regards the price of the beverages and similar establishments, the most pleased are the Spanish guests and the least pleased are those of a certain nationality which is not shown in Table 22. As far as the information given is concerned, people of 'other ethnicity,' a minority of visitors/tourists, are the most pleased. The English are the least happy of all. The French who frequent the nHVP are the least pleased in terms of access routes and in terms of roads and pedestrian paths, with the "other" group

Table 21. Visitor satisfaction with dependent variables, by nationality

Dependent variables	Sum of squares	df	Mean square	Z	Sig.
Level of satisfaction with the visit / experience in the HVP	1.126	5	0.225	0.640	0.669
Level of satisfaction with security in the territory of HVP	2.747	5	0.549	1.144	0.340
Level of satisfaction with the price of catering establishments	3.333	5	0.667	1.430	0.217
Level of satisfaction in relation to the quality of beverage and similar establishments	5.786	5	1.157	2.312	0.048**
Level of satisfaction in relation to the quality of the accommodation	4.185	5	0.837	1.776	0.123
Level of satisfaction in relation to the quality of sports activities practised	6.619	5	1.324	2.098	0.072*
Level of satisfaction in relation to the quality of cultural activities carried out	1.973	5	0.395	0.822	0.537
Level of satisfaction regarding cleanliness/hygiene of public spaces	1.388	5	0.278	0.506	0.771
Level of satisfaction with the information provided (transport, routes, monuments, cultural agenda, etc.)	5.743	5	1.149	2.428	0.039**
Level of satisfaction in relation to local/traditional commerce	0.540	5	0.108	0.215	0.955
Level of satisfaction in relation to tourist entertainment companies	1.476	5	0.295	0.471	0.797
Level of satisfaction in relation to places of interest (e.g. Interpretive infrastructures, interpretive centres, museum spaces, viewpoints, among others)	2.843	5	0.569	1.073	0.379
Level of satisfaction in relation to access routes	12.031	5	2.406	4.987	0.000***
Level of satisfaction with public transport	1.946	5	0.389	0.661	0.654
Level of satisfaction in relation to routes and pedestrian routes (associated information, signs, road status, …)	6.881	5	1.376	2.042	0.078*
Level of satisfaction in relation to communication networks (mobile, wireless)	12.106	5	2.421	2.480	0.036**
Level of satisfaction in relation to medical care in the tourist territory (health technicians, mobile units and proximity to a medical centre)	6.968	5	1.394	1.851	0.111

Note: *** level of statistical significance of 1%; ** level of statistical significance of 5%, * level of statistical significance of 10%.

Table 22. Level of satisfaction of nHVP visitors, by nationality (average)

	Visitors nationality				
	Portuguese	English	Spanish	French	Other
	Average	Average	Average	Average	Average
Level of satisfaction in relation to the quality of beverage and similar establishments	4.17	4.29	4.69	4.40	3.89
Level of satisfaction with the information provided (transport, routes, monuments, cultural agenda, etc.)	3.87	3.86	4.00	4.20	4.25
Level of satisfaction in relation to access routes	3.77	3.57	3.38	3.20	4.45
Level of satisfaction in relation to routes and pedestrian routes (associated information, signs, road status,…)	3.49	3.71	3.55	3.20	4.38
Level of satisfaction in relation to communication networks (mobile, wireless)	3.13	3.57	3.33	3.80	4.22

being the most pleased. With regard to contact networks, Portuguese tourists in nHVP are those who are least pleased, with those with the highest degree of satisfaction being those of 'other nationality'.

4. DISCUSSION AND FINAL CONSIDERATIONS

HVPs are located in designated low-density areas. These villages have lost their inhabitants, and their economic structure is very poor. Nevertheless, they have tremendous potential in terms of cultural, landscape and architectural heritage, and are highly desirable in terms of tourism. Tourism is expected to be a strategic driver for growth and sustainability in these villages, and the HVP will seek to be appealing to both residents and tourists. In this regard, it was considered necessary to understand the degree of satisfaction of both residents and visitors regarding HVP tourism.

As far as residents are concerned, we have tried to gauge their level of satisfaction with the impact of tourism in their area, their affiliation with the HVP brand, their pride in belonging to the HVP community and culture, their willingness to recommend other people to visit the HVP, and their cooperation in promoting tourist activities and events in the area. As far as tourists are concerned, we have tried to evaluate the degree of satisfaction with the HVP in relation to a number of variables.

The findings obtained indicate that the bulk of the nHVP population is elderly and without higher education. The number of people who make up the household is below the national average, which brings into question the revival of the population of these villages, reflecting, with aggravation, the national trend. In addition, about 1/5 of the population live on their own, creating problems in terms of defence, health and the protection of the territories from environmental disasters (such as fires).

In the general assessment of the results (based on the responses to the questionnaire surveys) and of the residents, it should be noted that the majority are satisfied or very satisfied with the impact of tourism on their community, as well as with the association with the HVP brand.

In general, the population living in the HVP has a medium/high level of satisfaction with the effect of tourism on the respective village and also with the association of the village with the HVP brand, which is very positive. Residents still feel proud to be part of the local community and culture, but just 33 per cent of them are interested in the growth of tourist events in the area, partially due to the advanced age of their residents.

As far as the profile of the residents is concerned, those with an average age of about 50 years are those who appear to be working together to facilitate tourism activities and events in the village. Those with more qualifications are those who choose to visit the HVP and those who engage most in the dynamization activities of the area.

As for tourists, the average age is about 40 years and those with the lowest level of education are the most likely to visit the nHVP, with the focus on tourists with a higher level of education. The majority of tourists in the HVP are Portuguese, followed by English and Spanish.

With regard to visitors, different behaviours were identified, namely:

- the main reason for the visit is the historical and cultural heritage of the region, followed by interaction with nature and rurality;
- The overall satisfaction with the nHVP experience is high, although in some respects the satisfaction is lower than the overall satisfaction (e.g. the security of the tourist destination, the quality

of catering and accommodation facilities, sports and cultural activities, access routes and communication networks);

- At the end of the experience, the opinions of the respondents are divided between those who would recommend a visit to the nHVP (50 per cent) and those who intend to repeat the visit (49 per cent). These data should be reflected by policymakers, serving as indicators to understand which areas to improve.

As far as visitor satisfaction is concerned, they are generally satisfied and very satisfied with the HVP visited. The highest level of satisfaction of HVP visitors is related to visits/experience in general in HVP, followed by security in the territory and quality of accommodation and drinking establishments and the like. Visitors are less satisfied in relation to the public means of transport available, the communication networks (mobile, wireless) and the medical care available in the tourist territory (health technicians, mobile units and proximity to medical offices).

As far as the profile of visitors is concerned, it appears that visitors who are less satisfied with the routes and walking routes and communication networks are those with an average age of around 50 years. Those with the 2nd cycle of basic education and higher education are the most satisfied in relation to these issues. In terms of medical care in the tourist area (health technicians, mobile units and proximity to medical centres), the youngest are the most dissatisfied.

As far as the quality of drinks and similar establishments is concerned, the most satisfied are the Spanish visitors and the least satisfied are those of 'other nationality'. As far as the information provided is concerned, visitors of 'other nationality' are the most satisfied, with the English being the least satisfied. The French who visit the nHVP are the least satisfied in terms of access routes and in terms of routes and pedestrian routes, with the "other nationality" remaining the most satisfied. With regard to communication networks, Portuguese visitors to nHVP are those who are least satisfied and those with the highest level of satisfaction are those of 'other nationality'.

In view of the results obtained, it is expected that this study on residents and visitors to the nHVP will be a useful tool for the development and design of tourism strategies and policies that, on the one hand, contribute to reversing the trend of desertification and, at the same time, provide the conditions for sustainable development. It may further contribute to the reversal of the trend of desertification, fostering the attractiveness of these territories and retaining the resources and surpluses that otherwise tend to leave the region.

ACKNOWLEDGMENT

This work is financed by the budget of the Regional Operational Program of the Centre in its ERDF component with the reference CENTRO-01-0145-FEDER-023984.

UA

Polytechnic of Guarda – Research Unit for Inland Development (IPG-UDI)

AAHP

REFERENCES

Almeida, A. M. (2017). Rural development and rural tourism: An institutional approach. *European Journal of Applied Business and Management, 3*(1), 101–117.

Bohlin, M., Brandt, D., & Elbe, J. (2016). Tourism as a vehicle for regional development in peripheral areas–myth or reality? A longitudinal case study of Swedish regions. *European Planning Studies, 24*(10), 1788–1805. doi:10.1080/09654313.2016.1194807

Çakar, K., & Uzut, İ. (2020). Exploring the stakeholder's role in sustainable degrowth within the context of tourist destination governance: The case of Istanbul, Turkey. *Journal of Travel & Tourism Marketing, 37*(8-9), 8–9, 917–932. doi:10.1080/10548408.2020.1782307

Carson, D. A. (2018). Challenges and opportunities for rural tourism geographies: A view from the 'boring'peripheries. *Tourism Geographies, 20*(4), 737–741. doi:10.1080/14616688.2018.1477173

Dupeyras, A., & Maccallum, N. (2013). *Indicators for Measuring Competitiveness in Tourism - a Guidance Document*. OECD Tourism Papers. doi:10.1787/5k47t9q2t923-

Gill, A. M., & Williams, P. (2011). Rethinking resort growth: Understanding evolving governance strategies in Whistler, British Columbia. *Journal of Sustainable Tourism, 19*(4/5), 629–648. doi:10.1080/09669582.2011.558626

Liu, Y. (2016). *Tourism Destination Governance The case study of Hemavan and Tärnaby* (Master Thesis). UMEA Universitet.

Melo, A. I., Ferreira, A. d., Santos, C. A., Augusto, D., Gomes, G., Azevedo, G. M., & Marques, R. (2020). *Projeto PlowDeR: Relatório Final*. UA Editora – Universidade de Aveiro.

Petrevska, B., & Gerasimova, V. M. (2012). Tourism in Regional Development: Empirical Evidence. *Innovative Issues and Approaches in Social Sciences., 5*(2), 6–20. doi:10.12959/issn.1855-0541.IIASS-2012-no2-art01

Ramos, G. M., & Fernandes, J. L. (2016). Tourism territories in low density areas: The case of Naturtejo geopark in Portugal. *Journal of Tourism. Heritage & Services Marketing, 2*(1), 14–21.

Salvatore, R., Chiodo, E., & Fantini, A. (2018). Tourism transition in peripheral rural areas: Theories, issues and strategies. *Annals of Tourism Research, 68*, 41–51. doi:10.1016/j.annals.2017.11.003

Santos, C., Melo, A., Ferreira, A., Augusto, D., Gomes, G., Azevedo, G., . . . Marques, R. (2019). PLowDeR - Framework para Análise do Impacto Económico e Social das Atividades Turísticas nos Territórios de Baixa Densidade: o Caso das Aldeias Históricas de Portugal. Aveiro.

Umbelino, J. (1998). Turismo em espaço rural: da utopia à realidade. In Turismo Horizontes Alternativos (pp. 175-188). Lisbon: Colibri.

Valeri, M. (2015). Sustainability development and competitiveness of Rome as a tourist destination. *Tourism and Hospitality Management, 21*(2), 203–217. doi:10.20867/thm.21.2.7

Vieira, I. C. (2014). *Desenvolvimento turístico sustentável em pequenas cidades históricas: O caso de Lamego* (Tese de Doutoramento). Vila Real: UTAD.

Zahra, A. (2011). Rethinking regional tourism governance: The principle of subsidiarity. *Journal of Sustainable Tourism, 19*(4), 535-552. doi:.576764 doi:10.1080/09669582.2011

ENDNOTE

1 ILDA – digital platform to support the purposes of the *PLowDeR* project. The ILDA platform allows the various stakeholders to collect information on the indicators that are most useful for monitoring, namely, the evolution of the business, the profile and the satisfaction of visitors, making it possible, if necessary, to adjust the strategy to their profile (Santos, et al., 2019).

Chapter 11
Role of Stakeholders and Perceived Destination Competitiveness Towards Ethnic Enclave Destination Management:
An Overview of "Little India" in Malaysia

Munieleswar Raju
Taylor's University, Malaysia

Puvaneswaran Kunasekaran
Taylor's University, Malaysia

ABSTRACT

The focus of this research is on the role of stakeholders and perceived destination competitiveness in ethnic enclave destination management. Ethnic enclave destination is one of the unique tourism products in terms of art, culture, heritage, and lifestyle. In Malaysia, Little India, Brickfields is one of the popular Indian enclave destinations significantly contributing to the development of tourism in Malaysia. The perceived destination competitiveness determines the destination attributes, and the destination management is ensured by the stakeholders. However, there are just a few pieces of research that look at the relationship between perceived destination competitiveness and destination management. As a result, this study used an integrated model of TDC to examine the mix of stakeholders and perceived destination competitiveness in Little India, Brickfields. This study will have a significant impact on ethnic enclave destination management by identifying stakeholders and their roles in boosting destination competitiveness at the destination of concern.

DOI: 10.4018/978-1-6684-3889-3.ch011

INTRODUCTION

Ethnic enclave destination is one of the unique tourism products in terms of art, culture, heritage, and lifestyle. This ethnic enclave represents the geographical space where the minority immigrants resided and settled at a specific location by showing their traditional and authentic lifestyle. In Malaysia, Little India is a famous tourist destination where visitors may experience ethnic cuisine, buy for unique goods, and learn about a minority culture. Little India, Brickfields is one of the popular Indian enclave destinations in Malaysia significantly contributing to the development of tourism in Malaysia. The popularity and success of this destination rely on some related destination attributes and strategic destination management. The perceived destination competitiveness determines the destination attributes, and the destination management is ensuring by the stakeholders.

AN OVERVIEW OF ETHNIC ENCLAVE TOURISM DESTINATION

One of the world's fastest-growing industries is tourism (Stankova and Amoiradis, 2019; Ahani et.al.,2019). This industry is vital not just for economic reasons but also for its social development and monitoring functions. The tourism sector contributes significantly to employment creation and foreign exchange profits (Bhatt, Pandya & Modi, 2019; Matthew et.al.,2021). In 2019, Malaysia welcomed 26,100,784 international tourists, bringing in a total of RM86.14 billion in tourism revenues (Chin, Razali & Yong 2020). According to a recent annual rating by MasterCard Inc, Kuala Lumpur is the world's fifth most-visited city (Reuters, 2019). Like in other large cities, tourism plays an integral part in Kuala Lumpur's economic life, generating income, jobs, and opportunities for the industry. The tourism industry has a wide range of infrastructure and amenities that it requires and jobs in several sectors of the community (DBKL, 2019).

"With global trends pointing to experiential tourism as a preferred method of travel, travellers want to experience a site that provides the local community a rich cultural history and excellent artistic expressions" says former Deputy Tourism, Arts, and Culture Minister Muhammad Bakhtiar Wan Chik. During the recent launch of Tourism Malaysia's Kuala Lumpur (KL) Heritage Food Trails package 2019, he said that new initiatives would continue to bring visitors to places like Kampung Baru Petaling Street, and Brickfields to pursue authentic local experiences. As a result, an ethnic enclave tourism destination is a good fit for meeting visitors' needs. This destination is also an essential element of the accessible tourism offerings and Malaysia's significant contributors. As a result, ethnic enclave tourism contributes significantly to the development of Kuala Lumpur (Samad et.al.,2021). Ethnic enclave attractions are regarded as one of the world's most popular tourist destinations (Valenzuela-Garcia, Parella, & Güell, 2017; Kim, 2018; Woosnam et.al.,2019). This ethnic enclave has grown into an intriguing tourism attraction that emphasizes the destination's cultural similarities through social and economic factors (Santos & Yan, 2008). The names "Little" or "Town" are commonly used to describe enclave attractions such as "Little India" or "Chinatown" (Khan, 2015).

This ethnic enclave has been the subject of numerous successful studies. According to George (2010), ethnic tourism refers to tourists visiting regions with intangible cultural heritages. He also stated that local intangible cultural assets provide significant economic benefits to the tourism industry. According to his research, culture is the most compelling driving force for a market segment that looks to be growing in both demand and supply in the global tourism sector, and these cultural features are being translated into commercial products for exchange. Ethnic enclave tourism can provide economic and social benefits that substantially impact ethnic groups' traditional values, lifestyles, and sense of identity (Yang and Wall, 2016; Wang, Shen & Ye, 2020). Ethnic tourism has expanded beyond the 'minority community' boundary geography and everyday life, where community culture and lifestyle are significant tourist draws (Oakes, 2016; Gao et.al.,2019). However, the majority of tourism research focuses on rural ethnic tourism (Chio, 2017; Lor, Kwa, & Donaldson, 2019; Solikhah, 2020). Ethnic studies can be carried out in urban areas as part of ethnic enclave tourism. In other words, ethnic enclave tourism refers to travel to an urban place where the local culture serves as an attraction and aids in the economic development of the destination.

According to Lin (2013), enclaves are secure resources since they provide a regular, consistent, and efficient source of revenue from tourism throughout the year. He investigates a land-use dispute between an ethnic enclave and a tourist enclave near Sun Moon Lake in Taiwan in his study. His findings are primarily focused on ethnic and tourist enclaves. As a tourist enclave with no defined borders in a fast-developing territory, he noted that social prejudice and exclusion between local Taiwanese and aborigines (Thao people) and their descendants are observable. He also stated that it is a classic example of how "ethnic enclaves" have evolved into "tourism enclaves," as evidenced by real estate investors, corporate associations such as hotel agencies, and city government outreach programs spiralling and altering the local landscapes of the Sun Moon Lake. Although this study focused on the distinctions between ethnic enclaves and tourism enclaves, there is no harm in sharing resources at a destination with ethnic enclave tourism destinations if sustainable practices are used.

Woosnam et al. (2019) proposed a new concept of Ethnic Neighborhoods Tourism, linked to the commodification of ethnic cultures that are not easily found in other city sections or countries. They went on to say that, while it allows people to have "genuine" ethnic experiences without having to fly abroad, it may also be a valuable tool for immigrants' socio-economic growth. As a result, increased support for tourism growth in general and recognition of tourism's value to the city translates into greater awareness of ethnic neighbourhood tourism, which creates jobs for minority inhabitants and the necessity to participate in tourism planning and development. As a result, tourist planners in ethnic enclaves should educate minority populations about the importance of their ethnicity and encourage cross-cultural awareness and understanding rather than constantly prioritizing economic development (Hardy, Beeton, & Pearson, 2002).

'LITTLE INDIA' BRICKFIELDS AS AN ETHNIC ENCLAVE DESTINATION

Ethnic enclave tourism contributes significantly to the development of Malaysia's tourism industry. In Malaysia, significant towns such as Kuala Lumpur, Penang, Ipoh, Malacca, and Johore have deliberately enforced ethnic enclave attractions in the form of urban ethnic tourism and known as "Little India" (Bachek, Zainudin, & Haron, 2014; Rahman, Halim, & Zakariya, 2018; Hassan & Ali, 2019). Little India is a top-rated attraction in all cities, captivating visitors with the allure of Indian cuisines, colourful apparel, beautiful jewellery, and a vibrant lifestyle (Henderson, 2008; Kurtz, 2017; McKercher & Koh, 2017). When locals and visitors of Indian ancestry from Malaysia and other countries continue to visit this attraction, the Little India ethnic enclave becomes a must-visit tourist destination (Campus, Long, & Long, 2017; Xiaomei and Daming, 2018).

Little India, Brickfields, in notably, have hundreds of years of history as residential districts for their respective ethnic communities, but they are today well-known for being commercial areas that cater to both the local community and visitors (Hall and Barrett, 2018; Nayan, Jones, & Ahmad, 2019). Little India, also known as Brickfields, dates to the British colonial administration when the area was famed for its brickmaking. It is just south of the city centre, near Kampung Attap to the north, Bangsar to the west, Taman Seputeh to the east, and Mid-Valley Megamall to the south. Many religious structures are visible in Brickfields. Some of them are well over a century old. (Bachek et al., 2014; Zainudin, Bachek, & Haron, 2016). Furthermore, Brickfields was the location of the central depot for 'Keretapi Tanah Melayu' (KTM). Workers from Sri Lanka and India were brought in to work for Malayan Railway and were accommodated in this destination's railway quarters (Yahya, Safian, & Burhan, 2020; Khoo, 2021. Moreover, Sri Kandaswamy Temple, a 115-year-old Hindu temple and 117-a year old Viveanda Ahramam are located here (Rajantheran, 2009; Rajantheran et.al.,2014; Basha, 2016; Willford, 2020)

In October of 2010, Brickfields was renamed Little India. Many Indian ex-pats live in high-rise condominiums and apartments along Jalan Sultan Abdul Samad, making Brickfields a desirable location for the local Indian population and many Indian expatriates (King, 2011; Zamri, 2020). In 2016, Brickfields was placed third on Airbnb's top trending locations list. The real estate value in this location increased by 7,900% to 9,900%, approximately from Rm40,000 to Rm4 million. Many of the shops in Brickfields are owned by Indian merchants. Jalan Tun Sambanthan, formerly known as Brickfields Road, is lined with businesses providing Indian groceries, traditional meals, sweets and snacks, textiles, jewellery, garlands, and whatever else an Indian family requires. Brickfields "Little India" has become a tourist attraction in Kuala Lumpur because of the items on offer and its proximity to KL Central Station (Baxstrom, 2009; Khalid et al., 2019; Baxstrom, 2020).

ISSUES IN 'LITTLE INDIA' BRICKFIELDS

Although 'Little India' is exciting, vibrant, and happening, the ethnic enclave's delight is limited due to its geographical location, where space is limited, and boundaries are narrow. This lack of space eventually leads to crowding and jamming, which leaves tourists disappointed (Khan, 2015; Kumar, 2020). In addition, visitors to 'Little India' have expressed concern that, although the destination has a unique traditional image and culture, there is a lack of greenery, parking, and tourist facilities in the neighbourhood, which is preventing them from returning (Ujang, Moulay and Zakaria, 2019). Without a suitable restoration plan, the area will deteriorate and lose its allure, making sustained ethnic enclave tourism challenging to achieve (Omar et.al.,2018; Ramlee et.al.,2018).

According to a study on cultural tourism in Malaysia, tourists are unaware of Malaysia's cultural wealth, the country's image as a destination for cultural tourism has not been fully realized, and non-culture attractions have attracted tourists' interest (Sudipta and Sarat, 2010; Wong & Lau, 2016; Murugan, Perumal, & Kamarudin, 2021). As a result of the lack of cultural exoticism, tourists' intentions and motivation to visit ethnic enclave sites have been influenced. Other South Asian migrants, such as Pakistanis, Bangladeshis, and Sri Lankans, are increasingly visible in the Little India enclave (Khan, 2015; Rajan, & Percot, 2020; Ullah et.al., 2021). This infiltration produces cultural conflict among local communities to promote the destination as a "Little India" by employing Indian cultural features explicitly.

All related stakeholders, such as destination providers in Malaysia's need to enhance and maintain their involvement in the tourism sector particularly in sustainable tourism practices (Mohaidin, Wei, and Murshid, 2017; Yunus, Samadi, & Omar, 2018). The local communities' participation in developing countries such as Malaysia continues to face institutional and organizational challenges, leading to limited opportunities for them to participate in tourism planning and development processes (Marzuki, Hay and James, 2012; Wondirad, Tolkach, & King, 2020). Although 'Little India' is known for ethnic enclave tourism, it is a place of work and business. The primary focus always on food and beverage and retail activities, sales offices and specialist suppliers such as remittance traders and travel agencies (Henderson, 2008; Monks-Husain, 2020). The promotional campaign always take charge by the local tourism board, municipal councils, local travel agencies and event management companies (Mokhtar, Yazid, & Shamsudin, 2020). Based on that, we can see there is lack of participation of local communities in promoting their place as tourist centric rather than busy with their business trading and works.

The strategic cornerstone of tourism growth is community empowerment (Khalid et.al.,2019; Vukovic et.al.,2021). In tourism development, local communities must have the knowledge, skills, and abilities to participate in long-term tourism development activities. According to Ramakrishnan (2011), the local government organized a task group in 2010 to look into the contentious concerns presented in 'Little India.' On the other hand, the task force lacks urban development professionals and planners to examine the physical and human traits required for Little India to succeed. If the government continue to overlook it, according to Lovrenciear, 2017, this Indian enclave may soon follow in the footsteps of several other historical, cultural business districts in the city. As a result, it demonstrated a lack of empowerment offered to local populations in making decisions at the attraction. The traders in 'Little India' want to see the enclave business centre revitalized because it has a lot of tourism potential, mainly because there are many heritage and pre-war buildings within walking distance, which can provide significant economic transformation by providing more job opportunities for locals in tourism. As a result, tourist planners must engage local people in decision-making on long-term tourism development initiatives.

Cleanliness leads to strength and health (Chaabane et.al.,2019; Awan, Shamim, & Ahn, 2020). Cleanliness and effective hygiene measures should become part of the brand promise to reposition the destination's image in the eyes of tourists, assuring visitors (Park et al.,2019; Rumere, Wiranatha, & Pujaastawa, 2020). In the year 2019, Brickfields received a "C" for overall cleanliness. "In brickfields, the storefront was obviously clean, but it was unpleasant at the back alleyways," says Kuala Lumpur mayor Datuk Nor Hisham Ahmad Dahlan. Furthermore, restaurant owners are reckless and irresponsible by failing to supervise their employees who are improperly disposing of trash. The area's back lanes were constantly clogged with garbage and stench. Nor Hisham claimed that cleaning up the township was not a one-person task. It must be a coordinated and integrated effort of all stakeholders (Tawil et.al.,2021; Rumere, Wiranatha, & Pujaastawa, 2020). However, business owners in Little India Brickfields are rarely seen participating to solve this issue.

ROLE OF STAKEHOLDERS IN ENHANCING DESTINATION COMPETITIVENESS OF 'LITTLE INDIA'

Anyone who is involved with a project, organization, or industry is referred to as a stakeholder (Jamal & Getz, 1995; Bouwen & Taillieu, 2004; Hovelynck et.al.,2020). To put it another way, they have a metaphorical "stake" in the management (Orts & Strudler,2009; Kovačić, Jovanović & Šagovnović, 2019). A tourism stakeholder can be a single person, such as a visitor or a taxi driver. They could be a group of people, such as a family or a student group and could be a business or an organization (Murtagh et.al.,2017; Knecht, 2017 Su, Wall & Ma, 2019). Stakeholder analysis required for any firm, whether in tourism or another industry, to ensure that it is successfully considering the needs and requirements of stakeholders. Stakeholder analysis identifies and categorises significant stakeholders based on their level of involvement, interest, and influence (Abdullah et.al.,2018; Acuti at.al.,2019). The organization must then figure out how to work with and satisfy these stakeholders. Theoretically, if an organization works with all of its stakeholders' interests and requirements in mind, it will be more likely to satisfy them and hence conduct effective business operations.

One of the most important industry in the world is tourism (Sofronov, 2018; Kyrylov et.al.,2020; Khurramov, 2020). Means that many people, groups, and organizations, such as the government, tourism organizations and operators, small and medium businesses (SMBs), non-governmental organizations (NGO), tourists or visitors, communities, and suppliers, are all involved in some way (Kapera, 2018; Vu & Ngo, 2019; Pagliara et.al.,2021). Stakeholders have been characterized by as external, secondary, or primary, depending on the type of participation (Cardwell, Williams, & Pyle, 2017; Van der Zee, Gerrets, & Vanneste, 2017; Urbinati et.al.,2020). Tourism industry operators, developers, investors, communities, and visitors are the primary stakeholders. Government, state government, municipalities, non-governmental organizations, and domestic tourism suppliers are examples of the secondary category. International organizations, financial institutions, international traders, and suppliers are the final external stakeholders. Thus, Ministry of Tourism, Local Tourism Board of Kuala Lumpur, Kuala Lumpur City Hall (DBKL), local business traders, local communities, and visitors are the stakeholders related to Little India Brickfields.

Malaysian government play a vital role in Little India, Brickfields since year 2010. Former Malaysian Prime Minister Datuk Seri Najib Tun Razak's objective of developing and conserving the community's interests and ensuring that Malaysian Indians are not left out of mainstream growth spurred the makeover of Brickfields. Since Brickfields was renamed Little India eight years ago, the Indian community's socioeconomic status has improved dramatically, owing to the government's attempts to solve challenges that have plagued the community (Singh, 2011; Nayan, Jones, & Ahmad, 2019). According to one of the restaurant owners, many tourists flock to Little India because they appreciate Indian food, the spices, and the authenticity of the dish. Furthermore, one of Brickfields' vendors remarked that there is no good location to conduct business and that Brickfields is the best option. A store owner selling prayer materials and other necessities also commended the government besides giving loans and help to enable Indians to become businesses and enhance their level of living in Little India. The government has showed us some options to raise the Little India's local communities living standards. They also provide some financial aids with loans and help through the Indian Community Entrepreneur Development Scheme (SPUMI), 1Malaysia People's Aid (BR1M), and Amanah Saham 1Malaysia, all of which are part of the National Entrepreneurial Group Economic Fund (TEKUN) (AS1M) (Zainol et.al.,2017; Thuraisingham & Valu, 2020).

Apart from that, in 2017, Indian Prime Minister Narendra Modi and former Malaysian Prime Minister Najib Razak jointly opened the Torana Gate in Kuala Lumpur's "Little India." The Torana Gate project, which was started by former Indian Prime Minister Manmohan Singh and was expected to be finished by 2010, was five years late. The gate was given as a gift from India to commemorate the opening of Little India in the Brickfields neighborhood of Kuala Lumpur as a sign of friendship between the two nations. Torana Gate, according to Prime Minister Modi, would be a landmark of India-Malaysia relationship. This gate serves as a link between two countries. He went on to say that Indian visitors go to

Kuala Lumpur in big numbers and that they invariably visit Little India. As a result, Torana Gate has remained a key attraction of the location to this day. (Avineshwaran, 2015; NST Online. 2015).

The local authority at the destination is also plays a vital role in developing Little India as tourist destination. Brickfields is a Kuala Lumpur's neighborhood designated as a historic site in the Kuala Lumpur City Plan 2020 (Isa et.al.,2020; Yin & Abdullah, 2020). The colonial administration, an Indian population has lived here and transformed it into an Indian ethnic enclave destination (Chung, 2018; Baxstrom, 2020). Brickfields has emerged as a prospective urban development region due to its strategic position. The government's improvement in enhancing Brickfields' image in 2010 has also aided in marketing its tourism elements. The Kuala Lumpur City Hall attempted their local plan in 2008 by considering heritage zones by the National Heritage Act 2005 to protect old structures that have architectural and historical importance. Moreover, Kuala Lumpur City Hall (Dewan Bandaraya Kuala Lumpur) (DBKL) oversees city planning, transportation and traffic management, economic development, street lighting, river and drainage maintenance, construction and maintenance of public housing, provision of public facilities, cultural and sporting activities, environmental protection, and disbursement of government funds.

Traffic management is one of DBKL's frequent activities in Little India, Brickfields. DBKL is always enforcing parking rules in that region, although an appropriate parking facility are already available. As previously discussed in the issues, the space in Brickfields is extremely crowded, with just a little amount of space accessible for transit accessibility and movement (Zamri, N. A. (2020; Yap, Chua, & Skitmore, 2021). The double-parking problem in Brickfields is one of those annoying behaviors that may cause traffic congestion and inconvenience other vehicles (NST Online.,2017; Ravindran, 2017). Since, additional parking spaces have been created near Tun Sambanthan Complex and Kompleks TLK Brickfields, DBKL has always taken strong action against these double parkers. Apart from that, DBKL is also constructing adequate space for shops, restaurants, and small markets to serve the local Indian population and others who visit Jalan Tun Sambanthan to dine or shop. As a result, DBKL is doing its part to alleviate traffic congestion and make this location more appealing for seamless operations.

Meanwhile, local business traders, residents, and visitors play an important part in the functioning of Little India, Brickfields as major stakeholders. Local business traders guarantee that Indian culture's uniqueness and authenticity are constantly preserved. They founded the Brickfields Business Community Society (BBCS) to look after the local traders' welfare and seek justice in Little India' Brickfields (Priya., 2010; The Star, 2015). This society has resolved many complications between the authorities and the traders at this location. A one-day protest against illegal foreign traders at the specified destination was one of the latest actions performed by the society. More than a hundred businesses along Jalan Tun Sambanthan were shuttered for the day because local shopkeepers in 'Little India' alleged that foreign dealers were hurting their local business. The shopkeepers also protested against foreigners participating in Indian shopping carnivals and trade fairs organized by a Kuala Lumpur-based events management firm.

In addition, BBCS president Barath Maniam also stated that the merchants and shopkeepers in the Indian business enclave were angry that no action had been taken against illegal immigrants selling clothes and fashion accessories openly. He added that the foreigners, particularly from India, Pakistan, and Bangladesh, offer similar items with local vendors; however, some of their items are of lesser quality and are somewhat cheap. This declaration demonstrates that the society is concerned about the quality of business and the well-being of local traders. This initiative taken by the society will allow the ethnic enclave idea to be preserved for the long term, and tourists will have a better experience. The maintenance of authenticity is critical at this enclave location, and the generated resources may be one of the elements that encourage people to visit and return.

PERCEIVED DESTINATION COMPETITIVENESS

According to numerous researches, tourists and their requirements are the ultimate driving force influencing tourism destination competition and competitiveness. Today, destinations compete for visitors' attention based on the quality of their tourism experiences (Zainuddin, Mohd Radzi,& Mohd Zahari, 2018). In this context, tourism competitiveness refers to a destination's capacity to attract visitors, increase expenditure, and give enjoyable and memorable experiences. It also adds to the well-being of destination people and the long-term preservation of the destination's natural resources (Patti, 2019; Fuentes, 2019).

Measuring tourism competitiveness is problematic because it is an artificial construct that is not immediately observable and has a multifaceted structure emerging from the total of factors, many of which are linked to other tourist destinations as a point of comparison (Martinez, Martín, & Fernández, 2020; Roman et.al., 2020). There are around five main competitiveness models that have been identified as being among the most important. These five separate models show the perceived destination competitiveness variables (Novais, Ruhanen, & Arcodia, 2018). The attempts of destinations to attract more tourists are indisputable proof of stated competitiveness; therefore, it is critical to employ destination competitiveness characteristics. As a result, the Integrated Model (IM) of Competitiveness will be suitable model to explain the perceived destination competitiveness as the primary reference framework (Dwyer et al., 2003; Nadalipour, Khoshkhoo & Eftekhari, 2019). In contrast to the Crouch and Ritchie model (1999), this IM theory depicts the relationships between various variables in destination competitiveness more accurately.

After a series of seminars in both Korea and Australia in 2001, which members of important tourist sector groups attended, government officials, and academics, the Integrated Model of TDC shown in Figure 1 was developed in a collaborative effort by experts from both countries (Dwyer & Kim, 2003). The model includes resources, destination management, demand conditions, and situational conditions, all essential contributors to perceived destination competitiveness. The elements of a destination that attract visitors and allow tourists to visit are referred to as Resources or Core Resources in the Integrated Model. It can be divided into three groups; Endowed (Inherited), Created and Supporting (Armenski et al., 2012; Azzopardi & Nash,2017; Neto et.al.,2020).

Destination management variables in integrated model are those that can improve the quality and efficiency of supporting elements and resources, as well as respond best to the limits imposed by the environment" (situational conditions). In terms of destination management, the Integrated Model distinguishes between governmental and private sector actions. The public sector has implemented a national tourist plan, national destination marketing, national and regional recruiting programs, environmental restrictions, and climate change conservation and management policies. Industry participation in and support of destination marketing initiatives, industry support services, industry acceptance of "green" tourism activities, and environmental certification schemes, among other things, are all considered private sector activities (Minguzzi, 2006; Boley & Perdue, 2012; Fyall & Garrod, 2019).

Figure 1.

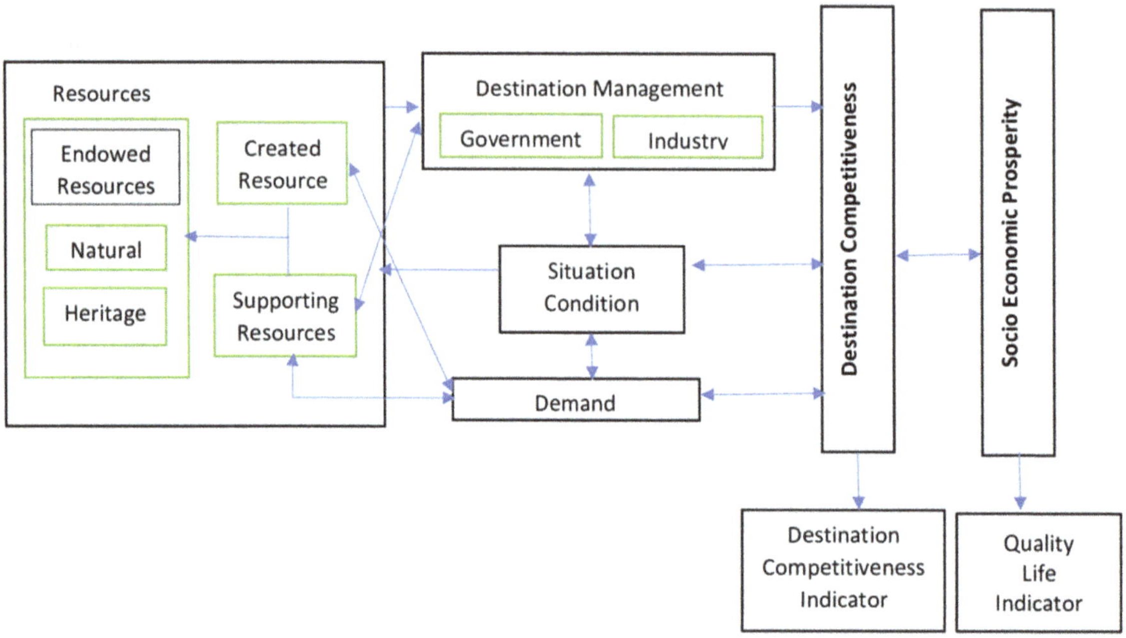

Furthermore, tourism demand is made up of three essential components: awareness, perception, and preferences. Various strategies, including destination marketing and market ties such as religious, sporting, ethnic, and so on, can be used to raise awareness. The perception of a place is influenced by its image, which has an impact on visitor numbers. The majority of tourism scholars think that destination image has a considerable impact on destination selection and general visitor purchasing and travel behaviour. Preferences are the links between visitors and the products that are thought to be available in their destination. Overall, demand conditions refer to the kinds of items and services developed anywhere in a location due to visitor preferences and travel motivations (Lopes, 2011; Pike, 2015; Pavlovi et al.,2016; Milićević et.al,.2020). Finally, Situational Conditions are extrinsic factors that influence visitor flows and are influenced by destination competitiveness. Organizations work in the destination's operational environment and remote environment, two interconnected settings. (Mulec & Wise, 2013; Dwyer et al., 2014; Ghaderi & Aarabi, 2020).

The perceived destination competitiveness of Little India is divided into three resources as described in IM Tourist Destination Competitiveness. Endowed (inherited), Created, and Supporting resources are the variables. Endowed resources can be classed as either natural (mountains, lakes, beaches, rivers, climate, and more) or cultural/heritage (art, cuisine, handicrafts, language, customs, architectural heritage and belief systems). Since this location's nature is located in a city, cultural or heritage-based resources has been chosen to represent inherited resources. Created resources include lodging, restaurants, guided tours, transportation, special events, and various activities, entertainment, and shopping accessible. Meanwhile, Supporting Resources are enabling the visitation by adding value to the visitors' experience, such as local people's hospitality, telecommunication systems, local government management, destination accessibility, and relationships with residents (Dwyer et al., 2004; Crouch 2011; Armenski et al., 2012; Fernández et al.,2020).

Finally, the Situational Conditions are extrinsic factors that influence visitor flows and are influenced by destination competition. The destination's operational environment and remote environment are two interconnected settings in which organizations operate. The operational environments of various private and public sector institutions at a destination are critical because the general structure of the industry in which they are based has a significant impact on the behaviour and performance of these institutions. This recognizes that the overall structure of the business environment impacts a company's conduct and performance. The remote environment refers to the external relationship between variables that limit organizations' or decision makers' strategic options but over which they have no control. As a result, situational conditions refer to economic, social, cultural, demographic, environmental, political, legal, governmental, regulatory, technological, and competitive developments and events that can impact how businesses and other organizations in the destination operate, posing both opportunities and challenges to their operations. Destination's ability to react to such changes has a substantial impact on their competitiveness as a destination. (Italian) (Dwyer et.al, 2009; Mulec & Wise, 2013; Dwyer et.al., 2014). All these variables are can be tested to understand the destination competitiveness level of Little India Brickfields and the evaluate he socio-economic prosperity of the local communities at this Indian Enclave.

THEORETICAL AND PRACTICAL IMPLICATIONS

This integrated model (IM) of tourism destination competitiveness (TDC) was widely used in past literate to identify the destination competitiveness based on demand perspective. This theory has also been widely employed in previous tourism and destination studies. The focus of this research is on the role of stakeholders and perceived destination competitiveness in ethnic enclave destination management. However, there are just a few researches that look at the relationship between perceived destination competitiveness and destination management. Furthermore, no precise research on the role of stakeholders at ethnic enclave destinations have been established. As a result, this study used integrated model of TDC to examine the mix of stakeholders and perceived destination competitiveness in Little India, Brickfields. Furthermore, comprehending the role of stakeholders in tourism is critical, as it will improve the destination management of a particular place. As a result, this study will have a significant impact on ethnic enclave destination management by identifying stakeholders and their roles in boosting destination competitiveness at the destination in concern. This study will give an indication to local authorities such as Kuala Lumpur City Hall, the Kuala Lumpur Tourism Bureau, and the Ministry of Tourism, Arts, and Culture to focus on ethnic enclave tourist destinations such as Little India, Brickfields as one of Malaysia's most important tourism products. Furthermore, this research is important because it will guide responsible parties such as the government, non-governmental organizations (NGOs), and the private sector in paying greater attention to ethnic enclave sites. As a result, it will directly or indirectly contribute to the promotion of 'Little India' in Malaysia, as well as provide a good opportunity for vendors at 'Little India' and tourism players to contribute to the national GDP.

CONCLUSION

The review shows that this Indian enclave is highly competitive with all resources available at the destination. All three destination resources are very important. The destination management operators should prioritize developing and promoting this place to attract more visitors and improve local communities (business traders and residents) lifestyle and socioeconomic well-being. The future research should also emphasize destination image and place attachment and how these factors influence the visitors' intention to this destination. Moreover, the research also should focus on both demand and supply side of the destination towards sustainable development.

REFERENCES

Abdullah, A. R., Weng, C. N., Afif, I., & Fatah, A. (2018). Ecotourism in Penang National Park: A multi-stakeholder perspective on environmental issues. *Journal of Business and Social Development*, *6*(1), 70–83.

Acuti, D., Grazzini, L., Mazzoli, V., & Aiello, G. (2019). Stakeholder engagement in green place branding: A focus on user-generated content. *Corporate Social Responsibility and Environmental Management*, *26*(2), 492–501. doi:10.1002/csr.1703

Ahani, A., Nilashi, M., Ibrahim, O., Sanzogni, L., & Weaven, S. (2019). Market segmentation and travel choice prediction in Spa hotels through TripAdvisor's online reviews. *International Journal of Hospitality Management*, *80*, 52–77. doi:10.1016/j.ijhm.2019.01.003

Armenski, T., Gomezelj, D. O., Djurdjev, B., Ćurčić, N., & Dragin, A. (2012). Tourism destination competitiveness-between two flags. *Economic research- Ekonomska Istrazivanja*, *25*(2), 485–502. doi:10.1080/1331677X.2012.11517519

Avineshwaran, T, (2015). *Modi to inaugurate 'Torana' Gate at Brickfields*. Available at: https://www.thestar.com.my/news/nation/2015/11/23/modi-torana-gate/

Awan, M. I., Shamim, A., & Ahn, J. (2020). Implementing 'cleanliness is half of faith'in re-designing tourists, experiences and salvaging the hotel industry in Malaysia during COVID-19 pandemic. *Journal of Islamic Marketing*.

Azzopardi, E., & Nash, R. (2017). A Review of Crouch and Ritchie's, Heath's, and Dwyer and Kim's Models of Tourism Competitiveness. *Tourism Analysis*, *22*(2), 247–254. doi:10.3727/108354217X14888192562483

Bachek, S. H., Zainudin, H., & Haron, N. A. (2014). Preservation of culture and built heritage in new urban development: A case study on Little India Brickfields, Kuala Lumpur. *International Proceedings of Economics Development and Research*, *74*, 1.

Basha, B. (2016). *Viability of adaptive reuse with social approach: case study of Asia Heritage Row, Kuala Lumpur* (Master's thesis). Gombak, Selangor: International Islamic University Malaysia.

Baxstrom, R. (2009). *Transforming Brickfields: Development and governance in a Malaysian city*. NUS Press.

Baxstrom, R. (2020). 1. The Founding of Brickfields and the Prewar Development of Kuala Lumpur. In *Houses in Motion* (pp. 25–53). Stanford University Press. doi:10.1515/9780804775861-003

Bhatt, P. J., Pandya, H. A., & Modi, N. R. (2019). *A review of possible impacts on tourism by climate change*. Academic Press.

Boley, B. B., & Perdue, R. R. (2012). Destination management, competitiveness, and quality-of-life: a review of literature and research agenda. Handbook of tourism and quality-of-life research, 515-528.

Bouwen, R., & Taillieu, T. (2004). Multi-party collaboration as social learning for interdependence: Developing relational knowing for sustainable natural resource management. *Journal of Community & Applied Social Psychology*, *14*(3), 137–153. doi:10.1002/casp.777

Campus, S. L., Long, J. S., & Long, B. S. (2017). *Malaysia in Indian Filmography: The Case of "Don" and "Kabali"*. Management (STHEM), Universiti Utara Malaysia.

Cardwell, L. A., Williams, S., & Pyle, A. (2017). Corporate public relations dynamics: Internal vs. external stakeholders and the role of the practitioner. *Public Relations Review*, *43*(1), 152–162. doi:10.1016/j.pubrev.2016.11.004

Chaabane, W., Nassour, A., Bartnik, S., Bünemann, A., & Nelles, M. (2019). Shifting towards sustainable tourism: Organizational and financial scenarios for solid waste management in tourism destinations in Tunisia. *Sustainability*, *11*(13), 3591. doi:10.3390u11133591

Chin, C., Razali, A., & Yong, J. (2020, September 12). *Malaysians Reveal Best Travel and Food Spots in Their Hometowns*. Academic Press.

Chio, J. (2017). Rendering rural modernity: Spectacle and power in a Chinese ethnic tourism village. *Critique of Anthropology*, *37*(4), 418–439. doi:10.1177/0308275X17735368

Chung, S. S. Y. (2018). The Coliseum Theatre, Kuala Lumpur: A Site for Collective Engagement, A Space of Urban Imagination. *Fabrications*, *28*(2), 212–234. doi:10.1080/10331867.2018.1446657

Crouch, G. I. (2011). Destination competitiveness: An analysis of determinant attributes. *Journal of Travel Research*, *50*(1), 27–45. doi:10.1177/0047287510362776

Dwyer, L., Cvelbar, L. K., Mihalič, T., & Koman, M. (2014). Integrated destination competitiveness model: Testing its validity and data accessibility. *Tourism Analysis*, *19*(1), 1–17. doi:10.3727/108354214X13927625340073

Dwyer, L., Edwards, D., Mistilis, N., Roman, C., & Scott, N. (2009). Destination and enterprise management for a tourism future. *Tourism Management*, *30*(1), 63–74. doi:10.1016/j.tourman.2008.04.002

Dwyer, L., & Kim, C. (2003). Destination competitiveness: Determinants and indicators. *Current Issues in Tourism*, *6*(5), 369–414. doi:10.1080/13683500308667962

Dwyer, L., Mellor, R., Livaic, Z., Edwards, D., & Kim, C. (2004). Attributes of destination competitiveness: A factor analysis. *Tourism Analysis*, *9*(1-2), 91–101. doi:10.3727/1083542041437558

Fernández, J. A. S., Azevedo, P. S., Martín, J. M. M., & Martín, J. A. R. (2020). Determinants of tourism destination competitiveness in the countries most visited by international tourists: Proposal of a synthetic index. *Tourism Management Perspectives*, *33*, 100582. doi:10.1016/j.tmp.2019.100582

Fuentes, Z. (2019). *Environment as a Key Factor of Health and Well-Being Tourism Destinations in Five European Countries*. Academic Press.

Fyall, A., & Garrod, B. (2019). Destination management: A perspective article. *Tourism Review*.

Gao, J., Ryan, C., Cave, J., & Zhang, C. (2019). Tourism border-making: A political economy of China's border tourism. *Annals of Tourism Research, 76*, 1–13. doi:10.1016/j.annals.2019.02.010

Ghaderi, E., & Aarabi, S. S. (2020). *Urban Tourism Competitiveness in Iran*. Applying Dwyer's Integrated Model.

Hall, T., & Barrett, H. (2018). Planning, regeneration and urban policy. In *Urban Geography* (pp. 122–155). Routledge. doi:10.4324/9781315652597-6

Hassan, A. S., & Ali, A. (2019). *Colonial influence on Indian Muslim religious monuments in Penang*. Academic Press.

Henderson, J. C. (2008). Managing urban ethnic heritage: Little India in Singapore. *International Journal of Heritage Studies, 14*(4), 332–346. doi:10.1080/13527250802155851

Hovelynck, J., Craps, M., Dewulf, A., Sips, K., Taillieu, T., & Bouwen, R. (2020). Relational Practices for Generative Multi-Actor Collaboration. The Sage Handbook of Social Constructionist Practice, 258.

Isa, M., Goh, A. T., Abdul Majid, H., Yuen, A., Abdul Rahman, F., Shahidan, S. I., ... Sharuddin, S. N. (2020). *The Kuala Lumpur Heritage Agenda: Reviews and Recommendations for the Kuala Lumpur Draft Structure Plan 2040 & Kuala Lumpur City Plan 2020 by ICOMOS Malaysia*. Academic Press.

Jamal, T. B., & Getz, D. (1995). Collaboration theory and community tourism planning. *Annals of Tourism Research, 22*(1), 186–204. doi:10.1016/0160-7383(94)00067-3

Kapera, I. (2018). Sustainable tourism development efforts by local governments in Poland. *Sustainable Cities and Society, 40*, 581–588. doi:10.1016/j.scs.2018.05.001

Khalid, N. S., Othman, R. R., Abdullah, I. C., Marzukhi, M. A., & Nasrudin, N. (2019, November). The Regaining of Traditional Street: Combining Street Network Configuration with Temporary Urban Interventions. *IOP Conference Series. Earth and Environmental Science, 385*(1), 012019. doi:10.1088/1755-1315/385/1/012019

Khalid, S., Ahmad, M. S., Ramayah, T., Hwang, J., & Kim, I. (2019). Community empowerment and sustainable tourism development: The mediating role of community support for tourism. *Sustainability, 11*(22), 6248. doi:10.3390u11226248

Khan, S. (2015). Otherness" of ethnic enclave attractions in multicultural cities: A study of Chinatown and Little India. *Asia-Pacific Journal of Innovation in Hospitality and Tourism, 4*(1), 63–76. doi:10.760340930-015-0004-8

Khoo, S. L. (2021). *Creative City as an Urban Development Strategy: The Case of Selected Malaysian Cities*. Springer Nature. doi:10.1007/978-981-16-1291-6

Khurramov, O. K. (2020). The role of the tourism sector in the digitalization of the service economy. *Economics and Innovative Technologies, 2020*(1), 6.

Kim, J. (2018). Manhattan's Koreatown as a Transclave: The Emergence of a New Ethnic Enclave in a Global City. *City & Community*, *17*(1), 276–295. doi:10.1111/cico.12276

King, R. (2011). *Transforming brickfields: Development and governance in a Malaysian city Houses in Motion: The Experience of Place and the Problem of Belief in Urban Malaysia*. Academic Press.

Knecht, S. (2017). Exploring different levels of stakeholder activity in international institutions: Late bloomers, regular visitors, and overachievers in Arctic Council Working Groups. In *Governing Arctic Change* (pp. 163–185). Palgrave Macmillan. doi:10.1057/978-1-137-50884-3_9

Kovačić, S., Jovanović, T., & Šagovnović, I. (2019). Exploring the gap in destination image and destination personality perception between tourists and stakeholders: Case of Roman heritage sites in Serbia. *Turizam*, *23*(3), 133–144. doi:10.5937/turizam23-22773

Kumar, V. (2020). *Morality in the Making of Urban Space* (Doctoral dissertation). State University of New York at Buffalo.

Kurtz, J. (2017). *The happy traveler: Unpacking the secrets of better vacations*. Oxford University Press.

Kyrylov, Y., Hranovska, V., Boiko, V., Kwilinski, A., & Boiko, L. (2020). International Tourism Development in the Context of Increasing Globalization Risks: On the Example of Ukraine's Integration into the Global Tourism Industry. *Journal of Risk and Financial Management*, *13*(12), 303. doi:10.3390/jrfm13120303

Little India @ Brickfields - VisitKL. (n.d.). Retrieved August 3, 2021, from http://www.visitkl.gov.my/visitklv2/document/files/thingstodo/little_india.pdf

Lopes, S. D. F. (2011). Destination image: Origins, developments and implications. *PASOS. Revista de Turismo y Patrimonio Cultural*, *9*(2), 305–315. doi:10.25145/j.pasos.2011.09.027

Lor, J. J., Kwa, S., & Donaldson, J. A. (2019). Making ethnic tourism good for the poor. *Annals of Tourism Research*, *76*, 140–152. doi:10.1016/j.annals.2019.03.008

Martinez, J. M. G., Martín, J. M. M., & Fernández, J. A. S. (2020). Innovation in the measurement of tourism competitiveness. In *Analyzing the relationship between innovation, value creation, and entrepreneurship* (pp. 268–288). IGI Global. doi:10.4018/978-1-7998-1169-5.ch013

Matthew, O. A., Ede, C., Osabohien, R., Ejemeyovwi, J., Ayanda, T., & Okunbor, J. (2021). Interaction effect of tourism and foreign exchange earnings on economic growth in Nigeria. *Global Business Review*, *22*(1), 7–22. doi:10.1177/0972150918812985

McKercher, B., & Koh, E. (2017). Do attractions "attract" tourists? The case of Singapore. *International Journal of Tourism Research*, *19*(6), 661–671. doi:10.1002/jtr.2138

Milićević, S., Petrović, J., Kostić, M., & Lakićević, M. (2020). Tourism product in the function of improving destination competitiveness: Case of Vrnjačka Banja, Serbia. *Quality - Access to Success*, *21*(178), 133–138.

Minguzzi, A. (2006). Destination competitiveness and the role of destination management organization (DMO): An Italian experience. In Tourism local systems and networking (pp. 211-222). Routledge.

Mokhtar, M. R., Yazid, M. F. M., & Shamsudin, M. F. (2020). Sustainability of tourism industry in Malaysia. *Journal of Postgraduate Current Business Research, 1*(1).

Monks-Husain, A. (2020). "Better Me Than Someone Else": Making Sense of Multiple Publics in Little India. *The American Papers, 2019-2020*, 41.

Mulec, I., & Wise, N. (2013). Indicating the competitiveness of Serbia's Vojvodina Region as an emerging tourism destination. *Tourism Management Perspectives, 8*, 68–79. doi:10.1016/j.tmp.2013.07.001

Murtagh, M. J., Minion, J. T., Turner, A., Wilson, R. C., Blell, M., Ochieng, C., Murtagh, B., Roberts, S., Butters, O. W., & Burton, P. R. (2017). The ECOUTER methodology for stakeholder engagement in translational research. *BMC Medical Ethics, 18*(1), 1–12. doi:10.118612910-017-0167-z PMID:28376776

Murugan, I., Perumal, V., & Kamarudin, K. M. (2021). *Challenges in the practice of traditional Kolam among Indian women in the Klang Valley, Malaysia*. Academic Press.

Nadalipour, Z., Khoshkhoo, M. H. I., & Eftekhari, A. R. (2019). An integrated model of destination sustainable competitiveness. *Competitiveness Review, 29*(4), 314–335. doi:10.1108/CR-12-2017-0086

Nayan, N. M., Jones, D. S., & Ahmad, S. (2019). Historic Open Space: The Identity of [Padang] Merdeka Square, Kuala Lumpur, Malaysia. *Asian Journal of Behavioural Studies, 4*(17), 29–39. doi:10.21834/ajbes.v4i17.182

Neto, A. Q., Dimmock, K., Lohmann, G., & Scott, N. (2020). Destination competitiveness: How does travel experience influence choice? *Current Issues in Tourism, 23*(13), 1673–1687. doi:10.1080/13683500.2019.1641070

Novais, M. A., Ruhanen, L., & Arcodia, C. (2018). Destination competitiveness: A phenomenographic study. *Tourism Management, 64*, 324–334. doi:10.1016/j.tourman.2017.08.014

Oakes, T. (2016). Ethnic tourism in China. In *Handbook on ethnic minorities in China*. Edward Elgar Publishing. doi:10.4337/9781784717360.00020

Omar, D., Ramlee, M., Yunus, R. M., & Samadi, Z. (2018). Urban Public Space Revitalisation Strategies for George Town World Heritage Site. *Environment-Behaviour Proceedings Journal, 3*(7), 285–291. doi:10.21834/e-bpj.v3i7.1235

Omar, S. R., & Omar, S. N. (2018). Malaysian heritage food (MHF): a review on its unique food culture, tradition and present lifestyle. *Int J Heritage Art Multimedia, 1*(3), 1-15.

Online, N. S. T. (2015). *Najib, Modi jointly launch Torana Gate in Brickfields*. Available at: https://www.nst.com.my/news/2015/11/113341/najib-modi-jointly-launch-torana-gate-brickfields

Online, N. S. T. (2017). *Illegal parking in Brickfields*. Available at: https://www.nst.com.my/action-line/2017/11/304902/illegal-parking-brickfields

Orts, E. W., & Strudler, A. (2009). Putting a stake in stakeholder theory. *Journal of Business Ethics, 88*(4), 605–615. doi:10.100710551-009-0310-y

Pagliara, F., Aria, M., Russo, L., Della Corte, V., & Nunkoo, R. (2021). Validating a theoretical model of citizens' trust in tourism development. *Socio-Economic Planning Sciences*, *73*, 100922. doi:10.1016/j.seps.2020.100922

Patti, S. (2019). Determinants of tourist destination competitiveness in a low-carbon tourism. In *Positioning and Branding Tourism Destinations for Global Competitiveness* (pp. 74–99). IGI Global. doi:10.4018/978-1-5225-7253-4.ch004

Pavlović, D., Avlijaš, G., & Stanić, N. (2016). Tourist perception as key indicator of destination competitiveness. *TEME: Casopis Za Društvene Nauke, 40*(2).

Pike, S. (2015). *Destination marketing: essentials*. Routledge. doi:10.4324/9781315691701

Priya. (2010). *Brickfields business community oppose setting up of kiosks*. Available at: https://www.thestar.com.my/news/community/2010/09/09/brickfields-business-community-oppose-setting-up-of-kiosks

R. (2019, September 4). *KL 5th world's most-visited city, Bangkok takes top spot - MasterCard*. Retrieved August 1, 2021, from https://www.nst.com.my/news/nation/2019/09/518721/kl-5th-worlds-most-visited-city-bangkok-takes-top-spot-mastercard

Rahman, N. A., Halim, N., & Zakariya, K. (2018, September). Architectural value for urban tourism placemaking to rejuvenate the cityscape in Johor Bahru. *IOP Conference Series. Materials Science and Engineering*, *401*(1), 012010. doi:10.1088/1757-899X/401/1/012010

Rajan, S. I., & Percot, M. (Eds.). (2020). *Dynamics of Indian migration: Historical and current perspectives*. Taylor & Francis. doi:10.4324/9780367818043

Rajantheran, M. (2009). *Siddha philosophy in the Malaysian Indian Culture: A historical outlook*. Academic Press.

Rajantheran, M., Gill, S. S., Muniapan, B., Silllalee, K., & Manimaran, S. (2014). A critical analysis of siddha tradition in the context of Malaysian Hindu culture. *Life Science Journal*, *11*(7).

Ramlee, M., Omar, D., Mohd Yunus, R., & Samadi, Z. (2018). Attributes of successful public spaces through users perception. *Asian Journal of Quality of Life*, *3*(11), 21–28. doi:10.21834/ajqol.v3i11.118

Ravindran, S. (2017). *DBKL: No more double-parking in Brickfield's Little India*. Available at: https://www.thestar.com.my/metro/metro-news/2017/10/14/dbkl-no-excuse-for-doubleparking-city-hall-to-conduct-more-enforcement-activities-in-brickfields-to/

Roman, M., Roman, M., Prus, P., & Szczepanek, M. (2020). Tourism competitiveness of rural areas: Evidence from a region in Poland. *Agriculture*, *10*(11), 569. doi:10.3390/agriculture10110569

Rumere, N., Wiranatha, A. S., & Pujaastawa, I. B. G. (2020). Planning of Country Border Area as A Tourism Destination in Jayapura City. *E-Journal of Tourism*, 384-398.

Samad, N. S. A., Abdul-Rahim, A. S., Afandi, S. H. M., & Johari, M. M. (2021, May). Factors influencing the public park use in Kuala Lumpur, Malaysia. *IOP Conference Series. Earth and Environmental Science*, *756*(1), 012085. doi:10.1088/1755-1315/756/1/012085

Santos, C. A., & Yan, G. (2008). Representational politics in Chinatown: The ethnic other. *Annals of Tourism Research*, *35*(4), 879–899. doi:10.1016/j.annals.2008.06.006

Singh, A. (2011). India–Malaysia Strategic Relations. *Maritime Affairs: Journal of the National Maritime Foundation of India*, *7*(1), 85–105.

Sofronov, B. (2018). The development of the travel and tourism industry in the world. *Annals of Spiru Haret University. Economic Series*, *18*(4), 123–137.

Solikhah, N. (2020, April). Ethnic tourism and sustainable of vernacular settlement in Cibeo Village, Baduy Dalam. *IOP Conference Series. Earth and Environmental Science*, *452*(1), 012021. doi:10.1088/1755-1315/452/1/012021

Stankova, M., & Amoiradis, C. (2019). Effects Of Promotion Practices On The Sustainable Development Of Tourist Destinations. *Entrepreneurship*, *7*(1), 84–96.

Su, M. M., Wall, G., & Ma, Z. (2019). A multi-stakeholder examination of destination image: Nanluoguxiang heritage street, Beijing, China. *Tourism Geographies*, *21*(1), 2–23. doi:10.1080/14616688.2017.1385031

Sudipta, K. S., & Sarat, L. (2010). Cultural tourism in Malaysia in the perspective of Indian tourists: A study. *Revista de turism-studii si cercetari in turism*, (10), 48-53.

Tawil, M., Tawil, Y. P., Rahmarini, G. M., & Salmon, I. P. (2021). *Collaborative Governance as a Strategy for Optimizing the Potential of Coastal Areas and Local Tourism: A Study in Tangerang Regency*. Academic Press.

The Star. (2015). *Police launch pilot project to 'clean up' Brickfields*. Available at: https://www.thestar.com.my/Metro/Community/2015/09/28/Focused-sweep-in-Brickfields

Thuraisingham, J., & Valu, V. V. (2020). 10. The financial vulnerability of Indian micro and small entrepreneurs in Malaysia. In *Vulnerable Groups in Malaysia* (pp. 140–151). De Gruyter Oldenbourg. doi:10.1515/9783110610970-010

Ullah, A. A., Akram Hossain, M., & Huque, A. S. (2021). Non-Conventional Migration: An Evolving Pattern in South Asia. *Journal of Asian and African Studies*. doi:10.1177/00219096211008468

Urbinati, A., Landoni, P., Cococcioni, F., & De Giudici, L. (2020). Stakeholder management in open innovation projects: A multiple case study analysis. *European Journal of Innovation Management*.

Valenzuela-Garcia, H., Parella, S., & Güell, B. (2017). *Revisiting the 'ethnic enclave economy': Resilient adaptation of small businesses*. Academic Press.

Van der Zee, E., Gerrets, A. M., & Vanneste, D. (2017). Complexity in the governance of tourism networks: Balancing between external pressure and internal expectations. *Journal of Destination Marketing & Management*, *6*(4), 296–308. doi:10.1016/j.jdmm.2017.07.003

Vu, H. M., & Ngo, V. M. (2019). *Strategy Development from Triangulated Viewpoints for a Fast Growing Destination Toward Sustainable Tourism Development–A Case Of Phu Quoc Islands in Vietnam*. Academic Press.

Vu, H. M., & Ngo, V. M. (2019). Strategy Development from Triangulated Viewpoints for a Fast Growing Destination Toward Sustainable Tourism Development–A Case Of Phu Quoc Islands in Vietnam. *Journal of Tourism and Services, 10*(18), 117–140. doi:10.29036/jots.v10i18.86

Vukovic, D. B., Petrovic, M., Maiti, M., & Vujko, A. (2021). Tourism development, entrepreneurship and women's empowerment–Focus on Serbian countryside. *Journal of Tourism Futures.*

Wang, Y., Shen, H., Ye, S., & zhou, L. (2020). Being rational and emotional: An integrated model of residents' support of ethnic tourism development. *Journal of Hospitality and Tourism Management, 44*, 112–121. doi:10.1016/j.jhtm.2020.05.008

Willford, A. (2020). 8. The Figure of the Tamil in Modern Malaysia. In *Clio/Anthropos* (pp. 223–272). Stanford University Press. doi:10.1515/9780804772402-009

Wondirad, A., Tolkach, D., & King, B. (2020). Stakeholder collaboration as a major factor for sustainable ecotourism development in developing countries. *Tourism Management, 78*, 104024. doi:10.1016/j.tourman.2019.104024

Wong, N. L., & Lau, K. L. (2016). Voices of Third Generation Malaysians: Malaysian or Malaysian Indian'. *Sarjana, 31*(1), 45–54.

Woosnam, K. M., Maruyama, N. U., Ribeiro, M. A., & Joo, D. (2019). Explaining minority residents' attitudes of ethnic enclave tourism from general perceptions of tourism impacts. *Journal of Tourism and Cultural Change, 17*(4), 467–484. doi:10.1080/14766825.2019.1601407

Xiaomei, W., & Daming, X. (2018). The mismatches between minority language practices and national language policy in Malaysia: A linguistic landscape approach. *Kajian Malaysia: Journal of Malaysian Studies, 36*(1).

Yahya, M. S. S., Safian, E. E. M., & Burhan, B. (2020). The Development and Distribution Pattern of Railway Network for Urban Public Transport Using GIS from 1990 Until 2019 in The Klang Valley and Kuala Lumpur, Malaysia. *Journal of Social Transformation and Regional Development, 2*(2), 1–10.

Yang, L., & Wall, G. (2016). *Planning for ethnic tourism*. Routledge. doi:10.4324/9781315600741

Yap, J. B. H., Chua, C. Y., & Skitmore, M. (2021). Towards Sustainable Mobility with Transit-Oriented Development (TOD): Understanding Greater Kuala Lumpur. *Planning Practice and Research, 36*(3), 314–336. doi:10.1080/02697459.2021.1883249

Yin, I., & Abdullah, J. (2020). The development control of urban centre in Kuala Lumpur. *Planning Malaysia, 18*(13).

Yunus, R. M., Samadi, Z., & Omar, D. (2018). Community's Participation: A strategic approach in revitalizing heritage streets. *Journal of ASIAN Behavioural Studies, 3*(10), 29–37. doi:10.21834/jabs.v3i10.301

Zainol, N. R., Al Mamun, A., Hassan, H., Rajennd, A., & Muniady, L. (2017). Examining the effectiveness of micro-enterprise development programs in Malaysia. *Journal of International Students, 10*(2).

Zainuddin, Z., Zahari, M. S. M., Radzi, S. M., Hanafiah, M. H. M., & Ishak, N. (2018). Tourism images and the international tourists response behaviour: A case of Langkawi Island, Malaysia. *International Journal of Academic Research in Business & Social Sciences*, *8*(16), 60–82.

Zainudin, H., Bachek, S. H., & Haron, N. A. (2016). Preservation of Culture and Built Heritage in New Urban Development: A Case Study on Little India Brickfields. Academic Press.

Zamri, N. A. (2020). Walkways for visually impaired person in Brickfields. Academic Press.

Chapter 12
Veszprém–Balaton 2023 European Capital of Culture Title for Sustainable Cooperation:
The Key Role of Local and Regional Stakeholder Networks

Edit Kővári
University of Pannonia, Hungary

Katalin Lőrincz
University of Pannonia, Hungary

Marco Valeri
https://orcid.org/0000-0002-9744-506X
Niccolò Cusano University, Italy

ABSTRACT

The European Capital of Culture (ECoC) title enables European citizens to learn about one another's culture, to admire the relics of their common past, and to experience the wealth of European values and the sense of belonging together. Winner cities and regions wish to sustain the effect of the title in order to maintain economic, social and cultural prosperity. Veszprém and Balaton region in Hungary won the title for 2023, which influences local and regional destination management. The aim of the research is to introduce the key role, factors, and pitfalls of local and regional stakeholders' network system and to highlight the challenges of sustainable cooperation.

DOI: 10.4018/978-1-6684-3889-3.ch012

INTRODUCTION

The title of European Capital of Culture was created more than 30 years ago on the proposal of Greek-born actress and minister of culture Melina Mercouri (until 2001 the title was called "European City of Culture"), which has been one of the most important and well-known projects in European Union cultural policy ever since. The title itself is awarded by the European Union's Directorate-General for Education, Youth, Sport and Culture in order to highlight the cultural richness of its Member States. The aim of the programme is to increase the sense of belonging to a common (European) cultural area for European citizens and thus to improve their understanding of each other.

In fact, the title of European Capital of Culture (ECoC) is to raise awareness that the common field of European culture is created by the joint contribution of different countries and cities. In the language of the arts it conveys that European culture is common and is always the result of the cultural activity of that region. The city holding the title will be the focus of European attention with the cultural products and concrete projects that are created there. The project's explicit aim is to improve and strengthen the image of the ECoC initiation and the ECoC city, increasing the efficiency of the local creative industry and increasing its tourist attractiveness and the cooperation of the locals (Albattat et al., 2020; Chemli et al., 2020; Toanoglou et al., 2021; Valeri, 2021a).

The Veszprém-Balaton 2023 (VEB2023) project places great emphasis on cooperation and regional relations, so the city of Veszprém is scheduled to work with almost 260 municipalities in Balaton (biggest Lake of Central Europe) and Bakony (mountain range near Veszprém city) in the ECoC 2023 programme series. The proposal and the project, as the slogan *Beyond* suggests, go beyond limits, geographical boundaries and are about to bring together municipalities, businesses, NGOs and to cooperate in a network. The phrase "move beyond" covers a wide spectrum initiations to change attitude towards "boring Tuesdays, trouble and isolation, mistrust, prejudice, tradition, abandoned barns, obstacles and indifference" (BidBook Veszprém2023 2018).

The study briefly presents the history and objectives of the ECoC title and puts this programme in the context of cooperation, contacts and networks. In the course of the national and international outlook of the literature, it deals with the network and tourism relationship system, and the selection process of ECoC 2023. The focus of the study will be on the Veszprém-Balaton 2023 programme itself, highlighting the challenges of cooperation and networking – from winning the title to the leading years. By presenting specific programmes and operational solutions, a networked cooperation model of regional municipal and cultural actors, the University of Pannonia, civil communities and the management of the ECoC is presented. The study looks for the answer to the following question:

- How can (regional) cooperation and networking be linked to European Capital of Culture applications?
- How does this idea apply in the Veszprém-Balaton 2023 (VEB2023) project?
- What concrete steps have been taken during the preparation of the project (until the manuscript is completed by 2020)?

DESIGN/METHODOLOGY/APPROACH

The study exhibits the stakeholders' network of some ECoC bidders and winners especially focusing on Veszprém and Balaton region, Hungary as a future ECoC city. The research applies two methodology: the secondary data analysis is applied based on bid books and reports of ECoC cities; and the primary data is based on regular project meetings and their minutes between the ECoC management, the representative of the University of Pannonia, head of institution, members of local and regional government and other stakeholders such as tour boards in Veszprém and the Balaton region.

Results

The results highlight the challenges of sustainability, cooperation and networking of stakeholders ECoC cities, especially Veszprém-Balaton 2023. It provides an overview of the most important directions, content and the planned operating model of stakeholders' network system regarding how the ECoC programmes can be successful.

Originality/Value

The value of the paper is that the researchers identifies key factors and pitfalls of ECoC stakeholders' network system in order to create successful ECoC programme series. The second value is that the development of the ECoC programme in the process is presented in real-time based on minutes due to the involvement of the authors.

Limits

The study cannot illustrate all ECoC cities and bidders. It is limited mainly some previous winners and the Hungarian winner and bidders. Also this study applies written documents as interviews had to be postponed due to the pandemic situation.

** The research is supported by the Research Centre (PE-GTK-GSKK A095000000-3 sz.).at Faculty of Business and Economics of University of Pannonia (Veszprém, Hungary).*

1. THE ECOC PROGRAMME

1.1. History and Values of the European Capital of Culture Programme

The title of European Capital of Culture allows European citizens to learn more about each other's cultures, to admire the memories of our common history, to experience the richness of European values and the sense of European togetherness. A total of 62 cities have so far been honoured with the title (including 2020), which has brought long-term development to the city's cultural infrastructure, increased citizens' self-esteem and pride in the city, improved the image of the city, resulted in more active participants in culture, increased tourism, greater recognition and economic progress.

The ECoC programme itself has been constantly changing since its inset in 1985; and moving together with the challenges of the present age. For the applicant city, the European Union has compiled a manual containing the criteria and deadlines for the application documentation. In addition to the specific objectives, the ECoC programme highlighted four values globally (Figure 1); these are *openness*, *introduction*, *involvement* and *model*. The philosophy of the ECoC programme builds on the concrete realisation of relations and cooperation at different levels.

Figure 1. Global values of the European Capital of Culture programme
Source: Own figure based on https://veszprembalaton2023.hu/en/about-the-programme2

Openness: strengthens the cohesion among the population of the European Union. It creates a system modelled on European free trade that facilitates the movement of cultural creations instead of products, emphasizing its own diversity and fostering better mutual understanding.

Introduction: offers an opportunity to introduce lesser-known regions to other European states.

Involvement: generates interest among local people to participate in the development of the area, thus increasing the bond between the community and the municipality.

Model: the development of the winning cities can serve as a positive model for the development of the cultural industry in other regions.

Along with the global ECoC values the Veszprém-Balaton 2023 programme has its own significant ones which are the followings:

- **Community building:** contributes to the connection of fragmented and non-communicative social strata.
- **Volunteering:** By involving the community it helps residents of the region to gain new experiences.
- **Tradition:** it supports project development with the awareness of local values and plans to make improvements considering these.
- **Innovation:** stimulates cultural and economic demand for the region by producing new market goods and intellectual products.
- **Sustainability:** we implement developments that help modernise the region in ways that serve the long-term environmental and economic interests of the region.

The ECoC programmes present a particular challenge for the candidate and winner cities. It expects cooperation both vertically and horizontally in the management. The host city should cooperate with the EU and national institutions, while horizontal partnerships are expected between local creative and professional communities and NGOs.

1.2. Networks and Cooperation in Tourism and ECoC Programmes

After the regime change, the main focus of the development of Hungarian cities was the transformation of the economic structure, which was mostly driven by external energies (global economy) and external opportunities. Today, small and medium-sized enterprises and internal energies of the city and its region are taking the lead in the transformation (decentralised management, creation of cooperation, networking, innovative institutions, NGOs). In addition to creating jobs and diversifying the economic structure, a complex programme for improving quality of life is the focus. In addition to improving the quality of the physical environment it also means providing an attractive cultural and service environment and access to it (Lőrincz 2017). Morvai et al. (2020) examined the effects generated by the European Capitals of Culture eastern European programme series. The impact of the projects implemented the relationship between the ECoC and urban development, which were based on the cooperation of actors involved in urban development.

Over the past decades, the need for networks and cooperation has increased in economic sectors, including tourism (Flower 2016; Valeri, 2016; Valeri and Baggio, 2020a; 2020b). On the one hand, meeting the ever-changing needs of visitors clearly makes cooperation between the cultural institutions and the tourism sector of cities a clear necessity (Michalkó-Rátz 2005). On the other hand, tourism is characterised by a number of independent, often small players and market fragmentation, making networking particularly important, based on financial stability, a clear programme, an effective planning structure and flexible organisational units (Soisalon-Soininen - Lindroth, 2006; Baggio and Valeri, 2020; Valeri, 2021b; Valeri and Baggio, 2021). A workable cooperation can be established if we can bring the right partners into a network, but it is also of paramount importance to coordinate the person or staff with the right competences and capabilities to establish clear agreements and to manage external relations. In the regional dimension, tourism destination management organizations can be considered as a kind of vertical networks. (Panyor et al., 2011; Valeri and Katsoni, 2021).

In the context of the ECoC programme, networks and cooperation, there have been numerous publications. Rátz (2014) stressed that "product development in niche tourism is based on uniqueness, the development of customized services, the phenomenon of co-creation (Kővári et al, 2021), i.e. the cooperation between the guest and the service provider in order to personalise the service, plays an increasingly important role" (Rátz 2013:25). Bognar – Kovacs (2013) stressed in his study that the ECoC project in Liverpool had contributed to the strengthening of cooperation. It concluded that "partnerships, networks and cultural cooperation before the ECoC only existed formally, but during the programme they became more close and efficient. ... Partnerships with civil society and creative communities have developed a series of long-term partnerships to ensure that the cultural sector's new role is maintained. The Liverpool approach, which set out to reposition the city, has led to a complex network of multi-player collaborations. For this reason, the Liverpool programme is often mentioned as one of the most successful ECoC projects" (Bognar – Kovacs 2013:34).

2. THE VESZPREM-BALATON2023 PROJECT

Winning the ECoC title is a huge opportunity for every city. Many examples in Europe show that the programme allows neighbourhoods or even entire cities to have a completely new, modern image, or even to introduce the old, traditional one to a wider audience. ECoC-related developments, urban rehabilitation actions, assuming sustainable planning and implementation will bring long-term positive change to the local population. During the selection of the tender, the independent jury examines the candidate municipalities in several circles in order to put on the table a more complex and colourful concept that will determine the future of the city for decades.

2.1. Profile of Veszprém and the Balaton region

Veszprém Hungary's 16th most populous city, but in view of its cultural impact, as described in the competition, it is in the top 5. This is due especially to the cultural and spiritual importance of Lake Balaton. In addition to the emblematic cultural attractions of Veszprém (i.e. Veszprém Street Music Festival, VeszprémFest, Auer Festival,Vass Collection, House of Arts, Dezső Laczkó Museum, Agóra Cultural Centre), the project focused on the cultural profile of the region (Lőrincz – Raffay 2019).

The diverse cultural offer of the Balaton area is built on a huge number of threads by professional relationships and networks. Events of the region (i.e. Valley of Arts, Paloznaki Jazzpiknik, Salvatore Quasimodo International Poetry Competition, Balaton Sound); the emblematic attractions (Benedictine Abbey of Tihany, Festetics Castle, Bakonyi Natural History Museum, Pannon Starda, House of Jewish Excellence) and its special venues (Hungarian Translation House, Peace Cultural Center, Snétberger Music Talent Center, Lower Ore Amphitheater, the largest former pioneer camp, Liliomkerti market) serve the needs of guests with different profiles and motivations across a wide spectrum.

Criticism of the current cultural offer is the idea that no formal level of cultural administration covers the area, and consequently there is a lack of an organisation necessary for the proper organisation and diversity of regional culture. By taking over the well-established culture model in Veszprém, which encourages entrepreneurial initiatives that can follow needs, there is also a chance to build a "unified, inclusive, likeable cultural community" in the Balaton area (BidBook Veszprém2023 2018).

2.2. Metaphorical Sense and "Shine Veszprém-Balaton" Image

In the competition of Veszprém and Balaton, the slogan summarising the art program – "Cross your borders!" – is closely linked to the concept of "overrun" (or "beyond"). The slogan itself calls for a metaphorical and direct journey: to overcome barriers, to cross and exceed our own borders. The proposal states that "we want to address the individual, the creativity of the individual, helping to exceed his own boundaries" (BidBook Veszprém2023:94).

After the announcement of the ECoC title, the image was revised in 2019. The winner of the image manual competition was the creative work "Shine Veszprém-Balaton" which also made the official slogan of the Veszprém-Balaton2023 programme: "Shine!". The image builds on the motifs and ideas of awareness, creativity, enlightenment and interaction. In this way, it expresses the main objectives of the programme: to invigoration of the cultural life of the city and the region and the dissemination of its values. The concepts of "interaction" and "regional dimension" expressed in the basic idea are consistent with the emphasis on cooperation and relationships.

2.3. BidBook Veszprém2023 in the Context of Networks and Collaborations

The application document summarising the Veszprém-Balaton2023 programme, the so-called BidBook Veszprém2023, refers in countless places to the implementation of contacts (84), cooperation (61), networks (40) and regional thinking (56). In connection with networking, the idea of the Barn programme and the setting up of a regional team can be highlighted.

In the "beyond charm and isolation" chapter, the Barn programme, a solution to the cultural sites of the regional project, is introduced in the "beyond charm and isolation" chapter. According to this, "barns currently empty can be great places for community life. This is what our villages lack most, according to the people who live there: a place to gather, where they can be together as a community, dance, have fun – even in winter. This is the solution of our project, in which we create community centres of different sizes with vibrant life" (BidBook Veszprém2023 2018:26). During the project, ten locations are planned: empty barns will be renovated or new ones will be created and built. The planning pays particular attention to local communities, i.e. "... the first step will be to bring together the alliances that ensure operation, not architectural design, which will only start after the former. Sustainability is key" (BidBook Veszprém2023 2018:26).

The entire area of the Veszprém-Balaton2023 programme can be described as a diverse region with more than 260 independent municipalities. The existence of a regional team is of paramount importance, so that, on the one hand, regional programme planning and implementation are carried out on several levels and, on the other hand, directly reach the population. In the course of implementation, it is planned to establish 12 ECoC bases in municipalities that already act as hubs for rural cultural life. In addition, at least one person or organisation will be needed as an official ECoC contact point in the smallest villages. These bases and contact points will both be under the professional management of the regional coordinator (Kumar et al., 2021).

2.4. Support for the Application Phase

In connection with the Veszprém-Balaton2023 competition, a personal network of supporting, outstanding professionals and public figures with ties to Veszprém has also been established. In 2017-2018, the artistic assistance was carried out by the Arts Council, lobbying and the personal support of the program were carried out by the Veszprém-Balaton2023 Ambassadors – in close cooperation with the tender writers and management.

The Arts Council consisted of artists and professionals who had long been active in various fields of cultural life in Veszprém and the Balaton region, with long-standing experience and outstanding artists. The task of the members of the Arts Council was to put together a series of programmes related to the ECoC competition and to promote the success of the competition through their expertise, ideas, proposals and implementation.

The Veszprém-Balaton 2023 ambassadors carried out lobbying activities during the application phase. The ambassadors with Veszprém roots or local ties were invited from among the artists, athletes and professionals respected in their profession and also known to the general. They supported the Veszprém-Balaton 2023 competition with their own person and whole heart.

3. IMPLEMENTATIN OF THE PROJECT: ECONOMIC, ORGANISATIONAL AND REGIONAL COOPERATION

3.1. Financing

There is a common misconception about the ECoC programme that development and implementation are financed by the European Union. With regard to funding, it can be said that the role of the EU is small: in fact, the ECoC gives the winning city the title and tender opportunities. The investments and programme development linked to the ECoC project themselves are largely made from the resources of the Hungarian State and the municipalities of Veszprém and the region. Veszprém-Balaton 2023 Plc's tender, sponsorship and own sales revenues contribute to the revenue side. In 2020 the national state resources were fixed in a decision (1468/2020 (VIII. 5) entitled "Government decision on measures related to the title of European Capital of Culture 2023" and amounting to 88.5 billion HUF.

3.2. Organisations and Management

The management organization behind the Veszprém-Balaton2023 project is also special in its structure and operation in many ways. The ECoC was awarded to the city of Veszprém county law (December 2018), involving stakeholders such as the institutions, cultural actors and NGOs of Veszprém and the region. One of the guarantees of regionality and joint coordination is the project management organization itself, Veszprém-Balaton 2023 Plc.

Figure 2. The organizers and partner stakeholders in the Veszprém-Balaton2023 programme
Source: https://veszprembalaton2023.hu/en/our-partners

On the one hand, Veszprém-Balaton 2023 Plc as a company is owned by major regional and micro-regional organisations as well as the the local municipalities. After winning the ECoC title, Plc. became the main designer and organizer of project development, international relations, cultural and artistic activities. On the other hand, VEB 2023, as a project organisation is planned to take a kind of developmental curve. It will be integrated into the existing public, municipal and civil sectors in order to play its main coordinator role by 2023. This is an ongoing organizational development process. The relationship of the stakeholders are illustrated by Figure 3. The main players in the "inner circle" of the institutions are Veszprém city, Veszprém Utility Service Co., Veszprém Government Commissioner's Office and Veszprém-Balaton2023 Plc. One of the key partner institution is the University of Pannonia.

3.3. Networking in the Veszprém-Balaton Region

The Veszprém-Balaton 2023 program names several specific networks, and BidBook Veszprém2023 itself presents this ambition in details. One of the starting points is the regional thinking that "we plan to implement our programme as a joint venture between Veszprém and the Balaton region. The leader of the competition is Veszprém, but our program extends to a much wider area of the Balaton region: with nature and NATURA 2000 sites, UNESCO's Bakony-Balaton Geopark and the Balaton Highland cultural landscape marked with the UNESCO World Heritage List. This is also central to our natural wealth programme" (BidBook Veszprém2023 2018:6).

The implementation of this project element is the so-called 'partnership at regional level' (internal document VEB 2023 2020), which clarifies the cooperation base and economic and professional objectives of the municipalities in the ECoC region. During the July 2020 initiative, Veszprém also entered into an economic alliance with other local authorities in order to successfully and sustainably implement the ECoC project at regional level. In this partnership, the territorial defining takes place "from the southern slopes of Bakony to the southern shore of Lake Balaton", nearly 260 municipalities have been invited. VEB 2023 notified the municipalities of the conditions of the regional partnership and the specific direction of cooperation both in the face-to-face consultation and in a partnership letter.

The essence of the network operation is that municipalities who wish to participate actively in the ECoC programme can express their commitment by concluding a cooperation contract. The contract is concluded with the Public Foundation for the Culture of the Veszprém-Balaton Region, which was set up to support the work of Veszprém-Balaton 2023 Plc. ECoC membership, together with all the cultural, economic and tourist benefits of the programme, involves a financial contribution from the municipalities. For district centres this is a contribution of 2 EUR per inhabitant per year, while for other municipalities this amount is 1 EUR 1 per inhabitant per year.

With the payment of the fee due between 2021 and 2023, the intention is that the participating municipalities, NGOs, institutions, associations and local market players truly feel VEB 2023 programmes to their own. They appreciate together all the positive benefits, while being responsible for the implementation and quality of the programmes undertaken. Through the conclusion of the contract with the Public Foundation, they want to ensure the full use of the funds transparently and channel the amount paid from the region into a so-called regional aid fund.

The regional partnership can be formulated along the lines of 'community building, tradition, innovation and sustainability'. With the partnership, regional municipalities realize the following benefits:

- **Providing a professional background:** communication and marketing at domestic and European level; professional and management support, professional mentoring if necessary
- **Possibilities for joining the programme:** potential beneficiary of small-scale community proposals; specific support for productions and projects under tender schemes; a dedicated application framework for district centres; possible locations for projects and productions carried out by Veszprém-Balaton 2023 Plc.
- **Communication cooperation:** for existing events and initiatives

The communication and coordination of the regional partnership is carried out from the management side by a so-called regional coordinator working in the plc, whose task will be to contact the municipalities, inform them and channel the various needs in the right place.

3.4. Specific Directions for Cooperation

The two concrete institutional cooperation of the ECoC programme in 2020 are important milestones: the channelling of regional higher education University of Pannonia and the ECoC voluntary programme, the Shiners.

The regionally connected University of Pannonia and Veszprém-Balaton 2023 Plc. have entered into a strategic agreement to achieve the European Capital of Culture project as successfully as possible. The key concept is that the two partners values such as community building, volunteering, sustainability, traditional and innovation are alike. As a priority partner, the University of Pannonia aims is to work together with VEB 2023 Plc. in a number of areas:

- Writing up tenders with external stakeholders to develop and realize VEB 2023 programs
- Knowledge sharing and cooperation in building up the VEB 2023 management team
- Student and professors involvement in initiating and realizing the ECoC programmes: The University of Pannonia has announced as an ECoC themed course in which students, teachers and ECoC partners work together to implement and evaluate project ideas.
- One of the key strategic objectives is to establish one of the planned ECoC Points on the campus that can host events, civil organisation projects and serves as the University Volunteer Base.
- The strategic agreement also includes joint development of the ECoC monitoring system, the preparation and submission of collective tenders in the field of creative industry and sustainability, and the professional support of the university ECoC research group.

3.5. Concrete Events for the 2020 Preparation Year

2023 is pre-emanating from the so-called pre-years 2020-2022. The direct aim of this effort is to establish the priority year, and the organizers and programme promoters should have the opportunity to test some programme elements. The strategic goal is that when the whole of Europe's attention is focused on Veszprém and the Balaton area, it can really provide a range of value to the locals and those who come here.

The year 2020 was extremely hectic in terms of festivals and big events, as the Covid-19 virus basically rewrote and changed the plans. Yet there were some events through which we can demonstrate and emphasising network cooperation.

- Filmpicnic in networked form was organized first in 2020 September in Veszprém and Balatonfüred in cooperation with several organisations. The program is a film festival involving and addressing the audience. Award-winning movies, short films, exciting documentaries and animated works could be seen, as well as audience meetings, professional and music programmes. The event was an excellent example of network cooperation: home and outdoor venues, as well as non-profit and profit-making organisations provided the location for film screenings. The official communication also verified stakeholders' collaboration: "The Veszprém-Balaton Filmpicnic initiative under the European Capital of Culture 2023 programme supports the aim of strengthening the cultural-creative region of Veszprém and Lake Balaton and increasing its European visibility. The new location selection for the annual inspection of the Hungarian motion picture production was born out of the close cooperation of the Veszprém-Balaton2023 programme, the National Film Institute and the Hungarian Academy of Motion Picture Arts and Sciences. Veszprém-Balaton Filmpicnik is a place and time choice for the profession and audience to watch Hungarian films, meet Hungarian stars and filmmakers in late summer Balaton, discover or re-watch contemporary and classical masterpieces, audience films and movie delicacies" (www.veszprembalaton2023.hu/hu/events/filmpiknik 2020).
- The other program was Kőfeszt (Stonefest) where six villages in the Káli Basin within the Balaton Upland joined the program beside the initiating village of Kővágóörs. The programs entwined the villages thematically and chronological. Folklore was the main character of the all-art event along with light music, theatre and literary programs. The Veszprém-Balaton2023 programme contributed to successful management primarily with the volunteering during the programmes.
- Despite COVID19 several ECoC-supported events were implemented to strengthen quality development and internationalism. In the summer visitors could enjoy the Auer Festival, the Veszprém Street Music Festival, the Rosé, Riesling and Jazz Days. Among the autumn programs were the Festival of Dance and then the Salvatore Quasimodo Festival in Balatonfüred. The UNESCO's City of Music summit and the World Music Day took place though half online in October.

4. CONCLUSION

The study examines the emergence of cooperation and networking in tourism in European capital of culture applications on the basis of literature resources and minutes of the ECoC regular meetings. In addition to the ECoC objectives established and operated by the European Union, it highlighted four values in global terms, openness, introduction, inclusion and model. It can be said that the philosophy of the ECoC programme builds on the implementation of relations and cooperation at different levels. Based on the literature, it can be said that one of the keys to successful ECoC programmes is the quality, depth of cooperation and thinking in the network.

The Veszprém-Balaton 2023 application material relies on the construction of cooperation and networks in both quantitative and qualitative terms. The implementation of the Barn programme or regional actors and regional coordinator system supports this. The new image of the program ("Shine Veszprém-Balaton!") builds on the motif world of interaction and the regional dimension. And the slogan "Shine" symbolizes the sparking-up the city and the region cultural life of and the dissemination of its values.

The preparation of the 2023 project started in 2020 in a number of concrete steps. As regards financing, the Hungarian Government fixed the level and timing of ECoC support in a government decree.

The management organization Veszprém-Balaton 2023 Plc. contacted nearly 260 regional municipalities with the possibilities and framework for joining the programme in order to establish a partnership at regional level. During the preparation year of 2020 several programs have already been organized with local and regional stakeholders' cooperation. If in this hard pandemic year this could be successfully implemented then it is a hopeful sign that by 2023 cooperation will even be stronger.

REFERENCES

Albattat, A., Jamaludin, A., Zuraimi, N. S. M., & Valeri, M. (2020). Visit Intention and Destination Image in Post- Covid- 19 Crisis Recovery. *Current Issues in Tourism*. Advance online publication. doi :10.1080/13683500.2020.1842342

Baggio, R., Valeri, M. (2020). Network science and sustainable performance of family businesses in tourism. *Journal of Family Business Management*. doi:10.1108/JFBM-06-2020-0048

Bognár, S., & Kovács, L. (2013). A világ egy városban. Liverpooli EKF tapasztalatok. *Modern Geográfia, 1*, 23-36.

Chemli, S., Toanoglou, M., & Valeri, M. (2020). The impact of Covid-19 media coverage on tourist's awareness for future traveling. *Current Issues in Tourism*, 1–8. Advance online publication. doi:10.1080/13683500.2020.1846502

Criticism of the Shine logo: https://vehir.hu/cikk/53239-kivalasztottak-a-veszprem-balaton-2023-zrt-uj-ekf-arculatat

European Capital of Culture Application Manual. (n.d.). https://ec.europa.eu/programmes/creative-europe/sites/creative-europe/files/capitals-culture-candidates-guide_en_vdec17.pdf

European Capital of Culture information publication in English. (n.d.). https://ec.europa.eu/programmes/creative-europe/sites/creative-europe/files/library/capitals-culture-30-years-brochure_en.pdf

European Capital of Culture. (n.d.). https://ec.europa.eu/programmes/creative-europe/actions/capitals-culture_en

European Commission. (2013). *Pécs 2010, European Capital of Culture. Report*. https://ec.europa.eu/regional_policy/hu/projects/best-practices/hungary/2672

Film picnic 2020 program and information. (n.d.). www.veszprembalaton2023.hu/hu/events/filmpiknik

Government No 1468/2020 (VIII. 5) Korm.'s Decision on measures relating to the title of European Capital of Culture 2023. Hungarian Gazette 2020 Issue 183: https://magyarkozlony.hu/dokumentumok/07d96a0272f2a6c08461de0ecbf7bcb2abd31aa7/megtekintes

Kővári, E., Pásztor, J., & Raffay-Danyi, Á. (2021). A pannon egyetem közösségének kultúrafogyasztása, Az érzelmi és kulturális intelligencia összefüggése – veszprém 2023 európa kulturális fővárosa. *Vezetéstudomány, 52*(4), 48-62. doi:10.14267/VEZTUD.2021.04.05

Kumar, S., Valeri, M., & Shekhar. (2021, August 13). Understanding the relationship among factors influencing rural tourism: A hierarchical approach. *Journal of Organizational Change Management*. Advance online publication. doi:10.1108/JOCM-01-2021-0006

Lőrincz, K. (2017). A fenntarthatóság társadalmi vetületei az Európai Kulturális Főváros projektben. Comitatus különszám, 64-72.

Lőrincz, K., & Raffay, Á. (2019). "Beyond, azaz túllépni saját magunkon" - a turizmus szerepe a Veszprém2023 Európai Kulturális Főváros projektben. Turisztikai és Vidékfejlesztési Tanulmányok, 4(2), 18-38.

Melina Mercouri Award. (n.d.). https://ec.europa.eu/programmes/creative-europe/actions/capitals-culture_en

Michalkó, G., & Rátz, T. (2005). A kulturális turizmus élménygazdaságtani szempontjai. In A magyar városok kulturális gazdasága. MTA Társadalomkutató Központ.

Morvay, S., Rechnitzer, J., & Fekete, D. (2020). Európa Kulturális Fővárosai Kelet-Közép-Európában. *Tér és Társadalom, 34*(1).

Panyor, Á., Győriné Kiss, E., & Gulyás, L. (2011). *Bevezetés a desztináció menedzsmentbe*. Keszthely. http://www.tankonyvtar.hu/en/tartalom/tamop425/0034_bevezetes_a__desztinacio_men_jegyzet/bevezetes_a%20_desztinacio_menedzsmentbe_jegyzet.pdf

Rátz, T. (2014). Az EKF-turizmus mint lehetséges niche termék. A kulturális turizmus egy különleges területének vizsgálata. Turizmus Bulletin, 15(2), 24-33.

Soisalon-Soininen, T., & Lindroth, K. (2006). Regional Tourism Co-operation in Progress. In Tourism Local Systems And Networking. Elsevier Science & Technology.

Strategic cooperation of Veszprém-Balaton2023 Plc. with the University of Pannonia. (n.d.). https://veszprembalaton2023.hu/hu/news/veszprem-balaton-2023-egyuttmukodes-felsofokon

Toanoglou, M., Chemli, S., & Valeri, M. (2021). The organizational impact of COVID-19 crisis on travel perceived risk across four continents. *Journal of Organizational Change Management*. Advance online publication. doi:10.1108/JOCM-12-2020-0369

Valeri, M. (2016). Networking and cooperation practices in the Italian tourism business. *Journal of Tourism, Heritage & Services Marketing*, *2*(1), 30–35. doi:10.5281/zenodo.376333

Valeri, M. (Ed.). (2021a). Tourism risk. Crisis and recovery management. Emerald Publishing.

Valeri, M. (2021b). *Organization Theories. Implications for the Strategic Management*. Springer.

Valeri, M., & Baggio, R. (2020a). Social network analysis: Organizational implications in tourism management. *The International Journal of Organizational Analysis*. Advance online publication. doi:10.1108/IJOA-12-2019-1971

Valeri, M., & Baggio, R. (2020b). Italian tourism intermediaries: A social network analysis exploration. *Current Issues in Tourism*, *24*(9), 1270–1283. doi:10.1080/13683500.2020.1777950

Valeri, M., & Baggio, R. (2021). Increasing the efficiency of knowledge transfer in an Italian tourism system: A network approach. *Current Issues in Tourism*, 1–16. Advance online publication. doi:10.1080/13683500.2021.1937960

Valeri, M., & Katsoni, V. (Eds.). (2021). Gender and tourism: challenges and entrepreneurial opportunities. Emerald Publishing. doi:10.1108/9781801173223

Veszprém-Balaton2023 Image Manual. (n.d.). http://archiv.veszprembalaton2023.hu/arculat

Veszprém-Balaton2023 official website. (n.d.). www.veszprembalaton2023.hu

Virág, Á. (2016). A turisztikai együttműködések gazdasági szerepe. Studia Mundi – Economica, 3(1), 115-124.

Chapter 13
Legacy Economics and Meta-Management of Creativity in Destinations:
Legacy Economics for Destination Management

Angela Besana
IULM University, Italy

Maria Cristina Vannini
IULM University, Italy

ABSTRACT

The concept of "Legacy" might have many interpretations, going from tangible to intangible, and accordingly, it can include a series of items that can be bequeathed to different kind of "heirs." Either the giver or the community as main role players involved, the legacy is a bond for public administrations and destination managers. Legacy economics will be here connected to the innovative management of cultural destinations, where several and multi-goals-holders connect in order to provide resilience of villas, museums, and cultural assets together with increasing audiences and tourists. The collection of data about resilient creativity in Como (in the North of Italy) will show connections between legacy and management of public-private organizations for the last decade. A versatile and wide range of public and private interests together with national and international relevance will emerge. Community-type destinations will result as an original and innovative governance for resilience of culture and creativity connected with legacies and meta-management.

DOI: 10.4018/978-1-6684-3889-3.ch013

1. LEGACY: CONTENTS AND MISSIONS

Legacy is strictly connected to heritage and in its turn, heritage is to be distinguished from history, the former being a celebration of the latter, as the late Professor Emeritus David Lowenthal put it in his important publication (Lowenthal, 1998).

For many decades, experiential marketing, especially that applied to tourism, has established heritage as one of the pillars of the marketing mix connected to the storytelling of a place or of a product in general (Schmitt, 2010; Mossberger, 2008). According to this definition, heritage can be seen as a holder of stratified values connected to a piece of crystallised history and therefore sometimes it is made up or "corrected" in order to embellish a particularly unattractive storytelling or to add allure where this is missing (Lowenthal, 1998). This magical treatment derives from heritage being the content of a legacy, and therefore it gains the distinctiveness of richness. In fact, legacy is usually what our ancestors pass down to us: the patrimony or the assets, either tangible, in cash or in kind, or intangible, being what has been collected over one or many lifetimes and bequeathed upon posterity.

From a legal point of view, legacy derives from a will, a statement, a donation or any such document or action by which somebody (individual or legal persons, not-for-profit organisations included) wishes to transmit goods, assets, patrimony or heritage to one person or a group of people, aiming at preserving or increasing the assets. In most cases, the patrimony collected by a family, a corporation or an institution of any kind may include, as mentioned above, tangibles (i.e. money, buildings, land, collections of goods or art: materials, tools, machines, written pages, musical scores), but in many other cases, the patrimony is mainly intangible and includes know-how, expertise, education, cultural traditions, creative uniqueness in doing things, or the unique maintenance of a landscape, the so called "genius loci" (Prokopis, Farmaki, Saveriades & Spanou, 2019; Geng-Qing Chi, Pan & Del Chiappa, 2018; Govers & Go, 2009; Kavaratzis & Hatch, 2013).

The rationale of the legacy calls for a strict consistency between the "inherited items" and the recipients since in many cases, the bestower lists a series of requirements and restrictions the beneficiaries and receivers must follow to maintain hold of the heritage, receive the benefits and exploit its potential. In order to guarantee this consistency, Boards of Directors have the tendency to include as many stakeholders involved as possible, depending on the chosen legal form: trust, foundation, association, institution, legato, or even mere pre-emption (Lugosi, 2014).

Being able to manage the relationship between the community, the tourist and the legacy is no longer enough; what has become imperative is the ability to control the relationships among all the stakeholders in communities and the tourism sector who in one way or another make a contribution to the potential management of the 'inherited items'. This necessarily means legacies should be provided with some form of governing body, either public or private or a mixture of both (Coles et al., 2013), which can draw up a strategy to increase the quality of both the uniqueness of the heritage and any other local tourist attractions (Valeri, 2021b; Thees, Pechlaner, Olbrich & Schuhbert, 2020).

First of all, heritage must be strictly connected with the identity/patrimony of the giver and must fulfil the identity of the beneficiaries with its endorsement to communities and tourists.

The reasons someone wishes to bequeath a legacy might range from psychological and personal factors on the one hand, to economic, financial or fiscal ones on the other (Brewer, 2001).

Depending on both the motivation of the giver and the kind of assets, the giver might decide to leave the patrimony to a single person or to a small number of individuals – preferably through a trust that manages the legacy on their behalf; to a small or wider community; or to an institution.

In all these cases, the beneficiaries – individuals or communities – find themselves in the situation of accepting, sometimes forcefully, the will of the giver. It will be on the chosen legal entity established with the will and any enforcing and legal documents, to preserve and foster the original mission of the legacy, to manage the assets, to exploit possible development and distribute benefits. There is a second type of legacy which is intimately connected to the identity of a place – the "genius loci" which presents an inception that is definitely bottom-up (Cutcher, 2010; Saraniemi & Komppula, 2019). In this case, the beneficiaries have a much more relevant and active role because it is up to them to recognise a common identity and determine the means in which the identity manifests itself and reverberates on them. In some way, it can be stated that they work to create their own legacy. The constitution of this kind of legacy is what has led to the creation of the European Capitals of Culture of which Matera is an interesting example, and later on to the Italian Capitals of Culture (Palmer & Richards, 2004; Pepe & Percoco, 2020).

This process is the answer to a need of a local community, and it is what happens when a community asks for the recognition and the valorisation of its own identity. In this case, there is not just one giver but a multitude of givers who share their assets with the community in a sort of mutual adoption of the concept of legacy. Sometimes, it can even prompt the local administration, or the legal entity established to manage this sort of bottom-up legacy, to purchase the assets that the community considers an essential part of their cultural and historical legacy, i.e. Cernobbio, in the vicinity of Como, and the purchase and valorisation of Villa Bernasconi as a civic museum, (Vannini, Nasciuti, Suzzani & Tavecchio 2019).

In this second type of legacy two further aspects play a relevant role, one is cultural and the second one is the touristic exploitation of a cultural heritage asset. The final aim of this process, beyond the preservation of the heritage, its history and the related knowledge, is usually the economic well-being and improvement of a certain area, as the UNESCO Heritage list declares on a world-wide level, and as the diffusion of the concept of "ecomuseum" or "diffuse museum" demonstrates on a smaller and local level (Davis, 2009).

Differently from the "top-down" legacy where the assets are bestowed, generally, along with a financial endowment (i.e. the National Trust in the UK) to allow for the management and maintenance of the bequeathed heritage, in the "bottom-up" legacy it is essential to find a substantial and constant flow of money. Legacy economics are connected here to the innovative management of cultural destinations, where several and multi-goals-holders connect in order to ensure the resilience of villas, museums and cultural assets together with the growth of audiences and tourists.

An analysis of the experience of Como (in the north of Italy) will provide evidence on how much legacies are connected with the public spending of public administrations, who are leaders in the project management and meta-management, and with tourist flows in order to maximise the competitiveness and attractiveness of the destination. Meta-management can be meaningful during the start-up phase and implementation of a 'resilient project'. Nevertheless, from a long-term perspective, partnerships must be maintained and protected against unfulfilled commitments.

A versatile and wide range of public and private interests together with national and international interests emerged for the destination of Como from 2010 to 2017. This community-type destination resulted in an original and innovative public and private governance for resilience of culture and added value of creativity connected with legacies and meta-management. This paper is to be considered a starting point for wider investigations on this topic and is useful for managers of cultural foundations who want to explore the links between legacy, community and tourism.

Legacy economics here relates to the innovative management of cultural destinations, where several and multi-goals-holders connect in order to ensure the resilience of villas, museums and cultural assets together with increasing audiences and tourists. The collection of data about resilient creativity in Como will allow to show the connections between the legacy, governance and management of public-private assets over the last decade. A versatile and wide range of public and private interests together with national and international relevance will emerge. Community-type destinations will result in an original and innovative governance for the resilience of culture and creativity connected with legacies and meta-management.

2. LEGACY ECONOMICS. GOVERNANCE AND MANAGEMENT ISSUES

A legacy is both an asset and a commitment. This asset is evidence of the past and different stakeholders commit themselves to recalling, re-discovering, appreciating and integrating the legacy in a community, a destination and a landscape. As a consequence, an integrated and multidisciplinary framework that fosters the consistency of the legacy is needed.

Above all, it is necessary to contextualise and emphasise a valuable legacy and its resilience within the urban environment from a community (local) perspective to a tourist's (national and international) one. This approach should include distinctive attitudes, strategies and visions, estimating and shaping the environment and stakeholders in connection with the legacy. The governance should include communities, especially for bottom-up legacies and once communities have become aware of their legacies, tourists can be involved in creating an enduring and attractive destination, where all stakeholders are available for management and branding of this uniqueness (Thees, Pechlaner, Olbrich & Schuhbert, 2020; Butowski, 2019; De Lucia, Pazienza, Balena & Caporale, 2019; Biagi, Ladu, Meleddu & Royuela, 2019).

A holistic and inclusive strategy is here aimed at promoting social awareness of the legacy, and as a consequence, its sustainability and offer for citizens and tourists. The complexity of contents, stakeholders, governance and management could result in a chaotic and impossible challenge (Valeri, 2021b). This can be faced with an upgrading from local to international perspectives, from principal owners to private and public stakeholders, from marketing and fundraising to project management and partnerships. When the legacy has achieved full awareness, it is suitable for reuse and a cumulative and upgrading engagement of roles, resources, relationships, co-marketing and branding can be considered. It is a matter of a multitasking process with the goal of achieving a unique governance, management and branding of the legacy, wherever it is located in the urban landscape or not (Kumar et al., 2021; Geng-Qing Chia, Panc & Del Chiappa, 2018; Ávila-Robinson & Wakabayashi, 2018; Ashworth, 1988).

The landscape can be interpreted as a holistic network of natural, physical and socio-cultural elements (such as empathy and *nostalgia*) that interact and evolve with the qualities of the past, the present, and the future around unique assets like legacy (Gaoa, Linb & Zhang, 2020). If the legacy conveys local identity and offers points of reference to citizens, communities and visitors, the matching of multiple grading and upgrading is a product of continuous dynamic interaction between this asset, its distinctiveness and human activity and this combination is simultaneously produced, changed and consumed by different stakeholders at different locations and levels, from city to destination, from destination to landscape (Kim, Stylidis & Oh, 2019). Because of its pluralistic and diversified nature, it is crucial to acknowledge and engage all relevant types of stakeholders, their networks, their partnerships, with a focus on their types and skills during the re-creation of a legacy in order to lead and plan its comprehensive development (legacy governance and management).

Such multiple governance can result in the best match of public and private managers, together with fund-givers and sponsors, who can lead and pool funds for which they play managing roles. In Table 1, stakeholders increase and evolve in the destination with a focus on a legacy and, if stakeholders are usually understood as a hierarchical governance, they are here playing poliarchy roles with public administrations and private associations, foundations, clubs, citizens' committees, whose matching and partnerships can both maximise resources and international grants. The best collection and allocation of resources can result in a competitive and sustainable destination for community and tourists (both national and international tourists) with different performances and impacts on citizens and tourist flows (Yanga, Chenb, Sud & Yu-Xia, 2020; Vareiro, Ribeiro & Remoaldo, 2019).

Table 1. Evolution and upgrading of a competitive and sustainable destination with a focus on legacy

	STAKEHOLDERS **FROM hierarchy → TO poliarchy**			
TO multiple and matched GOVERNANCE ↑ FROM single governance			DESTINATION with FOCUS ON LEGACY	TO international tourists ↑ FROM community
	↑ .... →			
	DESTINATION with no focus on legacy			
	RESOURCES FROM local → TO international grants			

Source: own

This presents an opportunity to link the past with the present in an innovative and challenging manner, while stimulating future development within the themed landscape. This happens thanks to international grants much more than local funding and thanks to a poliarchy much more than a hierarchy. Themed urban landscapes can link the elements of legacy (above all, if a heritage) throughout the cityscape and in a business model where a multiple and matched governance is better than a single one and it can be enabled with a community full of empathy, advocacy and vision for local growth and tourism growth.

Including architecture as a central theme within a landscape, every stakeholder is aware of the urban identity as well as the protection of the heritage through its positioning within a sustainable cultural context. The latter needs to be an inherent part of the tourism strategy. A themed landscape is not merely the sum of the present and historical elements. Selection and packaging are crucial in the creation of a themed tourism product, whose focus can be on legacy. It is therefore important to find the balance between a condensed and manageable theme versus the implications of offering different creative contents. Failing to acknowledge all stakeholders involved increases the risk of presenting a simplified and selective thematisation and will result in a lack of sense of place and community (Albattat et al., 2020; Chemli et al., 2020; Toanoglou et al., 2021; Valeri, 2021a). If the community is not engaged, the legacy will fail to achieve sustainability and attractiveness for tourists.

Because of the uniqueness and the specific identity of the legacy, it is not possible to apply a 'one size fits all' blueprint regarding the development of the tourism product. Implementing a themed landscape approach implies more than merely adding up similar elements. It requires engagement, stakeholders, story-telling, planning and a holistic approach. The lack of an integrated approach may result in a lack of coordination, structured network actors and systematic dialogue whilst increasing further simplification, spot co-marketing and fragmentation (Jansen-Verbeke, 2009; McKercher & Ho, 2006; Markwell, Stevenson & Rowe, 2004; Burmil & Enis, 2004; Valeri, 2016; Valeri & Baggio, 2020a; 2020b; 2020c; 2021; Baggio & Valeri, 2020). This may hinder the positioning of the destination in a broader context.

What is the core of a multiple and efficient management of a legacy, without fragmentation and simplification?

It is a matter of meta-management both of the legacy and the destination. Marketing (and social media marketing) and fundraising can be a pivotal strategy. Project management can include them for every event and plan of reuse, in order to signal competitiveness in a very keenly competitive context of cultural and creative offers. Nevertheless, fragmentation in the governance and resources can be a weakness, as it is in destinations where the focus is either forgotten or misplaced.

Local administrations can be leaders and an efficient node of a unique management of the poliarchy. They collect opinions, agreements, votes, resources, including both public and private grants, and both national and international funding. They can enable legacies, firstly for communities and secondly, they can enable them for tourists. For this goal, they can collect partners around a unique branding of the legacy, so that resources are not wasted on spot and fragmented promotion.

As a consequence, branding is a pivotal strategy. Distinctiveness through branding is of primary importance, in order to succeed with marketing and fundraising. A destination with a focus on a legacy is primarily for the community and not for mass tourism. Targets refer to strong and keen interests in that history and in that past, with brand awareness and connections with the original contents of the legacy, even if it is transformed and changed to be transmitted to next generations (sustainable legacy). Otherwise, future generations would rapidly disengage and lose their advocacy. Branding can strengthen community engagement and attract tourists. The brand may be the unique story-teller, around which stakeholders can commit themselves for long-term and binding partnerships.

In conclusion, a competitive and sustainable destination with a focus on a legacy is in need of multiple governance and multiple strategies: branding for uniqueness is one of the leading strategies.

3. THE DESTINATION OF COMO AND ITS LEGACIES FOR COMMUNITIES AND TOURISTS

Bounded by mountains and lake, Como is a destination for cultural tourism with fairs, events and meetings of several forms and motivations of tourism. In Como and in the vicinity (province), legacies have been enabled for decades, with the bottom-up approach of communities becoming aware and fully engaged in the valorisation process.

In Lombardy, Como ranks third in tourist flows after Milan and Brescia, and it is the first for international tourist flows, In 2019, Como was the seventh province for the share of international tourist spending (Rome, 16.9%; Venice, 9.3%; Milan, 9.1%; Florence, 6.6%; Naples, 4.1%; Verona, 2.9%; Como, 2.7%; Bolzano, 2.7%; Turin, 2.5%; Trento, 2% according to *Federazione delle Associazioni Italiane Alberghi e Turismo*).

Legacy includes heritage, from monuments to churches, from museums to archaeology, from villas to gardens. Public administrations collect resources from different sources (both national and international) and they can be leaders of pooled funds together with grant-making foundations (banks and community led ones), sponsors and citizens, so that multi-stakeholder fundraising results in multiple and matched governance, which positively affects legacy restitution and tourism. Como and the vicinity benefit from bottom-up legacies.

For example, the Cernobbio local administration has been recently involved in the restoration and re-opening of Villa Bernasconi and the poliarchy included citizens, public and private grant-makers (from grant-making foundations to the chamber of commerce). For Cernobbio, the municipality planned the recovery of an Art Nouveau Villa as a "city museum". The villa had belonged to Davide Bernasconi, an entrepreneur at the end of the 19th century who made a fortune for his family and for the town out of silk and textile industries for almost one century. His action, though, gave birth to an industrial town (similar to the UNESCO Crespi Village at Crespi d'Adda) in the paternalistic entrepreneurship vogue following the second Industrial Revolution. Afterwards his memory vanished rapidly. Nevertheless, memories and connections with local multi-stakeholders re-emerged as soon as the municipality initiated the museum project based on the engagement of the residents – relatives of former workers/suppliers/clients of the Bernasconi industry – to collect materials, stories and objects to create the narrative of the museum. The Villa, now not only bears witness to Bernasconi's business but it is the teller of the history of the town for two centuries. While also attracting foreign visitors – panels and labels are written in both Italian and English – the main target is local and cross-generations like the multi-stakeholders (chamber of commerce, local design and arts and craft schools…) who were involved in the project.

As demonstrated in the Villa Bernasconi example, the community plays a role in the co-creation of value, and the Como municipality in investing. Community engagement is demonstrated in meetings, when the municipality looks for opinions, approval and involvement of citizens, who are taxpayers and are, as a consequence, giving resources as stakeholders in a poliarchy. The community is given evidence also in groups, committees, associations and foundations like FAI (*Fondo Ambiente Italiano*), which is in charge of monuments, villas, gardens and other kinds of heritage around Lake Como.

Revenues and grants from public administrations of Como have been constantly increasing, though the focus will be here dedicated to expenditures. This results in a meta-management of private and public stakeholders for different types of legacies, for different and available resources, for versatile efforts and expenditures and for the benefit of citizens and tourists. Como legacies include different forms of heritage: museums, villas, gardens, where multiple events take place. The public administration collects resources, delivering grants for current spending on to culture and tourism events and delivering funds for investing in assets to be restored, maintained and returned in the form of a bottom-up legacy for communities and tourists.

If restoration and valorisation expenditures are included in expenses for investments (for culture), along with these expenses, current expenditures for culture mainly concern events relating to these heritage sites. Nevertheless, current expenditures for tourism can relate to fairs, meetings and promotion of these heritage sites, and marketing and branding are included. Branding can include different efforts and multimedia, while combining different legacies around different landscapes and their stakeholders. The Lake Como Restart Video is an example of Como and Lecco Chamber of Commerce as a stakeholder with strong commitments to branding around a collection of multiple heritages and uniqueness around the values and visions of this destination. Both kinds of expenditure can include expenditure for staffs, creation of consortia and intermediate or ancillary governances and administrative collaterals for partnerships who are in charge of fundraising, grant-making, marketing and branding.

3.1. Research Questions

The multiple and efficient management of Como destination can be to the responsibility of the municipality, whose governance is matched with other public and private stakeholders in order to collect and maximise resources, from local to international ones. Thanks to these funds, the municipality enables legacies for different audiences, from community to national and international tourists. This poliarchy is the governance, who manages legacies, whose cultural sustainability and competitiveness are both founded on expenditure.

- How much do local expenditures for culture and tourism affect Italian and international tourism?
- What kind of expenditure can affect legacy for tourism?

3.2. Methodology and results

First of all, we collected data from different datasets: Italian public administrations' open data, statements of revenues and expenses of the Como local administration, and data from Consorzio Como Turistica and Polys Observatory. These data include the number of tourists in Como (the territory of the town and province), the shares of tourists opting for hotel accommodation or not, shares of Italian and international tourists in the town of Como and in the province of Como, public expenditures of the local administration for culture and for tourism, both current spending and investments. Expenditures are supported by resources from the local administration, with the transfer of funds from central administrations, national and international grants, local taxes, private grants and any other revenue. These data were collected from 2010 to 2017. The time series refers to the years before and soon after the Milan World Expo, a world event with an impact on Milan and the nearby landscape. Above all, Como could profit from the externality of this world event: 2,569,997 tourists in 2014 both in Como and in the province of Como; 2,760,074 in 2015 and 2,874,038 in 2016.

If legacy is to be included in expenditures for the destination of Como, whose managers pooled funds and grants from different sources, cultural investments increased during the whole period, with the top amount being more than 1.6 million euro in 2017. During the same period the current expenditure for culture diminished, though it was more than 2.6 million euro in 2017. For tourism, the current expenditure was at its highest at the beginning of the period (648,616 euro in 2010) and was 414,919 in 2017. Tourism investments were zero in 2017 and they topped 304,568 euro in 2014, just before Expo.

In order to connect the expenditures of a poliarchy of resources and stakeholders, we interpolated the shares of Italian and international tourists with current expenditures and investments, in order to provide evidence on how much the efforts of the public leader of poliarchy affect flows. International tourism dominated throughout the investigated period, between 68.7% in 2010 and 80.3% in 2017. Italian tourists made up 31.3% in 2010 reducing to 19.7% in 2017. Current expenditures for culture are positively correlated with the share of Italian tourists and they are negatively correlated with international (foreign) tourists. If the current expenditure refers to legacies with strong distinctiveness and identity in this themed landscape, where communities are empowered, engaged and resilient, the current expenditure and public efforts can positively affect citizens and national tourism much more than international tourism.

Figure 1. Current Expenditure for Culture and International Tourists

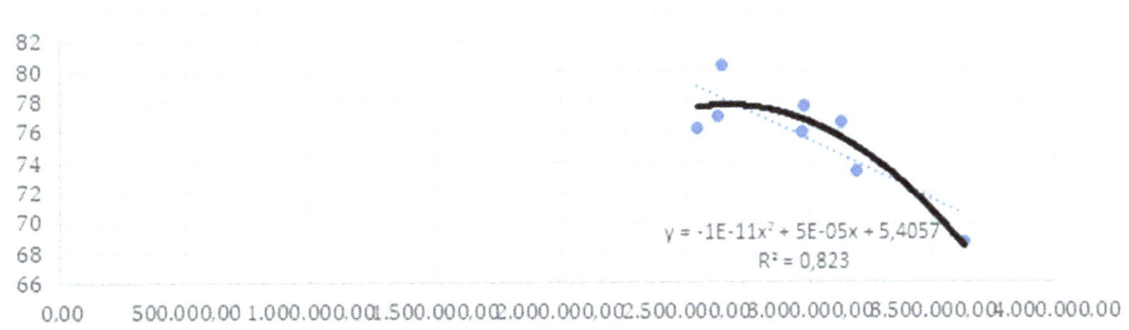

At the same time, current expenditure for tourism is positively correlated with the share of Italian tourists and is negatively correlated with international (foreign) tourists (Figure 3 and Figure 4).

Figure 2. Current Expenditure for Culture and Italian Tourists
Source: own

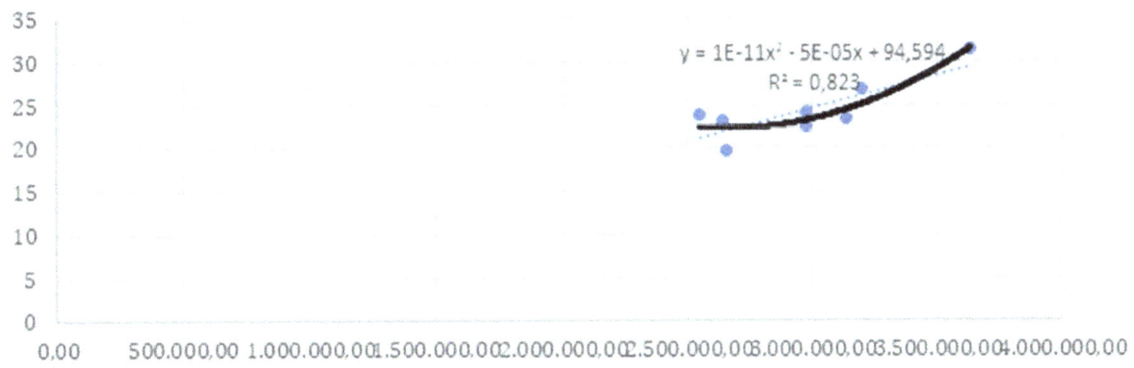

Figure 3. Current Expenditure for Tourism and International Tourists

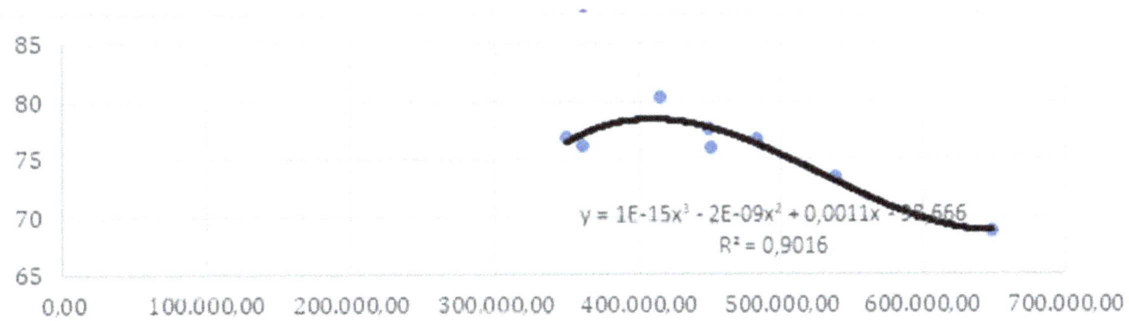

Figure 4. Current Expenditure for Tourism and Italian Tourists

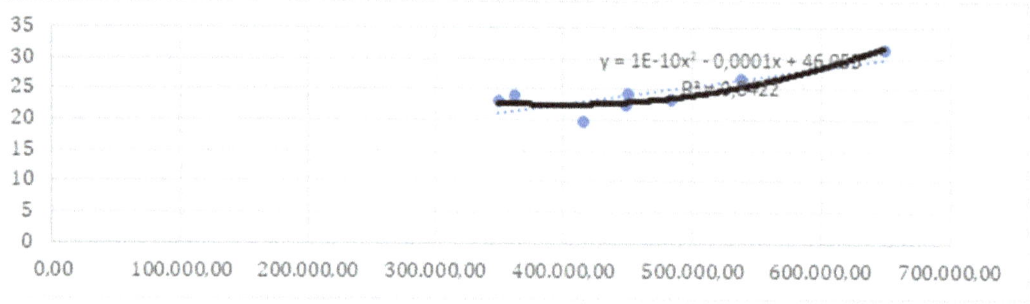

If shares of tourist flows are still separated and analysed for their correlation with public expenditures for cultural investments, there is no evidence of a strong and unique correlation between investments for culture and Italian and international tourism (Figure 5 and Figure 6). This can confirm that investments are, firstly, for citizens and, only as a subsequent goal and impact, tourists are a target. The poliarchy is supplying cultural investments for citizens. National and international tourist flows can profit from these investments and beautification, after communities have played their roles in the restitution process.

Figure 5. Investments for Culture and International Tourists

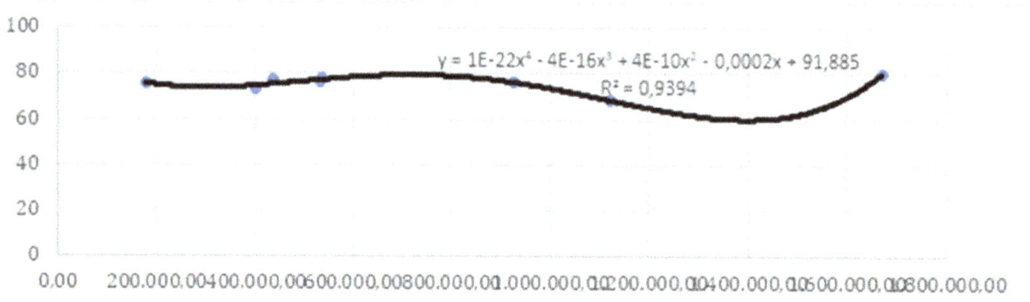

Figure 6. Investments for Culture and Italian Tourists
Source: own

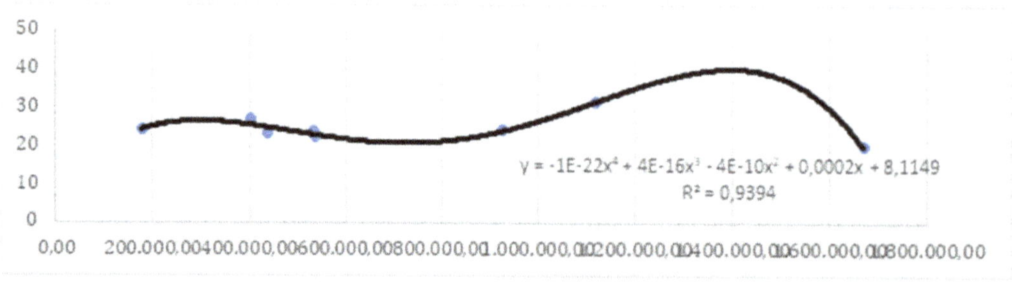

These polynomial regressions were significant for separate correlations between two series of data, for shares of national and international tourism. In order to investigate how much public expenditures are correlated with tourist flows, we implemented a multiple regression. If the legacy reuse increases with cultural investments, this kind of expenditure has a positive impact on tourists, as for the multiple regression we implemented this stepwise with Jump Statistics. At the same time, current expenditure for tourism is positively correlated with tourist flows, both national and international.

Tourists = 4,579,240.9 +

-0.94 CultureCurrentExpenditure +

+ 2.04 TourismCurrentExpenditure +

+ 0.09 CultureINVESTMENTS +

- 0.58 TourismINVESTMENTS

Table 2 Output Summary of multiple regression.

Summary of Fit					
RSquare	0.994736				
RSquare Adj	0.987716				
Root Mean Square Error	19956.11				
Mean of Response	2730253				
Observations (or Sum Wgts)	8				
Analysis of Variance					
Source	DF	SS	Mean Square	F Ratio	
Model	4	2.258e+11	5.64e+10	141.7169	
Error	3	1194739130	398246377	Prob > F	
Total	7	2.269e+11		0.0010	
Parameter Estimates					
Term	Estimate	Standard Error	T Ratio	Prob > |t|	
Intercept	4579240.9	207400	22.08	0.0002	
Current Expenditure for Culture	-0.94365	0.148277	-6.36	0.0079	
Current Expenditure for Tourism	2.0490577	0.548056	3.74	0.0334	
Investments for Culture	0.0980485	0.025071	3.91	0.0297	
Investments for Tourism	-0.581085	0.048693	-11.93	0.0013	
Effect Tests					
Source	Nparm	DF	Sum Squares	F Ratio	Prob > F
Current Expenditure for Culture	1	1	1.613e+10	40.5018	0.0079
Current Expenditure for Tourism	1	1	5566861405	13.9784	0.0334
Investments for Culture	1	1	6090819360	15.2941	0.0297
Investments for Tourism	1	1	5.671e+10	142.4117	0.0013

Source. Elaboration with Jump Software.

For the variables in this equation, tourists increase both with increasing current expenditure for tourism and with increasing investments for culture. If the poliarchy is firstly giving legacies back to communities, tourist flows are not indifferent to events like fairs and festivals (tourism current expenditures) and to investments in monuments, heritage, cultural and creative assets in a destination and can include the supply of versatile and diversified bundles. Tourism current expenditures can include multimedia efforts (videos, social media, etc.) with different emphasis on different legacies (gardens/lake/mountains, tangible/intangible, culture/creativity, craftsmanship/manufacturing memory ...) and branding. This branding can result in a *versioning* of the same uniqueness to accommodate different tastes and the motivations of national and international tourists. Cultural tourists are the targets of this branding, though business tourists are often attracted to conferences and meetings in Como as they appreciate its legacies.

4. CONCLUSION. META-MANAGEMENT WITH A FOCUS ON LEGACY

When communities and local administrations collaborate in order to restore and reopen heritage sites, whose creativity is founded on identity, memory and history, the co-created value is available for tourists. With increasing tourist flows, the destination of Como has aggregated stakeholders and funds around legacies and there is a positive correlation between national tourist flows and current expenditures both for culture and for tourism. If data on tourists are summed up for national and international flows, the positive correlation is tested with current expenditure for tourism and investments for culture. These investments give back and enable legacies for citizens and, as a consequence, for tourists. When a destination is paying attention to legacies, the management is a poliarchy of different stakeholders, and expenditures can positively affect tourism.

Figure 7 shows the tree of legacy restitution. The roots are in a legacy which is firstly a co-created value of communities and enabling administrations. Multiple and worthwhile projects are then financed thanks to partnerships with multiple stakeholders like grant-making foundations, sponsors, public administrations and public funding from national and regional administrations. From roots to branches, stakeholders and their funds, they both support a poliarchy that can protect and brand the legacy to the benefit of not only communities but also for national and international tourists.

Creativity and legacy are bound together when media (and social media, too) are helpful for story-telling and innovative delivery of contents of what is memory (the legacy).

Branches of this tree for Como are both efficient and valuable for communities and tourists. On one side, the positive correlation of tourists with investments for culture, it shows that investments were available for citizens and they have also gained the attention of tourists. On the other side, the positive correlation with the current expenditure for tourism shows that once legacy is available, the importance of promotion, branding and events is not dismissed in order to grant tourists' advocacy and the sustainability of legacy both for communities and tourists.

Branding can be very useful where partnerships are falling away and commitments need to be strengthened. Branding can be approached with different efforts and engagement of the poliarchy: from the chamber of commerce to associations of hotels, from consortia to municipalities. Como stakeholders have jointly developed and enhanced mixed governance, co-marketing and branding for at least a decade.

Multimedia have been combined with traditional advertising, in order to reach international audiences, so that their flows still prevail. Cultural foundations, who are owners and/or managers of legacies, have been profiting from this poliarchy for their story-telling of culture and creativity.

A versatile and wide range of public, private, national and international interests and relevance is evidenced for Como from 2010 to 2017. This community-type destination results in an original and innovative mixed governance (poliarchy) for resilience of culture and contribution of creativity connected with legacies and meta-management.

This paper is to be considered a starting point for wider investigations on this topic and it is useful for managers of cultural foundations who want to explore the links between legacy and community. Once community is fully aware of the legacy, community is engaged and engaging cultural foundations (both operating and grant-making) with multiple stakeholders. All features of legacies can become relevant for communities and tourists, if governance grants multi-stakeholders' engagement. Thanks to meta-management, multiple governance, poliarchy, funds and grants, resilience is guaranteed together with increasing international tourist flows. From the roots of the legacy, branches of a growing poliarchy are delivering memory, contents and visions of heritage, which are given back to communities and are available for tourists who are looking for multiple versions of themed landscapes.

Figure 7. The tree of legacy restitution
Source: own

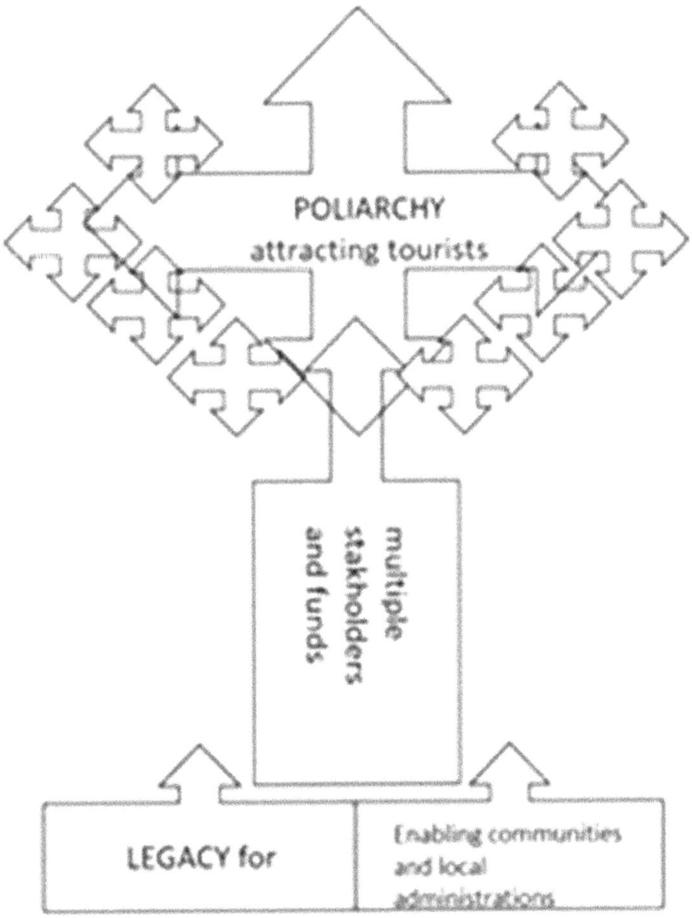

REFERENCES

Aitchison, C., MacLeod, E. N., & Shaw, J. S. (2000). *Leisure and Tourism Landscapes. Social and cultural geographies*. Routledge.

Albattat, A., Jamaludin, A., Zuraimi, N. S. M., & Valeri, M. (2020). Visit Intention and Destination Image in Post- Covid- 19 Crisis Recovery. *Current Issues in Tourism*. Advance online publication. doi :10.1080/13683500.2020.1842342

Antrop, M. (2006). Sustainable landscapes: Contradiction, fiction or Utopia? *Landscape and Urban Planning*, *75*(34), 187–197. doi:10.1016/j.landurbplan.2005.02.014

Ashworth, G. (1988). Marketing the historic city for tourism. In B. Goodall & G. Ashworth (Eds.), *Marketing in the Tourism Industry The Promotion of Destination Regions*. Routledge.

Ávila-Robinson, A., & Wakabayashi, N. (2018). Changes in the structures and directions of destination management and marketing research: A bibliometric mapping study, 2005–2016. *Journal of Destination Marketing & Management*, *10*, 101–111.

Baggio, R., & Valeri, M. (2020). Network science and sustainable performance of family businesses in tourism. *Journal of Family Business Management.* doi:10.1108/JFBM-06-2020-0048

Biagi, B., Ladu, M. G., Meleddu, M., & Royuela, V. (2019). Tourism and the city: The impact on residents' quality of life. *International Journal of Tourism Research*, *22*(2), 168–181. doi:10.1002/jtr.2326

Brewer, M. B. (2001). The many faces of social identity: Implications for political psychology. *Political Psychology*, *22*(1), 115–125. doi:10.1111/0162-895X.00229

Burmil, S., & Enis, R. (2004). An Integrated Approach to Landscape and Planning. *Journal of Architectural and Planning Research*, *21*, 140–151.

Butowski, L. (2019). Tourist sustainability of destination as a measure of its development. *Current Issues in Tourism*, *22*(9), 1043–1061. doi:10.1080/13683500.2017.1351926

Chemli, S., Toanoglou, M., & Valeri, M. (2020). The impact of Covid-19 media coverage on tourists' awareness for future traveling. *Current Issues in Tourism*, 1–8. Advance online publication. doi:10.1080/13683500.2020.1846502

Coles, T., Fenclova, E., & Dinan, C. (2013). Tourism and corporate social responsibility: A critical review and research agenda. *Tourism Management Perspectives*, *6*, 122–141. doi:10.1016/j.tmp.2013.02.001

Cutcher, L. (2010). Local heroes: Co-producing history, 'community' and the self. *Consumption Markets & Culture*, *13*(1), 79–90. doi:10.1080/10253860903346773

Davis, P. (2009). Ecomuseums and the representation of place. *Rivista Geografica Italiana*, *116*, 483–503.

De Lucia, C., Pazienza, P., Balena, P., & Caporale, D. (2019). Exploring local knowledge and socio-economic factors for touristic attractiveness and sustainability. *International Journal of Tourism Research*, *22*(1), 81–99. doi:10.1002/jtr.2320

Du Cros, H., & McKercher, B. (2001). A New Model to Assist in Planning for Sustainable Cultural Heritage Tourism. *International Journal of Tourism Research*, *3*(2), 165–170. doi:10.1002/jtr.297

Gaoa, J., Linb, S., & Zhang, C. (2020). Authenticity, involvement, and nostalgia: Understanding visitor satisfaction with an adaptive reuse heritage site in urban China. *Journal of Destination Marketing & Management*, *15*, 100404. doi:10.1016/j.jdmm.2019.100404

Geng-Qing Chi, C., Pan, L., & Del Chiappa, G. (2018). Examining destination personality: Its antecedents and outcomes. *Journal of Destination Marketing & Management*, *9*, 149–159. doi:10.1016/j.jdmm.2018.01.001

Govers, R., & Go, F. (2009). *Place branding: Glocal, virtual and physical identities, constructed, imagined and experienced*. Palgrave Macmillan. doi:10.1007/978-0-230-24702-4

Jansen-Verbeke, M. (2009). The territoriality paradigm in cultural tourism. *Tourism (Zagreb)*, *19*(1-2), 25–31.

Jansen-Verbeke, M. (2009). The territoriality paradigm in cultural tourism. *Tourism (Zagreb)*, *19*(1-2), 25–31.

Kavaratzis, M., & Hatch, M. J. (2013). The dynamics of place brands: An identity-based approach to place branding theory. *Marketing Theory*, *13*(1), 13. doi:10.1177/1470593112467268

Kim, S., Stylidis, D., & Oh, M. (2019). Is perception of destination image stable or does it fluctuate? A measurement of three points in time. *International Journal of Tourism Research*, *21*(4), 447–461. doi:10.1002/jtr.2273

Kumar, S., Valeri, M., & Shekhar. (2021, August 13). Shekhar (2021). "Understanding the relationship among factors influencing rural tourism: A hierarchical approach. *Journal of Organizational Change Management*. Advance online publication. doi:10.1108/JOCM-01-2021-0006

Lowenthal, D. (1998). *The heritage crusade and the spoils of history*. Cambridge University Press. doi:10.1017/CBO9780511523809

Lugosi, P. (2014). Mobilising identity and culture in experience co-creation and venue operation. *Tourism Management*, *40*, 165–179. doi:10.1016/j.tourman.2013.06.005

Markwell, K., Stevenson, D., & Rowe, D. (2004). Footsteps and memories: Interpreting an Australian urban landscape through thematic walking tours. *International Journal of Heritage Studies*, *10*(5), 457–473. doi:10.1080/1352725042000299063

McKercher, B., & Ho, S. Y. P. (2006). Assessing the Tourism Potential of Smaller Cultural and Heritage Attractions. *Journal of Sustainable Tourism*, *14*(4), 473–488. doi:10.2167/jost 620.0

Mossberg, L. (2008). Extraordinary Experiences through Storytelling. *Scandinavian Journal of Hospitality and Tourism*, *8*(3), 195–210. doi:10.1080/15022250802532443

Palmer, R., & Richards, G. (2004). European Cities and Capitals of Culture. In *Study prepared for the European Commission*. Palmer-Rae Associates.

Pepe, A., & Percoco, A. (2020). *Quel che resta della Legacy di Matera 2019 dopo il Covid-19. Fondazione E.* Mattei.

Prokopis, A. C., Farmaki, A., Saveriades, A., & Spanou, E. (2019). The "genius loci" of places that experience intense tourism development. *Tourism Management Perspectives*, *30*, 19–32. doi:10.1016/j.tmp.2019.01.002

Saraniemi, S. (2011). From destination image building to identity-based branding. *International Journal of Culture, Tourism and Hospitality Research*, *5*(3), 247–254. doi:10.1108/17506181111156943

Saraniemi, S., & Komppula, R. (2019). The development of a destination brand identity: A story of stakeholder collaboration. *Current Issues in Tourism*, *22*(9), 1116–1132. doi:10.1080/13683500.2017.1369496

Schmitt, B. (2010). Key concepts of experience marketing. *Foundations and Trends in Marketing*, *5*(2), 55–112. doi:10.1561/1700000027

Thees, H., Pechlaner, H., Olbrich, N., & Schuhbert, A. (2020). The Living Lab as a Tool to Promote Residents' Participation in Destination Governance. *Sustainability*, *12*(3), 1–27. doi:10.3390u12031120

Toanoglou, M., Chemli, S., & Valeri, M. (2021). The organizational impact of COVID-19 crisis on travel perceived risk across four continents. *Journal of Organizational Change Management*. Advance online publication. doi:10.1108/JOCM-12-2020-0369

Valeri, M. (2016). Networking and cooperation practices in the Italian tourism business. *Journal of Tourism, Heritage & Services Marketing*, *2*(1), 30–35. doi:10.5281/zenodo.376333

Valeri, M. (Ed.). (2021a). Tourism risk. Crisis and recovery management. Emerald Publishing.

Valeri, M. (2021b). *Organization Theories. Implications for the Strategic Management*. Springer.

Valeri, M., & Baggio, R. (2020a). Social network analysis: Organizational implications in tourism management. *The International Journal of Organizational Analysis*. Advance online publication. doi:10.1108/IJOA-12-2019-1971

Valeri, M., & Baggio, R. (2020b). Italian tourism intermediaries: A social network analysis exploration. *Current Issues in Tourism*, *24*(9), 1270–1283. doi:10.1080/13683500.2020.1777950

Valeri, M., & Baggio, R. (2020c). A critical reflection on the adoption of blockchain in tourism. *Journal Information Technology and Tourism*, *23*(2), 121–132. doi:10.100740558-020-00183-1

Valeri, M., & Baggio, R. (2021). Increasing the efficiency of knowledge transfer in an Italian tourism system: A network approach. *Current Issues in Tourism*, 1–16. Advance online publication. doi:10.1080/13683500.2021.1937960

Vannini, M.C., Nasciuti, G., Suzzani, V., & Tavecchio, G. (2019). Brand building analysis report Villa Bernasconi, Internal report. *Proceedings of Progetto LIBERTYTUTTI – Fondazione Cariplo and Progetto Maison Bernasconi – Regione Lombardia.*

Vareiro, L., Ribeiro, J. C., & Remoaldo, P. C. (2019). What influences a tourist to return to a cultural destination? *International Journal of Tourism Research*, *21*(2), 280–290. doi:10.1002/jtr.2260

Yanga, Y., Chenb, M.-H., Sud, C.-H., & Yu-Xia, L. (2020). Asymmetric effects of tourist arrivals on the hospitality industry. *International Journal of Hospitality Management*, *84*, 102323. doi:10.1016/j.ijhm.2019.102323

Chapter 14
Reaching the New Tourist Through Creativity:
Sustainable Development Challenges in Croatian Coastal Towns

Christian Stipanović
University of Rijeka, Croatia

Elena Rudan
Faculty of Tourism and Hospitality Management, University of Rijeka, Croatia

Vedran Zubović
Faculty of Tourism and Hospitality Management, University of Rijeka, Croatia

ABSTRACT

In today's modern world, creative expression is opening up new dimensions of business and new opportunities for economic development. One field of economic activities in which this is evident is tourism. Creativity in tourist destinations can be viewed in different ways, for example, through creative action (undertaken by destination management, residents, entrepreneurs, and tourists) and through creative spaces and creative events. Creativity plays a vital role in all elements involved in the creation of a destination's offering, regardless of which form of tourism is the focus of development efforts. Given the growing role of self-actualisation of individuals in society and the displaying of social status, creativity has in the past 20 years begun to positively impact on economy activities taking place in tourist destinations. Creativity is especially important in developing cultural tourism in all its sub-types, where it is seen as a means of animating and adding value to cultural heritage locations.

INTRODUCTION

In today's modern world, creative expression is opening up new dimensions of business and new opportunities for economic development. One field of economic activities in which this is evident is tourism.

DOI: 10.4018/978-1-6684-3889-3.ch014

Creativity in tourist destinations can be viewed in different ways, for example, through creative action (undertaken by destination management, residents, entrepreneurs and tourists) and through creative spaces and creative events. Creativity plays a vital role in all elements involved in the creation of a destination's offering, regardless of which form of tourism is the focus of development efforts. Given the growing role of self-actualisation of individuals in society and the displaying of social status, creativity has in the past 20 years begun to positively impact on economy activities taking place in tourist destinations (Hartley et al., 2012). Creativity is especially important in developing cultural tourism in all its sub-types, where it is seen as a means of animating and adding value to cultural heritage locations (Richards, 2020).

Creativity in tourism has still not been conceptually defined because creativity in tourism focuses on a broad range of tourism and hospitality service sectors and related industries. In light of this, a question arises as to the role of developed creative industries as platforms for achieving creativity in tourist destinations, in particular in urban spaces that until recently have primarily based their offerings on static resources.

1. CREATIVITY IN CITY TOURISM DEVELOPMENT

Creativity in tourism can best be described as an opportunity for tourists to develop their creative potential by actively participating in learning processes that are characteristic of the destination in which they are staying (Richards & Raymond, 2000). Bacon (2005) sees creativity as originality and usefulness, while Virfinija (2016) argues that creativity in tourism is a part of cultural tourism; it can greatly contribute to developing new experiences thus driving social, cultural and economic growth, and its focus is on intangible resources such as learning, developing experiences and traditions. The growing role of creativity in tourism is further bolstered by the generation of the value-added of tourism products, which is reflected in the creative economy through the aspects of creative space, creative events and creative tourism (Richards & Wilson, 2006). Creative space gathers creative producers and artisans to make a creative environment for tourism consumption (Marques & Richards, 2014). Creative environments encourage the development of creative industries. The United Nations Conference on Trade and Development (UNCTAD) views creativity in the context of the creative industries as industries that make, produce and commercialise intangible goods and services of a cultural nature. Creativity can also be viewed from the perspective of sustainability that ensures the authenticity of experiences of local cultures (Raymond, 2007) or from the perspective of participative learning in the field of art or cultural heritage (Garcês, Pocinho & Jesus, 2018). Enabling creativity helps to ensure sustainability and, vice versa, sustainability helps to ensure future creativity. The future of tourism development is not only about the number of attractions and how they should be market managed. It is also about the creativity of projects and programmes and the ability to bring together and coordinate sustainable development-based projects (Valeri, 2015).

In recent years, the term "creativity" has often emerged in the strategic thinking of tourism development aimed at designing new tourism offerings (Rudan, 2012). The broad spectrum of diverse directions of creativity in tourism represents a challenge to entrepreneurial initiatives in innovating and managing the tourism product and the sustainable development of a destination. The creativity of entrepreneurial initiatives in any destination is determined by spatial dynamics, which is defined by the specific processes taking place within a geographical region, processes that involve creating an offering and managing the destination product and market (Berg & Hassink, 2014). Turok (2009) notes that in seeking new ways to stand out in a global economy, tourist destinations are increasingly turning to the creative industries for help. Such development often hinges on a combination of assets, knowledge and skills on which the creative economy is based. In Croatia, creativity began to be widely applied to tourism facilities such as design hotels, specialised museums, art galleries and wineries. The incorporation of creativity into a tourism context has made the diversification strategy of vital importance, in particular in the sphere of cultural tourism. In modern market conditions, cultural tourism in all its sub-forms uses creativity as an element in creating distinctiveness and improving competitiveness.

Cities are where creative economies are developing at the fastest pace and, as such, cities should become creative tourist destinations, reinforced through their cultural policies. Krstinić Nižić, Rudan and Trinajstić (2019) emphasise that in the past decade the cultural and creative industries have become a popular line of local city development. Many cities are using these industries as tools, particularly in synergy with tourism, to boost their economic development and become more visible in the broader environment. According to Richards (2020), growing cities seek to keep abreast of tourist trends focused on contemporary creativity, innovativeness and intangible creations. Cities are transforming into creative industry centres, with each city possessing its own specific creative power. Cities are using the creativity of their entrepreneurs and their spaces to join forces with other cities to foster new creative impulses. For example, creative cities can build networks such as the UNESCO Creative Cities Network. Such networks are created to promote cooperation among those cities that have incorporated creativity as a strategic factor into their sustainable urban development (Valeri, 20216; Valeri and Baggio, 2020a; 2020b; 2020c; 2021; Baggio and Valeri, 2020). The Cultural and Creative Cities Monitor, developed by the European Commission, underscores culture and creativity as vital concepts of European Union development, with creativity providing ways to reinterpret heritage (European Commission, 2019). According to Oliveira & Paulino (2017), the elements based on which the level of creativity in cities is defined are talent, technology and tolerance; culture, communication, cooperation; cleanliness, safety; learning, innovation; comfortable; justice, equality, diversity; sustainability, mobility, solidarity; free citizenship's exercise, socially complex environments, areas culturally dynamic, and quality of democratic life.

In the development of cities that encourage creativity, a special form of cultural tourism has emerged: creative tourism. Creative tourism began as a concept linked to small-scale courses and workshops displaying the creativity of a destination and then gradually spread to cover a broad range of experiences that include situations and encounters in informal learning (Richards & Raymond, 2000). Creative tourism provides tourists with an opportunity to develop their creative potential and skills through active participation in creative processes. Active participation in the creative processes of a destination is becoming increasingly important to tourist destination visitors because it provides them with a unique tourist experience and enables communication with residents with whom they engage in local cultural exchange (Richards, 2020). This type of thinking and these development trends have brought about a new definition of creative tourism as a convergence of tourism and a creative economy that generates knowledge-based creative activities connecting producers, consumers and places using technology, talent or skills to create important intangible cultural products, creative content and experiences (OECD, 2014). Strange (2016) points out, however, that one outcome of the policies and practices that might lead cities to use creativity in search of tourists is the argument that the arrival of the events and iconic structures and buildings that have been deployed in the search for this creative tourism has produced places that are similar to each other (Adamo, G. E., Ferrari, S., & Gilli, M., 2019). This is where cultural-policy makers and launchers of creative initiatives must be on their guard, given than creativity should be the result of the synergy of space, individuals and the community, and cannot be copied. One of the ways to develop creativity is to join the Creative Tourism Network, an international organisation that fosters creative tourism development worldwide. Established in 2010 as a non-profit organisation, its aim is to promote all types of destinations whose offerings are based on creativity as a differential element to attract new generations of travellers and to create value chains (Creative Tourism Network, 2020). These destinations, which include rural areas, islands, mid-size cities and metropolises from around the world, benefit from the organisation's promotion and are officially designated "Creative Friendly Destinations" by the organisation.

The tourism market, reflecting the preferences of new tourists and marked by the pandemic crisis caused by the COVID-19 virus, has set the development of creativity and creative industries as one of the drivers of new economic activities in tourism of the future, that could provide a significant contribution to the economy (Putrivi, 2020; Albattat et al., 2020; Chemli et al., 2020; Toanoglou et al., 2021; Valeri, 2021a). To address the new challenges facing creativity in tourism as a result of the pandemic, the rapid growth of the Internet and technologies is being used to generate new forms of tourism reality and tourist experiences. Creative institutions have provided "smart" Internet solutions, with good examples of creativity in the e-environment being animated films, and online exhibitions, concerts, book fairs, etc. The creative industries have begun to introduce virtual reality tools that can provide potential tourists, regardless of where they are, a realistic view of the tourism experience through interaction and visualisation in simulated tourism environments (Lo & Cheng, 2020).

2. DESTINATION STAKEHOLDERS AS DRIVERS OF A CREATIVE TOURISM OFFERING

Creativity in cities is based on the creative thinking of all destination stakeholders as well as on the concepts and strategies of city development. According to Della Lucia and Segre (2017), the synergistic action of the cultural and creative industries value chains and city tourism can result in the accelerated and enhanced economic development of regions. Developed cultural and creative industries are a platform from which new directions can be forged in the tourism development of city centres, in which stakeholders such as public administration enterprises, theatres, tourist boards, cultural associations, entrepreneurs and residents generate creative offerings. Cities are specific spaces of creative individuals who need to encourage the creative involvement of residents in city development, in particular in creating improved and better-designed living spaces. Public authorities have a potentially important role in adopting various cultural policy development lines thus affecting the development of creativity which such policies should enable. The task of the local government is to put in place a development framework and strategic development concepts to valorise the resource basis to create a distinctive product and optimise the value chain (Valeri, 2021b). The local government also drives creative tourism development in synergy with other economic branches, and brings together all stakeholders to ensure the destination's sustainable development focused on the recognition and conservation of the residents' autochthonous culture and the natural resource base.

Local or regional tourist boards create frameworks for synergistic action among stakeholders, thus facilitating adjustments to the market and communication with entrepreneurs. Tourist boards act as centralised bodies, undertaking concrete measures in creating an environment conducive to the promotion and development of tourism by different sectors, including the creative industries sector (Upadhya & Vij, 2020). Local entrepreneurs are individuals, sharing common values, beliefs, behaviours, heritage and linguistic resources, which they use to generate entrepreneurial activity through creativity (Rae, 2020). The creative offering is spread outside of the local community and into regional and international settings by hoteliers, caterers and accommodation providers.

The role of residents in the development of creativity is also important, because residents are the generators, creators and performers of the creative economy (either by participating in it or creating it). They are the strength behind the generation of a destination's creative offering, not only in terms of tourism but much broader; however, residents are enabled to act only if the cultural and other policies of their cities have created a platform for creative expression. One of the sources of creativity is the innovative thinking of individuals who shape creativity through entrepreneurial initiatives. Interaction between the concentration of various talents and the economic sector that encompasses creativity as entrepreneurial potential is the basis for developing new value in tourism (Lorenzen & Frederiksen, 2008).

The creative offering of stakeholders is based on several strategic determinants, that is, concepts, such as active participation, co-creation, authentic experiences, involvement with residents, and the development of skills (Remoaldo et al. 2020). Competition is also important because it represents an opportunity to develop the creative potential of a tourist destination. Such an approach calls for the creators of an offering (private and public) to develop new entrepreneurial initiatives through which the tourism product is differentiated and tailored to the demands of tourists. Stipanović, Rudan and Zadel (2018) point out that the strategic management of cultural policy development, based on the synergistic action of entrepreneurs in the cultural and creative industries and those in the tourism sector, as well as on a culture of collaboration and the horizontal and vertical integration of all stakeholders, is becoming a necessity in modern cities. When taking such action, however, it is important to remember that uniqueness is at the heart of creativity because copying positive creative practices is not creativity.

3. CREATIVITY IN DEVELOPING THE TOURISM OFFERINGS OF CROATIAN COASTAL CITIES

The creative ideas found in the tourism of Croatian cities can be compared with the parameters set by previous studies that show creativity to be one of the possible solutions in a shift away from the mass production of tourism products and services. The Adriatic coastal zone makes up 22% of the total land area of Croatia and accounts for 25.6% of the country's population (1,411,000 inhabitants), living in 350 – 400 cities and towns located along about 700 km of the coast - 9% of the total coastline (Ministry of Economy and Sustainable Development of Croatia, 2020). Today Croatian coastal towns are becoming increasingly important tourist destinations with more or less distinctive tourism resources. According to Florida (2003), cities that attract much creative talent and creative production have huge potential for developing creative entrepreneurship.

Creativity in Croatian coastal cities is tied to management, planning, the economy, social inclusion, culture and local identity, providing a strategic basis for a creative environment (Krivačić & Janković, 2017; Perinić, 2010). Creativity has a non-monetary value that contributes towards inclusive social development, dialogue and understanding among people. It can be seen as a driver of sustainable as well as social development as it encourages tourism development based on the preservation and promotion of the existing resource base through creative individuals. Managing the sustainability of creative cities is faced with difficulties stemming from the number of potential tourists, the number of inhabitants, and the location of the city, factors that often call into question the development of creative potential in a creative economy, as a driver of economic development aligned with sustainable development (Vivant, 2013). Marked seasonality in Croatian tourism, in particular in the tourism of coastal towns, is a major problem and, at the peak of the season, the streets and public facilities of these towns are overrun with tourists. The tourism offering of Croatian coastal towns is typically developed based on their rich cultural and historical heritages and on events (for example, the Dubrovnik Summer Festival – established in 1950, this cultural festival stages musical, theatrical and dance performances in historical locations throughout the town; Opatija, the Imperial Town – the festival brings to life traditions from the beginning of the twentieth century, and presents a variety of music and dance performances; *Rapska Fjera* – the first and largest Medieval-themed summer festival in Croatia, based on a tradition that began in 1364. Small artisan shops and craftsman workshops provide visitors with a glimpse of life in the Middle Ages.). While the historical heritage of Croatian coastal towns gives them very similar starting points for the use of creativity in tourism valorisation, their specific historical events, diverse cultural heritage and the impact of modern society have caused them to develop and implement creativity in different ways. In recent decades, Croatian coastal towns have been shifting away from static cultural tourism towards a cultural tourism offering that employs creativity as a means of tourism service/product innovation. This creativity has pervaded not only the cultural industry but also the gastronomy, music, film, and design industries and many others. Creativity is clearly visible in the technologies and digitalisation that intermediate in and help to shape the tourism offering in the faster mobility of people, goods and services. The tourism experience is affected by new technologies in which creativity plays a crucial role.

3.1. Examples of Creativity in the Tourism of Croatian Coastal Cities

In Croatia, initiatives have been launched by cities seeking to boost their tourism and economic development by designing offerings based on creativity and the creative industries. In 2013 the Croatian Cluster of Competitiveness of Creative and Cultural Industries (CCCCCI) was formed at the national level. The CCCCCI aims to foster the urban development of cities based on branding creativity in cities by promoting entrepreneurship, the cultural and creative industries, and tourism. As part of the CCCCCI, Creative Cities of Croatia is an innovative network that brings together the representatives of local self-government and communities, as well as experts in the creative and cultural industries, to jointly design models for city cultural and creative resource management to help drive local economies and create new investment opportunities. To this end, the Creative Cities of Croatia network is mapping creative and cultural potential, educating local stakeholders, and cooperating with them to find the essential identity of each city, based on which a city brand will be designed to later serve as a starting point for each city to plan its future tourism and economic development (Creative Cities of Croatia, 2017). The development of a management model involves identifying a city's strengths and opportunities and relies on the participation of its citizens as representatives of different interest spheres that benefit from and contribute to driving the local economy. The model aims to create the unique identity of cities, based on which targeted investments will be made to boost development and create jobs.

To interpret creativity in city tourism and its effect on sustainable city development, the following cities and towns were selected because, in their development designs, they have made improvements to conventional tourism through a combination of creativity and culture.

3.1.1. Rijeka

With 128,624 inhabitants (2011), Rijeka is the third largest city in Croatia. In 2019, Rijeka recorded 162,162 tourist arrivals and 407,677 overnights (The Croatian Bureau of Statistics, 2020) and in 2020 it became a European Capital of Culture – EcoC. The city places strong emphasis on the development of both creativity in tourism and the creative industries by establishing creative industry incubators (Rijeka 2020). An important example of creativity in Rijeka's tourism development is the repurposing of abandoned industrial spaces, such as the former Rikard Benčić Motor and Tractor Factory in the heart of the city, which is being developed into three buildings that will have cultural value: the City of Rijeka Museum, the City of Rijeka Library, and the Children's House aimed at developing creativity in children. In 2017, the Museum of Modern and Contemporary Art was moved into the premises of the former Sugar Refinery, located in the vicinity. Creativity is expressed through the repurposing of abandoned spaces for cultural purposes, which is an indication of sustainability in development policies focused on spaces that have lost their original purpose. In the spirit of this type of creativity, the M/V Galeb is also being renovated. Built in 1938, the vessel took on the role of a floating residence in 1953. The M/V Galeb was sailable until 1979, after which time it fell into disrepair. Rijeka is currently renovating the vessel, a symbol of twentieth-century political and social history, to transform it into a museum ship and cultural tourism attraction. The vessel is to be moored at one of the Port of Rijeka quays. In 2020, Rijeka planned to open a large number of creative events to showcase its cultural and creative assets. Unfortunately, the COVID-19 pandemic has prevented many of these events from reaching their full potential. Nevertheless, the most important exhibitions are "Fiume Fantastika: Phenomena of the City" (held in the industrial premises of Export-Drvo) and "51000 Balthazargrad" (the largest exhibition within the Children's House programme of the Rijeka 2020 project). The most important programme spurring local community creativity is "27 Neighbourhoods", which brings together 27 neighbourhoods throughout Primorje-Gorski Kotar County (Rijeka is the County capital) and connects each of them with a neighbourhood in one of the 27 EU member states. The ongoing importance of this programme lies in motivated and active residents who are aware that culture can be an engine of development in their environments and make them distinctive. In addition to giving new meaning to abandoned spaces, creativity is also expressed through the various creative events and workshops that have been promoted more intensively as part of the ECoC Rijeka 2020 project (Stipanović, Rudan, & Zubović 2019).

3.1.2. Pula

Pula is Croatia's eighth largest city, with a population of 57,460 (2011). In 2019, it received 439,541 tourists, who accounted for 2,067,041 overnights (The Croatian Bureau of Statistics, 2020). That same year the Cultural and Creative Cities Monitor named Pula the top-performing Croatian city with the highest cultural and creative potential (The Cultural and Creative Cities Monitor, 2019). Using creativity, Pula has succeeded in developing its rich cultural heritage into tourism products such as the Museum of Istrian Olive Oil, and the interactive multimedia Memo Museum (or Museum of Good Memories), and events such as the Spectacula Antiqua, featuring gladiators, and the Lighting Giants (illuminated port cranes). The city has positioned creativity throughout its entire economy thanks to the mediation of numerous companies and partners in creativity-based entrepreneurship. It was also among the first to map a creative cluster of its creative actors in the creative industries to bring together stakeholders and encourage collaboration on new projects and in new partnerships through international funding and platforms. Special importance is attached to the valorisation of Pula's bay and coast in which the Uljanik Shipyard plays a pivotal role, as well as to the execution of numerous ideas in the field of culture.

3.1.3. Bale

Bale is a small historical town with a population of 1,127 residents located in the south of Istria (2011). In 2019, it had 54,402 tourist arrivals and 335,846 tourist overnights (The Croatian Bureau of Statistics, 2020). Bale's tourism development vision is focused on the development of the town as a hotel-town (Creative Cities of Croatia, 2017). The reception office would be located at the entrance to the town, and the white stone-flagged streets would become the corridors of the "hotel". All private accommodation renters, caterers, and other tourism workers are involved in the hotel-town project, and the creation of a common Bale brand also brings together all stakeholders engaged in the town's tourism offering. The town's creativity is demonstrated in the joint offering of its accommodation facilities and in its integrated tourism offering. Bale is on the World List of Paleontological Sites. As far as it is known, the Bale site is the only one in the world where the fossilised remains of dinosaurs are found under the sea. Some of the bones are estimated to be up to 200 million years ago and are on display in the Dinosaur Museum. The town's creative offering also defines Bale as a tourist destination for butterfly-spotting and offers a variety of creative workshops for pre-schoolers and school-age children (Tourism Office Bale, 2020).

3.1.4. Dubrovnik

Thirteenth in size in Croatia, Dubrovnik has a population of 42,615 inhabitants in 2011. In 2019 it recorded 443,971 tourist arrivals and 4,295,071 overnights (The Croatian Bureau of Statistics, 2020). The town's creativity in tourism is demonstrated through its best-known tourist event, the Dubrovnik Summer Festival, and through the project "Lazareti – the Creative District of Dubrovnik", comprising the renovation of three naves in which cultural and tourism programmes will be held (Creative Cities of Croatia, 2017). The aim of the Dubrovnik Development Agency (DURA) is to facilitate the association, networking, integration and joint action of all public sector stakeholders in the creative industries, thus encouraging entrepreneurship in the spheres of culture and the arts – from architecture, design, film-making, music, and information and communication technology to theatre and dance, the visual arts, photography, publishing, fashion design, robotics and model-making (DURA, 2014). Creativity is visible in the wide variety of workshops organised. For example, as part of the project "*Kultura u điru*" (Culture-to-Go), the Natural History Museum of Dubrovnik and the Art Gallery of Dubrovnik joined forces to organise creative workshops called "Artistic Virus". The workshops were held in various locations in the outdoors and the participants created small sculptures or installations of "artistic viruses" to encourage responsible behaviour during the pandemic (Art Gallery of Dubrovnik, 2020)

3.1.5. Zadar

The fifth-largest city in Croatia, Zadar has a population of 170,053 people (2011). In 2019, it recorded 609,785 tourist arrivals and 2,020,302 overnights (The Croatian Bureau of Statistics, 2020). The Sea Organ makes Zadar one of the best-known examples of creativity in tourism (Creative Cities of Croatia, 2017). The Sea Organ is a natural music instrument carved into 70-metre-long stone steps, and musical chords are created by the sea pushing air through organ pipes. The Sea Organ Festival "Mitra Zonata" represents the synergy of the sound of the sea organ and the singing of traditional songs performed by *klapa* (harmony-singing) choirs on board sailboats. Another distinctive example of the town's creativity is the urban installation Greetings to the Sun, consisting of 300 multi-layer glass plates, placed in a 22-metre-diameter circle at the same level as the waterfront. Absorbing sunlight during the day, after sunset the lighting elements built into the installation turn on, creating a light show. Zadar has also founded a Centre for Creative Industries, as a specialised entrepreneurial incubator for specialised entrepreneurial education to help the formal education system keep abroad of market trends in the creative industries (Centre for Creative Industries Zadar, 2018).

3.1.6. Biograd na Moru

Biograd na Moru has 5,569 inhabitants in 2011 and in 2019 it had 182,803 tourist arrivals and 964,548 overnights (The Croatian Bureau of Statistics, 2020). Located on the central Adriatic coast, this tourist town expresses its creativity through a number of well-known events such as the Biograd Street Music Festival with street performances, entertainers and musicians; the Biograd Night, presenting local customs and culture; creative food-and-wine tours; etno and eco experiences of the Biograd hinterland; and Fun Park Biograd, the largest themed park in Croatia (with attractions such as Tornado, Simulator, Alien, XD Dark Drive, Western Express, Pirate Battle, and others). Biograd also offers an application called Adventure a la Carte, an imaginative and interactive guide that uses the solving of riddles to take tourists on a tour of five destinations along the Biograd Riviera (Tourism Office Biograd na Moru, 2020).

These positive examples are confirmation of the growing awareness of the opportunities that exist for implementing creativity in the sustainable development of coastal cities but are also indications of the absence of a development model that would more closely bring together stakeholders to create new and original solutions in transforming heritage resources into a distinctive creative offering and brand.

4. DISCUSSION

The new lines of tourism development in Croatia's coastal cities show that each city is seeking to build its own, particular tourism offering based on the available natural resources, cultural and historical heritage, as well as creativity. In the cities and town considered in this paper, creativity refers to different elements within a tourist destination, such as creative spatial features, and the creativity of stakeholders such as destination managers, residents and entrepreneurs, together with the creativity of tourists who have a need to stay in such spaces and take part in creative events. Every city and town is creative in its own way. With the support of local authorities, creative communities can deliver projects and programmes in the field of culture, thus helping to design a novel, creative tourism offering. The analysis of cities and towns also reveals two courses of action for achieving sustainable tourism development based on creativity. The first course of action involves repurposing physical objects (tangible heritage) through creative content and programmes (sustainable development) and the second, involves using the creative industries to imbue intangible assets with novel value and content focused on the new tourist. Sustainability is reflected in the repurposing of abandoned and dilapidated spaces for new cultural purposes (for example, the industrial heritage of Rijeka that has been repurposed into a library and museum) as well as in the creativity of presenting and reviving forgotten crafts and traditions of the past. Della Lucia and Trunfio (2018) claim that the intangible factors of the urban transformation of creativity include the innovative thinking of individuals, the formation of the creative class and the extraction of maximum value from cultural heritage.

Importantly, creative content and projects in tourism help to create novel tourist attractions, which are adjusted to the needs and interests of the modern tourist and visitor through a variety of offerings. It is a fact, however, that the development of creative tourism in a community should not focus exclusively on visitors (tourists) arriving from other cities or countries but also on residents to ensure sustainable development (socio-economic effects).

Creativity in Croatia's coastal cities is making new qualitative advances in the presentation of cultural and historical heritage, with residents being a primary source of creative initiatives. The creative entrepreneurial initiatives of residents often provide the basis for the sustainable management of a city's heritage. By ensuring the adequate presentation and use of old elements in novel purposes, creative expression helps to promote the sustainable development of Croatian coastal cities and towns. Although the creation of creative value in these coastal cities and towns needs to be based on their cultural heritage, it should not be limited to cultural tourism only. Instead, creative value should be created based on the original ideas of individuals as well as on a winning combination of cultural heritage and creative content, knowledge, technology and education.

Knowledge, digitalisation and technological achievements (Colletta, 2018) help to develop stakeholder synergy and interdependence (foremost through the application of information technologies), based on which creativity is more easily shaped into parts of a tourism product, becoming more accessible to a wider market (Trinchini et al. 2019). While the observed cities and towns have no creative-tourism development documents incorporated into their strategic documents, they recognise the need to develop a creative offering targeting modern tourists and their needs. Creative destinations are not built on isolated entrepreneurial initiatives but rather on the strategic thinking and integrated action of all stakeholders involved in development (under the leadership of tourist boards or local governments). With the globalisation of tourism compelling cities to become creative to ensure their economic development, creativity needs to become a catalyst in achieving differentiation and gaining competitive advantages for cities while preserving their authenticity. This type of thinking will help many cities and towns on Croatia's coast to achieve their goal, which is to attract the new tourist, who is focused on creative technologies and views heritage in the context of the creative industries. At any rate, everyone – the local community, destination management, local governments, entrepreneurs and, ultimately, tourists – stands to profit from such solutions for city tourism development. To that end, the development of competitive creativity in tourism should be fostered by building entrepreneurship based on educated human resources and knowledge development because knowledge and creativity are the main production factors within the creative industries in which all stakeholders act in synergy to accomplish a common goal (Mangifera, 2019).

CONCLUSION

Creativity provides opportunities for the transformation of Croatia's coastal cities and towns from tourist destinations of marked seasonality whose tourism development hinges on natural preconditions (sea, sun, sand) into distinctive destinations with creative offerings. Analysis of the current state of the creative offering of coastal Croatian cities and towns reveals that it is confined to isolated entrepreneurial endeavours and creative examples (with the exception of Rijeka as a 2020 European Capital of Culture).

There is also no coordinated action taken to valorise resources in a creative way that fosters sustainable development.

Preconditions to development are awareness of the opportunities provided by a creative offering and keeping abreast of new trends in tourism, and putting in place a dynamic model of development that will spur and unlock the creative potential of stakeholders (supply providers, destination managers, residents) through synergistic action (horizontal and vertical integration) to optimise the value chain and generate new value. The vision of a creative city, as a distinctive tourist destination, must bring together all stakeholders and partial interests to generate new value for tourists and to improve the quality of life of residents as the promotors of local culture and traditions. Creativity that knows no boundaries uses the resource base to generate new experiences, thus producing a unique and distinctive product (transforming visual experiences in multi-sensory experiences). Through digitalisation and new technologies, creativity enables heritage and tradition to be interpreted in novel ways that are focused on the preservation of indigenous values and on sustainable development. Croatia's coastal cities and towns can differentiate themselves from a globalised offering by creatively bringing together the past, the present and the future and by linking traditional values and modern technologies focused on the new modern tourists and their needs. Ultimately, creativity helps to ensure sustainability and, in turn, sustainability ensures the future of creativity.

REFERENCES

Adamo, G. E., Ferrari, S., & Gilli, M. (2019). Creativity as a source of differentiation in urban tourism: The case of Torino city. *International Journal of Tourism Research*, *21*(3), 302–310. doi:10.1002/jtr.2261

Albattat, A., Jamaludin, A., Zuraimi, N. S. M., & Valeri, M. (2020). Visit Intention and Destination Image in Post- Covid 19 Crisis Recovery. *Current Issues in Tourism*. Advance online publication. doi :10.1080/13683500.2020.1842342

Bacon, S. (2005). Positive Psychology's Two Cultures. *Review of General Psychology*, *9*(2), 181–192. doi:10.1037/1089-2680.9.2.181

Baggio, R., & Valeri, M. (2020). Network science and sustainable performance of family businesses in tourism. *Journal of Family Business Management*. doi:10.1108/JFBM-06-2020-0048

Berg, S. H., & Hassink, R. (2014). Creative industries from an evolutionary perspective: A critical literature review. *Geography Compass*, *8*(9), 653–664. doi:10.1111/gec3.12156

Centar za kreativne industrije Zadar. (2018). Retrieved April 18, 2020, Available at: https://www.inovacija-zadar.hr/novosti/provedba-projekta-centar-za-kreativne-industrije/161.html

Chemli, S., Toanoglou, M., & Valeri, M. (2020). The impact of Covid-19 media coverage on tourist's awareness for future traveling. *Current Issues in Tourism*, 1–8. Advance online publication. doi:10.1080/13683500.2020.1846502

Colletta, T. (2018). *To promote urban heritage culture in the age of information revolution: a new creativity.* In ICOMOS 19th General Assembly and Scientific Symposium "Heritage and Democracy", New Delhi, India.

Creative Tourism Network. (2020). Retrieved April 22, 2020, Available at: http://www.creativetourismnetwork.org/presentation/

Della Lucia, M., & Segre, G. (2017). Intersectoral local development in Italy: The cultural, creative and tourism industries. *International Journal of Culture, Tourism and Hospitality Research, 11*(3), 450–462. doi:10.1108/IJCTHR-03-2016-0032

Della Lucia, M., & Trunfio, M. (2018). The role of the private actor in cultural regeneration: Hybridizing cultural heritage with creativity in the city. *Cities (London, England), 82*, 35–44. doi:10.1016/j.cities.2018.05.003

Državni zavod za statistiku. (2011). *Stanovništvo u najvećim gradovima i općinama*. Retrieved September 13, 2020, Available at: www.dzs.hr

European Commission. (2019). *The Cultural and Creative Cities Monitor*. Retrieved October 10, 2020, Available at: https://ec.europa.eu/jrc/en/publication/eur-scientific-and-technical-research-reports/cultural-and-creative-cities-monitor-2019-edition

Florida, R. (2003). Cities and the creative class. *City & Community, 2*(1), 3–19. doi:10.1111/1540-6040.00034

Garcês, S., Pocinho, M., & Jesus, S. N. D. (2018). Review of optimism, creativity and spirituality in tourism research. *Tourism and Hospitality Management, 24*(1), 107–117. doi:10.20867/thm.24.1.6

Hartley, J. (2012). *Key Concepts in Creative Industries*. Sage.

Hrvatski klaster konkurentnosti kreativnih i kulturnih industrija. (2013). Retrieved April 19, 2020, Available at: https://www.udruga-gradova.hr/wordpress/wp-content/uploads/2016/07/5.-HKKKI.pdf

Krivacic, D., & Jankovic, S. (2017). Managerial attitudes on environmental reporting: Evidence from Croatia. *Journal of Environmental Accounting and Management, 5*(4), 327–341. doi:10.5890/JEAM.2017.12.005

Krstinić Nižić, M., Rudan, E., & Trinajstić, M. (2019). The role of creative cities in regional development. *Poslovna Izvrsnost, 13*(1), 35–55. doi:10.22598/pi-be/2019.13.1.35

Lo, W. H., & Cheng, K. L. B. (2020). Does virtual reality attract visitors? The mediating effect of presence on consumer response in virtual reality tourism advertising. *Information Technology & Tourism, 22*(4), 537–562. doi:10.100740558-020-00190-2

Lorenzen, M., & Frederiksen, L. (2008). Why do cultural industries cluster? Localization, urbanization, products and projects. In P. Cooke & L. Lazzeretti (Eds.), *Creative cities, cultural clusters and local economic development* (pp. 155–179). Edward Elgar.

Mangifera, L. (2019). Stakeholders Analysis for Creative Industry Development. *Jurnal Ekonomi Pembangunan: Kajian Masalah Ekonomi dan Pembangunan, 19*(2), 156-162.

Marques, L., & Richards, G. (2014). Creative districts around the world (4th ed.). Breda: NHTV.

Ministarstvo gospodarstva i održivog ravoja, Središnji državni portal. (2020). Retrieved April 23, 2020, Available at: https://mzoe.gov.hr/o-ministarstvu-1065/djelokrug-4925/vode-i-more/more-i-priobalje-opcenito/1437

Mreža kreativni gradovi Hrvatske. (2017). Retrieved May 22, 2020, Available at: https://kreativnigradovi.net/kreativni-gradovi-hrvatske/

Mreža za razvoj i kreativnost. (2013). Retrieved May 18, 2020, Available at: https://hkkkki.eu/

Oliveira, A., & Paulino, F. (2017). European creativity and urban regeneration. *The Journal of Public Space, 2*(2), 127–140. doi:10.5204/jps.v2i2.98

Organisation for Economic Co-operation and Development. (2014). Tourism and the creative economy. Paris: OECD.

Perinić, L. (2010). Kreativni gradovi: uzroci i posljedice. *Drugost: časopis za kulturalne studije,* (1), 110-115.

PutriviN. (2020). *How Digital Innovation and Dynamic Organization Concept Can Help Creative Industries to Survive COVID-19 Global Pandemic*. https://ssrn.com/abstract=3590926 doi:10.2139/ssrn.3590926

Rae, D. (2020). Intercultural Entrepreneurship in Creative Place-Making. In *Value Construction in the Creative Economy*. Palgrave Macmillan. doi:10.1007/978-3-030-37035-0_7

Raymond, C. (2007). Creative Tourism New Zealand: The practical challenges of developing creative tourism. In G. Richards & J. Wilson (Eds.), *Tourism, Creativity and Development* (pp. 145–157). Routledge.

Razvojna agencija Grada Dubrovnika. (2014). Retrieved November 14, 2020, Available at: https://dura.hr/news/dani_kreativnih_kulturnih_industrija/index.html

Remoaldo, P., Matos, O., Freitas, I., Gôja, R., Araújo Alves, J., Ribeiro, V., Pereira, M., & Xavier, C. (2020). An international overview of certified practices in creative tourism in rural and urban territories. *Journal of Hospitality & Tourism Research (Washington, D.C.)*, 1–29. doi:10.1177/1096348020950792

Richards, G. (2020). Designing creative places: The role of creative tourism. *Annals of Tourism Research, 85*, 1–11. doi:10.1016/j.annals.2020.102922

Richards, G. & Raymond, C. (2000). Creative tourism. *Atlas News, 23*(8), 16-20.

Richards, G., & Wilson, J. (2006). Developing creativity in tourist experiences: A solution to the serial reproduction of culture? *Tourism Management, 27*(6), 1209–1223. doi:10.1016/j.tourman.2005.06.002

Rijeka. (2020). *Europska prijestolnica kulture*. Retrieved November 16, 2020, Available at: https://rijeka2020.eu/en/

Rudan, E. (2012). Razvojne perspektive kreativnoga turizma Hrvatske. *Ekonomska Misao i Praksa*, (2), 713–730.

Stipanović, C., Rudan, E., & Zadel, Z. (2018). Possibilities of entrepreneurial development in cultural tourism of Croatian cities. In *24th Biennial International Congress: Tourism and hospitality industry*. Opatija: Faculty of Tourism and Hospitality Management Opatija, University of Rijeka.

Stipanović, C., Rudan, E., & Zubović, V. (2019). Cultural and creative industries in urban tourism innovation – the example of city of Rijeka. *Tourism in South East Europe – ToSEE*, (5), 655-666.

Strange, I. (2016). Urban planning, architecture and the making of creative spaces. In P. Long & N. D. Morpeth (Eds.), *Tourism and the creative industries: theories, policies and practice* (pp. 100–113). Routledge. doi:10.4324/9781315735870-6

Toanoglou, M., Chemli, S., & Valeri, M. (2021). The organizational impact of COVID-19 crisis on travel perceived risk across four continents. *Journal of Organizational Change Management*. Advance online publication. doi:10.1108/JOCM-12-2020-0369

Trinchini, L., Kolodii, N. A., Goncharova, N. A., & Baggio, R. (2019). Creativity, innovation and smartness in destination branding. *International Journal of Tourism Cities*, *5*(4), 529–543. doi:10.1108/IJTC-08-2019-0116

Turistička zajednica grada Biograd na Moru. (2020). Retrieved November 16, 2020, Available at: https://discover-biograd.com

Turistička zajednica općine Bale. (2020). Retrieved November 16, 2020, Available at: https://www.bale-valle.hr/

Turok, I. (2009). The distinctive city: Pitfalls in the pursuit of differential advantage. *Environment & Planning A*, *41*(1), 13–30. doi:10.1068/a37379

Umjetnička galerija Dubrovnik. (2020). Retrieved November 15, 2020, Available at: https://www.ug-dubrovnik.hr/?file=naslovna

United Nations Conference on Trade and Development. (2020). Retrieved November 28, 2020, Available at: https://unctad.org/webflyer/world-investment-report-2020

Upadhya, A., & Vij, M. (2020). Creative tourist experience: Role of destination management organizations. In *Destination Management and Marketing: Breakthroughs in Research and Practice* (pp. 763–783). IGI Global. doi:10.4018/978-1-7998-2469-5.ch042

Valeri, M. (2015). Sustainability development and competitiveness of Rome as a tourist destination. *Tourism and Hospitality Management*, *21*(2), 203–217. doi:10.20867/thm.21.2.7

Valeri, M. (2016). Networking and cooperation practices in the Italian tourism business. *Journal of Tourism, Heritage & Services Marketing*, 2(1), 30–35. doi:10.5281/zenodo.376333

Valeri, M. (2021). *Organization Theories. Implications for the Strategic Management*. Springer.

Valeri, M. (Ed.). (2021a). Tourism risk. Crisis and recovery management. Emerald Publishing.

Valeri, M., & Baggio, R. (2020a). Social network analysis: Organizational implications in tourism management. *The International Journal of Organizational Analysis*. Advance online publication. doi:10.1108/IJOA-12-2019-1971

Valeri, M., & Baggio, R. (2020b). Italian tourism intermediaries: A social network analysis exploration. *Current Issues in Tourism*, *24*(9), 1270–1283. doi:10.1080/13683500.2020.1777950

Valeri, M., & Baggio, R. (2020c). A critical reflection on the adoption of blockchain in tourism. *Journal Information Technology and Tourism*, *23*(2), 121–132. doi:10.100740558-020-00183-1

Valeri, M., & Baggio, R. (2021). Increasing the efficiency of knowledge transfer in an Italian tourism system: A network approach. *Current Issues in Tourism*, 1–16. Advance online publication. doi:10.1080/13683500.2021.1937960

Virfinija, J. (2016). Interaction between Cultural/Creative Tourism and Tourism/Cultural Heritage Industries. In L. Butowski (Ed.), *Tourism – From Empirical Research Towards Practical Application* (pp. 137–156). Tech.

Vivant, E. (2013). Creatives in the city: Urban contradictions of the creative city. *City. Cultura e Scuola*, *4*(2), 57–63. doi:10.1016/j.ccs.2013.02.003

Chapter 15
The Contribution of Female Creativity to the Development of Gastronomic Tourism in Greece:
The Case of the Island of Naxos in the South Aegean Region

Panoraia Poulaki
University of the Aegean, Greece

Antonios Kritikos
University of West Attica, Greece

Nikolaos Vasilakis
Ionian University, Greece

Marco Valeri
https://orcid.org/0000-0002-9744-506X
Niccolò Cusano University, Italy

ABSTRACT

Gastronomy as an important part of the culture, morals, customs, and traditions of the countries meets the needs of visitors and at the same time contributes to tourism development through the rich gastronomic heritage. Gastronomic tourism offers the possibility of increasing the income criteria of a population, increasing jobs, and developing the national economy. The purpose of this study is to highlight the role of female creativity in the development of a gastronomic destination with a case study of Naxos in the South Aegean in Greece. At the same time, it aims to highlight the great challenges and opportunities brought about by gastronomic tourism with the contribution of women's entrepreneurship. Women's creativity and entrepreneurship create new job opportunities. The dynamic presence and activity of the women of Naxos highlights the tourism destination of Naxos as a destination of high gastronomy. Secondary sources present gender discrimination in entrepreneurship, women's entrepreneurship in tourism, and the connection of tourism with gastronomic heritage.

DOI: 10.4018/978-1-6684-3889-3.ch015

1. INTRODUCTION

Visitors to a tourism destination due to their growing desire to experience a variety of tourism experiences have placed gastronomy in their high preferences as an important part of their daily lives, being an important tourism motivation for them (Kim, Park, Lamb, 2019). Tasting a food is a gastronomic experience that represents the heritage and the characteristic idiom of a tourism destination and its inhabitants, reflecting its economic, social, environmental physiognomy (Brunori, Rossi, 2000) and experiencing the visitor other cultures through food (Long, 2004).

Gender and female creativity play a major role in the promotion of gastronomic culture, since in the world of gastronomic entrepreneurship, women entrepreneurs create in an innovative way, use their imagination and take advantage of the combined and unique resources they have (Chiles, Bluedorn, Gupta 2007). On the one hand, several scholars have approached the specific research field of female entrepreneurship from the optical of female creative perspectives, focusing on their creative imagination (Lachmann, 1977) and on the other hand, the possible limitations in the integration of women have been examined in business life (Al-Dajani, Marlow, 2013).

The specific contradictions that arise from the various research approaches that have been carried out do not invalidate the idea that the emotional and reminiscent criteria of local gastronomy turn a holiday place into an attractive and special resort (Kivela, Johns, 2002; Santos et al., 2021a; 2021b). They also reveal the key ingredients that highlight the cultural heritage of a destination (Björk, Kauppinen-Räisänen, 2016) and provide special personal experiences for the visitor (Finkelstein, 1989), gastronomic culture experiences that remain unforgettable (Wolf, 2002). Many times, social systems are gender-based due to the integration of institutionalized structures preventing women from realizing new ideas and creating businesses (Welter, 2011) within a wide range of contexts that include gender differences (García, Welter, 2013).

The unforgettable gastronomic experiences experienced by the visitor in Greece with the various culinary traditions and local food through female creativity and gastronomy are examined in this primary research effort by collecting primary data through interviews with women entrepreneurs located on the island of Naxos, located at the southeastern end of Greece and European Union. The primary research emphasizes the contribution of women entrepreneurship in the promotion of this island as a gastronomic destination. Successful women entrepreneurs enjoy social appreciation through their creativity (Baron, 2005). Conclusions and suggestions encourage female gastronomic creativity, as it offers the opportunity to progress by acquiring resources (McKeever, Jack, Anderson, 2015), examining the hypothesis that social systems are often gender-based and prevent women entrepreneurs from implementing new ideas (Welter, 2011) because of social forces (Shahriar, 2008).

2. LITERATURE REVIEW

The dynamic presence of female creativity in Greece and in the world of entrepreneurship appears in the 1980s with women entrepreneurs who are engaged in agricultural life and entrepreneurship by founding the first women's cooperatives, with a great example of the island of Lesvos and specifically the Municipality of Petra (Giagou, Apostolopoulos, 1996).It highlights the dynamic presence of women in the business world where tourism plays an important role that creates jobs and contributes fundamentally to the development and competitive part of several areas (Camisón, Forés, Navarro, 2017) with huge social and cultural benefits in addition to economic development for a tourism destination (Hieu, Rasovska, 2017). The specific involvement of Greek women entrepreneurs with agricultural life and their dynamic presence in the 1980s is not accidental. Previous studies highlight the relationship between tourism and employment in agricultural life and the production of gastronomic products (Telfer, Wall, 1996: 645-653).

In a destination and a "creative city" is developed a creative environment through which people develop their talent and creativity (Cohendet, Grandadam, Simon, 2010). The characteristics of a region play an important role in the preferences of visitors who love the local gastronomic cuisine, and these should be especially considered and not only healthy local food (Boatto, Gennari, 2011). Visitors enjoy a variety of interconnected activities that are constantly evolving thanks to tourism (Hall, Mitchell, 2000). The experience of gastronomy describes the approach of authenticity with unique features (Heidegger, 1996) as gastronomic tourism helps the visitors of a tourism destination to get to know further the culture and local cuisine of a tourism resort (Diaconescu, Moraru, Stănciulescu, 2016). It also contributes to the full fight against seasonality (World Tourism Organization, 2017) and beyond. For the complete satisfaction of the needs of the guests with the term "gastronomy" in the field of hospitality at the international level it is mentioned that it includes food and drink. The view that tourism and gastronomy are associated with realistic and real features from the stage of accommodation to the provision of food and beverage services for the visitor is explored (Scarpato, 2002).

Entrepreneurship provides the possibility of an identity (Anderson, 2000) and action by acquiring the characterization of the "process of socialized character" (Drakopoulou, Anderson, 2007). Women's entrepreneurship and gastronomy are theoretically part of a globalized economic context, an economic context that in recent years has caused a homogenization in the productive and cultural sector at the international level as well as a relevance of characteristics in each country or destination (Bonanno, 1992). The international literature has identified several concerns in general about how much a woman benefits compared to the man in a managerial position from the creation of social networks and their investment in human capital (Baines, 2000; Valeri, 2016; Valeri and Baggio, 2020a; 2020b; 2020c; 2021; Baggio and Valeri, 2021).

In areas of Europe or even further south there is a diversity in environmental and cultural level that offers the possibility of developing new job opportunities not only in areas located in the "city" (Whatmore, 1988) but also in business tourism entities where the mild tourism is located, in areas where organic farming is flourishing where women acquire new roles with important professional activities (Ventura, 1994: 80-90).

Many historians mention the lost professional status of women in areas dominated by agriculture where men have acquired a role model provided by them as an alternative form of recreation (Dahlstrom, 1996). The main concern of the European Union is the balance in professional and personal level for everyone, with women entrepreneurs representing 30% of all entrepreneurs in Greece and in the EU 31% of entrepreneurs estimated at 10.3 million (European Commission, 2014).

International tourism due to its flexibility, capacity and flexible features strengthens the unequal distribution of employment and the income criteria of women (Peeters, Ateljevic, 2009). The motivation is big enough for women to be involved in gastronomic entrepreneurship. More and more food, both locally and regionally, offers further value to a tourism destination (Tefler, Wall, 1996), as in addition to Greece, countries such as Spain, Italy, France invest in promoting their local cuisine with evaluation axis the fact that there are visitors who are motivated for their trip to participate in activities related to gastronomic interests (SETE, 2010).

Various studies have shown that the traditional role of women has changed due to their participation in the tourism sector and they have been given a new opportunity to create income criteria and autonomy (Tucker, Boonabaana, 2011). The international literature focuses on the gender pay gap that takes place in different spatial and temporal domains (Gray, Benson, 2003) as gender is considered the "most widespread organizational principle" of society (Ahl, Nelson, 2010; Valeri and Katsoni, 2021; Valeri, 2021b). The specific references to international examples of women entrepreneurship are related to the correlation between gender and tourism development, how their prosperity occurs and therefore their "empowerment" through development programs (Rowlands, 1997). With the development intervention that takes place significantly in areas that are mainly dominated by rural life, the possibility of women's participation in the development of the national economy is offered, bringing prosperity both to the community itself and to their family (Baden, Milward, 2000; Kumar et al., 2021).

Gastronomy is directly related to the manners, customs, traditions, history of the place, climatic conditions and the geographical location of a tourism destination (Gyimothy, Rassing, Wanhill, 2000) and agritourism (Kastenholz, Davis, Paul, 1999). The cultivation of local products has a special dynamic. The preparation of a local product takes place in parallel with the preparation of another local product from another country. This is true in several countries, a typical example is the "moussaka" of Greece with the "pasta" of Italy as the tasting of a food contributes at the same time to the development of learning and socialization (Tikkanen, 2007). Gastronomy is original in terms of dishes and wines of local culture (Green, Dougherty, 2008). People travel from country to country to try a new cuisine and get to know the local culture (Kercher, Okumus, 2008).

The review of the current literature therefore highlights the aspirations of women entrepreneurship that contributes to the development of gastronomy both internationally and in Greece. With the term "embeddedness" it tries to be integrated in the Greek data of gastronomic tourism (Korsgaard, Ferguson, Gaddefors, 2015) shaping the specific business process through the extent of the relationship it acquires with the environment it wishes to join (McKeever, Jack, Anderson, 2015). The study of women entrepreneurship in gastronomic tourism with a case study in Naxos highlights the process of integration of women entrepreneurs in male gender fields as previous studies have pointed out (Kristoferson, Aggestam, 2017).

3. GASTRONOMY IN THE ISLAND OF NAXOS IN THE SOUTH AEGEAN REGION

The South Aegean Region has managed to integrate cultural elements with flavors. It is a model of balanced island development between the primary, secondary and tertiary sectors. The revenues of the South Aegean Region reached 4.41 billion euros in 2018, an amount that represents 28.2% of the country. In the period 2016-2018, travel receipts in the South Aegean increased by 40.8%, i.e. at almost twice the rate compared to the national average, which was 22.8%. The share of the South Aegean in the total travel receipts of the country, increased from 24.6% (2016) to 28.2% (2018). Travel receipts come from tourists from Germany and the United Kingdom, with significant contributions from the United States, Italy and France. The overnight stays reached 51 million in 2018, a price that represents 22.5% of the whole country (South Aegean Region, 2019, Bank of Greece, 2019). Most overnight stays are for tourists from Germany, the United Kingdom, Italy, France and the United States. The South Aegean Region is the most expensive tourism destination in the country, as the cost per visit to it is 665.8 euros, much higher than the national average which does not exceed 450 euros. In recent years, the emphasis on the development of the hotel potential has been given to their quality upgrade (Bank of Greece, 2019). According to the South Aegean Region, citing the official data of the Bank of Greece, for the 9 months January-September 2019, the South Aegean is at the top among the 13 Regions of the country, in all indicators related to arrivals, overnight and mainly revenue. In the South Aegean, the arrivals of visitors from the main countries of origin mentioned above, in the 9 months of January-September 2019, amounted to 6.025 million compared to 5.816 million visitors in the corresponding period of 2018 and 5.083 million in 2017. It is worth noting that expenditure in the South Aegean, according to the final data of the Bank of Greece for 2018, amounted to 599.4 euros (South Aegean Region, 2019, Bank of Greece, 2019). Both the islands of the Aegean Sea and the Cyclades are well-known tourism destinations with international recognition.

"Aegean Gastronomy" participates in major international events. Characteristic is the program "Taste the seasons" which was implemented in the context of the "Gastronomic Region of Europe 2019" and was presented on Gastronomic Tourism Day at the International Tourism Fair BIT 2018 in Milan. The South Aegean Region as the "Gastronomic Region of Europe 2019" presented at the James Beard Foundation (Temple of American Gastronomy) in New York the unique gastronomic tradition of the South Aegean for the third consecutive year. Exquisite flavors of the Cyclades and the Dodecanese stole the show at the event "International: Southaegean Feast" in May 2019 in the USA. There was also a "gastronomic tour of the flavors of the Aegean" from antiquity to modern cuisine. The gastronomy of the South Aegean has participated in the International Food & Beverage Exhibition ANUGA 2019 in Cologne, Germany in October 2019. It is the world's largest international food and beverage exhibition and is the most important business and communication platform for the industry. The islands of Cyclades and Dodecanese presented their excellent quality products, attracting the interest of buyers. Bee products, olives, pastries and cheese products gave the opportunity to the visitors of ANUGA to get a "taste" of the special gastronomic identity of each region. Also, the producers of the South Aegean Region have participated in major exhibition events such as SIAL in Paris and Tutto Food in Milan. The South Aegean Region organized important activities abroad from Berlin to New York. A very impressive and tasty dish is the sea bream plate with plenty of tomatoes, capers and potatoes. Also, the lamb of long hours of slow cooking with "revithada", (with chickpeas), baked in the wood oven, in a hull (South Aegean Region, 2019) is of exceptional taste.

Since antiquity with the worship of the god Dionysus to the architecture of the Venetians, every corner of the island of Naxos is a cultural treasure. Naxos is evolving into a top gastronomic destination, "marrying" the local products of the island with the tourism industry. Impressive and enchanting are many villages in Naxos, especially the village of Apeiranthos, a stone-built village, a real gem for the island. Naxos is an independent branded and recognizable destination. It has particularly developed maritime tourism, sports, cultural/religious, agritourism, ecotourism and gastronomy, offering quality tourism. It has a rich cultural reserve of monuments, churches, monasteries, traditional settlements and archeological sites, capable poles of attraction for tourism (Stadium, Koutoulas, 2019). Most companies (82.2%) are sole proprietorships in Naxos, 29 are anonymous and 18 are of limited liability. Tourism companies make up 25.2% of the total (South Aegean Region, 2019). The processing of agricultural products and the production of food have an important role in craft/processing enterprises. The presence of the following cooperatives is very important: Agricultural/Olive Cooperative of Damariona Naxos, Urban Cooperative of Naxos Butchers, Urban Cooperative of Handmade Traditional and non-Products and Items of Apeiranthos Naxos and the Limited Liability Cooperative. The great position of the Union of Agricultural Cooperatives of Naxos is underlined, which consists of 12 cooperatives with 3,654 members with intense activity. The Union of Agricultural Cooperatives of Naxos employs 135 permanent employees and about 40 seasonal workers. Using method and program, it has brought the products of the island to a high level of quality (Stadium, Koutoulas, 2019), South Aegean Region, 2019). Naxos, thanks to its natural wealth and rich tradition is self-sufficient in a huge variety of products, which are guaranteed and certified. Naxos produces agricultural (olive, olive oil, potato, wine, citrus) and livestock products (milk, cheese, meat). The history of the potato begins in the early 19th century and its cultivation was an important source of income for the farmers of Naxos. In 1953, the Greek state established the Potato Seed Production Center on the island and thus Naxos is the only potato seed production center in Greece. The Naxos potato is now a Product of Geographical Indication.

The peasant woman, the rural woman, is a heroine and is officially honored on October 15 based on the decision of the UN General Assembly. In the European Union 42% of the rural population are women. A small number of women in Naxos are owners of agricultural production units. In the past, there was a devaluation of women's skill in the kitchen, reproducing the stereotype "Only men have the technique, discipline and passion that turn cooking into art". The women's cooperatives of many villages of Naxos, as in other parts of Greece, have joined forces and their techniques to offer unique traditional products to locals and tourists. These are products that are part of the island culture (South Aegean Region, 2019). It is worth mentioning the Women's Traditional Art Cooperative of the village of Apeiranthos, which deals with both textiles and local products. Also active is the Women's Cooperative of the village of Filoti, Naxos, which produces a variety of local flavors, homemade jams, liqueurs, flavored olive oil with herbs of the island, pastries and local "tsipouro" drink.
Special recipes of Naxos are:

- "Salatouri": Steamy salad with cooked skate, parsley, lemon oil and plenty of onion.
- "Monk": A dish that combines eggplants, beef, tomatoes and gruyere in a delicious pyrgaki.
- Pork with "provatses": Pork cooked with a very special local herb, the "provatses".
- "Rosto": Sunday dish with casserole pork with plenty of garlic and salt and pepper, red wine and tomato sauce with bay leaf.
- "Patoudo": Easter dish with lamb cooked on the vines and stuffed with "sefkoula", "koutsounades" (poppy leaves before flowering), spring onion, fennel, dill, spinach, myrons and cauliflower, rice and raisins.
- "Galaktoboureko" dessert in the village of Halki.

The Union of Agricultural Cooperatives of Naxos has a modern cheese unit with a dynamic presence in the field of dairy. The products they produce are Naxos gruyere, the famous kefalotyri, also known as arsenic, hard cheese, "xinomizithra", white brine cheese, butter and strained yogurt. Naxos gruyere made from cow's milk and sheep and goat is a special and top cheese. This is an exquisite Product of Designation of Origin that has been distinguished with a bronze taste award at the Greek Cheese Festival in 2007, with a silver taste award in 2009 at the same festival and gold in 2011. It also won a gold award and two taste stars in 2011 in Brussels. The famous Naxos gruyere (from cow's and goat's milk) is the protagonist in the effort to make Naxos an attractive destination in the domestic and international market. Naxos, although a small and arid island with rich biodiversity, has managed to become a distinct gastronomic destination. Its products also include aromatic citrus that give unique liqueurs, olive groves, vineyards, wine and, of course, the best fishing from the boats on the island. The beekeeping association of the island consists of 71 members who produce about 50 tons of thyme honey per year (https://

www.greekgastronomyguide.gr/odoiporiko-sti-naxo/). The Naxians are the ones who manage the natural resources and the infrastructure of the island, they are the ones who produce goods and services, they are the ones who offer hospitality to the visitors. They are distinguished for their energy, ingenuity and entrepreneurship. They have created the success of Naxos as a tourism destination. Naxos has its own gastronomic tradition and culture with a special identity and character. Typical local products, in addition, are "ham" (pork sausage reminiscent of prosciutto), honey, herbs, citrus fruits, citrus (used in liqueurs), "papilona" (fruit like bergamot and nerazzi), (https:/www.greekgastronomyguide.gr/odoiporiko-sti-naxo/). There is an amalgam of local, often strictly endemic products, and influences from the cuisines of mainland Greece, Egypt and Asia Minor. The island has received many influences from various conquerors, such as: Franks, Turks and Venetians. Also, a special local product is the "gouna" made from some oily fish such as mackerel or kolio, smeared with dry oregano and other herbs, plenty of salt and pepper and dried in the sun for hours. Citrus is a famous local liqueur made from the aromatic leaves of citrus by distillation, according to old family recipes. This is one of the three Greek liqueurs recognized by the state. There are many spoon sweets of Naxos, such as: citrus, lemon, orange, elm, quince, frappe, pumpkin, watermelon, grape, wild apple, etc. The olive oil of the island that is produced from 400,000 olive trees is also exceptional.

4. METHODOLOGICAL FRAMEWORK OF THE RESEARCH

Qualitative primary research was conducted using interviews with institutions on the island of Naxos during the period August-October 2019. The institutions (public and private) that participated in this research were involved in tourism in general and gastronomy. Particularly:

- Regional Unit of Naxos
- Municipality of Naxos
- Cycladic Chamber
- Naxos Hoteliers Association
- Naxos Rooms to Let Association
- Naxos Mass Catering Association
- Naxos Commercial Association
- Cooker Club of the South Aegean
- Tourism Agencies
- Association of Agricultural Cooperatives of Naxos
- Urban Cooperative of Handmade Traditional and Non-Products and Infinity Items
- "Melimilon" (catering business with traditional products of Naxos)
- "Vioma" - Culinary Culture Experience
- Distilleries - Citrus museums
- "Philema" Food Tours (gastronomic cooking lessons / gastronomic routes)
- "Mothers of the Aegean Sea"
- "Platia" Restaurant (restaurant offering cooking lessons)
- "Basiliko" Cooking Lessons
- "Paradise" Tavern (family tavern with traditional local ingredients, made in the family fields)

- "Stou Lefteri" Tavern (traditional tavern village "Apeiranthos" with dishes completely adapted to the culture of the village)
- Oinochoros"-Wine Cava

It is worth mentioning the good practices of gastronomic experiences from the private stakeholders, such as:

1. "Melimilon" - Catering business with Brunch from traditional Naxian materials. Production of jams and sweets of the spoon. https://www.e-kyklades.gr/travel/showroom/tourism/MELIMILON?lang=en.
2. Distilleries - Citrus Museums. They can be visited by tourists aiming to know how to distill and create the drink. https://www.greekgastronomyguide.gr/item/potopoieio-vallindra-naxos/
3. "Vioma" - Culinary Culture Experience. Experiential gastronomic experiences, activities and events in order to highlight the gastronomic wealth of the island. https://www.vioma.gr/en/
4. "Philema" Food Tours: Gastronomic cooking lessons, gastronomic tours on the island, to get acquainted with the local cuisine and culture. https://philemafoodtours.com/
5. "Mothers of the Aegean Sea" Action of the South Aegean region with the aim of promoting the local traditional "mama's" cuisine. Https://www.argiro.gr/2019/05/aegean-mamas-know-best-aegean-gardeners-giorti-gastronomias-tis-afthentikes-gefseis-tis/
6. "Platia" Restaurant. Restaurant that offers cooking lessons. http://www.platianaxos.com/greece.php
7. "Oinochoros"-Wine Cava. A wine cellar that offers wine tastings from both Naxos varieties and famous varieties. https://www.naxospress.gr/arthro/exodos/naxos-oinohoros-mia-diaforetiki-protasi-apolamvanontas-ena-poly-kalo-krasi
8. "Basiliko" Cooking Lessons: Cooking lessons oriented to the use of traditional raw materials, from the land of Naxos. https://basiliko.gr/
9. Tavern "Paradise": Family tavern of Naxos, with traditional local raw materials, made in the fields of the family. Large variety of traditional dishes daily. A pole of attraction for many famous "chefs", "food bloggers" but also for people with high gastronomic requirements, who want to taste traditional Naxian dishes. https://www.paradisonaxos.gr/index.php?option=com_content&view=article&id=76&Itemid=81&lang=en
10. Tavern "Stou Lefteri": Special tavern in the traditional village "Apeiranthos" in Naxos. Dishes adapted to the culture of the village and consequently to the gastronomic profile of Naxos. The mountainous Naxos is also famous for its rich livestock, a fact that implies excellent quality animal raw materials. https://travelfoodpeople.com/blog/2017/8/24/naxos-tradition-and-good-food-in-apiranthos-at-lefteris

5. CONCLUSION

The conclusions that emerged from the interviews of the institutions on the island of Naxos in the South Aegean, are as follows:

- Intense gastronomic presence of the South Aegean Region in Greece and internationally.

- Strategic actions for extroversion of women entrepreneurship in the field of gastronomy.
- Consolidation of the position of the South Aegean Region as an international gastronomic destination on the world map of gastronomy utilizing international trends.
- Promoting the value and history of "Aegean Gastronomy" as a model of healthy eating.
- Increasing of the number of labeled and certified local products.
- Expansion of the use of local products throughout the tourism chain and relevant consumer education.
- Introduction of incentives for young people to return to the traditions of their place.
- Promotion of local products in the international markets, support of the local economy of the islands.
- Connection of tourism with gastronomy, which is a strategic goal of the South Aegean Region.
- Recognition of the nutritional tradition and uniqueness of the products and recipes of Naxos.
- Organized efforts of the South Aegean Region for the promotion of the rare gastronomic wealth.
- Contribution of the women of the Agricultural Cooperatives in the production and distribution of the local products but also of the cooks.
    - The decisive role of the Federation of Cycladic Women's Associations.
- Strong relationship of women with the agri-food sector, the food service and promotion of culture and history.
- Connection of the preparation of good food with the culture, manners and customs.
- Promotion of gastronomy as part of the cultural capital, which is inherited, acquired and activated individually and collectively.
- Inextricable link between women's activity and local gastronomy.
- Strong authenticity in local products and cooking.
- Connecting the local traditional cuisine with ingenuity, self-sufficiency and hard work.
- Promotion of the cuisine of Naxos as an integral part of the Mediterranean cuisine and diet.
- Integration of the gastronomic tradition of Naxos and the Cyclades with multicultural elements from merchants, conquerors and travelers.
- Evolution of the gastronomy of Naxos from women's activity to art, empirical science and necessity.
- Constant evolution of the inexhaustible local cuisine that is constantly evolving with pride, ingenuity and relationship building.
- Intense female activity in gastronomy, agriculture, animal husbandry and in the production of local products.
- Increasing number of women employed in recent years with more sensitivity to environmental and healthy eating issues.
- Promoting the role and position of women in the production and distribution of food.
- Contribution of the woman of Naxos to the eating habits and cooking practices and her contribution to gastronomy.
- The kitchen is very important for the woman of Naxos beyond a simple place for cooking.
- Contribution of cooking to the stabilization and strengthening of social ties.
- Connecting the good meal with recognition, support and reward.
- The main role of flavor festivals for the promotion of local gastronomy and the rescue and promotion of local products.

In conclusion, female creativity plays a leading role on the island of Naxos in the South Aegean Region. It contributes to the creation of many positive social, cultural and environmental influences and is inextricably linked to the development of gastronomic tourism. The special traditional cuisine of Naxos develops the local community, contributes to the continuation of the production of traditional products and transmits customs, traditions and events that are a means of communication between the cultures and peoples of modern globalized society. The specific island gastronomy promotes the cultural heritage and the uniqueness of the villages of the island. It offers the visitor the opportunity to get to know the local products, the Cycladic cuisine and the daily life of the inhabitants with respect for the environment and tradition.

REFERENCES

Aggestam, M., & Kristoferson, C.W. (2017). How women entrepreneurs build embeddedness: A case study approach. *International Journal of Gender and Entrepreneurship.*

Ahl, H., & Nelson, T. (2010). Moving forward: Institutional perspectives on gender and entrepreneurship. *International Journal of Gender and Entrepreneurship, 2*, 5–9.

Al-Dajani, H., & Marlow, S. (2013). Empowerment and entrepreneurship: A theoretical framework. *International Journal of Entrepreneurial Behaviour & Research, 19*(5), 503–524.

Anderson, A. (2000). Paradox in the periphery: An entrepreneurial preconception. Entrepreneurship and Regional Development. *International Journal (Toronto, Ont.), 12*(2), 91–110.

Baden, S. & Milward, K. (2000). *Gender inequality and poverty: trends, linkages, analysis and policy implications.* BRIDGE Report.

Baggio, R., & Valeri, M. (2020). Network science and sustainable performance of family businesses in tourism. *Journal of Family Business Management.* DOI . doi:10.1108/JFBM-06-2020-0048

Baines, S., & Wheelock, J. (2000). Work and employment in small businesses: Perpetuating and challenging gender traditions. *Gender, Work and Organization, 7*(1), 45–56.

Bank of Greece. (2019). https://www.bankofgreece.gr/

Baron, R. A., & Shane, S. A. (2005). *Entrepreneurship: A process perspective.* Thomson.

Björk, P., & Kauppinen-Räisänen, H. (2016). Exploring the multi-dimensionality of travellers' culinary-gastronomic experiences. *Current Issues in Tourism, 19*(12), 1260–1280.

Boatto, V., & Gennari, A.J. (2011). *La roadmap del turismo enologico.* Franco Angeli.

Bonanno, A. (1992). *The Agricultural and Food Sector in the New Global Era.* Concept Publishing Company.

Brunori, G., & Rossi, A. (2000). Synergy and coherence through collective action: Some insights from wine routes in Tuscany. *Sociologia Ruralis, 40*(4), 409–423.

Camisón, C., Forés, B., & Navarro, B.M. (2017). Current issues in tourism cluster and firm-specific antecedents of organizational innovation. *Current Issues in Tourism, 20*(6), 617–646.

Chiles, T. H., Bluedorn, A. C., & Gupta, V. K. (2007). Beyond creative destruction and entrepreneurial discovery: A radical Austrian approach to entrepreneurship. *Organization Studies, 28*, 467–493.

Cohendet, P., Grandadam, D., & Simon, L. (2010). The Anatomy of the Creative City. *Industry and Innovation, 17*(1), 91–111.

Dahlstrom, M. (1996). Young Women in a Male Periphery-Experiences from the Scandinavian North. *Journal of Rural Studies, 12*(3), 259–271.

Diaconescu, D., M., Moraru, R., Stănciulescu, G. (2016). Considerations on Gastronomic Tourism as a Component of Sustainable Local Development. Amfiteatru Economic, 18(10), 999-1014.

Drakopoulou-Dodd, S., & Anderson, A. (2007). Mumpsimus and the mything of the individualistic entrepreneur. *International Small Business Journal, 25*(4), 341–360.

EC (European Commission). (2014). *Statistical Data on Women Entrepreneurship in Europe.* Directorate-General for Enterprise and Industry. B-1049 Brussels.

Finkelstein, J. (1989). *Dining Out: A Sociology of Modern Manners.* Polity Press.

García, M. C. D., & Welter, F. (2013). Gender identities and practices: Interpreting women entrepreneurs' narratives. *International Small Business Journal, 31*(4), 384–404.

Giagou, D., & Apostolopoulos, C. (1996). Rural Women and the Development of Agrotouristic Co-operatives in Greece: The case of Petra, Lesvos. *Journal of Rural Cooperation, 1996*(24), 143–155.

Gray, S.R., & Benson, P.G. (2003). Determinants of executive compensation in small business development centers. *Nonprofit Management & Leadership, 13*(3).

Green, G.P., & Dougherty, M.L. (2008). Localizing linkages for food and tourism: Culinary tourism as a community development strategy. *Community Development Journal: An International Forum, 39*(3), 148–158.

Gyimothy, S., Rassing, C., & Wanhill, S. (2000). Marketing Works: A Study of Restaurants on Bornholm. Denmark. *International Journal of Contemporary Hospitality Management, 12*, 371–379.

Hall, C., M., & Mitchell, R. (2000). Wine Tourism in the Mediterranean: A Tool for Restructuring and Development. *Thunderbird International Business Review, 42*(4), 445–465.

Heidegger, M. (1996). *Being and Time.* State University of New York Press.

Hieu, V.M., & Rasovska, I. (2017). Craft villages and tourism development, a case study in Phu Quoc island of Vietnam. *Management, 21*(1), 223–236.

Kastenholz, E., Davis, D., & Paul, G. (1999). Segmenting Tourism in Rural Areas: The Case of North and Central Portugal. *Journal of Travel Research, 37*, 353–363.

Kercher, M., Okumus, F., & Okumus, B. (2008). Food Tourism as a Viable Market Segmet: It's All How You Cook the Numbers. *Journal of Travel and Tourism Marketing, Volume, 25*(2), 137–146.

Kim, S., Park, E., & Lamb, D. (2019). Extraordinary or ordinary? Food tourism motivations of Japanese domestic noodle tourists. *Tourism Management Perspectives*, *29*, 176–186.

Kivela, J., & Johns, N. (2002). A mythological approach in analyzing dining out experiences: Implications for remaking of tourist destinations. *An International Tourism Research Conference*.

Korsgaard, S., Ferguson, R., & Gaddefors, J. (2015). The best of both worlds: How rural entrepreneurs use placial embeddedness and strategic networks to create opportunities. *Entrepreneurship and Regional Development*, *27*(9-10), 574–598.

Koutoulas, D. (2019). *Assignment of study - masterplan for the Cultural Route of Naxos*. Stadium.

Kumar, S., & Valeri, M. (2021). Understanding the relationship among factors influencing rural tourism: A hierarchical approach. *Journal of Organizational Change Management*. Advance online publication. doi:10.1108/JOCM-01-2021-0006

Lachmann, L. (1977). *The Market as an Economic Process*. Basil Blackwell.

Long, M. L. (2004). *Culinary Tourism*. University Press of Kentucky.

McKeever, E., Jack, S., & Anderson, A. (2015). Embedded entrepreneurship in the creative re-construction of place. *Journal of Business Venturing*, *30*(1), 50–65.

Peeters, L. W. J., & Ateljevic, I. (2009). Women empowerment entrepreneurship nexus in tourism: Processes of social innovation. In S. Page & J. Ateljevic (Eds.), *Tourism and entrepreneurship: International perspectives* (pp. 75–89). Routledge.

Rowlands, J. (1997). *Questioning empowerment: working with women in Honduras. Academic Press.*

Santos, V., Ramos, P., Sousa, B., Almeida, N., & Valeri, M. (2021a). Factors influencing touristic consumer behavior. *Journal of Organizational Change Management*. Advance online publication. doi:10.1108/JOCM-02-2021-0032

Santos, V., Sousa, B., Ramos, P., & Valeri, M. (2021b). Emotions and Involvement in Tourism Settings. *Current Issues in Tourism*. Advance online publication. doi:10.1080/13683500.2021.1932769

Scarpato, R. (2002). Gastronomy Studies in Search of Hospitality. *Journal of Hospitality and Tourism Management*, *9*(2), 152–163.

SETE. (2010). *The Gastronomy in the Greek Tourism Marketing*. Available in Http://www.sete.gr/files/Media/Ebook/2010/100222%20gastronomy%20f.pdf

Shahriar, A. Z. M. (2018). Gender differences in entrepreneurial propensity: Evidence from matrilineal and patriarchal societies. *Journal of Business Venturing*, *33*(6), 762–779.

South Aegean Region. (2019). http://www.pnai.gov.gr/

Telfer, D., & Wall, G. (1996). Linkages between tourism and food production. *Annals of Tourism Research*, *23*(3), 635–653.

Tikkanen, I. (2007). Maslow's hierarchy and food tourism in Finland: Five cases. *British Food Journal*, *109*(3), 635–653.

Tucker, H., & Boonabaana, B. (2011). A critical analysis of tourism, gender and poverty reduction. *Journal of Sustainable Tourism.*

Valeri, M. (2016). Networking and cooperation practices in the Italian tourism business. *Journal of Tourism, Heritage & Services Marketing*, *2*(1), 30–35. https://doi.org/10.5281/zenodo.376333

Valeri, M. (2021). *Organization Theories. Implications for the Strategic Management.* Springer.

Valeri, M., & Baggio, R. (2020a). Social network analysis: Organizational implications in tourism management. *The International Journal of Organizational Analysis*. Advance online publication. doi:10.1108/IJOA-12-2019-1971

Valeri, M., & Baggio, R. (2020b). Italian tourism intermediaries: A social network analysis exploration. *Current Issues in Tourism*, *24*(9), 1270–1283. https://doi.org/10.1080/13683500.2020.1777950

Valeri, M., & Baggio, R. (2020c). A critical reflection on the adoption of blockchain in tourism. *Journal Information Technology and Tourism*, *23*(2), 121–132. doi:10.100740558-020-00183-1

Valeri, M., & Baggio, R. (2021). Increasing the efficiency of knowledge transfer in an Italian tourism system: A network approach. *Current Issues in Tourism*. Advance online publication. doi:10.1080/13683500.2021.1937960

Valeri, M., & Katsoni, V. (Eds.). (2021). *Gender and tourism: challenges and entrepreneurial opportunities*. Emerald Publishing.

Ventura, F. (1994). Women in Italian Agriculture: New Roles, New Problems. In L. van der Plas & M. Fonte (Eds.), *Rural Gender Studies in Europe* (pp. 80–90). Van Gorcum.

Welter, F. (2011). Contextualizing entrepreneurship-conceptual challenges and ways forward. *Entrepreneurship Theory and Practice*, *35*(1), 165–184.

Whatmore, S. (1988). *Farming Women: Gender, Work and Family Enterprise*. Macmillan.

Wolf, E. (2002). *Culinary tourism: A tasty economic proposition*. http//: www.culinarytourism.org/faq.php

World Tourism Organization. (2017). *Affiliate Members Report. Volume sixteen. In Second Global Report on Gastronomy Tourism*. UNWTO.

Chapter 16
Assessing the Potential for Tourism Development:
Northern Portugal as a Surf Destination

Goretti Silva
https://orcid.org/0000-0001-6661-0555
Polytechnic Institute of Viana do Castelo, Portugal

Susana Rachão
Polytechnic Institute of Viana do Castelo, Portugal

Alexandra I. Correia
Polytechnic Institute of Viana do Castelo, Portugal

Luís André Soares
Polytechnic Institute of Viana do Castelo, Portugal

ABSTRACT

Increasing awareness of the relevance of natural resources for the practice of outdoor leisure and recreation purposes has contributed to the increasing attention upon the assessment and monitoring of the quality of sports spots. As such, evaluating the existing natural conditions and leisure-related infrastructures is crucial, helping public and private stakeholders in decision-making processes. This chapter aims at examining the conditions for surf tourism in the north of Portugal, enabling the monitoring and positioning as an international flagship destination for surf practitioners. This is done through a mixed-methods approach comprising the demand and supply perspectives and supported by an assessment tool developed to evaluate the conditions for the practice of surfing activities within a framework of sustainable tourism development. The relevance of assessing the potential and monitoring of the existing conditions for sport and outdoor activities, namely surf tourism, according to a defined set of attributes, is also discussed.

DOI: 10.4018/978-1-6684-3889-3.ch016

INTRODUCTION

The increasing awareness of the relevance of natural resources for the practice of outdoor leisure and recreation activities has contributed to the increase of attention upon the monitoring of the quality of outdoor spots in the context of tourism. Surf tourism (surf tour operators packages including transport, accommodation and food) has come into prominence only in the 1990s (Buckley, 2002a). Nevertheless, surfing consisting of wave surfing, windsurfing, and kite surfing is a growing sports activity within the worldwide adventure outdoor tourism sector, representing 35 million surfers in 2018 (International Surfing Association, 2014).

Schools specialised in surfing activities (bodyboard, surf, kitesurf, windsurf and stand-up paddle) are widespread (Fadda, 2019), and surf tourism has become a significant factor influencing regional economic, social and environmental processes (Buckley, 2002a). At the same time, the number of surf lifestyle entrepreneurs has grown worldwide, contributing to local destination development, namely through surfing festivals, and creating bounds and networks with local communities (Marchant & Mottiar, 2011; Ratten, 2018, Valeri, 20216; Valeri and Baggio, 2020a; 2020b; 2020c; 2021; Baggio and Valeri), which are key issues in sustainable approaches to tourism management (Martin & Assenov, 2012; Valeri, 2021).

Clearly, as for other tourism destination in general, surf destinations' competitiveness arises from many factors, which include, at a first level of priority, a sustainable approach to tourism. Sustainable tourism and responsible tourism are multi-dimensional concepts which include environmental protection, economic development, social equity and ecological effects (Ritchie and Crouch, 2003). In order to develop a sustainable tourist destination, and for it to remain competitive and sustainable, it is pertinent to gain insight into the concept of sustainable and responsible tourism from a more holistic perspective, in addition to the assessment of destination competitiveness attributes and the key tourism destination components (Ritchie and Crouch, 2003), which are often complex, subjective and situational natural environment and geographic location, climate, scenery, but also man-made attractions, tourism infrastructure and supporting facilities.

Thus, sustainable tourism development refers to a process which envisages a responsible management process of resources, integrating the natural, cultural and human environments and considers tourism effects on the cultural heritage and traditional elements, and the dynamics of each local community (Chan, 2010). Responsible tourism shifts responsibility towards individual, organizations and businesses. Equally important in attracting people to visit the destination, in a responsible and successful tourism approach, the key element in is planning and governance (Rasdi et al., 2019; Valeri and Fadlon, 2018). Planning at the local community level needs be led by an integrated culture which implies the cooperation of all stakeholders involved in the tourism decision making process (Rasdi, et al. 2019). Moreover, tourist destination needs to define a governance body, including private and public actors/individuals and the local community, which implies a broad shared, participatory, decision-making process (Valeri and Fadlon, 2018). In this context, cooperation and networking can represent an organizational model suitable to develop adaptive socio-economic systems, enabling busineses and communities to face today's complex competitive context (Valeri and Baggio, 2020a; 2020b; 2021; Valeri and Fadlon, 2018).

As acknowledged by Brochado et al. (2018), there is still limited research on surf tourism in Portugal compared to other surfing destinations. As such, assessing the existing natural conditions and leisure-related infrastructures for residents and visitors' fulfilment is crucial in tourist destinations' planning and development processes, leading to better-informed decision-making processes. This chapter aims at examining the conditions for surf tourism in the north of Portugal. This is done through a mixed-methods approach comprising the demand and supply perspectives and supported by an assessment tool developed to evaluate the conditions for the practice of surfing activities within a framework of sustainable tourism development.

Despite the Surf Resource Sustainability Index constructed by Martin and Assenov (2015), it does not classify the attributes according to the different dimensions identified at the surf destination. In order to address this gap, an original assessment tool comprising a defined set of attributes was developed. The proposed tool includes three categories of attributes, each comprising different indicators: 1) the qualifying/ structural attributes; 2) the subjective attributes (which are inherent to individuals – surfing tourists), and (3) the intensifiers (adding extra value to the surfing experience). Though it can be employed in any other surf tourism destination, it has been submitted to a solid construction and validation process.

The results of this research will reinforce the idea that assessing the resources and conditions for the practice of surf is important particularly significant for Portugal as it enables it to monitor its performance and find ways to improve its positioning as an international flagship destination. This chapter is structured as follows: the following section reviews the literature on the main themes guiding the study (section 2); the methods employed and the description of the collected data (section 3); the results and discussion (section 4); the main conclusions (section 5), including the relevance of assessing the potential and monitoring of the existing conditions for sport and outdoor activities, namely surf tourism according to a defined set of attributes. Limitations and recommendations for future research (section 6) are also presented.

LITERATURE REVIEW

Surfing as a Tourism Product

Surfing tourism is a social phenomenon (Buckley, 2002a) that revolves around specific consumption behaviours (Ponting & McDonald, 2013). As a tourism activity, surfing can be defined as travelling individuals outside their usual residence, travelling at least 40kms, staying overnight, and whose main motivation is surfing (Buckley, 2002a; Ponting, 2009; Towner, 2018), either for recreational or commercial surfing. Buckley (2002a) distinguishes these two types of surfing, referring to the first as independent surfers (they plan their trips independently, use their transport and equipment and stay overnight in local accommodation), whereas commercial surfing is when individuals buy surf tour packages. The increasing demand has culminated with the development of several businesses and the creation of surf schools (Fadda, 2019); thus, surf packages include services such as the sale of sports equipment and classes.

The literature on the surf tourism demand is comprehensive exploring surfing tourists' sociodemographic profiles and travelling behaviours. In general and according to several studies, worldwide surfing practitioners are predominantly male, between 30 and 35 years old, with high educational and income levels, and characterised as adventurous individuals (Barbieri & Sotomayor, 2013; Dolnicar & Fluker, 2003a; Kruger & Saayman, 2017; Porter & Usher, 2019). Table 1 illustrates distinct typologies of surfing tourists as identified in the literature.

Table 1.Surfing tourists' typologies

Typology	Explanation	Authors
Radical adventurers	Importance is given to the local surf season and secret locations	(Dolnicar & Fluker, 2003a)
Ambivalent	Anything is important at all	
Luxury surfers	Importance is given to the accommodation, food and safety	
Price-conscious safety seekers	Surf-related items, personal safety and health, family facilities and the quality of accommodation	
Price-conscious adventurers	Surf-related items, personal safety and health, search for new locations and the lack of crowd	
Incidental surf tourists	First time practising with local surf instructor	Martin & Assenov (2012)
Soft surf tourists	Stay in *surf camps* with access to surf instructors and surf material.	
Hard surf tourists	Individuals who undertake the travel with the sole purpose of surfing (leisure or competition)	
Hard-core surf-rider tourists	Depending on the level of individuals' experience	Orams & Towner (2013)
Recreational surf-rider tourists		
Casual surf-rider tourists		
Kooks (beginner/wannabe surfers		
Beginners	Depending on the level of individuals' experience	Kruger & Saayman (2017)
Weekenders		
Amateurs		
Professionals		
Source: authors' elaboration		

Existing studies (Sotomayor & Barbieri, 2016) suggest that further attention is needed to explore destination preferences across different types of surfers. The destination preferences are influenced by the quality of services provided and perceived value (Ko & Pastore, 2004), and, therefore, will impact the likelihood to recommend the surfing experience to others and repeated visits (Brochado et al., 2018).

There is a wide range of factors that influence the choice of surfing spots. Factors that relate to specific aspects of surfing (e.g., wave consistency, wave quality, and crowd) and the attributes of the surfing destination, such as information, accessibility, among others. For instance, seasonality and high peak of tourism flows (e.g. crowded surf spots) have a strong influence on surf tourists' satisfaction (Buckley, 2002b; Mach et al., 2018) (Buckley, 2002a). Experienced surfers who travel with their equipment, and sometimes with their own transport, value more natural conditions such as perfect waves (Portugal et al., 2017), whereas destination attributes such as surfing infrastructures (local surf schools/surf camps) are highly valued by novice surfers (Barbieri & Sotomayor, 2013). However, currently, surfing tourists (both experienced and novice) are relying on tour operators for their travel arrangements (e.g. transport within the destination), leading surf business operators to provide adequate services (Buckley, 2002a).

For these reasons, planning for the sustainable development of surf tourism requires multiple stakeholders and multiple indicators to assess and monitor the conditions for the practice of the activity and the attractiveness of destinations. Moreover, as Buckley (2002a) stated, surf tourism produces social, economic and environmental impacts for the host communities. Economic impacts include creating jobs in accommodation, restaurants, and surf-related businesses while the environmental impacts are related to increased resource consumption to satisfy tourist needs.

Therefore, identification and management of surfing tourism attributes are particularly critical to the sustainability of the tourism destination. Significant attributes influencing the choice of a surfing destination are accessibility and infrastructure, as argued by different authors (Barbieri & Sotomayor, 2013; Buckley, 2002a; Martin & Assenov, 2014). However, as Buckley (2002a) claimed, the most easily accessible surf spots are crowded, and most uncrowded surf spots are inaccessible. Dolnicar and Fluker (2003b) stated that the lack of crowd is one of the essential features of the surf destination, but the more popular and crowded it becomes, the less attractive it gets. The regional governance must foresee surf destination management plans to avoid overcrowding of surfers and surf tour operators (Mach & Ponting, 2018) and regulate their recreation capacity management activity (Buckley, 2002b).

Equally significant attributes are beach safety and supporting facilities. Physical hazards (e.g. dangerous areas, conflicting uses) at surf sites can be addressed with signage such as safety information panels (Martin & Assenov, 2014) and with information panels regarding local surf etiquette (Mach & Ponting, 2018). In addition, beach safety and supporting facilities are criteria for awarding a coastal zone with the blue flag, a label that distinguishes coastal zones with outstanding natural values (Halkos & Matsiori, 2012; Lazarow et al., 2008; Martin & Assenov, 2015).

Thus, combining the attributes mentioned above in one assessment tool is challenging, but it provides tourism planners seeking to achieve higher effectiveness levels a governance tool, enabling them to prioritise actions or investments.

METHODOLOGY

The surf tourism destination assessment was based on the combination of qualitative and quantitative research methods through a systematic assessment of the attributes of the surf destination with the help of specific measurement instruments. Three instruments have been selected to measure the surf destination attributes, combining different but complementary perspectives according to the principle of methodological triangulation (Decrop, 1999).

Previous studies have emphasised the value of mixed-method research for producing relevant results (Imran et al., 2014), particularly in outdoor recreation activities (Bichler & Peters, 2020). The mixed-methods approach was chosen due to its suitability regarding the holistic perspective undertaken within this study, employing a systematic integration of quantitative and qualitative data (Creswell, 2009) within a single study. First, data on websites and blogs were collected, identifying main surf spots in the North of Portugal and information concerning its baseline attributes. Secondly, semi-structured interviews and questionnaires were performed with different kinds of participants and in two distinct research phases, as displayed in Table 2. Following the work of Bichler & Peters (2020), the concurrent mixed-method approach was employed, which combines the integration of quantitative and qualitative data in the data interpretation and analysis stages (Khoo-Lattimore et al., 2019; Teddlie & Yu, 2007).

Table 2 .Sources for methodological triangulation of the surfing tourism destination assessment

Research type	Instruments	Research object	Data analysis	Population	Sample size
Quantitative	Questionnaire	Satisfaction, perceived value and behavioural attitude towards surfing	IPA	Surfers	32
Quantitative	Assessment matrix	Set of attributes to assess the conditions of surfing spots	Descriptive statistics (means)	Surf spots	34
Qualitative	Semi-structured interview	Public administration; participation in regional administration.	Thematic content analysis	Surf-related business owners	4

Source: authors' elaboration

Semi-structured interviews were employed with four surf-related business owners: (i) surf tourism animation firm; (ii) accommodation unit specialised in surfing activities; (iii) surf school and (iv) a surf club. Surf-related business owners were asked to fulfil the assessment matrix developed by the research team, were questioned about further improvements that needed to be implemented, and their perception of public organisations' participation in surf tourism development at the regional level. The assessment matrix was developed into a broader research process focusing on outdoor tourism. It was based on an extant literature review and validated by a group of academics and practitioners involved in surfing tourism-related activities. It is a robust instrument comprising different destination attributes (mostly quantitatively rated scores) and capable of objectively assessing and evaluating surfing tourism-related activities' conditions, as illustrated in Table 3.

Questionnaires in the form of face-to-face structured interviews were used to inquiry surf practitioners concerning their level of satisfaction, perceived value and behavioural attitude towards the surfing practice in the research area. The questionnaire was thus divided into three sections. The first section aimed to assess the behavioural consumption patterns of surfers, asking about their travel arrangements. The second section was designed to evaluate respondents' satisfaction and perceived value regarding surfing conditions in Northern Portugal, different rating items on a 5-point Likert scale ranging from "1 = very unsatisfied" to 5 = "very satisfied". Content analysis was applied for the qualitative part of the study; concerning quantitative analysis, descriptive analyses were performed using the SPSS software. Items included in the surf spots assessment tool were submitted to an Importance-Performance analysis. The third section collected data on respondents' sociodemographic profiles (gender, age, country of residence, marital status, level of education, main occupation).

Quantitative and qualitative data started to be collected in May 2019 and are in progress. However, given the current situation provoked by the COVID-19, qualitative data collection was interrupted.

Table 3. Surfing tourism destination assessment

Attributes	Attribute rating scores	References
Classified area	1 – No 5 – Yes	(Gamarra et al., 2019; Scheske et al., 2019)
Accessibility	1 – On foot; 2 – On foot, non-motorised vehicles and two-wheel motorised vehicles; 3 – On foot, non-motorised vehicles, two-wheeled motor vehicles and light motorised four-wheel vehicles without organised parking; 4 – On foot, non-motorised vehicles, motorised two-wheeled vehicles, light motorised four-wheeled vehicles with organised parking, caravans, motorhomes and heavy vehicles with organised parking; 5 – On foot, non-motorised vehicles, motorised two-wheeled vehicles, light motorised four-wheeled vehicles with organised parking, caravans, motorhomes, heavy vehicles with organised parking and is served by public transport networks.	(Alaeddinoglu & Can, 2011; Priskin, 2001)
Infrastructure	1 – Does not have any type of built infrastructure; 2 – Has infrastructures for access to the beach (if necessary); 3 – Has infrastructures for access to the beach (if necessary) and parking; 4 – It has infrastructures for access to the beach (if necessary), parking and public showers; 5 – It has infrastructures for access to the beach (if necessary), parking, public showers and changing rooms.	(Halkos & Matsiori, 2012)
Supporting facilities	1 – The beach has no supporting facilities; 2 – The beach has bar/restaurant/mini-market; 3 – The beach has bar/restaurant/mini market and accommodation services; 4 – The beach has bar/restaurant/mini-market, accommodation services and an information desk; 5 – The beach has bar/restaurant/mini-market, accommodation services, information desk and sports equipment rental shops / Surf School.	(Barbieri & Sotomayor, 2013)
Signalling	1 – The beach does not have any signage; 2 – The beach has signs of the conditions of the sea; 3 – The beach has signs of the conditions of the sea and signs indicating dangerous areas; 4 – The beach has signs of the conditions of the sea, signs indicating dangerous areas and signs of conflicting uses; 5 – The beach has signs of the conditions of the sea, signs indicating dangerous areas, signs of conflicting uses and has a health and safety information panel.	(Martin & Assenov, 2015)
Surveillance	1 – Does not have any surveillance service during the bathing season; 2 – The beach has permanent surveillance during the bathing season; 3 – The beach has permanent surveillance and has rescue equipment (namely buoys) available and easily accessible during the bathing season; 4 – The beach has permanent surveillance, rescue equipment (namely buoys) available and easily accessible and allows easy access to rescue and emergency means during the bathing season; 5 –The beach has permanent surveillance, rescue equipment (namely buoys) available and easily accessible, allows easy access to emergency means, a first aid kit equipped with first aid material, and staff qualified to provide first aid during the bathing season.	(Lazarow et al., 2008; Martin & Assenov, 2014)
Universal accessibility	1 – The beach is not accessible to people with reduced mobility; 2 – The beach has easy access to the bathing area, with an accessible route that allows access to people with reduced mobility; 3 – The beach has easy access to the bathing area, with an accessible path that allows access for people with reduced mobility and space for parking vehicles for people with reduced mobility; 4 – The beach has easy access to the bathing area, with an accessible route that allows access for people with reduced mobility, parking for people with reduced mobility and easy access to support structures with an accessible route that allows access people with reduced mobility; 5 – The beach has easy access to the bathing area, with an accessible path that allows access for people with reduced mobility, parking for vehicles with people with reduced mobility and easy access to support structures with an accessible path that allows access people with reduced mobility as well as a school/club/organisation capable of providing adapted services for people with reduced mobility.	(UNWTO, 2016)
Crowd	1 – Very crowded; 2 – Enough crowd; 3 – Neither too little nor too much crowd; 4 – Little crowd; 5 – Nothing crowd;	(Dolnicar & Fluker, 2003b) (Barbieri & Sotomayor, 2013)
Wave quality	1 – Reasonable; 3 – Good; 5 – Very good;	(Barbieri & Sotomayor, 2013; Dolnicar & Fluker, 2003b)
Wave consistency	1 – One to two days a week; 3 – Three to four days a week; 5 – Five or more days a week.	(Barbieri & Sotomayor, 2013; Dolnicar & Fluker, 2003b)
Source: authors' elaboration		

DATA ANALYSIS

For the qualitative part of the study, content analysis was applied. The interview data was submitted to a qualitative thematic analysis (Fox et al., 2010; Neuman, 2014), and deductive content analysis was employed as the themes were grounded on the theory developed in the literature review section. To ensure the validity of the results was employed the triangulation of the researchers (Decrop, 1999; Dwyer et al., 2012). The first author of the paper coded transcriptions individually, and the remaining authors verified the coding to have systematic coding reliability. Whereas quantitative analysis was supported using SPSS (Statistical Package for the Social Sciences) software, and descriptive analyses were performed. Importance-Performance Analysis, developed by Martilla and James in 1977, was carried out, as it is an effective technique to determine which destination or product attributes need improvement, examining the difference between "Performance (P)" and "Importance (I)" and identifying areas where to focus on, when resources are limited (Mimbs et al., 2020). Martilla and James proposed interpretations for the four quadrants as follows: Q1 – high importance and high performance (Keep Up Good Work); Q2 – high importance and low performance (Concentrate Here); Q3 – low importance and low performance (Low priority); Q4 – low importance and high performance (Possible Overkill) (Marasinghe et al., 2021).

RESEARCH STUDY AREA: NORTHERN PORTUGAL

According to the *Federação Portuguesa de Surf* (Portuguese Surfing Federation), in 2016, there were 82 surf clubs and 218 surf schools in total in Portugal (Brochado et al., 2018). Despite the growing demand for surf tourism in Portugal, research studies are limited to social and economic impacts in local communities focusing primarily in Central and Southern regions, such as Ericeira, Costa da Caparica, Cascais, Algarve, Nazaré and Peniche (Brochado et al., 2018; Frank, 2015; Machado et al., 2018; Moutinho et al., 2007; Portugal et al., 2017; Silva & Ferreira, 2014). For Brochado et al. (2018), these are the most important surfing regions in Portugal, based on the number and scope of international competition circuits organised, such as the World Surf League (WSL), the world's top competitive surfing organisation (e.g. WSL in Peniche, Cascais, Azores and Ericeira). However, some international surfing events were also organised in Northern Portugal, namely the World Bodyboard Championship in 2018, hosted in Praia da Arda, Viana do Castelo, and repeated for five years (Correia et al., 2021). Despite this, and the existence of various relevant surf spots, before this research, little was known about the conditions for surfing along Northwest Coast, and in particular in the nine municipalities analysed, namely, Caminha, Viana do Castelo, Esposende, Póvoa de Varzim, Vila do Conde, Matosinhos, Porto, Vila Nova de Gaia e Espinho.

RESULTS AND DISCUSSION

Stakeholders' Perceptions of Regional Governance

Restricted access to specific surf breaks is crucial for its environmental maintenance and prevents accidents among practitioners, as described by one of the interviewees (ATS02), as illustrated in Table 4. It is evident that some surfing spots are more appealing given their natural and infrastructural conditions and, therefore, more crowded (Dolnicar & Fluker, 2003b; Mach & Ponting, 2018). Thus, recognising the increasing demand for specific surfing spots, regulating access conditions is vital for their natural preservation (Buckley, 2002a).

Table 4. Thematic content analysis of interviews data

	Interviewees insights			
Interview themes	**ATS02**	**AM21**	**AM27**	**AM29**
Improvements to be implemented	Creation of restricted areas for SUP (Stand-Up Paddle) activities. Rio Paiva is not navigable, but boats and jet skis are sailing. Boat drivers disrespect those who are performing non-motorised nautical activities (e.g., SUP and canoeing).	Public transportation is lacking. Tourists have to travel by taxi or car rental within the region.	It should be improved the international communication and promotion of the Northern region and the creation of infrastructures.	Since 2001, a facility like *Viana Beach Centre* has been needed. There was a need for infrastructure to support surfing activities. The Portuguese Environmental Agency has to analyse the different situations.
Regional public administration	The public administration only promotes Porto and Douro and does not look at the remaining territories of Northern Portugal. There is no strategy for the rest of the territory.	The scarce public transportation is a major problem for municipalities like Viana do Castelo.	There is no sufficient coordination between the public sector actors'.	The public administration is addressing regional problems with slow responses to surf tourism stakeholders.
Participation in regional management	Informal meetings with local accommodation operators.	Creation of collaboration/ partnerships with other surf businesses. The organisation of the *Windsurfing World Championship* in partnership with the local municipality.	I am currently teaching Kitesurfing in a higher education course (unique in the world) at the local university. Collaboration with music festivals.	The organisation of water sports events (*BlueWays*) and meetings with the municipality and with Local Development Associations (LDA).
Source: authors' elaboration				

One crucial component of surfing tourism is the ability for surfers to autonomously travel between spots, particularly for novice surfers, as suggested in the literature (Barbieri & Sotomayor, 2013; Buckley, 2002a). Therefore, a good public transportation network is a crucial attribute and key issue in the regional development of surfing for the study area destinations, as one of the interviewees (AM21) highlighted.

Qualitative data results show that national and international promotion of Northern Portugal as an outdoor destination as a whole and as a surf destination, in particular, is scarce (AM27). This idea was reinforced by interviewees when asked about the intervention of the regional public administration, as explained:

The public administration only promotes Porto and Douro and does not look at the remaining territories of Northern Portugal. There is no strategy for the rest of the territory. (ATS02)

Moreover, the interviews also suggest that problem-solving capacity is still a challenge for local governance:

The public administration is addressing regional problems with slow responses to surf tourism stakeholders. (AM29)

Reviewed literature on surf tourism emphasised the importance of involving multiple stakeholders in destination governance processes and its key role in promoting more effective communication, investment activities and regulation strategies (Buckley et al., 2017; Mach & Ponting, 2018). Interviewees gave the region a positive note on that, referring to the organisation of a few recent surfing events (e.g. *Windsurfing World Championship*), as a change in the mind-set of regional governance, emphasising their commitment to the development of strategic events (AM21; AM29), also reflected in the implementation of kitesurfing lessons in a higher education degree (AM27).

Surfing Tourism Destination Assessment

The complete 'surfing tourism destination assessment is provided in Table 5. The thirty-four surf spots are listed, and the assessed mean values are provided for some attributes (ordinal scales). Concerning the thirty-four surf spots as a whole, the quality of the wave has the lowest mean score attribute (2.76), and surveillance has the highest (4.65). In terms of ranking the surf spots based on the 'surfing tourism destination assessment', '*Cabedelo do Douro*' and '*Cagalhotos*' surf spots have the worse conditions and rank at 2.00 and 2.44 respectively, and the sites with the highest scores are '*Vila Praia de Âncora*' and '*Azurara*' surf spots at 4.56 and 4.67 respectively.

Surf Practitioners' Profile and Travel Characteristics

Questionnaire results showed that over 70% of respondents were male (*n*=28), and the average age was approximately 30 years old. The majority (56%) reported that they were single and mostly (over 75%) residents in Northern Portugal. The majority (65.6%) have a higher education degree and are employees (53.1%). Overall, the respondents' sociodemographic and surf travel preferences are in line with previous research (Barbieri & Sotomayor, 2013; Porter & Usher, 2019; Portugal et al., 2017).

Regarding the level of expertise, a large proportion (40.6%) of the respondents were untrained practitioners, followed by trained practitioners (33.3%), federated athletes (18.8%), and 6.5% were professional athletes.

Concerning surfing consumption behaviour, only 37.5% acquired the activity in advance through the surf association/club, to which they belong, and performed it autonomously. These results are not surprising since a significant part of the sample (58.6%) are autonomous surf practitioners with training and residents in the northern region of Portugal, consistent with Correia et al. (2021) results.

The majority (65.5%) travelled to the surf spot using their vehicle (car), with friends and/or colleagues (50%), while 15.6% of respondents reported surfing with an association/club or alone, followed by another companion (12.5%), and only 6.3% with their family. These results about preferred surf travel companions are similarly found in various studies (e.g. Dolnicar & Fluker, 2003a; Porter & Usher, 2019; Portugal et al., 2017). Table 6 displays the evaluation of the destination as a whole, through mean rating for each attribute, using the Importance-Performance Analysis (IPA).

Table 5. Surfing tourism destination assessment

Surf spots	Accessibility	Infrastructure	Supporting facilities	Signalling	Surveillance	Universal accessibility	Crowd	Wave quality	Wave consistency	*Mean*
Atlântica	5.00	4.00	4.00	4.00	4.00	4.00	3.00	3.00	5.00	3.89
Azul	5.00	5.00	4.00	5.00	5.00	5.00	3.00	3.00	5.00	4.22
Aguçadoura	4.00	4.00	4.00	5.00	5.00	4.00	3.00	3.00	4.00	3.89
Apúlia	5.00	5.00	4.00	5.00	5.00	4.00	3.00	3.00	5.00	4.11
Arda	5.00	5.00	5.00	4.00	5.00	5.00	3.00	5.00	5.00	4.44
Azurara	5.00	5.00	5.00	5.00	5.00	5.00	3.00	5.00	5.00	4.67
Baía	5.00	5.00	4.00	5.00	5.00	4.00	3.00	5.00	5.00	4.22
Granja	4.00	4.00	4.00	4.00	5.00	4.00	1.00	1.00	4.00	3.44
Luz	2.00	3.00	4.00	5.00	4.00	1.00	3.00	1.00	2.00	3.11
Torreira	5.00	5.00	3.00	4.00	4.00	5.00	3.00	3.00	5.00	3.78
Afife	5.00	5.00	5.00	5.00	5.00	4.00	3.00	5.00	5.00	4.44
Aver-o-mar	5.00	4.00	4.00	5.00	5.00	4.00	3.00	3.00	5.00	3.89
Cabedelo do Douro	5.00	3.00	1.00	1.00	1.00	1.00	1.00	1.00	5.00	2.00
Canide Sul	5.00	4.00	3.00	5.00	5.00	5.00	3.00	3.00	5.00	4.00
Cortegaça	4.00	4.00	5.00	5.00	5.00	4.00	3.00	3.00	4.00	4.00
Esmoriz	5.00	4.00	5.00	5.00	5.00	4.00	3.00	5.00	5.00	4.22
Esposende	5.00	5.00	4.00	5.00	5.00	5.00	1.00	3.00	5.00	4.00
Funtão	4.00	4.00	2.00	5.00	5.00	4.00	3.00	3.00	4.00	3.78
Leça	5.00	5.00	5.00	5.00	5.00	5.00	3.00	3.00	5.00	4.22
Mindelo (Sul)	4.00	5.00	4.00	5.00	5.00	4.00	3.00	5.00	4.00	4.33
Moledo	4.00	4.00	4.00	5.00	5.00	4.00	3.00	3.00	4.00	4.00
Ofir	5.00	4.00	5.00	5.00	5.00	4.00	3.00	3.00	5.00	4.00
Ourigo	4.00	4.00	4.00	4.00	4.00	4.00	3.00	1.00	4.00	3.44
Paramos	4.00	3.00	2.00	5.00	5.00	4.00	3.00	3.00	4.00	3.67
Silvalde. Espinho	4.00	4.00	4.00	5.00	5.00	4.00	3.00	1.00	4.00	3.78
Vila Praia de Âncora	5.00	5.00	5.00	5.00	5.00	4.00	3.00	5.00	5.00	4.56
Aterro	5.00	5.00	4.00	5.00	5.00	5.00	3.00	3.00	5.00	4.22
Cabedelo	5.00	5.00	5.00	5.00	5.00	4.00	3.00	5.00	5.00	4.44
Carvalhido	5.00	5.00	4.00	5.00	5.00	4.00	3.00	3.00	5.00	4.22
Rock	5.00	5.00	4.00	5.00	5.00	3.00	3.00	3.00	5.00	4.11
Rodanho	5.00	5.00	4.00	5.00	5.00	4.00	3.00	3.00	5.00	4.22
Cagalhotos	4.00	3.00	2.00	1.00	1.00	1.00	3.00	3.00	4.00	2.44
Ingleses	4.00	4.00	4.00	5.00	5.00	1.00	3.00	1.00	4.00	3.56
Praia Internacional	5.00	5.00	5.00	5.00	5.00	4.00	1.00	5.00	5.00	4.11
Mean	**4.59**	**4.38**	**3.97**	**4.62**	**4.65**	**3.85**	**3.32**	**2.76**	**3.18**	**3.92**
N	**34**	**34**	**34**	**34**	**34**	**34**	**34**	**34**	**34**	**34**
Standard Deviation	**0.657**	**0.697**	**1.000**	**0.985**	**0.981**	**1.158**	**0.878**	**0.654**	**1.336**	**0.93**
Source: authors' elaboration										

Table 6. IPA of surfing destination attributes response means

Attribute description	Attribute Code	Mean Importance Rating *	Mean Performance Rating **
Degree of accessibility	DA	4.25	3.72
Transport	T	3.88	3.53
Quality of infrastructure	QI	4.22	4.09
Availability of support services	ASS	4.00	3.75
Existence of support equipment	ESE	4.09	3.75
Existence of signalling	ES	3.97	3.56
Safety	S	4.50	3.94
Assistance	A	3.78	3.75
Universal accessibility	UA	4.00	3.38
Total average		4.08	3.72

* Rating obtained from a 5-point Likert scale ranging from 1 " Not important" to 5 "Extremely important";
** Rating obtained from a 5-point Likert scale ranging from 1 "Very unsatisfied" to 5 "Very satisfied".

The results of the IPA suggest that all nine attributes studied are placed in Quadrant 1: "Keep Up Good Work" (see Figure 1) which corresponds to high-performance assessments. This indicates that surf practitioners placed high importance on the destination attributes, and the performance of those attributes met their expectations.

Figure 1. IPA of surfing destination attributes in Northern Portugal

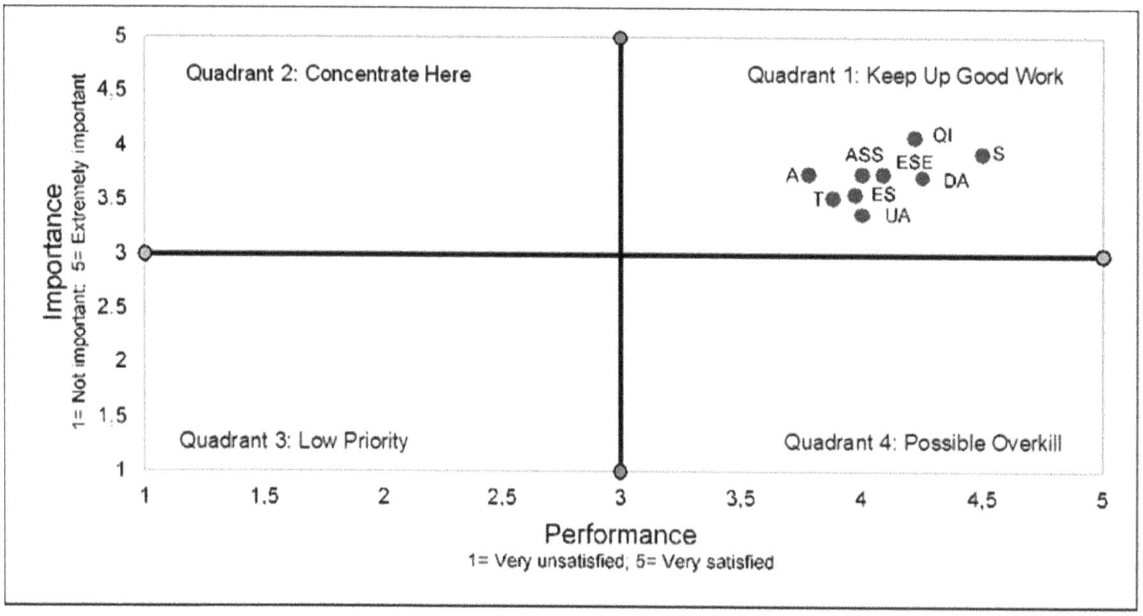

From the IPA results, it is possible to argue that surf practitioners perceived the "safety", "degree of accessibility", and "quality of infrastructure" as significant attributes of the surf destination. However, the perceived performance of these attributes is at a moderate level, particularly the "degree of accessibility" performance (mean score = 3.72). These results reflect previous studies on the issue of personal and water "safety" as it is a significant attribute for surfers (Dolnicar & Fluker, 2003b; Martin & Assenov, 2014; Portugal et al., 2017; Sotomayor & Barbieri, 2016).

CONCLUSION AND IMPLICATIONS

Surfing represents an adventure activity particularly relevant for some tourist destinations, given its attractive potential, and the current demand for experiential tourism, from most visitors (Brochado et al., 2018). This study aimed to understand the relevance of assessing the potential and monitoring of the existing conditions for surf tourism in Northern Portugal, according to a defined set of attributes. Using a concurrent mixed-method approach, this study, part of a more extensive, ongoing investigation process, provided insights about the regional governance and assessment of specific surf spots within Northern Portugal, which has recently been positioned as a surfing destination. Despite recent investments (e.g., surf infrastructures as the High-Performance Centres), interviews' results suggest a lack of an articulated regional strategy coherent with the identified potential and opportunities. Additionally, based on surfers' individual perceptions of the surfing spots, the results of an Importance-Performance Analysis results show that surfers perceived the performance of the studied destination attributes at a moderate level.

Thus, this study has some important implications for destinations' regional governance. The creation of surf schools/clubs and high-performance sports centres increased the recognition of Northern Portugal as a surfing destination and has demonstrated the role and commitment of regional and local stakeholders in the development process. Nevertheless, different destination attributes require further attention and investment. Public transportation must be revisited as it is one of the attributes with lower scores and an essential aspect for the competitiveness of any surfing destination. Additionally, as many destinations focusing on attracting diversified surfers (experienced and novices) (Brochado et al., 2018), the promotion of Northern Portugal as an outdoor tourism destination is still a challenge and, its image and positioning as a surf destination, in particular, needs to be reinforced and improved.

LIMITATIONS

Although this study is believed to have significantly contributed to the analysis of North Portugal as a surfing destination, some limitations and opportunities for further research are identified. First, different stakeholders (public, private, third sector and surf practitioners) could be invited to evaluate each surf spot. Different viewpoints would reinforce the validation process and assessment tool's reliability, providing even more accurate directions for regional governance. Secondly, although the study provides preliminary results from ongoing research, the number of stakeholders interviewed and the number of surfers should be enlarged for further and more firm conclusions.

REFERENCES

Alaeddinoglu, F., & Can, A. S. (2011). Identification and classification of nature-based tourism resources: Western Lake Van basin, Turkey. *Procedia: Social and Behavioral Sciences, 19*, 198–207. doi:10.1016/j.sbspro.2011.05.124

Barbieri, C., & Sotomayor, S. (2013). Surf travel behavior and destination preferences: An application of the Serious Leisure Inventory and Measure. *Tourism Management, 35*, 111–121. doi:10.1016/j.tourman.2012.06.005

Bichler, B. F., & Peters, M. (2020). Soft adventure motivation: An exploratory study of hiking tourism. *Tourism Review.* ahead-of-print. doi:10.1108/TR-10-2019-0403

Brochado, A., Stoleriu, O., & Lupu, C. (2018). Surf camp experiences. *Journal of Sport & Tourism, 22*(1), 21–41. doi:10.1080/14775085.2018.1430609

Buckley, R. (2002a). Surf tourism and sustainable development in indo-pacific islands. I. The industry and the islands. *Journal of Sustainable Tourism, 10*(5), 405–424. doi:10.1080/09669580208667176

Buckley, R. (2002b). Surf tourism and sustainable development in indo-pacific islands. II. Recreational capacity management and case study. *Journal of Sustainable Tourism, 10*(5), 425–442. doi:10.1080/09669580208667177

Buckley, R. C., Guitart, D., & Shakeela, A. (2017). Contested surf tourism resources in the Maldives. *Annals of Tourism Research, 64*, 185–199. doi:10.1016/j.annals.2017.03.005

Chan, J. K. L. (2010). Building sustainable tourism destination and developing responsible tourism: conceptual framework, key issues and challenges. Tourism Development Journal- An International Research Journal, 8(1), 24-32.

Correia, A., Nunes, A., Silva, G., Fernandes, P., Moreira, J., & Soares, L. (2021). Sport tourism event and perceived economic impacts: The case of World Bodyboard Championship 2018, Viana do Castelo, Portugal. In Advances in Tourism, Technology and smart systems (Vol. 208, pp. 223–233). Academic Press.

Creswell, J. W. (2009). *Research design - Qualitative, quantitative, and mixed methods approaches* (3rd ed.). SAGE Publications.

Decrop, A. (1999). Triangulation in qualitative tourism research. *Tourism Management, 20*(1), 157–161. doi:10.1016/S0261-5177(98)00102-2

Dolnicar, S., & Fluker, M. (2003a). Behavioural market segments among surf tourists: Investigating past destination choice. *Journal of Sport & Tourism, 8*(3), 186–196. doi:10.1080/1477508031000169O503

Dolnicar, S., & Fluker, M. (2003b). Who's riding the wave? An investigation into demographic and psychographic characteristics of surf tourists. *Proceedings of the 13th International Research Conference for the Council for Australian University Tourism and Hospitality Education (CAUTHE 2003).*

Dwyer, L., Gill, A., & Seetaram, N. (Eds.). (2012). *Handbook of research methods in tourism quantitative and qualitative approaches.* Edward Elgar. doi:10.4337/9781781001295

Fadda, N. (2019). Tourism from water boardsports: Evidence from the emerging destination of Sardinia. *Journal of the Knowledge Economy*, *10*(4), 1720–1737. doi:10.100713132-019-0587-8

Fox, D., Edwards, J., & Wilkes, K. (2010). Employing the Grand Tour approach to aid understanding of garden visiting. In G. Richards & W. Munsters (Eds.), *Cultural tourism research methods* (pp. 75–86). CABI Publishing. doi:10.1079/9781845935184.0075

Frank, F. (2015). Environmental awareness of surf tourists: A case study in the Algarve. *Journal of Spatial and Organizational Dynamics*, *3*(2), 102–113.

Gamarra, N. C., Correia, R. A., Bragagnolo, C., Campos-Silva, J. V., Jepson, P. R., Ladle, R. J., & Mendes Malhado, A. C. (2019). Are Protected Areas undervalued? An asset-based analysis of Brazilian Protected Area Management Plans. *Journal of Environmental Management*, *249*(July), 109347. doi:10.1016/j.jenvman.2019.109347 PMID:31419672

Halkos, G., & Matsiori, S. (2012). Determinants of willingness to pay for coastal zone quality improvement. *Journal of Socio-Economics*, *41*(4), 391–399. doi:10.1016/j.socec.2012.04.010

Imran, S., Alam, K., & Beaumont, N. (2014). Environmental orientations and environmental behaviour: Perceptions of protected area tourism stakeholders. *Tourism Management*, *40*, 290–299. doi:10.1016/j.tourman.2013.07.003

International Surfing Association. (2014). *ISA 50th Anniversary World Surfing Games*. https://www.isasurf.org/international-surfing-association-celebrate-50-years-success-isa-50th-anniversary-world-surfing-games-punta-rocas-peru/

Khoo-Lattimore, C., Mura, P., & Yung, R. (2019). The time has come: A systematic literature review of mixed methods research in tourism. *Current Issues in Tourism*, *22*(13), 1531–1550. doi:10.1080/13683500.2017.1406900

Ko, Y. J., & Pastore, D. L. (2004). Current issues and conceptualizations of service quality in the recreation sport industry. *Sport Marketing Quarterly*, *13*(2), 158–166.

Kruger, M., & Saayman, M. (2017). Sand, sea and surf: Segmenting South African surfers. *S.A. Journal for Research in Sport Physical Education and Recreation*, *39*(2), 115–132.

Lazarow, N., Miller, M. L., & Blackwell, B. (2008). The value of recreational surfing to society. *Tourism in Marine Environments*, *5*(2–3), 145–158. doi:10.3727/154427308787716749

Mach, L., & Ponting, J. (2018). Governmentality and surf tourism destination governance. *Journal of Sustainable Tourism*, *26*(11), 1845–1862. doi:10.1080/09669582.2018.1513008

Mach, L., Ponting, J., Brown, J., & Savage, J. (2018). Riding waves of intra-seasonal demand in surf tourism: Analysing the nexus of seasonality and 21st century surf forecasting technology. *Annals of Leisure Research*, *5398*, 1–19. doi:10.1080/11745398.2018.1491801

Machado, V., Carrasco, P., Contreiras, J. P., Duarte, A. P., & Gouveia, D. (2018). Governing Locally for Sustainability: Public and Private Organizations' Perspective in Surf Tourism at Aljezur, Costa Vicentina, Portugal. *Tourism Planning & Development*, *15*(6), 692–704. doi:10.1080/21568316.2017.1415958

Marasinghe, S., Perera, P., Simpson, G. D., & Newsome, D. (2021). Nature-based tourism development in coastal wetlands of Sri Lanka: An Importance–Performance analysis at Maduganga Mangrove Estuary. *Journal of Outdoor Recreation and Tourism, 33*(May), 100345. doi:10.1016/j.jort.2020.100345

Marchant, B., & Mottiar, Z. (2011). Understanding lifestyle entrepreneurs and digging beneath the issue of profits: Profiling surf tourism lifestyle entrepreneurs in Ireland. *Tourism Planning & Development, 8*(2), 171–183. doi:10.1080/21568316.2011.573917

Martin, S. A., & Assenov, I. (2012). The genesis of a new body of sport tourism literature: A systematic review of surf tourism research (1997-2011). *Journal of Sport & Tourism, 17*(4), 257–287. doi:10.1080/14775085.2013.766528

Martin, S. A., & Assenov, I. (2014). Investigating the importance of surf resource sustainability indicators: Stakeholder perspectives for surf tourism planning and development. *Tourism Planning & Development, 11*(2), 127–148. doi:10.1080/21568316.2013.864990

Martin, S. A., & Assenov, I. (2015). Measuring the conservation aptitude of surf beaches in Phuket, Thailand: An application of the Surf Resource Sustainability Index. *International Journal of Tourism Research, 17*(2), 105–117. doi:10.1002/jtr.1961

Mimbs, B. P., Boley, B. B., Bowker, J. M., Woosnam, K. M., & Green, G. T. (2020). Importance-performance analysis of residents' and tourists' preferences for water-based recreation in the Southeastern United States. *Journal of Outdoor Recreation and Tourism, 31*(April), 100324. doi:10.1016/j.jort.2020.100324

Moutinho, L., Dionísio, P., & Leal, C. (2007). Surf tribal behaviour: A sports marketing application. *Marketing Intelligence & Planning, 25*(7), 668–690. doi:10.1108/02634500710834160

Neuman, W. L. (2014). *Social research methods: Qualitative and quantitative approaches* (7th ed.). Pearson.

Orams, M. B., & Towner, N. (2013). Riding the wave: History, definitions, and a proposed typology of surf-riding tourism. *Tourism in Marine Environments, 8*(4), 173–188. doi:10.3727/154427313X13631129554901

Ponting, J. (2009). Projecting paradise: The surf media and the hermeneutic circle in surfing tourism. *Tourism Analysis, 14*(2), 175–185. doi:10.3727/108354209789116510

Ponting, J., & McDonald, M. G. (2013). Performance, agency and change in surfing tourist space. *Annals of Tourism Research, 43*, 415–434. doi:10.1016/j.annals.2013.06.006

Porter, B. A., & Usher, L. E. (2019). Sole surfers? Exploring family status and travel behaviour among surf travellers. *Annals of Leisure Research, 22*(4), 424–443. doi:10.1080/11745398.2018.1484782

Portugal, A. C., Campos, F., Martins, F., & Melo, R. (2017). Understanding the relation between serious surfing, surfing profile, surf travel behaviour and destination attributes preferences. *European Journal of Tourism Research, 16*, 57–73.

Priskin, J. (2001). Assessment of natural resources for nature-based tourism: The case of the Central coast region of Western Australia. *Tourism Management, 22*(6), 637–648. doi:10.1016/S0261-5177(01)00039-5

Rasdi, A. L. M., Som, A. P. M., & Usolludin, H. H., & Nordin, N. (2019). Responsible Tourism and Quality of Life: Perspective of Local Community. Journal of Tourism. *Hospitality and Environment Management*, *4*(13), 49–56.

Ratten, V. (2018). Entrepreneurial intentions of surf tourists. *Tourism Review*, *73*(2), 262–276. doi:10.1108/TR-05-2017-0095

Ritchie, J. R., & Crouch, G. I. (2003). *The Competitive Destination: A Sustainable Tourism Perspective*. CABI Publishers. doi:10.1079/9780851996646.0000

Scheske, C., Arroyo Rodriguez, M., Buttazzoni, J. E., Strong-Cvetich, N., Gelcich, S., Monteferri, B., Rodríguez, L. F., & Ruiz, M. (2019). Surfing and marine conservation: Exploring surf-break protection as IUCN protected area categories and other effective area-based conservation measures. *Aquatic Conservation*, *29*(S2), 195–211. doi:10.1002/aqc.3054

Silva, S. F., & Ferreira, J. C. (2014). The social and economic value of waves: An analysis of Costa de Caparica, Portugal. *Ocean and Coastal Management*, *102*, 58–64. doi:10.1016/j.ocecoaman.2014.09.012

Sotomayor, S., & Barbieri, C. (2016). An exploratory examination of serious surfers: Implications for the surf tourism industry. *International Journal of Tourism Research*, *18*(1), 62–73. doi:10.1002/jtr.2033

Teddlie, C., & Yu, F. (2007). Mixed methods sampling: A typology with examples. *Journal of Mixed Methods Research*, *1*(1), 77–100. doi:10.1177/1558689806292430

Towner, N. (2018). Surfing tourism and local stakeholder collaboration. *Journal of Ecotourism*, *17*(3), 268–286. doi:10.1080/14724049.2018.1503503

UNWTO. (2016). *Highlights of the 1st UNWTO Conference on Accessible Tourism in Europe (San Marino, 19-20 November 2014)*. 10.18111/9789284417902

Valeri, M. (2016). Networking and cooperation practices in the Italian tourism business. *Journal of Tourism, Heritage & Services Marketing*, *2*(1), 30–35. doi:10.5281/zenodo.376333

Valeri, M. (2021). *Organization Theories. Implications for the Strategic Management*. Springer.

Valeri, M., & Baggio, R. (2020a). Social network analysis: Organizational implications in tourism management. *The International Journal of Organizational Analysis*. Advance online publication. doi:10.1108/IJOA-12-2019-1971

Valeri, M., & Baggio, R. (2020b). Italian tourism intermediaries: A social network analysis exploration. *Current Issues in Tourism*, *24*(9), 1270–1283. doi:10.1080/13683500.2020.1777950

Valeri, M., & Baggio, R. (2021). Increasing the efficiency of knowledge transfer in an Italian tourism system: A network approach. *Current Issues in Tourism*, 1–16. Advance online publication. doi:10.1080/13683500.2021.1937960

Valeri, M., & Fadlon, L. (2018). Co-evolutionary prospects in tourism. *International Journal of Business and Management*, *13*(9), 95–105. doi:10.5539/ijbm.v13n9p95

Chapter 17
The Urban Sensescapes and Sensory Destination Branding

Irma Imamović
University of Minho, Portugal

António Joaquim Araújo Azevedo
University of Minho, Portugal

Bruno Miguel Barbosa de Sousa
https://orcid.org/0000-0002-8588-2422
Campus do Instituto Politécnico, Portugal

ABSTRACT

The sensory attractiveness of destinations has recently been one of the most important focal points when considering the impact of senses on positive and memorable travel experiences. Specifically, urban areas reflect different multisensory dimensions that can be analyzed and included in marketing efforts to stimulate positive tourists' experience and memory of a place, but also to encourage local sustainable development. Despite the fact that tourism studies center tourist experience mainly on visual impressions, recent researches stress the importance of recognizing and understanding the role of the human body in forming positive experiences, which is in tourism literature known as sensescapes (i.e., tactile, visual, gustatory, olfactory, and auditory senses).

INTRODUCTION

The sensory attractiveness of places has recently been recognized as an important aspect in the management of tourist destinations since studies suggest the positive impact of senses on memorable travel experience (Agapito et al., 2014). Even though tourism studies center tourist experience on visual impressions (Pan & Ryan, 2009), recent research stresses the importance of recognizing and understanding the role of the human body in the formation of positive experiences, which is in tourism literature known under the term sensescapes, i.e tactile, visual, gustatory, olfactory, and auditory senses (Dann

DOI: 10.4018/978-1-6684-3889-3.ch017

& Jacobsen, 2003; Pan & Ryan, 2009; Portheus, 1985; Agapito et al., 2017). It has been argued that destinations reflect authentic multisensory experiences, can be branded based on their unique sensory features, and in doing so stimulate tourists' positive emotions and memory of a destination. However, studies in tourism that empirically recognize the role of five human senses in the creation of a positive experience are still scantly researched, especially in the context of urban areas where there is still a need for more theoretical and empirical discussion.

Urban areas reflect different multisensory dimensions that can be analyzed and included in marketing efforts to stimulate positive tourists' experience and memory of a place, but also to encourage local sustainable development. The approach to sustainable tourism indeed depends on effective local governance and stakeholders (Beaumont & Dredge, 2010). Governance plays a key role in the creation of a stable environment for households, the private sector, and investors (Detotto et al., 2021). For instance, tourism stakeholders should focus on the creation of experiencescapes, operational spaces where they are interconnected, and where experiences are formed and consumed in a meaningful way (O'Dell, 2005). Furthermore, servicescapes can be designed by the man, physically controlled and managed, and consist of signs, tools, artifacts, and so forth (Bitner, 1992). All-natural and artificial objects of a large urban setting can provide a symbolic representation of a place where each human sense plays an important role in experiencing the city (Borer, 2013). Thus, the ability of a government to produce the services that visitors expect is important (Detotto et al., 2021). Finally, both experiencescapes and servicescapes should be coordinated to produce bettered strategies, convey meaningful sensory impressions, and stimulate consumer reactions, persuading certain consumption activities that are going to happen in a destination.

BACKGROUND

Urban Tourism

Defining urban tourism remains unspecified in the literature (Romero-García et al., 2019). However, scholars defined urban tourism in a variety of ways, such as being referred to as economic activity (Brouder & Ioannides, 2014), complex social phenomenon (Bulatova & Ul'chickij, 2015) as a short-term trip to urban settings with a large population or as a complex system (Romero-García, et al., 2019). However, it is deemed as a complex network, firstly because it is an open setting and can develop along with other cities, has a connection among urban tourism elements and other close cities, and finally, has diversified dimensions of tourism activities (Romero-García et al., 2019; Valeri, 2016; Valeri and Baggio, 2020a; 2020b; 2020c; 2021; Baggio and Valeri, 2020).

As a more simplified explanation, urban tourism refers to a variety of activities performed by a tourist where the city is the central destination and place of interest (Ye et al., 2019). It is dissimilar from other tourism forms due to its complexity and effect on the tourists' behavior (Edwards et al., 2008). In this context, tourists' experiences are formed based on different urban features, and accessibility of the city's space together with other attractions such as hospitality services, and public transportation (Ashworth & Page, 2011; Xu & Zhang, 2016). In general, tourism is mostly urban in its nature and is developing faster than tourism of other types of destinations across Europe (Nilsson, 2019).

Moreover, some academics argue that the complexity of urban tourism makes certain obstacles in creating tourists' satisfaction, since several urban dimensions clash, such as the city's features, sectors, and functions (Edwards et al., 2008). Wearing and Foley (2017) suggest that overall tourist satisfaction in the urban tourism context is based on summing different satisfactory characteristics regarding some tourism sectors like attractions and accommodation. Regarding that, accommodation and attraction sectors highly influence tourists' satisfaction (Yuan et al., 2018), where other scholars assume that tourism sectors play important role in investigations of overall satisfaction, as it could refer to upgrading the quality in particular sectors (Li et al., 2015; Shi et al., 2017).

When talking about the sensory dimension of urban tourism, modern cities present multisensory environments rich in scents, sounds, landscapes, tastes, and tactile stimuli that represent symbolic interpretation when experiencing the city (Borer, 2013). However, the process of globalization is affecting the urban environment in a way that cities appear more uniform in their architectonic style (Liu, 2012), thus some researchers highlight the importance of sensory experience to preserve the sense and memory of the place (Ouf, 2001). The urban environment is an authentic and eventful area (Crang, 2001), where the urban experience is created based on the interaction between the city's places and spaces (Wearing & Foley, 2017).

According to some researchers, traveling enables tourists to learn about new places, and in this process, they face three learning styles: visual learners who react to visual stimulus and account for 30-40% of the population, auditory learners who react to sound stimulus and represent 20-30% of the population, and finally kinesthetic learners that are mostly involved with physical activities and account for 30-50% of the population (Luecke, 2003). Following the same idea, tourist experience should attract tourists by multisensory stimuli (Dann & Jacobsen, 2003), and introduce a multisensory portfolio in the promotional material (Pan & Ryan, 2009). Although urban experience is the result of individuals' interpretations and sensory information, research on multisensory dimensions in urban studies is still vague (Quercia, et al., 2015).
Sensescapes in urban tourism

Smellscapes

Of the five senses, the smell is considered to have the greatest impact on memory (Krishna et al., 2010; Aggleton & Waskett, 1999; Morrin & Ratneshwar, 2000; 2003), cognitive responses (Krishna et al., 2010; Morrin & Ratneshwar, 2000), emotions (Hirch, 1995), affective responses (Morrin & Ratneshwar, 2000; 2003), attitudes and perception (Spangenber et al., 1996), and behavioral responses i.e. spend more time and money in the scented environment (Hirsch, 1995; Spangenber et al., 1996; Morrin & Ratneshwar, 2003). It has also been recorded that scent in combination with auditory cues (music) results in a better assessment of the shopping environment (Morrison et al., 2011). Dann and Jacobsen (2003) also found out that olfactory cues are strongly connected with the gastronomic experience.

In the marketing field, scents are being used as a marketing tool, not just to create a desire for a particular product, but to create an emotional bond between customers and the brand (Hultén, 2011). Its essence is not in just spreading pleasant perfume in a space, but in the art of taking the firms' brand image and identity, target market, and all marketing communications and connecting them with a scent that boosts these branding aspects (Schmitt, 1999; Hultén, 2011; Albattat et al., 2020; Chemli et al., 2020; Obermayer et al., 2021; Toanoglou et al., 2021; Valeri, 2021a). This leads to happier customers who strongly remember the brand. Scent marketing has been recognized as a prime trend in the future of retail marketing; hence if the urban setting can be perceived as a service environment, then scents can have a similar role on the bigger spatial scale (Henshaw et al., 2015). According to the authors, if cities are recognized as servicescape settings, scents can serve the purposes of urban place marketing that are intentionally or unintentionally dispersed into the environment to become a synonym for specific urban locations.

In the tourism industry, Xiao et al. (2018) state that Porteous (1985) used the term "smellscape" for the first time, where he explains how people perceive and easily memorize the environment around them through the sense of smell. Later, Rodoway (1994) introduced the term olfactory geography to designate scents and their sources, spaces, and individuals in those spaces. Additionally, Henshaw (2013) mentioned the term smellscapes was used to describe the general olfactory environment where individuals' sensory and mental processes take place.

The urban planning and design seem to focus on the visual representation of the cities, i.e., buildings and lights; tactile, i.e., sidewalk texture and surface; auditory, i.e., city planning to divert sounds, but the olfactory dimension of cities still lack the analysis (Henshaw et al., 2015). However, there is a growing number of studies on the olfactory perception of a place, and they are certain in our everyday experience of the urban environment, i.e., smelling the freshly baked bread in the bakery or smell of the cigarettes and beer in local pubs that helps us to better experience the city (Xiao et al., 2018). Cars, horns, people's conversations, and nature are also examples of sensuous features of the urban environment (Low, 2015).

The scent, as an intangible legacy can prompt and influence visitor's cognitive abilities regarding memories, emotions, and boost awareness connecting smells with the specific urban location (Davis & Thys-Şenocak, 2017). Hence, smells have a significant influence on human perception and the way they perceive destinations. Identifying the place through its olfactory dimensions creates a person-place bond (Rodoway, 1994), place attachment (Najafi & Shariff, 2011), and place-making (Henshaw, 2013). Additionally, Quercia et al. (2015) made the first urban smell dictionary through the usage of geo-referenced social media tags, where people perceived the city of Barcelona as related to aromas of food and nature, but London related to odors of waste and emission. A similar study of McLean (2017) resulted in an urban smell-map of Amsterdam applying smell walks with local people. Finally, Dann and Jacobsen, (2003) assume that modern tourist sites can become successful if they apply strategies to attract visitors by other sensory cues besides the vision, for instance aromatically appealing strategies.

Rodoway (1994) stated that people usually misjudge the role of scents in their social and emotional experiences of the urban area. For example, scents in Arabic and Eastern societies are considered important for creating bonds between people. On the other hand, modern urban policies in the Western part of the world neutralize the smell of public places to conserve the social order (Xiao et al., 2018). Compared to western societies, the Eastern and Arabic world appreciates the role of scents in daily urban experiences (Lefebvre, 1991).

The application of scents in the management of tourist destinations is evident in few examples such as the "Jorvik Viking Center" (Dann & Jacobsen, 2003) museum in York, the United Kingdom that revives the city of York from the tenth century. They focus on landscapes and sounds, pleasant scents (fruit scent), and unpleasant scents (rubbish scent, burnt wood, fish market) to intensify tourists' experience. The museum also deliberately introduced scents in the promotional material to guide visitors' urban experience (Henshaw et al., 2015). There is also an example of Holland 5D cinema, where they attempt to invoke tourists' senses to see, hear and smell the country in 5D.

One can say that little attention has been paid to olfactory impressions in urban tourism and their role in enhancing the tourist experience (Buzova et al., 2019). However, some academics argue that smell in an urban environment is usually associated with the gastronomic experience of the place (Kim & Fesenmaier, 2017). This statement can be confirmed by Xiong et al. (2015) whose study found that the olfactory representation of the Chinese city of Phoneix was closely associated with the local urban food.

Tastescapes

One way to apprehend the external world is through the tongue and its ability to recognize different tastes and textures (Kabat-Zinn, 2014). Urban culture depends on the role of taste in recognizing specific urban spots, via the production and consumption of local food and beverages, thus taste (food) acts as a cultural mediator between urban locations and urban experience (Borer, 2013). Moreover, gastronomy and tourism are closely related and according to literature, food is divided into four categories: part of local culture; part of tourist attraction; as tourist product and tourist experience (Chen & Huang, 2015). The sense of taste is considered a crucial factor of urban taste experience and place-making (Pink, 2008), but also for managing and marketing gastronomy tourism.

Urban areas are usually centered on food consumption experience, which enables us to experiment with different local cuisine, and one way to promote local food tourism and assure place differentiation is by organizing food-themed festivals (Lau & Li, 2019). It represents the way of life of the local people and their culture (Zhang et al., 2019). Generally, destinations' gastronomy is a key element in tourists' experience as it reflects the local environment, culture, and tradition (Björk & Kauppinen-Räisänen, 2017). It can also support authentic visitors' experiences (Kim et al., 2019) and can encourage the social and economic growth of the destination if strategically planned (Baldacchino, 2015). After all, the term "local" nowadays is used as a synonym for "authentic" which became an essential part of the tourism industry (Taylor, 2001). In the cultural context, gastronomy includes food, beverages, but also cooking traditions such as preparation, cooking, and consumption, advice, and tips on how to consume food and beverages (Santich, 2004). Moreover, it also encompasses culinary innovation, haute cuisine, and fine dining, however with the focus on local culture and culinary traditions (Hegarty & O'Mahony, 2001). Some scholars suggest that various travel destinations are identified based on their cuisine and culture, such as Mediterranean cuisine which includes fresh fish, fruits, and vegetables, or Mexican tortillas, beans, and corns (Björk & Kauppinen-Räisänen, 2017).

Another way to explore urban tastescapes is visiting food markets that became a symbol of experiencing the place authenticity and sensing the place from cultural and visitor's perspective (Fusté-Forné, 2018). In other words, local food for visitors is a way to experience authenticity and sensory embodiment of local tradition (Zhan et al., 2019). Crespi and Pérez (2015) also argue that urban food markets are becoming advocates of branding the city, where tourists can connect to local people and their daily urban experiences.

Gastronomic experience includes visits to the food market (Hall & Sharples, 2003), where a growing number of tourists appreciate local products because of their direct connection with local and regional culture and traditions (Suárez Velasco et al., 2017). Additionally, gastronomic tourism diversifies destinations' offers and minimizes the seasonality effects (Fusté-Forné, 2018).

Food markets are considered tourist attractions, usually situated in city centers (Vallbona & Perez, 2016), with unique textures and sensory experiences (Coles & Crang, 2011). For urban areas, food markets are the center of economy, politics, culture, and tourism as well (Fusté-Forné, 2018). For Lee et al. (2018) however, gastronomic tourism is connected to cultural images and attractions, history of cuisine, activity promotions, and experience functions. After all, choosing a travel destination is very much affected by gastronomic motivation (Berbel-Pineda et al., 2019), and food experience consequently affects satisfaction, which may lead to destination loyalty, i.e., intention to revisit a destination (Berbel-Pineda et al., 2019).

Soundscapes

The soundscape concept was pioneered by Schafer (1977). It represents the acoustic environment which exposes us to different auditory cues such as voices, transportation, animals, wind, mobile phones, raindrops, and so forth (Brown et al., 2015). In the 1960s, Schafer established a "World Soundscape Project (WSP)" to educate residents to value and preserve the sounds and to perceive the acoustic environments as an aesthetic and cultural construct. He also mentioned that a sound symbol as an essential part of soundscapes represents a particular place or time and should be considered in urban planning or landscape design (Jia et al., 2020).

Hence, sounds have a great influence on the perception of our surroundings (Kinayoglu, 2009). Wissmann (2014) claims that three sound categories were initially developed by Schaffer: keynote which includes faint sounds like wind or traffic and refers to any auditory cue; signal which can be easily recognized such as siren of the ambulance vehicle or dog barking, and finally sound mark that explains the sound of the urban community or neighborhood, i.e., Big Ben bells as an auditory symbol of London, United Kingdom.

At first, it was referred to physical and behavioral features of creatures (different auditory cues in the environment), but later this term was substituted with people, highlighting the human understanding of the auditory environment which is closely linked to emotions and memory (He et al., 2018). Sound represents an important part of any culture; hence it makes a destination animated and vivid (Saldanha, 2002), and influences tourists' emotions and feelings toward a destination (Waitt & Duffy, 2010). During a trip, tourists usually sense different sounds such as human voices or sounds of nature and translate these sensations to create their own experiences (Kim & Fesenmaier, 2016). In an urban setting, Aletta et al. (2016) concluded that tourists experience three groups of sounds: traffic noise, sounds of nature, and human crowd's sounds. Furthermore, Xiong et al. (2015) found that visitors' soundscape image of the heritage destination embraces different sounds, i.e., folk music, tourist and local people voice, the sound of the river, and similar.

Individuals' sensuous perspective of their external world is varying, but the hearing sense is very much influenced by the visual features of the environment (He et al., 2018). Landscape quality can affect the soundscape assessment of the environment (Viollon et al., 2002). In other words, functions of the urban environment influence soundscape evaluation of the place (Hong & Jeon, 2015), such as peaceful sounds in the environment might favor the spots where people relax or meditate, but places with vibrant sound environment might be preferred for parks and playgrounds. Generally, studies showed that people prefer natural sounds like birds, wind, and water, over traffic and construction sounds (Hong & Jeon, 2015).

The literature on urban soundscapes focused mostly on soundscape evaluation in specific urban functions like parks (Liu et al., 2014), residential zones (Ge & Hokao, 2006), squares (Yang & Kang, 2005), and historical parts of the city (Liu et al., 2019). Thus, one can conclude that visual perceptive and sound perspective of environment closely correlates, i.e., soundscape and landscape consistency has significantly affected the experience of religious tourism (He et al., 2018). Similar findings were produced by Ivona and Privitera (2019) who stressed the importance of the multisensory aspects of religious travel destinations; soundscapes shaped by religious rhythms walks, visiting sanctuaries, and any other kind of musical events. Furthermore, soundscape perception also depends on the interplay between lightscape and soundscape (Malva et al., 2015). Thus, previous studies confirm the relationship between visual and auditory information coming from our surroundings. Pedersen and Larsman (2008) showed that communities living in a neighborhood close to wind turbines exhibited higher noise displeasure with their visual appearance. Similarly, Viollon et al. (2002) found out that people become more intolerant to different kinds of noise with a more urbanized visual appearance of their surroundings.

The auditory element of the urban landscape plays an important role in the well-being of the city and its residents (Skanberg & Ohrstrom, 2002). The features of the urban environment depend on individuals' perception, thus urban soundscape perception is influenced by different aspects, such as physiological, psychological, environmental, and social (Torija et al., 2011). For sustainable development of urban tourism managers and researchers need to focus on acoustic settings and urban soundscape development. In the planning and management of urban soundscapes, local lifestyles and traditional sounds need to be protected to be able to create an authentic auditory signature of each city.

Although travel experiences reflect the multisensory character, involving various sensescapes, sounds are rarely included in process of choosing a travel destination and still require more focus in research (Ivona & Privitera, 2019).

Haptiscapes

Borer (2013) suggests that the body is a key instrument for perceiving and apprehending the sensuousness of the urban environment. From busy streets to public transportation, the body constantly interacts with other bodies and the external world, i.e., it plays two roles, both as a physical and communicative mechanism to grasp different impressions and experiences (Borer, 2013). Most of the time, the body and urban area act together by using electronic and mechanical devices (Grosz, 1995) or by walking through the city's avenues and pavements to experience the urban environment (Borer, 2013). In the marketing field, studies have found that tactile sense is important when making judgments on products (Peck & Childers, 2003), making product trials (Peck & Wiggins, 2006), and creating the feeling of ownership (Peck & Shu, 2009). Moreover, Peck (2011) suggests two types of tactile sense in consumer behavior studies (Santo et al., 2021a; 2021b). The first one is instrumental which means that we touch the product to obtain certain information i.e., product's features, temperature, textures, and similar. Another one is hedonic where the main goal of touching an object is a sensory experience.

Destination managers as well use the tactile features of a place to promote destinations. The summer resorts highlight the information in their promotion material on weather and temperature conditions, either hot or cold because sensing the temperature is connected to the sense of touch (Medway, 2015). Hoven (2011) explained how tourists experienced the Great Bear Rainforest through the sense of touch by shivering when hearing a bear's roar, feeling the wind, or different weather conditions such as hot, cold, or moisture. Furthermore, touching deliberately or accidentally forest elements while walking such as spider webs, trees, berries, leaves, etc. However, they highlight the importance of a tourist group in experiencing the forest by tactile sense, i.e., touching each other while helping to cross the mossy logs, slippery rocks, mud, rivers, or sharing hugs after a successful tour.

Studies on urban heritage sites confirmed that tactile sensory experience may be mainly hedonic since visitors usually touch walls, ornaments, objects, and so forth (Xiong et al., 2015). In rural locations, however, the tactile sense means touching flora and fauna (Agapito et al., 2017; Son & Pearce, 2005) or feeling the warmth of the Sun, coolness, or the wind (Agapito et al., 2014; Son & Pearce, 2005; Xiong et al., 2015). After all, because of the sense of touch tourists become close and intimate with a destination by exploring complex meanings of a place that cannot be easily communicated by language (Ballantyne et al., 2011).

Generally, literature has neglected the sense of touch and its role in experiencing the external world (touch-place interaction), its materiality and universality we share with the rest of human beings, but when discussed it is usually in the context of other sensory abilities, such as vision, where it is recognized as a noteworthy tool in experiencing places (Medway, 2015). That is why urban tourism stakeholders should consider tactile experience when planning a destination offers.

Landscapes

European Landscape Convention (ELC) has recognized the landscape as an essential factor that influences the quality of life and richness of the society, but also as an economic, social, ecological, and cultural medium (Liberato et al., 2019). The full landscape potential is underpinned by its progress based on the reciprocal and long-term relationship between social needs, economic interest, and environment (Nučič, 2012; Kumar et al., 2021), which is caused by urban development and its construction pace nowadays (Liberato et al., 2019). However, the landscape as a universal resource goes through continuous changes, due to production, society, culture, or geographic factors, and because of that, it requires constant assessment, protection, planning, and management (Liberato et al., 2019). The landscape exists in any place with human and nature interplay, both in rural and urban areas; hence it is universal, dynamic, and is subjected to many transformations over time, considering the natural and human activity (Liberato et al., 2019).

Landscape in the tourism industry is perceived as valuable since nowadays tourists seek authentic destinations (Yázigi et al., 1996), it influences tourists' decision process and is a relevant motivator to tourists when choosing a destination to visit (Liberato et al., 2019).

Previous studies on multisensory tourist experience argue that sight is the most popular sense for experiencing the places (Trang & Lee, 2018; Agapito et al., 2017; Agapito et al., 2014; Pan & Ryan, 2009; Buzova et al., 2019; Xiong et al., 2015). This is also evident in marketing studies, with a bundle of promotional and communication efforts, design, and campaigns where the vision is predominant sense. This is also a case of mass tourism where destinations are marketed as visual performances, i.e., tourist experience is formed on the tour bus or any other manipulated conditions where other senses are neglected or reduced to a minimum (Larsen, 2001).

According to Wang and Gu (2020), urban design and place-making are crucial for the continuousness of local culture and for creating a people-place bond that is part of the tourism of cultural places. Moreover, in leisure and tourism studies, the landscape is closely related to the sense of place, i.e., Kianicka et al. (2006) suggested that both tourists and residents are situated in the Swiss Alps, a famous tourist destination declared that the landscape is essential for sensing the place. In the study of Phillips (2015), one of his students worked with visual features of New York City and suggests that in the case of big urban areas such as New York, it is desirable to focus on tiny details to open the door for new perspectives and experience the city from a different angle, such as different color, textures, shapes, and patterns. Similar findings were addresses by Breiby and Slåtten (2018) and Qiu et al. (2018) who stated that an attractive landscape positively affects tourists' satisfaction with the destination.

Landscapes are an important part of planned and managed tourist destination offer. Indeed, sustainable tourism requires landscape analysis and planning, constant supervision of landscape changes, and creating awareness of landscape values.

CONCLUSION

Urban areas reflect many multisensory dimensions that can be analyzed and incorporated into marketing efforts to promote the positive experience, memory of a place, and serve the purpose of sensory destination branding. The sensory branding strategy can strengthen a destination image and ensure a competitive advantage (Aziz et al., 2014), but it depends on the consolidated actions of all stakeholders in the tourism industry.

This way of experiencing urban areas could be used by tourism management organizations to plan and promote the destination, contributing to destination branding i.e., positioning and destination image (Agapito et al., 2017). Moreover, urban sensory impressions can be used in planning a destination offer and enhancing the overall tourists' experience. With a consolidated management approach, all urban tourism stakeholders can work towards this goal, targeting tourists' profiles that correspond to the most appropriate sensory impression that can intensify their experience. For example, the multisensory urban route can be planned and designed to address different target groups such as visually impaired or blind people who can experience the place by other sensory stimuli.

Technology as well can assist in the overall experience, by modifying both sensory information and stimuli from the sensory route to tackle tourists' emotions, memories, and special moments during the trip. Moreover, marketers should introduce appropriate strategies in their communication material and design sensory cues in a way that triggers a person's imagination process before a visit, due to the fact this process positively moderates the impact of sensory cues on destination emotion (Ghosh & Sarkar, 2016). Moreover, multisensory impressions can be used as unique selling propositions (USPs) that may serve the purposes of sensory resource allocation and help the local sustainable development of the national, regional, and local tourism industry (Xiong, et al., 2015).

Tourism managers and organizations can benefit from a strategically planned multisensory environment, introducing pictures, handmade crafts and fabrics, music, spices, food, and handmade fabrics to market the destination and influence visitor's decision-making process. Understanding visitors' source of perceived value, tourism marketers can easily manage destination resources and adapt sensory cues to different tourist segments (Lv et al., 2020; Valeri 2021b).

REFERENCES

Agapito, D., Pinto, P., & Mendes, J. (2017). Tourists' memories, sensory impressions and loyalty: In loco and post-visit study in Southwest Portugal. *Tourism Management, 58*, 108–118. doi:10.1016/j.tourman.2016.10.015

Agapito, D. L., Da, C. M., Valle, P. S., & Almeida, H. M. (2014). Um contributo do marketing sensorial para o marketing da experiência turística rural. Sensory marketing rural tourist experiences. *PASOS Revista De Turismo Y Patrimonio Cultural, 12*(3), 611-621. . doi:10.25145/j.pasos.2014.12.045

Aggleton, J. P., & Waskett, L. (1999). The ability of odours to serve as state-dependent cues for real-world memories: Can Viking smells aid the recall of Viking experiences? *British Journal of Psychology, 90*(1), 1–7. doi:10.1348/000712699161170 PMID:10085542

Albattat, A., Jamaludin, A., Zuraimi, N. S. M., & Valeri, M. (2020). Visit Intention and Destination Image in Post- Covid- 19 Crisis Recovery. *Current Issues in Tourism*. Advance online publication. doi :10.1080/13683500.2020.1842342

Aletta, F., Kang, J., & Axelsson, Ö. (2016). Soundscape descriptors and a conceptual framework for developing predictive soundscape models. *Landscape and Urban Planning, 149*, 65–74. doi:.landurbplan.2016.02.001 doi:10.1016/j

Ashworth, G., & Page, S. J. (2011). Urban tourism research: Recent progress and current paradoxes. *Tourism Management, 32*(1), 1–15. doi:10.1016/j.tourman.2010.02.002

Aziz, N., Kefallonitis, E., & Friedman, B. A. (2014). Turkey as a Destination: A Study of Sensory Brand Associations. *Tourism, Culture & Communication, 14*(2), 77–89. doi:10.3727/109830414X14133839512587

Baggio, R., & Valeri, M. (2020). Network science and sustainable performance of family businesses in tourism. *Journal of Family Business Management.* doi:10.1108/JFBM-06-2020-0048

Baldacchino, G. (2015). Feeding the Rural Tourism Strategy? Food and Notions of Place and Identity. *Scandinavian Journal of Hospitality and Tourism, 15*(1-2), 223–238. doi:10.1080/15022250.2015.1006390

Ballantyne, R., Packer, J., & Sutherland, L. A. (2011). Visitors' memories of wildlife tourism: Implications for the design of powerful interpretive experiences. *Tourism Management, 32*(4), 770–779. doi:10.1016/j.tourman.2010.06.012

Beaumont, N., & Dredge, D. (2010). Local tourism governance: A comparison of three network approaches. *Journal of Sustainable Tourism, 18*(1), 7–28. doi:10.1080/09669580903215139

Berbel-Pineda, J. M., Palacios-Florencio, B., Ramírez-Hurtado, J. M., & Santos-Roldán, L. (2019). Gastronomic experience as a factor of motivation in the tourist movements. *International Journal of Gastronomy and Food Science, 100171*. Advance online publication. doi:10.1016/j.ijgfs.2019.100171

Bitner, M. J. (1992). Servicescapes: The impact of physical surroundings on customers and employees. *Journal of Marketing, 56*(2), 57–71. doi:10.1177/002224299205600205

Björk, P., & Kauppinen-Räisänen, H. (2017). A destination's gastronomy as a means for holiday well-being. *British Food Journal, 119*(7), 1578–1591. doi:10.1108/BFJ-09-2016-0394

Borer, M. I. (2013). Being in the City: The Sociology of Urban Experiences. *Sociology Compass, 7*(11), 965–983. doi:10.1111oc4.12085

Breiby, M. A., & Slåtten, T. (2018). The role of aesthetic experiential qualities for tourist satisfaction and loyalty. *International Journal of Culture, Tourism and Hospitality Research, 12*(1), 1–14. doi:10.1108/IJCTHR-07-2017-0082

Brochado, A., Stoleriu, O., & Lupu, C. (2019). Wine tourism: A multisensory experience. *Current Issues in Tourism*, 1–19. doi:10.1080/13683500.2019.1649373

Brouder, P., & Ioannides, D. (2014). Urban Tourism and Evolutionary Economic Geography: Complexity and Co-evolution in Contested Spaces. *Urban Forum, 25*(4), 419–430. 10.100712132-014-9239-z

Brown, A., Gjestland, T., & Dubois, D. (2015). Acoustic Environments and Soundscapes. *Soundscape and the Built Environment*, 1–16. . doi:10.1201/b19145-2

Bulatova, Y. K., & Ul'chickij, O. A. (2015). Historical and theoretical background to the formation of a tourist center: international experience. *Scientific Herald of the Voronezh State University of Architecture and Civil Engineering Construction and Architecture*, 92–104.

Buzova, D., Cervera-Taulet, A., & Sanz-Blas, S. (2019). Exploring multisensory place experiences through cruise blog analysis. *Psychology and Marketing*, *37*(1), 131–140. doi:10.1002/mar.21286

Chemli, S., Toanoglou, M., & Valeri, M. (2020). The impact of Covid-19 media coverage on tourist's awareness for future traveling. *Current Issues in Tourism*, 1–8. Advance online publication. doi:10.1080/13683500.2020.1846502

Chen, Q., & Huang, R. (2015). Understanding the importance of food tourism to Chongqing, China. *Journal of Vacation Marketing*, *22*(1), 42–54. doi:10.1177/1356766715589427

Coles, B., & Crang, P. (2011). Placing alternative consumption: commodity fetishism in borough fine foods market, London. In T. Lewis & E. Potter (Eds.), *Ethical Consumption. A Critical Introduction* (pp. 87–102). Routledge.

Crang, M. (2001). Rhythms of the city: temporalised space and motion. In *Timespace: geographies of temporality* (pp. 187–207). Routledge.

Crespi, M., & Pérez, M. (2015). Selected studies from the Tourism Network Workshop of the Regional Studies Association. In *Food markets: new tourism icons.Illustrative cases of Barcelona and Madrid* (pp. 127–139). Department of Economic Geography and Futures Study.

Dann, G., & Jacobsen, J. K. S. (2003). Tourism smellscapes. *Tourism Geographies*, *5*(1), 3–25. doi:10.1080/1461668032000034033

Davis, L., & Thys-Şenocak, L. (2017). Heritage and scent: Research and exhibition of Istanbul's changing smellscapes. *International Journal of Heritage Studies*, *23*(8), 723–741. doi:10.1080/13527258.2017.1317646

Detotto, C., Giannoni, S., & Goavec, C. (2021). Does good governance attract tourists? *Tourism Management*, *82*, 104155. doi:10.1016/j.tourman.2020.104155

Edwards, D., Griffin, T., & Hayllar, B. (2008). Urban Tourism Research. *Annals of Tourism Research*, *35*(4), 1032–1052. doi:10.1016/j.annals.2008.09.002

Fusté-Forné, F. (2018). Drawing a gastronomy landscape from food markets' produce. *International Journal of Culture, Tourism and Hospitality Research*, *12*(3), 378–384. doi:10.1108/IJCTHR-05-2018-0064

Ge, J., & Hokao, K. (2006). Research on residential lifestyles in Japanese cities from the viewpoints of residential preference, residential choice and residential satisfaction. *Landscape and Urban Planning*, *78*(3), 165–178. doi:10.1016/j.landurbplan.2005.07.004

Ghosh, T., & Sarkar, A. (2015). "To feel a place of heaven": Examining the role of sensory reference cues and capacity for imagination in destination marketing. *Journal of Travel & Tourism Marketing, 33*(sup1), 25–37. . doi:10.1080/10548408.2014.997962

Grosz, E. (1995). *Space, time and perversion: essays on the politics of bodies*. Routledge.

Hall, C., & Sharples, L. (2003). The consumption of experiences or the experience of consumption? An introduction to the tourism of taste. *Food Tourism Around the World*, 1–24. . doi:10.1016/B978-0-7506-5503-3.50004-X

He, M., Li, J., Li, J., & Chen, H. (2018). A comparative study on the effect of soundscape and landscape on tourism experience. *International Journal of Tourism Research*. Advance online publication. doi:10.1002/jtr.2237

Hegarty, J. A., & O'Mahony, G. B. (2001). Gastronomy: A phenomenon of cultural expressionism and an aesthetic for living. *International Journal of Hospitality Management, 20*(1), 3–13. doi:10.1016/S0278-4319(00)00028-1

Henshaw, V. (2013). *Urban smellscapes: Understanding and designing city smell environments*. Routledge. doi:10.4324/9780203072776

Henshaw, V., Medway, D., Warnaby, G., & Perkins, C. (2015). Marketing the "city of smells.". *Marketing Theory, 16*(2), 153–170. doi:10.1177/1470593115619970

Hirsch, A. R. (1995). Effects of ambient odors on slot-machine usage in a las vegas casino. *Psychology and Marketing, 12*(7), 585–594. doi:10.1002/mar.4220120703

Hong, J. Y., & Jeon, J. Y. (2015). Influence of urban contexts on soundscape perceptions: A structural equation modeling approach. *Landscape and Urban Planning, 141*, 78–87. doi:10.1016/j.landurbplan.2015.05.004

Hoven, B. van. (2011). Multi-sensory tourism in the Great Bear Rainforest. *Landabréfið*, 31–49.

Hultén, B. (2011). Sensory marketing: The multi-sensory brand-experience concept. *European Business Review, 23*(3), 256–273. doi:10.1108/09555341111130245

Ivona, A., & Privitera, D. (2019). Places and Religious Bands explorations in spiritual tourism. *International Journal of Religious Tourism and Pilgrimage, 7*(4), 54–63.

Jia, Y., Ma, H., Kang, J., & Wang, C. (2020). The preservation value of urban soundscape and its determinant factors. *Applied Acoustics, 168*, 107430. doi:10.1016/j.apacoust.2020.107430

Kabat-Zinn, J. (2014). Tastescape. *Mindfulness, 5*(1), 102–104. doi:10.100712671-013-0274-y

Kianicka, S., Buchecker, M., Hunziker, M., & Müller-Böker, U. (2006). Locals and Tourists Sense of Place. *Mountain Research and Development, 26*(1), 55–63. doi:10.1659/0276-4741(2006)026[0055:LATSOP]2.0.CO;2

Kim, J. J., & Fesenmaier, D. R. (2016). Measuring Human Senses and the Touristic Experience: Methods and Applications. *Analytics in Smart Tourism Design Tourism on the Verge*, 47–63. doi:10.1007/978-3-319-44263-1_4

Kim, J. J., & Fesenmaier, D. R. (2017). Sharing Tourism Experiences: The post-trip experience. *Journal of Travel Research, 56*(1), 28–40. doi:10.1177/0047287515620491

Kinayoglu, G. (2009). Using Audio-Augmented Reality to Assess the Role of Soundscape in Environmental Perception: An Experimental Case Study on the UC Berkeley Campus. *Proceedings: ECAADe 2009 International Conference on Education and Research in Computer Aided Architectural Design in Europe*.

Krishna, A. (2010). *Sensory marketing: Research on the sensuality of products*. Routledge.

Kumar, S., Valeri, M., & Shekhar. (2021, August 13). Shekhar (2021). "Understanding the relationship among factors influencing rural tourism: A hierarchical approach. *Journal of Organizational Change Management*. Advance online publication. doi:10.1108/JOCM-01-2021-0006

Larsen, J. (2001). Tourism Mobilities and the Travel Glance: Experiences of Being on the Move. *Scandinavian Journal of Hospitality and Tourism, 1*(2), 80–98. doi:10.1080/150222501317244010

Lau, C., & Li, Y. (2019). Analyzing the effects of an urban food festival: A place theory approach. *Annals of Tourism Research, 74*, 43–55. doi:10.1016/j.annals.2018.10.004

Lee, T. H., Chao, W. H., & Lin, H.-Y. (2018). Cultural inheritance of Hakka cuisine: A perspective from tourists' experiences. *Journal of Destination Marketing & Management, 7*, 101–111. doi:10.1016/j.jdmm.2016.09.006

Lefebvre, H. (1991). *The Production of Space*. Basil Blackwtl~ Inc.

Li, M., Fang, L., Huang, X., & Goh, C. (2015). A spatial–temporal analysis of hotels in urban tourism destination. *International Journal of Hospitality Management, 45*, 34–43. doi:10.1016/j.ijhm.2014.11.005 PMID:32287860

Liberato, D., Ferreira, F. A., & Azevedo, A. F. (2019). Landscape as a Tourism Resource: A Case Study in the Northern Region of Portugal. *Advances in Tourism, Technology and Smart Systems Smart Innovation, Systems and Technologies*, 627–639. . doi:10.1007/978-981-15-2024-2_54

Liu, J., Kang, J., Behm, H., & Luo, T. (2014). Landscape spatial pattern indices and soundscape perception in a multi-functional urban area, Germany. *Journal of Environmental Engineering and Landscape Management, 22*(3), 1–11. doi:10.3846/16486897.2014.911181

Liu, J., Yang, L., Xiong, Y., & Yang, Y. (2019). Effects of soundscape perception on visiting experience in a renovated historical block. *Building and Environment, 165*, 106375. doi:10.1016/j.buildenv.2019.106375

Liu, X. P. (2012). Perspective on the Problem of the Chinese Urban Space Homogenization and Identity Creation in the Era of Globalization. *Applied Mechanics and Materials, 174-177*, 2226–2231. . doi:10.4028/www.scientific.net/AMM.174-177.2226

Low, K. E. Y. (2015). The sensuous city: Sensory methodologies in urban ethnographic research. *Ethnography, 16*(3), 295–312. doi:10.1177/1466138114552938

Luecke, R. (2003). *Business communication*. Harvard Business School Press.

Lv, X., Li, C., & McCabe, S. (2020). Expanding theory of tourists' destination loyalty: The role of sensory impressions. *Tourism Management*, *77*, 104026. doi:10.1016/j.tourman.2019.104026

Malva, F. L., Verso, V. R. L., & Astolfi, A. (2015). Livingscape: A Multi-sensory Approach to Improve the Quality of Urban Spaces. *Energy Procedia*, *78*, 37–42. doi:10.1016/j.egypro.2015.11.111

McLean, K. (2017). Smellmap: Amsterdam—Olfactory Art and Smell Visualization. *Leonardo*, *50*(1), 92–93. doi:10.1162/LEON_a_01225

Medway, D. (2015). Rethinking Place Branding and the "Other" Senses. *Rethinking Place Branding*, 191–209. . doi:10.1007/978-3-319-12424-7_13

Morrin, M., & Ratneshwar, S. (2000). The Impact of Ambient Scent on Evaluation, Attention, and Memory for Familiar and Unfamiliar Brands. *Journal of Business Research*, *49*(2), 157–165. doi:10.1016/S0148-2963(99)00006-5

Morrin, M., & Ratneshwar, S. (2003). Does it Make Sense to Use Scents to Enhance Brand Memory? *JMR, Journal of Marketing Research*, *40*(1), 10–25. doi:10.1509/jmkr.40.1.10.19128

Morrison, M., Gan, S., Dubelaar, C., & Oppewal, H. (2011). In-store music and aroma influences on shopper behavior and satisfaction. *Journal of Business Research*, *64*(6), 558–564. doi:10.1016/j.jbusres.2010.06.006

Najafi, M., & Shariff, M. K. B. M. (2011). The Concept of Place and Sense of Place In Architectural Studies. *World Academy of Science, Engineering and Technology*, 1100–1106.

Nilsson, J. H. (2019). Urban bicycle tourism: Path dependencies and innovation in Greater Copenhagen. *Journal of Sustainable Tourism*, *27*(11), 1648–1662. doi:10.1080/09669582.2019.1650749

Nučič, M. (2012). Is sustainability science becoming more interdisciplinary over time? *Acta Geographica Slovenica*, 215–236. doi:10.3986/AGS52109

O'Dell, T. (2005). Experiencescapes: Blurring borders and testing connections. In T. O'Dell & P. Billing (Eds.), *Experiencescapes: Tourism, culture, and economy* (pp. 11–33). Copenhagen Business School Press.

Obermayer, N., Kővári, E., Leinonen, J., Bak, G., & Valeri, M. (2021, August). "How social media practices shape family business performance: The wine industry case study. *European Management Journal*. Advance online publication. doi:10.1016/j.emj.2021.08.003

Ouf, A. M. S. (2001). Authenticity and the Sense of Place in Urban Design. *Journal of Urban Design*, *6*(1), 73–86. doi:10.1080/13574800120032914

Pan, S., & Ryan, C. (2009). Tourism Sense-Making: The Role Of The Senses And Travel Journalism. *Journal of Travel & Tourism Marketing*, *26*(7), 625–639. doi:10.1080/10548400903276897

Peck, J. (2011). Does touch matter? Insights from haptic research in marketing. In Sensory Marketing (pp. 17–31). Routledge.

Peck, J., & Childers, T. L. (2003). To have and to Hold: The Influence of Haptic Information on Product Judgments. *Journal of Marketing*, *67*(2), 35–48. doi:10.1509/jmkg.67.2.35.18612

Peck, J., & Shu, S. B. (2009). The Effect of Mere Touch on Perceived Ownership. *The Journal of Consumer Research, 36*(3), 434–447. doi:10.1086/598614

Peck, J., & Wiggins, J. (2006). It Just Feels Good: Customers' Affective Response to Touch and Its Influence on Persuasion. *Journal of Marketing, 70*(4), 56–69. doi:10.1509/jmkg.70.4.056

Pedersen, E., & Larsman, P. (2008). The impact of visual factors on noise annoyance among people living in the vicinity of wind turbines. *Journal of Environmental Psychology, 28*(4), 379–389. doi:10.1016/j.jenvp.2008.02.009

Phillips, R. (2015). Playful and multi-sensory fieldwork: Seeing, hearing and touching. *New York. Journal of Geography in Higher Education, 39*(4), 617–629. doi:10.1080/03098265.2015.1084496

Pink, S. (2008). An Urban Tour: The Sensory Sociality of Ethnographic Place-making. *Ethnography, 9*(2), 175–196. doi:10.1177/1466138108089467

Porteous, J. D. (1985). Smellscape. *Progress in Physical Geography, 9*(3), 356–378. doi:10.1177/030913338500900303

Qiu, M., Zhang, J., & Zheng, C. (2018). Exploring tourists' soundscape emotion and its impact on sustainable tourism development. *Asia Pacific Journal of Tourism Research, 23*(9), 862–879. Advance online publication. doi:10.1080/10941665.2018.1494614

Quercia, D., Schifanella, R., Aiello, L. M., & McLean, K. (2015). Smelly Maps: The Digital Life of Urban Smellscapes. *9th International Conference on Web and Social Media, ICWSM 2015*, 327-336.

Rodaway, P. (1994). *Sensuous geographies: body, sense and place*. Routledge. doi:10.4324/9780203306864

Romero-García, L. E., Aguilar-Gallegos, N., Morales-Matamoros, O., Badillo-Piña, I., & Tejeida-Padilla, R. (2019). Urban tourism: A systems approach – state of the art. *Tourism Review, 74*(3), 679–693. doi:10.1108/TR-06-2018-0085

Saldanha, A. (2002). Music tourism and factions of bodies in Goa. *Tourist Studies, 2*(1), 43–62. doi:10.1177/1468797602002001096

Santich, B. (2004). The study of gastronomy and its relevance to hospitality education and training. *International Journal of Hospitality Management, 23*(1), 15–24. doi:10.1016/S0278-4319(03)00069-0

Santos, V., Ramos, P., Sousa, B., Almeida, N., & Valeri, M. (2021a). Factors influencing touristic consumer behavior. *Journal of Organizational Change Management*. Advance online publication. doi:10.1108/JOCM-02-2021-0032

Santos, V., Sousa, B., Ramos, P., & Valeri, M. (2021b). Emotions and Involvement in Tourism Settings. *Current Issues in Tourism*, 1–6. Advance online publication. doi:10.1080/13683500.2021.1932769

Schafer, R. M. (1977). *The Tuning of the World*. Knopf.

Schmitt, B. (1999). Experiential Marketing. *Journal of Marketing Management, 15*(1-3), 53–67. doi:10.1362/026725799784870496

Shi, B., Zhao, J., & Chen, P. (2017). Exploring urban tourism crowding in Shanghai via crowdsourcing geospatial data. *Current Issues in Tourism*, *20*(11), 1186–1209. doi:10.1080/13683500.2016.1224820

Skånberg, A., & Öhrström, E. (2002). Adverse Health Effects In Relation To Urban Residential Soundscapes. *Journal of Sound and Vibration*, *250*(1), 151–155. doi:10.1006/jsvi.2001.3894

Son, A., & Pearce, P. (2005). Multi-Faceted Image Assessment. *Journal of Travel & Tourism Marketing*, *18*(4), 21–35. doi:10.1300/J073v18n04_02

Spangenberg, E. R., Crowley, A. E., & Henderson, P. W. (1996). Improving the Store Environment: Do Olfactory Cues Affect Evaluations and Behaviors? *Journal of Marketing*, *60*(2), 67–80. doi:10.1177/002224299606000205

Suárez Velasco, J. E., Huaraca Vera, L. E., & Paladines, G. V. (2017). Desarrollo turístico de los mercados municipales en centros históricos. Mercado San Francisco. Quito-Ecuador. *Revista Publicando*, *11*(2), 331–339.

Taylor, J. P. (2001). Authenticity and sincerity in tourism. *Annals of Tourism Research*, *28*(1), 7–26. doi:10.1016/S0160-7383(00)00004-9

Toanoglou, M., Chcmli, S., & Valeri, M. (2021). The organizational impact of COVID-19 crisis on travel perceived risk across four continents. *Journal of Organizational Change Management*. Advance online publication. doi:10.1108/JOCM-12-2020-0369

Torija, A. J., Ruiz, D. P., & Ramos-Ridao, Á. (2011). Required stabilization time, short-term variability and impulsiveness of the sound pressure level to characterize the temporal composition of urban soundscapes. *Applied Acoustics*, *72*(2-3), 89–99. doi:10.1016/j.apacoust.2010.09.011

Trang, N. T., & Lee, G. (2018). Senses in leisure tourism: Scale development and its relationship with emotions. *International Journal of Tourism and Hospitality Research*, *32*(4), 21–39. doi:10.21298/IJTHR.2018.4.32.4.21

Valeri, M. (2016). Networking and cooperation practices in the Italian tourism business. *Journal of Tourism, Heritage & Services Marketing*, *2*(1), 30–35. doi:10.5281/zenodo.376333

Valeri, M. (Ed.). (2021). Tourism risk. Crisis and recovery management. Emerald Publishing.

Valeri, M. (2021). *Organization Theories. Implications for the Strategic Management.* Springer.

Valeri, M., & Baggio, R. (2020a). Social network analysis: Organizational implications in tourism management. *The International Journal of Organizational Analysis*. Advance online publication. doi:10.1108/IJOA-12-2019-1971

Valeri, M., & Baggio, R. (2020b). Italian tourism intermediaries: A social network analysis exploration. *Current Issues in Tourism*, *24*(9), 1270–1283. doi:10.1080/13683500.2020.1777950

Valeri, M., & Baggio, R. (2020c). A critical reflection on the adoption of blockchain in tourism. *Journal Information Technology and Tourism*, *23*(2), 121–132. doi:10.100740558-020-00183-1

Valeri, M., & Baggio, R. (2021). Increasing the efficiency of knowledge transfer in an Italian tourism system: A network approach. *Current Issues in Tourism*, 1–16. Advance online publication. doi:10.1080/13683500.2021.1937960

Vallbona, M. C., & Pérez, M. D. (2016). Los mercados de abastos y las ciudades turisticas. *PASOS Revista De Turismo y Patrimonio Cultural, 14*(2), 401–416. doi:10.25145/j.pasos.2016.14.026

Viollon, S., Lavandier, C., & Drake, C. (2002). Influence of visual setting on sound ratings in an urban environment. *Applied Acoustics, 63*(5), 493–511. doi:10.1016/S0003-682X(01)00053-6

Waitt, G., & Duffy, M. (2010). Listening and tourism studies. *Annals of Tourism Research, 37*(2), 457–477. doi:10.1016/j.annals.2009.10.017

Wang, S., & Gu, K. (2020). Pingyao: The historic urban landscape and planning for heritage-led urban changes. *Cities (London, England), 97*, 102489. doi:10.1016/j.cities.2019.102489

Wearing, S. L., & Foley, C. (2017). Understanding the tourist experience of cities. *Annals of Tourism Research, 65*, 97–107. doi:10.1016/j.annals.2017.05.007

Wissmann, T. (2014). *Geographies of urban sound*. Routledge.

Xiao, J., Tait, M., & Kang, J. (2018). A perceptual model of smellscape pleasantness. *Cities (London, England), 76*, 105–115. doi:10.1016/j.cities.2018.01.013

Xiong, J., Hashim, N. H., & Murphy, J. (2015). Multisensory image as a component of destination image. *Tourism Management Perspectives, 14*, 34–41. doi:10.1016/j.tmp.2015.03.001

Xu, Z., & Zhang, J. (2016). Antecedents and consequences of place attachment: A comparison of Chinese and Western urban tourists in Hangzhou, China. *Journal of Destination Marketing & Management, 5*(2), 86–96. doi:10.1016/j.jdmm.2015.11.003

Yang, W., & Kang, J. (2005a). Acoustic comfort evaluation in urban open public spaces. *Applied Acoustics, 66*(2), 211–229. . 011 . doi:10.1016/j.apacoust.2004.07

Yázigi, E., Carlos, A. F. A., & Cruz, R. C. A. (1996). *Turismo: Espaço, paisagem e cultura.* Turismo cultural. Impacto ambiental. Geografia. Politica social. Conservacion de la naturaleza. São Paulo Hucitec.

Ye, S., Wu, J. S., & Zheng, C. J. (2019). Are tourists with higher expectation more sensitive to service performance? Evidence from urban tourism. *Journal of Destination Marketing & Management, 12*, 64–73. doi:10.1016/j.jdmm.2019.01.002

Yuan, J., Deng, J., Pierskalla, C., & King, B. (2018). Urban tourism attributes and overall satisfaction: An asymmetric impact-performance analysis. *Urban Forestry & Urban Greening, 30*, 169–181. doi:10.1016/j.ufug.2018.02.006

Zhang, T., Chen, J., & Hu, B. (2019). Authenticity, Quality, and Loyalty: Local Food and Sustainable Tourism Experience. *Sustainability, 11*(12), 3437. doi:10.3390u11123437

KEY TERMS AND DEFINITIONS

Experiencescapes: The purposeful places where experiences are produced, staged, and consumed by diverse groups in tourism industry.

Haptiscapes: The sense of touch that is related to specific place.

Landscapes: The sense of sight that is related to specific place.

Sensescapes: Sensory experience of the environment.

Smellscapes: The sense of smell that is related to specific place.

Soundscapes: The sense of sound that is related to specific place.

Tastescapes: The sense of taste that is related to specific place.

Compilation of References

Aalbers, R. (2012). Organizing intra-organizational networks for innovation. PhD Theses in Economics and Business, University of Groningen.

Aalbers, R. (2020). Rewiring the Intrafirm Network under Downsizing: The Role of Tie Loss on Discretionary Tie Formation. *Long Range Planning*, *53*(3), 101858. Advance online publication. doi:10.1016/j.lrp.2018.11.002

Aalbers, R., & Dolfsma, W. (2017). Improving the value-of-input for ideation by management intervention: An intra-organizational network study. *Journal of Engineering and Technology Management*, *46*, 39–51. doi:10.1016/j.jengtecman.2017.10.006

Aalbers, R., & Dolfsma, W. (2019). Resilience of Information Flow during Restructuring: Characterizing Information Value being Exchanged and the Structure of a Network under Turmoil. *Journal of Business Research*, *100*, 299–310. doi:10.1016/j.jbusres.2019.03.003

Aalbers, R., Dolfsma, W., & Koppius, O. (2006). *On and off the beaten path: Transferring knowledge through formal and informal networksR Aalbers, O Koppius, W Dolfsma*. Circle Electronic Work. Paper Series.

Abdullah, A. R., Weng, C. N., Afif, I., & Fatah, A. (2018). Ecotourism in Penang National Park: A multi-stakeholder perspective on environmental issues. *Journal of Business and Social Development*, *6*(1), 70–83.

Abdullah, R., Karim, N. B. A., Patah, M. O. R. B. A., Zahari, H., Nair, G. K. S., & Jusoff, K. (2009). The linkage of employee satisfaction and loyalty in hotel industry in Klang Valley, Malaysia. *International Journal of Business and Management*, *4*(10), 152–160. doi:10.5539/ijbm.v4n10p152

Abubakar, A. M., & Ilkan, M. (2016). Impact of online WOM on destination trust and intention to travel: A medical tourism perspective. *Journal of Destination Marketing & Management*, *5*(3), 192–201. doi:10.1016/j.jdmm.2015.12.005

Acquisti, A., Brandimarte, L., & Loewenstein, G. (2015). Privacy and human behavior in the age of information. *Science*, *347*(6221), 509–514. doi:10.1126cience.aaa1465 PMID:25635091

Acuti, D., Grazzini, L., Mazzoli, V., & Aiello, G. (2019). Stakeholder engagement in green place branding: A focus on user-generated content. *Corporate Social Responsibility and Environmental Management*, *26*(2), 492–501. doi:10.1002/csr.1703

Adamo, G. E., Ferrari, S., & Gilli, M. (2019). Creativity as a source of differentiation in urban tourism: The case of Torino city. *International Journal of Tourism Research*, *21*(3), 302–310. doi:10.1002/jtr.2261

Agapito, D. L., Da, C. M., Valle, P. S., & Almeida, H. M. (2014). Um contributo do marketing sensorial para o marketing da experiência turística rural. Sensory marketing rural tourist experiences. *PASOS Revista De Turismo Y Patrimonio Cultural*, *12*(3), 611-621. . doi:10.25145/j.pasos.2014.12.045

Agapito, D., Pinto, P., & Mendes, J. (2017). Tourists' memories, sensory impressions and loyalty: In loco and post-visit study in Southwest Portugal. *Tourism Management, 58*, 108–118. doi:10.1016/j.tourman.2016.10.015

Aggestam, M., & Kristoferson, C.W. (2017). How women entrepreneurs build embeddedness: A case study approach. *International Journal of Gender and Entrepreneurship.*

Aggleton, J. P., & Waskett, L. (1999). The ability of odours to serve as state-dependent cues for real-world memories: Can Viking smells aid the recall of Viking experiences? *British Journal of Psychology, 90*(1), 1–7. doi:10.1348/000712699161170 PMID:10085542

Ahani, A., Nilashi, M., Ibrahim, O., Sanzogni, L., & Weaven, S. (2019). Market segmentation and travel choice prediction in Spa hotels through TripAdvisor's online reviews. *International Journal of Hospitality Management, 80*, 52–77. doi:10.1016/j.ijhm.2019.01.003

Ahl, H., & Nelson, T. (2010). Moving forward: Institutional perspectives on gender and entrepreneurship. *International Journal of Gender and Entrepreneurship, 2*, 5–9.

Ahmad, A. T. (2015). Penang Museums, Culture and History. *Kajian Malaysia, 33*, 153–174.

Ahmad, F., Draz, M. U., Su, L., Ozturk, I., & Rauf, A. (2018). Tourism and environmental pollution: Evidence from the one belt one road provinces of western China. *Sustainability, 10*(10), 1–22. doi:10.3390u10103520

Ahmed, N. K., Atiya, A. F., Gayar, N. E., & El-Shishiny, H. (2010). An empirical comparison of machine learning models for time series forecasting. *Econometric Reviews, 29*(5-6), 594–621. doi:10.1080/07474938.2010.481556

Ahmed, T., & Abdul-Kadir, A. (2013). Impact of information sources on the decision makes process of travel to the Egyptian tourist destination after January 25. *Tourism and Hospitality Management, 61*, 395–423.

Aires, L. (2011). *Paradigma qualitativo e práticas de investigação educacional.* UniversidadeAberta.

Aish, A., & Jöreskog, K. (1990). A panel model for political efficacy and responsiveness: An application of LISREL 7 with weighted least squares. *Quality & Quantity, 19*, 716–723. doi:10.1007/BF00152013

Aitchison, C., MacLeod, E. N., & Shaw, J. S. (2000). *Leisure and Tourism Landscapes. Social and cultural geographies.* Routledge.

Al Mansoori, S. A. N. A., Ab Yazid, M. S., Khatibi, A., & Azam, S. F. (2018). Strategic Management Planning and the Development of Healthcare Sector in Abu Dhabi: Structural Equation Modeling (SEM) Approach. *European Journal of Social Sciences Studies.*

Al Mansoori, S. A. N. A., Ab Yazid, M. S., Khatibi, A., & Azam, S. F. (2018). Validating A Measurement Model for Strategic Management Planning and the Development of Healthcare Sector in Abu Dhabi (SEHA). *European Journal of Human Resource Management Studies.*

Alaeddinoglu, F., & Can, A. S. (2011). Identification and classification of nature-based tourism resources: Western Lake Van basin, Turkey. *Procedia: Social and Behavioral Sciences, 19*, 198–207. doi:10.1016/j.sbspro.2011.05.124

Albattat, A. R., & Amer, H. (2016). Backpackers Expectation and Satisfaction towards Budget Hotel: A Case Study in Penang. *Almatourism-Journal of Tourism, Culture and Territorial Development, 7*(14), 168–180.

Albattat, A., Jamaludin, A., Zuraimi, N. S. M., & Valeri, M. (2020). Visit Intention and Destination Image in Post- Covid- 19 Crisis Recovery. *Current Issues in Tourism.* Advance online publication. doi:10.1080/13683500.2020.1842342

Albet, A. (2016). *Urban governance in southern Europe.* Routledge. doi:10.4324/9781315548852

Al-Dajani, H., & Marlow, S. (2013). Empowerment and entrepreneurship: A theoretical framework. *International Journal of Entrepreneurial Behaviour & Research*, *19*(5), 503–524.

Alegre, I., Mas-Machuca, M., & Berbegal-Mirabent, J. (2016). Antecedents of employee job satisfaction: Do they matter? *Journal of Business Research*, *69*(4), 1390–1395. doi:10.1016/j.jbusres.2015.10.113

Aletta, F., Kang, J., & Axelsson, Ö. (2016). Soundscape descriptors and a conceptual framework for developing predictive soundscape models. *Landscape and Urban Planning*, *149*, 65–74. doi:.landurbplan.2016.02.001 doi:10.1016/j

Allen, D., Shore, L. M., & Griffeth, R. W. (2003). The role of perceived organizational support and supportive human resource practices in the turnover process. *Journal of Management*, *29*(1), 99–118. doi:10.1177/014920630302900107

Almeida, A. M. (2017). Rural development and rural tourism: An institutional approach. *European Journal of Applied Business and Management*, *3*(1), 101–117.

Al-Oun, S., & Al-Homoud, M. (2008). The potential for developing community-based tourism among the Bedouins in the Badia of Jordan. *Journal of Heritage Tourism*, *3*(1), 36–54. doi:10.1080/1743873X.2008.9701249

Al-Refaie, A. (2015). Effects of human resource management on hotel performance using structural equation modeling. *Computers in Human Behavior*, *43*, 293–303. doi:10.1016/j.chb.2014.11.016

Anastasiei, B., & Dospinescu, N. (2018). A model of the relationships between the Big Five personality traits and the motivations to deliver word-of-mouth online. *Psihologija (Beograd)*, *4*(2), 1–12. doi:10.2298/PSI161114006A

Anderson, A. (2000). Paradox in the periphery: An entrepreneurial preconception. Entrepreneurship and Regional Development. *International Journal (Toronto, Ont.)*, *12*(2), 91–110.

Anholt, S. (2016). *Places: Identity, Image and Reputation*. Springer.

Anitha, J. (2014). Determinants of employee engagement and their impact on employee performance. *International Journal of Productivity and Performance Management*, *63*(3), 308–323. doi:10.1108/IJPPM-01-2013-0008

Antón, C., Camarero, C., Laguna, M., & Buhalis, D. (2019). Impacts of Authenticity, degree of adaptation and cultural contrast on travellers' Memorable Gastronomy Experiences. *Journal of Hospitality Marketing & Management*, *28*(7), 1–22. doi:10.1080/19368623.2019.1564106

Antrop, M. (2006). Sustainable landscapes: Contradiction, fiction or Utopia? *Landscape and Urban Planning*, *75*(34), 187–197. doi:10.1016/j.landurbplan.2005.02.014

Arbuckle, J. L. (2013). *Amos 22 user's guide*. SPSS.

Armenski, T., Gomezelj, D. O., Djurdjev, B., Ćurčić, N., & Dragin, A. (2012). Tourism destination competitiveness-between two flags. *Economic research- Ekonomska Istrazivanja*, *25*(2), 485–502. doi:10.1080/1331677X.2012.11517519

Asero, V., Gozzo, S., & Tomaselli, V. (2020). Shaping and re-shaping tourism areas. A network approach. In Destination Management and Marketing: Breakthroughs in Research and Practice (pp. 483-503), Hershey, PA: IGI Global.

Asero, V., Gozzo, S., & Tomaselli, V. (2016). Building Tourism Networks through Tourist Mobility. *Journal of Travel Research*, *55*(6), 751–763. doi:10.1177/0047287515569777

Asero, V., & Skonieczny, S. (2017). Formal Networks among Tourism Enterprises: Peculiarities of the Italian Solution. *Eastern European Business and Economics Journal*, *3*(1), 48–66.

Ashworth, G. (1988). Marketing the historic city for tourism. In B. Goodall & G. Ashworth (Eds.), *Marketing in the Tourism Industry The Promotion of Destination Regions*. Routledge.

Ashworth, G., & Page, S. J. (2011). Urban tourism research: Recent progress and current paradoxes. *Tourism Management, 32*(1), 1–15. doi:10.1016/j.tourman.2010.02.002

Atari, D. O., Luginaah, I. N., & Fung, K. (2009). The Relationship between Odour Annoyance Scores and Modelled Ambient Air Pollution in Sarnia, "Chemical Valley", Ontario. *International Journal of Environmental Research and Public Health*, 6. PMID:20054461

Atikahambar, Y., Zainal, A., Rahayu, M., & Mokhtar, R. (2019). Quality of Food and Tourists ' Satisfaction of Penang Delicacies in Predicting Tourists ' Revisit Intention Quality of Food and Tourists ' Satisfaction of Penang Delicacies in Predicting Tourists '. *Revisit Intention., 8*(12), 1606–1618.

Ávila-Robinson, A., & Wakabayashi, N. (2018). Changes in the structures and directions of destination management and marketing research: A bibliometric mapping study, 2005–2016. *Journal of Destination Marketing & Management, 10*, 101–111.

Avineshwaran, T, (2015). *Modi to inaugurate 'Torana' Gate at Brickfields.* Available at: https://www.thestar.com.my/news/nation/2015/11/23/modi-torana-gate/

Awan, M. I., Shamim, A., & Ahn, J. (2020). Implementing 'cleanliness is half of faith'in re-designing tourists, experiences and salvaging the hotel industry in Malaysia during COVID-19 pandemic. *Journal of Islamic Marketing.*

Azam, M., Alam, M. M., & Haroon Hafeez, M. (2018). Effect of tourism on environmental pollution: Further evidence from Malaysia, Singapore and Thailand. *Journal of Cleaner Production, 190*, 330–338. doi:10.1016/j.jclepro.2018.04.168

Aziri. (2011). Job satisfaction: A literature review. *Management Research and Practice, 3*(4), 77-86.

Aziz, N., Kefallonitis, E., & Friedman, B. A. (2014). Turkey as a Destination: A Study of Sensory Brand Associations. *Tourism, Culture & Communication, 14*(2), 77–89. doi:10.3727/109830414X14133839512587

Azzopardi, E., & Nash, R. (2017). A Review of Crouch and Ritchie's, Heath's, and Dwyer and Kim's Models of Tourism Competitiveness. *Tourism Analysis, 22*(2), 247–254. doi:10.3727/108354217X14888192562483

Bachek, S. H., Zainudin, H., & Haron, N. A. (2014). Preservation of culture and built heritage in new urban development: A case study on Little India Brickfields, Kuala Lumpur. *International Proceedings of Economics Development and Research, 74*, 1.

Bacon, S. (2005). Positive Psychology's Two Cultures. *Review of General Psychology, 9*(2), 181–192. doi:10.1037/1089-2680.9.2.181

Baden, S. & Milward, K. (2000). *Gender inequality and poverty: trends, linkages, analysis and policy implications.* BRIDGE Report.

Badulescu, D., Hoffman, I., Badulescu, A., & Badulescu, R. (2016). Local authorities' involvement in fostering Hungarian-Romanian cross-border cooperation in tourism. *Lex Localis. Journal of Local Self-Government, 14*(3), 337–358. doi:10.4335/14.3.337-358(2016)

Baggio, R., & Valeri, M. (2020). Network science and sustainable performance of family businesses in tourism. *Journal of Family Business Management.* doi:10.1108/JFBM-06-2020-0048

Baggio, R. (2008). Symptoms of complexity in a tourism system. *Tourism Analysis, 13*(1), 1–20. doi:10.3727/108354208784548797

Baggio, R. (2017). Network science and tourism – the state of the art. *Tourism Review, 72*(1), 120–131. doi:10.1108/TR-01-2017-0008

Baggio, R. (2020a). The science of complexity in the tourism domain: A perspective article. *Tourism Review, 75*(1), 16–19. doi:10.1108/TR-04-2019-0115

Baggio, R. (2020b). Tourism destinations: A universality conjecture based on network science. *Annals of Tourism Research, 82*(April), 102929. doi:10.1016/j.annals.2020.102929

Baggio, R., & Cooper, C. (2010). Knowledge transfer in a tourism destination: The effects of a network structure. *Service Industries Journal, 30*(10), 1757–1771. doi:10.1080/02642060903580649

Baggio, R., & Del Chiappa, G. (2014). Real and virtual relationships in tourism digital ecosystems. *Information Technology & Tourism, 14*(1), 3–19. doi:10.100740558-013-0001-5

Baggio, R., Micera, R., & Del Chiappa, G. (2020). Smart tourism destinations: A critical reflection. *Journal of Hospitality and Tourism Technology, 11*(3), 407–423. doi:10.1108/JHTT-01-2019-0011

Baggio, R., & Sainaghi, R. (2011). Complex and chaotic tourism systems: Towards a quantitative approach. *International Journal of Contemporary Hospitality Management, 23*(6), 840–861. doi:10.1108/09596111111153501

Baggio, R., & Sainaghi, R. (2016). Mapping time series into networks as a tool to assess the complex dynamics of tourism systems. *Tourism Management, 54*, 23–33. doi:10.1016/j.tourman.2015.10.008

Baggio, R., & Scaglione, M. (2018). Strategic visitor flows and destination management organization. *Information Technology & Tourism, 18*(1–4), 29–42. doi:10.100740558-017-0096-1

Baggio, R., Scott, N., & Cooper, C. (2010). Improving tourism destination governance: A complexity science approach. *Tourism Review, 65*(4), 51–60. doi:10.1108/16605371011093863

Bagheri, N. (2013). *The relation between emotional intelligence and job satisfaction. European Journal.*

Bagozzi, R., & Yi, Y. (1988). On the evaluation of structural equation models. *Journal of the Academy of Marketing Science, 16*(1), 74–94. doi:10.1007/BF02723327

Baines, S., & Wheelock, J. (2000). Work and employment in small businesses: Perpetuating and challenging gender traditions. *Gender, Work and Organization, 7*(1), 45–56.

Bakacsi, G. (2002). Szervezeti magatartás és vezetés. Közgazdasági és Jogi Könyvkiadó.

Bakar, K. A., & Farinda, A. G. (2012). Consumers' Attitude Towards "Mamak" Food. *Malaysia 3rd International Conference on Business And Economic,* 1304–1316.

Baldacchino, G. (2015). Feeding the Rural Tourism Strategy? Food and Notions of Place and Identity. *Scandinavian Journal of Hospitality and Tourism, 15*(1-2), 223–238. doi:10.1080/15022250.2015.1006390

Balkundi, P., Kilduff, M., Barsness, Z. I., & Michael, J. H. (2007). Demographic antecedents and performance consequences of structural holes in work teams. *Journal of Organizational Behavior, 28*(2), 241–260. doi:10.1002/job.428

Ballantyne, R., Packer, J., & Sutherland, L. A. (2011). Visitors' memories of wildlife tourism: Implications for the design of powerful interpretive experiences. *Tourism Management, 32*(4), 770–779. doi:10.1016/j.tourman.2010.06.012

Ballis, A., Tsouka, D., Moschovou, T., & Kasselouris, G. (2018). The impact of airport development on tourism in the Greek islands of the South Aegean Sea. *Journal of Air Transport Studies, 9*(2), 95–109. doi:10.38008/jats.v9i2.25

Baloglu, S., & McCleary, K. (1999). A Model of Destination Image Formation. *Annals of Tourism Research, 26*(4), 868–897. doi:10.1016/S0160-7383(99)00030-4

Bambauer-Sachse, S., & Mangold, S. (2011). Brand equity dilution through negative online word-of-mouth communication. *Journal of Retailing and Consumer Services*, *18*(1), 38–45. doi:10.1016/j.jretconser.2010.09.003

Bank of Greece. (2019). https://www.bankofgreece.gr/

Bannò, M., Piscitello, L., & Varum, C. (2015). Determinants of the internationalization of regions: The role and effectiveness of public policy measures. *Regional Studies*, *49*(7), 1208–1222. doi:10.1080/00343404.2013.821570

Bao, T., & Chang, T. (2014). Finding disseminators via electronic word of mouth message for effective marketing communications. *Decision Support Systems*, *67*, 21–29. doi:10.1016/j.dss.2014.07.006

Barbieri, C., & Sotomayor, S. (2013). Surf travel behavior and destination preferences: An application of the Serious Leisure Inventory and Measure. *Tourism Management*, *35*, 111–121. doi:10.1016/j.tourman.2012.06.005

Barlas, A., Mantis, K., & Koustelios, A. (2010). Achieving positive word-of-mouth communication: The role of perceived service quality in the context of Greek ski centres. *World Leisure Journal*, *52*(4), 290–297. doi:10.1080/04419057.2010.9674654

Baron, R. A., & Shane, S. A. (2005). *Entrepreneurship: A process perspective*. Thomson.

Barton, K. (2018). Guyana's Linden to Lethem Road: A Metaphor for Conservation and Development. *Focus On Geography*, *61*, 1–1. doi:10.21690/foge/2018.61.3p

Basha, B. (2016). *Viability of adaptive reuse with social approach: case study of Asia Heritage Row, Kuala Lumpur* (Master's thesis). Gombak, Selangor: International Islamic University Malaysia.

Baxstrom, R. (2009). *Transforming Brickfields: Development and governance in a Malaysian city*. NUS Press.

Baxstrom, R. (2020). 1. The Founding of Brickfields and the Prewar Development of Kuala Lumpur. In *Houses in Motion* (pp. 25–53). Stanford University Press. doi:10.1515/9780804775861-003

Beaumont, N., & Dredge, D. (2010). Local tourism governance: A comparison of three network approaches. *Journal of Sustainable Tourism*, *18*(1), 7–28. doi:10.1080/09669580903215139

Bechky, B. A., & Okhuysen, G. A. (2011). Expecting the Unexpected? How SWAT Officers and Film Crews Handle Surprises. *Academy of Management Journal*, *54*(2), 239–261. doi:10.5465/amj.2011.60263060

Becken, S., Jin, X., Zhang, C., & Gao, J. (2017). Urban air pollution in China: Destination image and risk perceptions. *Journal of Sustainable Tourism*, *25*(1), 130–147. doi:10.1080/09669582.2016.1177067

Beerli, A., & Martin, J. D. (2004b). Factors influencing destination image. *Annals of Tourism Research*, *31*, 657–681. https://www.sciencedirect.com/science/article/abs/pii/S0160738304000349

Beerli, A., & Martin, J. D. (2004a). Tourists' characteristics and the perceived image of tourist destinations: A quantitative analysis—a case study of Lanzarote, Spain. *Tourism Management*, *25*(5), 623–636. doi:10.1016/j.tourman.2003.06.004

Belias, D., Vasiliadis, L., & Velissariou, E. (2020). Internal Marketing in Tourism: The Case of Human Resource Empowerment on Greek Hotels. *Cultural and Tourism Innovation in the Digital Era*, 559-573.

Belias, D., Vasiliadis, L., & Zaftis, E. (2020). The Impact of Cultural Routes on Traditional Settlements: The Case of Greece. In *Cultural and Tourism Innovation in the Digital Era* (pp. 123–134). Springer. doi:10.1007/978-3-030-36342-0_10

Belias, D., Velissariou, E., Kyriakou, D., Vasiliadis, L., Mantas, C., Sdrolias, L., Aspridis, G., & Koustelios, A. (2017). Customer Relationship Management and Social Media in Greek Tourism. In *International Conference on Strategic Innovative Marketing* (pp. 59-65). Springer.

Benson, M., & O'Reilly, K. (2018). Colonial Traces and Neoliberal Presents: Situating Malaysia and Panama. In Lifestyle Migration and Colonial Traces in Malaysia and Panama (pp. 41-80). Palgrave Macmillan.

Bentler, P. (1990). Quantitative methods in psychology: Comparative fit indices in structural models. *Psychological Bulletin*, *107*(2), 238–246. doi:10.1037/0033-2909.107.2.238 PMID:2320703

Berbel-Pineda, J. M., Palacios-Florencio, B., Ramírez-Hurtado, J. M., & Santos-Roldán, L. (2019). Gastronomic experience as a factor of motivation in the tourist movements. *International Journal of Gastronomy and Food Science*, *100171*. Advance online publication. doi:10.1016/j.ijgfs.2019.100171

Berezan, O., Raab, C., Yoo, M., & Love, C. (2013). Sustainable hotel practices and nationality: The impact on guest satisfaction and guest intention to return. *International Journal of Hospitality Management*, *34*, 227–233. doi:10.1016/j.ijhm.2013.03.010

Berg, S. H., & Hassink, R. (2014). Creative industries from an evolutionary perspective: A critical literature review. *Geography Compass*, *8*(9), 653–664. doi:10.1111/gec3.12156

Beritelli, P. (2011). *Tourist destination governance through local elites: Looking beyond the stakeholder level* (Doctoral dissertation). Universität St. Gallen.

Beritelli, P., & Laesser, C. (2017). The Dynamics of Destinations and Tourism Development. In D. R. Fesenmaier & Z. Xiang (Eds.), *Design Science in Tourism, Tourism on the VergeVerge* (pp. 195–214). doi:10.1007/978-3-319-42773-7_13

Beritelli, P., Bieger, T., & Laesser, C. (2007). Destination governance: Using corporate governance theories as a foundation for effective destination management. *Journal of Travel Research*, *46*(1), 96–107. doi:10.1177/0047287507302385

Beritelli, P., Bieger, T., & Laesser, C. (2014). The New Frontiers of Destination Management: Applying Variable Geometry as a Function-Based Approach. *Journal of Travel Research*, *53*(4), 403–417. doi:10.1177/0047287513506298

Beritelli, P., & Laesser, C. (2014). Getting the cash-cow directors on board: An alternative view on financing DMOs. *Journal of Destination Marketing & Management*, *2*(4), 213–220. doi:10.1016/j.jdmm.2013.10.001

Berkesné, R. N. (2018). *CSR EMAT – A vállalatok társadalmi felelősségvállalásának kiválósági menedzsment értékelési eszköze* (PhD thesis).

Berman, S. L., Down, J., & Hill, C. W. (2002). Tacit knowledge as a source of competitive advantage in the National Basketball Association. *Academy of Management Journal*, *45*(1), 13–31.

Bernabé, C., & Hernández, M. (2016). Turismo en Albarracín y Comarca: Acción pública local y dinâmica reciente en clave de desarrollo turístico sostenible. *Anales de Geografía de la Universidad Complutense*, *36*(1), 173–194. doi:10.5209/rev_AGUC.2016.v36.n1.52718

Bhatt, P. J., Pandya, H. A., & Modi, N. R. (2019). *A review of possible impacts on tourism by climate change*. Academic Press.

Biagi, B., Ladu, M. G., Meleddu, M., & Royuela, V. (2019). Tourism and the city: The impact on residents' quality of life. *International Journal of Tourism Research*, *22*(2), 168–181. doi:10.1002/jtr.2326

Bichler, B. F., & Peters, M. (2020). Soft adventure motivation: An exploratory study of hiking tourism. *Tourism Review*. ahead-of-print. doi:10.1108/TR-10-2019-0403

Bickart, B., & Schindler, R. M. (2001). Internet forums as influential sources of consumer information. *Journal of Interactive Marketing*, *15*(3), 31–40. doi:10.1002/dir.1014

Bickerdyke, I. (1996). Australia: The Evolving Structure and Strategies of Business Networks. In *Networks of Enterprises and Local Development: Competing and Co-operating in Local Productive Systems* (pp. 203–216). Organisation for Economic Co-operation and Development.

Bieger, T., Beritelli, P., & Laesser, C. (2009). Size matters! Increasing DMO effectiveness and extending tourism destination boundaries. *Tourism Review*, *57*(3), 309–327.

Bigne, J. E., Sanchez, M. I., & Sanchez, J. (2001). Tourism image, evaluation variables and after purchase behaviour: Inter-relationship. *Tourism Management*, *22*(6), 607–617. doi:10.1016/S0261-5177(01)00035-8

Bitner, M. J. (1992). Servicescapes: The impact of physical surroundings on customers and employees. *Journal of Marketing*, *56*(2), 57–71. doi:10.1177/002224299205600205

Björk, P., & Kauppinen-Räisänen, H. (2016). Exploring the multi-dimensionality of travellers' culinary-gastronomic experiences. *Current Issues in Tourism*, *19*(12), 1260–1280.

Björk, P., & Kauppinen-Räisänen, H. (2017). A destination's gastronomy as a means for holiday well-being. *British Food Journal*, *119*(7), 1578–1591. doi:10.1108/BFJ-09-2016-0394

Blackwell, R., Miniard, P., & Engel, F. (2001). *Consumer Behavior*. Harcourt College.

Blasco, D., Guia, J., & Prats, L. (2014a). Emergence of governance in cross-border destinations. *Annals of Tourism Research*, *49*, 159–173. doi:10.1016/j.annals.2014.09.002

Blasco, D., Guia, J., & Prats, L. (2014b). Tourism destination zoning in mountain regions: A consumer-based approach. *Tourism Geographies*, *16*(3), 512–528. doi:10.1080/14616688.2013.851267

Boatto, V., & Gennari, A.J. (2011). *La roadmap del turismo enologico*. Franco Angeli.

Bognár, S., & Kovács, L. (2013). A világ egy városban. Liverpooli EKF tapasztalatok. *Modern Geográfia*, *1*, 23-36.

Bohdanowicz, P., Zientara, P., & Novotna, E. (2011). International hotel chains and environmental protection: An analysis of Hilton's Wecare! programme (Europe, 2006-2008). *Journal of Sustainable Tourism*, *19*(7), 797–816. doi:10.1080/09669582.2010.549566

Bohlin, M., Brandt, D., & Elbe, J. (2016). Tourism as a vehicle for regional development in peripheral areas – myth or reality? A longitudinal case study of Swedish regions. *European Planning Studies*, *24*(10), 1788–1805. doi:10.1080/09654313.2016.1194807

Boley, B. B., & Perdue, R. R. (2012). Destination management, competitiveness, and quality-of-life: a review of literature and research agenda. Handbook of tourism and quality-of-life research, 515-528.

Bollen, K. A. (1989). *Structural equations with latent variables*. Wiley. doi:10.1002/9781118619179

Bonanno, A. (1992). *The Agricultural and Food Sector in the New Global Era*. Concept Publishing Company.

Boniface, B., Cooper, R., & Cooper, C. (2016). *Worldwide Destinations: The geography of travel and tourism*. Routledge. doi:10.4324/9781315697666

Booyens, I., & Rogerson, C. M. (2015). Creative tourism in Cape Town: An innovation perspective. *Urban Forum*, *26*, 405–424. 10.100712132-015-9251-y

Booyens, I. (2016). Global–local trajectories for regional competitiveness: Tourism innovation in the Western Cape. *Local Economy*, *31*(1–2), 142–157. doi:10.1177/0269094215618598

Booyens, I., & Rogerson, C. M. (2016). Unpacking the geography of tourism innovation in Western Cape Province, South Africa. *Bulletin of Geography. Socio-Economic Series, 31*(31), 19–36. doi:10.1515/bog-2016-0002

Borenstein, S., & Rose, N. L. (2014). How airline markets work... or do they? Regulatory reform in the airline industry. In *Economic regulation and its reform: What have we learned?* (pp. 63–135). University of Chicago Press.

Borer, M. I. (2013). Being in the City: The Sociology of Urban Experiences. *Sociology Compass, 7*(11), 965–983. doi:10.1111oc4.12085

Borgatti, S. P., Everett, M. G., & Freeman, L. C. (2002). *UCINET for Windows: Software for Social Network Analysis*. Harvard.

Bouwen, R., & Taillieu, T. (2004). Multi-party collaboration as social learning for interdependence: Developing relational knowing for sustainable natural resource management. *Journal of Community & Applied Social Psychology, 14*(3), 137–153. doi:10.1002/casp.777

Bowden, J. (2003). A Cross-national Analysis of International Tourist Flows in China. *Tourism Geographies, 5*(3), 257–279. doi:10.1080/14616680309711

Braun, R. (2015). *Vállalati társadalmi felelősségvállalás A vállalatok politikája*. Akadémiai Kiadó.

Breiby, M. A., & Slåtten, T. (2018). The role of aesthetic experiential qualities for tourist satisfaction and loyalty. *International Journal of Culture, Tourism and Hospitality Research, 12*(1), 1–14. doi:10.1108/IJCTHR-07-2017-0082

Brewer, M. B. (2001). The many faces of social identity: Implications for political psychology. *Political Psychology, 22*(1), 115–125. doi:10.1111/0162-895X.00229

Brey, P. A. E. (2012). Anticipating ethical issue in emerging IT. *Ethics and Information Technology, 14*(4), 305–317. doi:10.100710676-012-9293-y

Brochado, A., Stoleriu, O., & Lupu, C. (2018). Surf camp experiences. *Journal of Sport & Tourism, 22*(1), 21–41. doi :10.1080/14775085.2018.1430609

Brochado, A., Stoleriu, O., & Lupu, C. (2019). Wine tourism: A multisensory experience. *Current Issues in Tourism*, 1–19. doi:10.1080/13683500.2019.1649373

Brouder, P., & Ioannides, D. (2014). Urban tourism and evolutionary economic geography: Complexity and co-evolution in contested spaces. *Urban Forum, 25*, 419-430. 10.100712132-014-9239-z

Brown, A., Gjestland, T., & Dubois, D. (2015). Acoustic Environments and Soundscapes. *Soundscape and the Built Environment*, 1–16. . doi:10.1201/b19145-2

Brown, J., Broderick, A. J., & Lee, N. (2007). Word of mouth communication within online communities: Conceptualizing the online social network. *Journal of Interactive Marketing, 21*(3), 2–220. doi:10.1002/dir.20082

Brown, T. A. (2015). *Confirmatory factor analysis for applied research* (2nd ed.). Guilford Press.

Brunori, G., & Rossi, A. (2000). Synergy and coherence through collective action: Some insights from wine routes in Tuscany. *Sociologia Ruralis, 40*(4), 409–423.

Buckley, R. (2002a). Surf tourism and sustainable development in indo-pacific islands. I. The industry and the islands. *Journal of Sustainable Tourism, 10*(5), 405–424. doi:10.1080/09669580208667176

Buckley, R. (2002b). Surf tourism and sustainable development in indo-pacific islands. II. Recreational capacity management and case study. *Journal of Sustainable Tourism, 10*(5), 425–442. doi:10.1080/09669580208667177

Buckley, R. C., Guitart, D., & Shakeela, A. (2017). Contested surf tourism resources in the Maldives. *Annals of Tourism Research*, *64*, 185–199. doi:10.1016/j.annals.2017.03.005

Buhalis, D., & Amaranggana, A. (2015). Smart Tourism Destinations Enhancing Tourism Experience Through Personalisation of Services. In Information and Communication Technologies in Tourism 2015 (pp. 377–389). doi:10.1007/978-3-319-14343-9_28

Buhalis, D. (2000). Marketing the Competitive Destination of the Future. *Tourism Management*, *21*(1), 97–116. doi:10.1016/S0261-5177(99)00095-3

Buhalis, D., Harwood, T., Bogicevic, V., Viglia, G., Beldona, S., & Hofacker, C. (2019). Technological disruptions in services: Lessons from tourism and hospitality. *Journal of Service Management*, *30*(4), 484–506. doi:10.1108/JOSM-12-2018-0398

Buhalis, D., & Law, R. (2008). Progress in information technology and tourism management: 20 years on and 10 years after the Internet—The state of eTourism research. *Tourism Management*, *29*(4), 609–623. doi:10.1016/j.tourman.2008.01.005

Buhalis, D., & Sinatra, Y. (2019). Real-time co-creation and nowness service: Lessons from tourism and hospitality. *Journal of Travel & Tourism Marketing*, *36*(5), 563–582. doi:10.1080/10548408.2019.1592059

Bulatova, Y. K., & Ul'chickij, O. A. (2015). Historical and theoretical background to the formation of a tourist center: international experience. *Scientific Herald of the Voronezh State University of Architecture and Civil Engineering Construction and Architecture*, 92–104.

Burmil, S., & Enis, R. (2004). An Integrated Approach to Landscape and Planning. *Journal of Architectural and Planning Research*, *21*, 140–151.

Burnette, J. L., Pollack, J. M., & Forsyth, D. R. (2011). Leadership in Extreme Contexts: A Groupthink Analysis of the May 1996 Mount Everest Disaster. *The Journal of Leadership Studies*, *4*(4), 29–40. doi:10.1002/jls.20190

Burton, J. (2010). *WHO Healthy Workplace Framework and Model: Background and Supporting Literature and Practice*. World Health Organization

Burt, R. S. (2000). The network structure of social capital. *Research in Organizational Behavior*, *22*, 345–423. doi:10.1016/S0191-3085(00)22009-1

Burt, R. S. (2002). Bridge decay. *Social Networks*, *24*(4), 333–363. doi:10.1016/S0378-8733(02)00017-5

Butowski, L. (2019). Tourist sustainability of destination as a measure of its development. *Current Issues in Tourism*, *22*(9), 1043–1061. doi:10.1080/13683500.2017.1351926

Buzova, D., Cervera-Taulet, A., & Sanz-Blas, S. (2019). Exploring multisensory place experiences through cruise blog analysis. *Psychology and Marketing*, *37*(1), 131–140. doi:10.1002/mar.21286

Byrne, B. M. (2010). Structural equation modeling with AMOS: Basic concepts, applications and programming (2nd ed.). London: Lawrence Erlbaum.

Byun, G., Karau, S. J., Dai, Y., & Lee, S. (2018). A three-level examination of the cascading effects of ethical leadership on employee outcomes: A moderated mediation analysis. *Journal of Business Research*, *88*, 44–53. doi:10.1016/j.jbusres.2018.03.004

Cakar, K. (2018). Critical success factors for tourist destination governance in times of crisis: A case study of Antalya, Turkey. *Journal of Travel & Tourism Marketing*, *35*(6), 786–802. doi:10.1080/10548408.2017.1421495

Çakar, K., & Uzut, İ. (2020). Exploring the stakeholder's role in sustainable degrowth within the context of tourist destination governance: The case of Istanbul, Turkey. *Journal of Travel & Tourism Marketing*, *37*(8-9), 8–9, 917–932. doi:10.1080/10548408.2020.1782307

Camilleri, M. A. (2018). The promotion of responsible tourism management through digital media. *Tourism Planning & Development*, *15*(6), 653–671. doi:10.1080/21568316.2017.1393772

Camisón, C., Forés, B., & Navarro, B.M. (2017). Current issues in tourism cluster and firm-specific antecedents of organizational innovation. *Current Issues in Tourism*, *20*(6), 617–646.

Campus, S. L., Long, J. S., & Long, B. S. (2017). *Malaysia in Indian Filmography: The Case of "Don" and "Kabali"*. Management (STHEM), Universiti Utara Malaysia.

Cardwell, L. A., Williams, S., & Pyle, A. (2017). Corporate public relations dynamics: Internal vs. external stakeholders and the role of the practitioner. *Public Relations Review*, *43*(1), 152–162. doi:10.1016/j.pubrev.2016.11.004

Carson, D. A. (2018). Challenges and opportunities for rural tourism geographies: A view from the 'boring'peripheries. *Tourism Geographies*, *20*(4), 737–741. doi:10.1080/14616688.2018.1477173

Casaló, F., Flavián, C., Guinalíu, M., & Ekinci, Y. (2015). Do online hotel rating schemes influence booking behaviors? *International Journal of Hospitality Management*, *1*, 28–36. doi:10.1016/j.ijhm.2015.05.005

Cascio, W. F. (2003). Responsible restructuring: Seeing employees as assets, not costs. *Ivey Business Journal*, *68*, 1.

Castro, C., Armario, M., & Ruiz, M. (2007). The influence of market heterogeneity on the relationship between a destination's image and tourists' future behavior. *Tourism Management*, *28*(1), 175–187. doi:10.1016/j.tourman.2005.11.013

Cavico, F. J., Mujtaba, B. G., & Samuel, M. (2016). Code words and covert employment discrimination: Legal analysis and consequences for management. *International Journal of Organizational Leadership*, *5*(3), 231–253. doi:10.33844/ijol.2016.60365

Centar za kreativne industrije Zadar. (2018). Retrieved April 18, 2020, Available at: https://www.inovacija-zadar.hr/novosti/provedba-projekta-centar-za-kreativne-industrije/161.html

Chaabane, W., Nassour, A., Bartnik, S., Bünemann, A., & Nelles, M. (2019). Shifting towards sustainable tourism: Organizational and financial scenarios for solid waste management in tourism destinations in Tunisia. *Sustainability*, *11*(13), 3591. doi:10.3390u11133591

Chahal, D. K. (2017). *The Determinants of Customer Satisfaction and Behavioral Intention among Restaurants in Klang Valley area of Malaysia : A Conceptual Study*. Academic Press.

Chan, J. K. L. (2010). Building sustainable tourism destination and developing responsible tourism: conceptual framework, key issues and challenges. Tourism Development Journal- An International Research Journal, 8(1), 24-32.

Chanias, S., Myers, M. D., & Hess, T. (2019). Digital transformation strategy making in pre-digital organizations: The case of a financial services provider. *The Journal of Strategic Information Systems*, *28*(1), 17–33. doi:10.1016/j.jsis.2018.11.003

Chatterjee, P. (2001). Online reviews: Do consumers use them? *Advances in Consumer Research. Association for Consumer Research (U.S.)*, *28*, 129–133.

Chemli, S., Toanoglou, M., & Valeri, M. (2020). The impact of Covid-19 media coverage on tourist's awareness for future traveling. *Current Issues in Tourism*, 1–8. Advance online publication. doi:10.1080/13683500.2020.1846502

Chen, C., & Tsai, D. C. (2007). How destination image and evaluative factors affect behavioural intentions? *Tourism Management*, *28*(4), 1115–1122. doi:10.1016/j.tourman.2006.07.007

Chen, Q., & Huang, R. (2015). Understanding the importance of food tourism to Chongqing, China. *Journal of Vacation Marketing*, *22*(1), 42–54. doi:10.1177/1356766715589427

Chen, Y. C., Shang, R. A., & Li, M. J. (2014). The effects of perceived relevance of travel blogs' content on the behavioral intention to visit a tourist destination. *Computers in Human Behavior*, *30*, 787–799. doi:10.1016/j.chb.2013.05.019

Chiang, C., Back, K.-J., & Canter, D. D. (2005). The Impact of Employee Training Satisfaction and Intention to Stay in the Hotel Industry. *Journal of Human Resources in Hospitality & Tourism*, *4*(2), 99–118. doi:10.1300/J171v04n02_06

Chi, C., & Gursoy, D. (2009). Employee satisfaction, customer satisfaction, and financial performance: An empirical examination. *International Journal of Hospitality Management*, *28*(2), 245–253. doi:10.1016/j.ijhm.2008.08.003

Chiles, T. H., Bluedorn, A. C., & Gupta, V. K. (2007). Beyond creative destruction and entrepreneurial discovery: A radical Austrian approach to entrepreneurship. *Organization Studies*, *28*, 467–493.

Chin, C., Razali, A., & Yong, J. (2020, September 12). *Malaysians Reveal Best Travel and Food Spots in Their Hometowns*. Academic Press.

Chio, J. (2017). Rendering rural modernity: Spectacle and power in a Chinese ethnic tourism village. *Critique of Anthropology*, *37*(4), 418–439. doi:10.1177/0308275X17735368

Chung, S. S. Y. (2018). The Coliseum Theatre, Kuala Lumpur: A Site for Collective Engagement, A Space of Urban Imagination. *Fabrications*, *28*(2), 212–234. doi:10.1080/10331867.2018.1446657

Clavé, S., & Wilson, J. (2017). The evolution of coastal tourism destinations: A path plasticity perspective on tourism urbanisation. *Journal of Sustainable Tourism*, *25*(1), 96–112. doi:10.1080/09669582.2016.1177063

Clemente, F. M., & Martins, F. M. L. (2017). Network structure of UEFA Champions League teams: Association with classical notational variables and variance between different levels of success. *International Journal of Computer Science in Sport*, *16*(1), 39–50. doi:10.1515/ijcss-2017-0003

Clemente, F. M., Martins, F. M. L., Kalamaras, D., & Mendes, R. S. (2015b). Network analysis in basketball: Inspecting the prominent players using centrality metrics. *Journal of Physical Education and Sport*, *15*(2), 212.

Clemente, F. M., Martins, F. M. L., Wong, P. D., Kalamaras, D., & Mendes, R. S. (2015a). Midfielder as the prominent partici¬pant in the building attack: A network analysis of national teams in FIFA World Cup 2014. *International Journal of Performance Analysis in Sport*, *15*(2), 704–722. doi:10.1080/24748668.2015.11868825

Cobos, L. M., Mejia, C., Ozturk, A. B., & Wang, Y. (2016). A technology adoption and implementation process in an independent hotel chain. *International Journal of Hospitality Management*, *57*, 93–105. doi:10.1016/j.ijhm.2016.06.005

Cohendet, P., Grandadam, D., & Simon, L. (2010). The Anatomy of the Creative City. *Industry and Innovation*, *17*(1), 91–111.

Cohen, J. (1988). *Statistical power analysis for the behavioural sciences* (2nd ed.). Erlbaum.

Coleman, J. S. (1988). Social Capital in the Creation of Human Capital. *The American Journal of Sociology*, *94*, 95-120.

Coles, B., & Crang, P. (2011). Placing alternative consumption: commodity fetishism in borough fine foods market, London. In T. Lewis & E. Potter (Eds.), *Ethical Consumption. A Critical Introduction* (pp. 87–102). Routledge.

Coles, T., Fenclova, E., & Dinan, C. (2013). Tourism and corporate social responsibility: A critical review and research agenda. *Tourism Management Perspectives*, *6*, 122–141. doi:10.1016/j.tmp.2013.02.001

Colletta, T. (2018). *To promote urban heritage culture in the age of information revolution: a new creativity.* In ICOMOS 19th General Assembly and Scientific Symposium "Heritage and Democracy", New Delhi, India.

Comissão Europeia. (2000). *Para um turismo urbano de qualidade: Gestão integrada da qualidade (GIQ) dos destinos turísticos urbanos.* Serviço das Publicações Oficiais das Comunidades Europeias.

Comissão Europeia. (2003). *A manual for evaluating the quality performance of tourist destinations and services.* Office for Official Publications of the European Communities.

Comissão Europeia. (2016). *The European tourism indicator system: ETIS toolkit for sustainable destination management.* Available at https://ec.europa.eu/growth/sectors/tourism/offer/sustainable/indicators/index_en.htm

ComissãoEuropeia. (2013). *Caixa de ferramentas do sistema europeu de indicadores de turismo para destinos sustentáveis.* Serviço de Publicações da União Europeia.

Confente, I., & Russo, I. (2015). Why do People Talk? A Comparison Between Offline WOM and Online WOM. In *ECSM2015-Proceedings of the 2nd European Conference on Social Media: ECSM 2015* (p. 81). Academic Conferences Limited.

Correia, A., Nunes, A., Silva, G., Fernandes, P., Moreira, J., & Soares, L. (2021). Sport tourism event and perceived economic impacts: The case of World Bodyboard Championship 2018, Viana do Castelo, Portugal. In Advances in Tourism, Technology and smart systems (Vol. 208, pp. 223–233). Academic Press.

Costello, A. B., & Osborne, J. W. (2005). Best practices in exploratory factor analysis: Four recommendations for getting the most from your analysis. *Practical Assessment, Research & Evaluation, 10*(7), 1–9.

Council of the European Union. (2016). *General Data Protection Regulation.* Available at https://data.consilium.europa.eu/doc/document/ST-5419-2016-INIT/en/pdf

Cousin, S., Chareyron, G., & Jacquot, S. (2017). Big data and tourism. In L. Lowry (Ed.), *The SAGE international encyclopedia of travel and tourism* (pp. 151–155). SAGE Publications.

Coutinho, C. P. (2015). Avaliação da qualidade da investigação qualitativa: Algumas considerações teóricas e recomendações práticas. In *F. Souza, D. Souza & A. Costa (Orgs.), Investigação qualitativa: Inovação, dilemas e desafios* (Vol. 2, pp. 101–121). Ludomedia.

Crang, M. (2001). Rhythms of the city: temporalised space and motion. In *Timespace: geographies of temporality* (pp. 187–207). Routledge.

Creative Tourism Network. (2020). Retrieved April 22, 2020, Available at: http://www.creativetourismnetwork.org/presentation/

Crespi, M., & Pérez, M. (2015). Selected studies from the Tourism Network Workshop of the Regional Studies Association. In *Food markets: new tourism icons.Illustrative cases of Barcelona and Madrid* (pp. 127–139). Department of Economic Geography and Futures Study.

Creswell, J. W. (2009). *Research design - Qualitative, quantitative, and mixed methods approaches* (3rd ed.). SAGE Publications.

Criticism of the Shine logo: https://vehir.hu/cikk/53239-kivalasztottak-a-veszprem-balaton-2023-zrt-uj-ekf-arculatat

Crittenden, A. B., Crittenden, V. L., & Crittenden, W. F. (2017). Industry transformation via channel disruption. *Journal of Marketing Channels, 24*(1–2), 13–26. doi:10.1080/1046669X.2017.1346974

Crouch, G. I., & Ritchie, J. R. B. (2010). A model of destination competitiveness/sustainability: Brazilian perspectives. *Public Administration Review (RAP), 44*(5), 1049-1066.

Crouch, G. I. (2011). Destination competitiveness: An analysis of determinant attributes. *Journal of Travel Research, 50*(1), 27–45. doi:10.1177/0047287510362776

Crouch, G. I., & Ritchie, J. R. B. (1995). *Destination competitiveness and the role of the tourism enterprise. In Conference of the International Management Development Association.* IMDA.

Crouch, G. I., & Ritchie, J. R. B. (1999). Tourism, competitiveness, and societal prosperity. *Journal of Business Research, 44*(3), 137–152. doi:10.1016/S0148-2963(97)00196-3

Cutcher, L. (2010). Local heroes: Co-producing history, 'community' and the self. *Consumption Markets & Culture, 13*(1), 79–90. doi:10.1080/10253860903346773

Cvelbar, L. K., Mayr, M., & Vavpotic, D. (2018). Geographical mapping of visitor flow in tourism: A user-generated content approach. *Tourism Economics, 24*(6), 701–719. doi:10.1177/1354816618776749

Czernek, K. (2013). Determinants of cooperation in a tourist region. *Annals of Tourism Research, 40*(1), 83–104. doi:10.1016/j.annals.2012.09.003

d'Angella, F., & Go, F. M. (2009). Tale of two cities' collaborative tourism marketing: Towards a theory of destination stakeholder assessment. *Tourism Management, 30*(3), 429–440. doi:10.1016/j.tourman.2008.07.012

Dahlstrom, M. (1996). Young Women in a Male Periphery-Experiences from the Scandinavian North. *Journal of Rural Studies, 12*(3), 259–271.

Dann, G., & Jacobsen, J. K. S. (2003). Tourism smellscapes. *Tourism Geographies, 5*(1), 3–25. doi:10.1080/1461668032000034033

Davis, L., & Thys-Şenocak, L. (2017). Heritage and scent: Research and exhibition of Istanbul's changing smellscapes. *International Journal of Heritage Studies, 23*(8), 723–741. doi:10.1080/13527258.2017.1317646

Davis, P. (2009). Ecomuseums and the representation of place. *Rivista Geografica Italiana, 116*, 483–503.

de Bruyn, C., & Alonso, A. F. (2012). Tourism destination governance: Guidelines for implementation. In E. Fayos-Sola, J. A. M. da Silva, & J. Jafari (Eds.), Knowledge management in tourism: Policy and governance applications (pp. 221–242). Bingley: Emerald Group.

De Lucia, C., Pazienza, P., Balena, P., & Caporale, D. (2019). Exploring local knowledge and socio-economic factors for touristic attractiveness and sustainability. *International Journal of Tourism Research, 22*(1), 81–99. doi:10.1002/jtr.2320

De Noni, I., Orsi, L., & Zanderighi, L. (2014). Attributes of Milan influencing city brand attractiveness. *Journal of Destination Marketing & Management, 3*(4), 218–226. doi:10.1016/j.jdmm.2014.06.001

De Silva, A. D. A., Khatibi, A., & Azam, S. F. (2018). Can Parental Involvement Mitigate—Swing Away from Science? Sri Lankan Perspectives. *Cogent Education, 5*(1), 1467244. doi:10.1080/2331186X.2018.1467244

De Silva, A. D. A., Khatibi, A., & Azam, S. F. (2018). Do the Demographic Differences Manifest in Motivation to Learn Science and Impact on Science Performance? Evidence from Sri Lanka. *International Journal of Science and Mathematics Education, 16*(1), 47–67. doi:10.100710763-017-9846-y

Decrop, A. (1999). Triangulation in qualitative tourism research. *Tourism Management, 20*(1), 157–161. doi:10.1016/S0261-5177(98)00102-2

Deguen, S., Pédrono, G., Ségala, C., & Mesbah, M. (2008). Association Between Pollution and Public Perception of Air Quality-SEQAP, a Risk Perception Study in France. *Epidemiology (Cambridge, Mass.)*, *19*, S216.

Deguignet, M., Juffe-Bignoli, D., Harrison, J., MacSharry, B., Burgess, N., & Kingston, N. (2014). *2014 United Nations list of protected areas.* Retrieved from https://wdpa.s3.amazonaws.com/WPC2014/2014_UN_LIST_REPORT_EN.pdf

Del Chiappa, G., & Baggio, R. (2015). Knowledge transfer in smart tourism destinations: Analyzing the effects of a network structure. *Journal of Destination Marketing & Management*, *4*(3), 143–144. doi:10.1016/j.jdmm.2015.02.001

Della Lucia, M., & Segre, G. (2017). Intersectoral local development in Italy: The cultural, creative and tourism industries. *International Journal of Culture, Tourism and Hospitality Research*, *11*(3), 450–462. doi:10.1108/IJCTHR-03-2016-0032

Della Lucia, M., & Trunfio, M. (2018). The role of the private actor in cultural regeneration: Hybridizing cultural heritage with creativity in the city. *Cities (London, England)*, *82*, 35–44. doi:10.1016/j.cities.2018.05.003

Detotto, C., Giannoni, S., & Goavec, C. (2021). Does good governance attract tourists? *Tourism Management*, *82*, 104155. doi:10.1016/j.tourman.2020.104155

Diaconescu, D., M., Moraru, R., Stănciulescu, G. (2016). Considerations on Gastronomic Tourism as a Component of Sustainable Local Development. Amfiteatru Economic, 18(10), 999-1014.

Diamantopoulos, A., & Siguaw, J. A. (2000). *Introducing to Lisrel: A guide for the uninitiated.* Sage. doi:10.4135/9781849209359

Dollinger, M. J. (1990). The evolution of collective strategies in fragmented industries. *Academy of Management Review*, *15*(2), 266–285. doi:10.2307/258157

Dolnicar, S., & Fluker, M. (2003b). Who's riding the wave? An investigation into demographic and psychographic characteristics of surf tourists. *Proceedings of the 13th International Research Conference for the Council for Australian University Tourism and Hospitality Education (CAUTHE 2003).*

Dolnicar, S., & Fluker, M. (2003a). Behavioural market segments among surf tourists: Investigating past destination choice. *Journal of Sport & Tourism*, *8*(3), 186–196. doi:10.1080/14775080310001690503

dos Santos Júnior, A., Mendes Filho, L., Almeida García, F., & Simoes, J. M. M. (2017). Smart Tourism Destinations: a study based on the view of the stakeholders. *Revista Turismo Em Análise -RTA, 28*(3), 358–379.

Drakopoulou-Dodd, S., & Anderson, A. (2007). Mumpsimus and the mything of the individualistic entrepreneur. *International Small Business Journal*, *25*(4), 341–360.

Dredge, D. (1999). Destination place planning and design. *Annals of Tourism Research*, *26*(4), 772–791. doi:10.1016/S0160-7383(99)00007-9

Državni zavod za statistiku. (2011). *Stanovništvo u najvećim gradovima i općinama.* Retrieved September 13, 2020, Available at: www.dzs.hr

Du Cros, H., & McKercher, B. (2001). A New Model to Assist in Planning for Sustainable Cultural Heritage Tourism. *International Journal of Tourism Research*, *3*(2), 165–170. doi:10.1002/jtr.297

Duch, J., Waitzman, J. S., & Amaral, L. A. N. (2010). Quantifying the performance of individual players in a team activity. *PLoS One*, *5*(6), e10937. doi:10.1371/journal.pone.0010937 PMID:20585387

Dupeyras, A., & Maccallum, N. (2013). *Indicators for Measuring Competitiveness in Tourism - a Guidance Document.* OECD Tourism Papers. doi:10.1787/5k47t9q2t923-

Dwyer, L., Cvelbar, L. K., Mihalič, T., & Koman, M. (2014). Integrated destination competitiveness model: Testing its validity and data accessibility. *Tourism Analysis*, *19*(1), 1–17. doi:10.3727/108354214X13927625340073

Dwyer, L., Edwards, D., Mistilis, N., Roman, C., & Scott, N. (2009). Destination and enterprise management for a tourism future. *Tourism Management*, *30*(1), 63–74. doi:10.1016/j.tourman.2008.04.002

Dwyer, L., Gill, A., & Seetaram, N. (Eds.). (2012). *Handbook of research methods in tourism quantitative and qualitative approaches*. Edward Elgar. doi:10.4337/9781781001295

Dwyer, L., & Kim, C. (2003). Destination competitiveness: Determinants and indicators. *Current Issues in Tourism*, *6*(5), 369–414. doi:10.1080/13683500308667962

Dwyer, L., Mellor, R., Livaic, Z., Edwards, D., & Kim, C. (2004). Attributes of destination competitiveness: A factor analysis. *Tourism Analysis*, *9*(1-2), 91–101. doi:10.3727/1083542041437558

Eagly, A., & Chaiken, S. (1993). *The psychology of attitudes*. Harcourt Brace Jovanovich College Publishers.

EC (European Commission). (2014). *Statistical Data on Women Entrepreneurship in Europe*. Directorate-General for Enterprise and Industry. B-1049 Brussels.

Echtner, C., & Ritchie, J. (1993). The measurement of destination image: An empirical assessment. *Journal of Travel Research*, *31*(4), 3–13. doi:10.1177/004728759303100402

Edwards, D., Griffin, T., & Hayllar, B. (2008). Urban Tourism Research. *Annals of Tourism Research*, *35*(4), 1032–1052. doi:10.1016/j.annals.2008.09.002

Eguskitza, X., & Huey, R. B. (2000). Supplemental Oxygen and Mountaineering Deaths. *American Alpine Journal*, *42*, 135–138.

Eisenhardt, K., & Graebner, M. (2007). Theory Building From Cases: Opportunities And Challenges. *Academy of Management Journal*, *50*(1), 25–32. doi:10.5465/amj.2007.24160888

Elbe, J., Hallén, L., & Axelsson, B. (2009). The destination-management organisation and the integrative destination-marketing process. *International Journal of Tourism Research*, *11*(3), 283–296. doi:10.1002/jtr.695

El-Said, O., & Aziz, H. (2019). Egypt's Competitiveness: Empirical Examination of the Relationship Between Destination Attributes, Tourist Satisfaction, and Behavioral Intentions Among the Hotel Guests. *Tourism and Hospitality Management*, *25*(1), 25. doi:10.20867/thm.25.1.2

Errichiello, L., & Micera, R. (2015). Smart Tourism Destination Governance. In J. Spender, G. Schiuma, & V. Albino (Eds.), *FKAD 2015 International Forum on Knowledge Asset Dynamics. Culture, Innovation and Entrepreneurship: connecting the knowledge dots* (pp. 2179–2191). Bari: IKAM - Institution of Knowledge Asset Management.

Escach, N., & Vaudor, S. (2014). Réseaux de villes et processus de recomposition des niveaux: Le cas des villes baltiques. *Cybergeo: European Journal of Geography*, *679*, 1–24. doi:10.4000/cybergeo.26336

European Capital of Culture Application Manual. (n.d.). https://ec.europa.eu/programmes/creative-europe/sites/creative-europe/files/capitals-culture-candidates-guide_en_vdec17.pdf

European Capital of Culture information publication in English. (n.d.). https://ec.europa.eu/programmes/creative-europe/sites/creative-europe/files/library/capitals-culture-30-years-brochure_en.pdf

European Capital of Culture. (n.d.). https://ec.europa.eu/programmes/creative-europe/actions/capitals-culture_en

European Commission. (2013). *Pécs 2010, European Capital of Culture. Report.* https://ec.europa.eu/regional_policy/hu/projects/best-practices/hungary/2672

European Commission. (2016). *Towards a European Pillar of Social Rights.* Author.

European Commission. (2019). *Digital economy and society index 2019. Country report Slovakia.* Retrieved from https://ec.europa.eu/digital-single-market/en/news/digital-economy-and-society-index-desi-2019

European Commission. (2019). *The Cultural and Creative Cities Monitor.* Retrieved October 10, 2020, Available at: https://ec.europa.eu/jrc/en/publication/eur-scientific-and-technical-research-reports/cultural-and-creative-cities-monitor-2019-edition

Eusébio, C., Carneiro, M.J., Madaleno, M., Robaina M., Rodrigues, V., Russo, M., Relvas, H., Gama, C. Lopes, M., Seixas, V., Borrego, C. & Monteiro, A. (2020). The impact of air quality on tourism: a systematic literature review. *Journal of Tourism Futures.* doi:10.1108/JTF-06-2019-0049

Fadda, N. (2019). Tourism from water boardsports: Evidence from the emerging destination of Sardinia. *Journal of the Knowledge Economy, 10*(4), 1720–1737. doi:10.100713132-019-0587-8

Faida, R., Zabidi, A., & Jamaludin, M. (2018). The Validation of Halal Awareness and Halal Practice Instrument among Halal Restaurant Food Handlers. *Article in Asian Journal of Social Sciences & Humanities, 1*(3), 2600–9056. doi:10.26666/rmp.jssh.2018.3.1

Fakeye, P. C., & Crompton, J. L. (1991). Image differences between prospective, first-time, and repeat visitors to the lower Rio Grande valley. *Journal of Travel Research, 30*(2), 10–15. doi:10.1177/004728759103000202

Farajat, S. A., Liu, B., Pennington-Gray, L. (2017). Addressing travel writer's role as risk brokers: the case of Jordan. *Journal of Policy Research in Tourism Leisure and Events, 9*, 23-39. doi:.2016.1231194 doi:10.1080/19407963

Farmaki, A. (2015). Regional network governance and sustainable tourism. *Tourism Geographies, 17*(3), 385–407. doi:10.1080/14616688.2015.1036915

Farmaki, A., & Papatheodorou, A. (2015). Stakeholder perceptions of the role of low-cost carriers in insular tourism destinations: The case of Cyprus. *Tourism Planning & Development, 12*(4), 412–432. doi:10.1080/21568316.2015.1013566

Farzanegan, M. R., Gholipour, H. F., Feizi, M., Nunkoo, R., & Andargoli, A. E. (2020). International Tourism and Outbreak of Coronavirus (COVID-19): A Cross-Country Analysis. *Journal of Travel Research.*

Faulkner, B., & Russell, R. (1997). Chaos and complexity in tourism: In search of a new perspective. *Pacific Tourism Review, 1*(2), 93-102.

Ferdinand, N., & Williams, N. (2013). International festivals as experience production systems. *Tourism Management, 34*, 202–210. doi:10.1016/j.tourman.2012.05.001

Fernández, J. A. S., Azevedo, P. S., Martín, J. M. M., & Martín, J. A. R. (2020). Determinants of tourism destination competitiveness in the countries most visited by international tourists: Proposal of a synthetic index. *Tourism Management Perspectives, 33*, 100582. doi:10.1016/j.tmp.2019.100582

Fewell, J. H., Armbruster, D., Ingraham, J., Petersen, A., & Waters, J. S. (2012). Basketball teams as strategic networks. *PLoS One, 7*(11), e47445. doi:10.1371/journal.pone.0047445 PMID:23139744

Film picnic 2020 program and information. (n.d.). www.veszprembalaton2023.hu/hu/events/filmpiknik

Finkelstein, J. (1989). *Dining Out: A Sociology of Modern Manners.* Polity Press.

Fishbein, M., & Ajzen, I. (1975). *Beliefs, Attitude, and intention behavior.* Addison-Wesley Publishing Company.

Florida, R. (2003). Cities and the creative class. *City & Community, 2*(1), 3–19. doi:10.1111/1540-6040.00034

Floridi, L. (2010). Ethics after the information revolution. In L. Floridi (Ed.), *Cambridge Handbook of Information and Computer Ethics* (pp. 3–19). Cambridge University Press. doi:10.1017/CBO9780511845239.002

Foo, R., & Krishnapillai, G. (2018). Preserving the Intangible Living Heritage in the George Town World Heritage Site, Malaysia. *Journal of Heritage Tourism*, 1–13.

Fornell, C., & Larcker, D. F. (1981). Evaluating structural equation models with unobservable variables and measurement error. *JMR, Journal of Marketing Research, 18*(1), 39–50. doi:10.1177/002224378101800104

Fortezza, F., & Pencarelli, T. (2018). A comprehensive picture of the social media challenge for DMOs. *Anatolia, 29*(3), 456–467. doi:10.1080/13032917.2018.1439759

Fox, D., Edwards, J., & Wilkes, K. (2010). Employing the Grand Tour approach to aid understanding of garden visiting. In G. Richards & W. Munsters (Eds.), *Cultural tourism research methods* (pp. 75–86). CABI Publishing. doi:10.1079/9781845935184.0075

Framke, W. (2002). The Destination as a Concept: A Discussion of the Business-Related Perspective versus the Socio-Cultural Approach in Tourism Theory. *Scandinavian Journal of Hospitality and Tourism, 2*(2), 92–108. doi:10.1080/15022250216287

Frank, F. (2015). Environmental awareness of surf tourists: A case study in the Algarve. *Journal of Spatial and Organizational Dynamics, 3*(2), 102–113.

Frans, P. (2011). The Protection of Working Relationships in the Netherlands. In The Protection of Working Relationships: A Comparative Study. Leiden: Kluwer Law International.

Fransen, K., Van Puyenbroeck, S., Loughead, T. M., Vanbeselaere, N., De Cuyper, B., Vande Broek, G., & Boen, F. (2015). Who takes the lead? Social network analysis as a pioneering tool to investigate shared leadership within sports teams. *Social Networks, 43*, 28–38. doi:10.1016/j.socnet.2015.04.003

Freeman, L. C. (1979). Centrality in Social Networks. *Social Networks, 1*(3), 215–239. doi:10.1016/0378-8733(78)90021-7

Fuchs, M., Höpken, W., & Lexhagen, M. (2014). Big data analytics for knowledge generation in tourism destinations – A case from Sweden. *Journal of Destination Marketing & Management, 3*(4), 198–209. doi:10.1016/j.jdmm.2014.08.002

Fuentes, Z. (2019). *Environment as a Key Factor of Health and Well-Being Tourism Destinations in Five European Countries*. Academic Press.

Fusté-Forné, F. (2018). Drawing a gastronomy landscape from food markets' produce. *International Journal of Culture, Tourism and Hospitality Research, 12*(3), 378–384. doi:10.1108/IJCTHR-05-2018-0064

Fyall, A., & Garrod, B. (2019). Destination management: A perspective article. *Tourism Review*.

Gajdošík, T. (2019a). Big Data Analytics in Smart Tourism Destinations. A New Tool for Destination Management Organizations? In V. Katsoni & M. Segarra-Oña (Eds.), *Smart Tourism as a Driver for Culture and Sustainability* (pp. 15–33). doi:10.1007/978-3-030-03910-3_2

Gajdošík, T. (2019b). Towards a Conceptual Model of Intelligent Information System for Smart Tourism Destinations. In R. Silhavy (Ed.), *Software Engineering and Algorithms in Intelligent Systems* (pp. 66–74). doi:10.1007/978-3-319-91186-1_8

Gajdošík, T. (2020). Smart tourists as a profiling market segment: Implications for DMOs. *Tourism Economics*, *26*(6), 1042–1062. doi:10.1177/1354816619844368

Gajdošík, T., Gajdošíková, Z., Maráková, V., & Borseková, K. (2017). Innovations and networking fostering tourist destination development in Slovakia. *Quaestiones Geographicae*, *36*(4), 103–115. Advance online publication. doi:10.1515/quageo-2017-0039

Gajdošík, T., Gajdošíková, Z., Maráková, V., & Flagestad, A. (2017). Destination structure revisited in view of the community and corporate model. *Tourism Management Perspectives*, *24*, 54–63. Advance online publication. doi:10.1016/j.tmp.2017.07.009

Gamarra, N. C., Correia, R. A., Bragagnolo, C., Campos-Silva, J. V., Jepson, P. R., Ladle, R. J., & Mendes Malhado, A. C. (2019). Are Protected Areas undervalued? An asset-based analysis of Brazilian Protected Area Management Plans. *Journal of Environmental Management*, *249*(July), 109347. doi:10.1016/j.jenvman.2019.109347 PMID:31419672

Gani, A. A., Mahdzar, M., Abdullah, N., & Muhammad, R. (2017). *Linking Image and Satisfaction of Food Tourism in Penang, Malaysia*. Academic Press.

Gaoa, J., Linb, S., & Zhang, C. (2020). Authenticity, involvement, and nostalgia: Understanding visitor satisfaction with an adaptive reuse heritage site in urban China. *Journal of Destination Marketing & Management*, *15*, 100404. doi:10.1016/j.jdmm.2019.100404

Gao, J., Ryan, C., Cave, J., & Zhang, C. (2019). Tourism border-making: A political economy of China's border tourism. *Annals of Tourism Research*, *76*, 1–13. doi:10.1016/j.annals.2019.02.010

Garcês, S., Pocinho, M., & Jesus, S. N. D. (2018). Review of optimism, creativity and spirituality in tourism research. *Tourism and Hospitality Management*, *24*(1), 107–117. doi:10.20867/thm.24.1.6

García, M. C. D., & Welter, F. (2013). Gender identities and practices: Interpreting women entrepreneurs' narratives. *International Small Business Journal*, *31*(4), 384–404.

Ge, J., & Hokao, K. (2006). Research on residential lifestyles in Japanese cities from the viewpoints of residential preference, residential choice and residential satisfaction. *Landscape and Urban Planning*, *78*(3), 165–178. doi:10.1016/j.landurbplan.2005.07.004

Geng-Qing Chi, C., Pan, L., & Del Chiappa, G. (2018). Examining destination personality: Its antecedents and outcomes. *Journal of Destination Marketing & Management*, *9*, 149–159. doi:10.1016/j.jdmm.2018.01.001

Ghaderi, E., & Aarabi, S. S. (2020). *Urban Tourism Competitiveness in Iran*. Applying Dwyer's Integrated Model.

Ghosh, T., & Sarkar, A. (2015). "To feel a place of heaven": Examining the role of sensory reference cues and capacity for imagination in destination marketing. *Journal of Travel & Tourism Marketing, 33*(sup1), 25–37. . doi:10.1080/10548408.2014.997962

Giagou, D., & Apostolopoulos, C. (1996). Rural Women and the Development of Agrotouristic Co-operatives in Greece: The case of Petra, Lesvos. *Journal of Rural Cooperation*, *1996*(24), 143–155.

Gibson, L., Lynch, P. A., & Morrison, A. (2005). The local destination tourism network: Development issues. *Tourism and Hospitality Planning & Development*, *2*(2), 87–99. doi:10.1080/14790530500171708

Gill, A. M., & Williams, P. (2011). Rethinking resort growth: Understanding evolving governance strategies in Whistler, British Columbia. *Journal of Sustainable Tourism*, *19*(4/5), 629–648. doi:10.1080/09669582.2011.558626

Gorsuch, R. (1983). *Factor analysis*. Lawrence Erlbaum.

Government No 1468/2020 (VIII. 5) Korm.'s Decision on measures relating to the title of European Capital of Culture 2023. Hungarian Gazette 2020 Issue 183: https://magyarkozlony.hu/dokumentumok/07d96a0272f2a6c08461de0ecbf7bcb2abd31aa7/megtekintes

Govers, R., & Go, F. (2009). *Place branding: Glocal, virtual and physical identities, constructed, imagined and experienced.* Palgrave Macmillan. doi:10.1007/978-0-230-24702-4

Gray, S.R., & Benson, P.G. (2003). Determinants of executive compensation in small business development centers. *Nonprofit Management & Leadership, 13*(3).

Green, G.P., & Dougherty, M.L. (2008). Localizing linkages for food and tourism: Culinary tourism as a community development strategy. *Community Development Journal: An International Forum, 39*(3), 148–158.

Gretzel, U. (2011). Intelligent systems in tourism: A Social Science Perspective. *Annals of Tourism Research, 38*(3), 757–779. doi:10.1016/j.annals.2011.04.014

Gretzel, U., Ham, J., & Koo, C. (2018). Creating the City Destination of the Future: The Case of Smart Seoul. In Y. Wang, A. Shakeela, A. Kwek, & C. Khoo-Lattimore (Eds.), *Managing Asian Destinations* (pp. 199–214). Perspectives on Asian Tourism. doi:10.1007/978-981-10-8426-3_12

Gretzel, U., & Scarpino-Johns, M. (2018). Destination resilience and smart tourism destinations. *Tourism Review International, 22*(3), 263–276. doi:10.3727/154427218X15369305779065

Grigoroudis, E., & Siskos, Y. (2010). Customer Satisfaction Evaluation: Methods for Measuring and Implementing Service Quality. Springer.

Grosz, E. (1995). *Space, time and perversion: essays on the politics of bodies.* Routledge.

Gruen, T. W., Osmonbekov, T., & Czaplewski, A. J. (2006). EWOM: The impact of customer-to-customer online know how exchange on customer value and loyalty. *Journal of Business Research, 59*(4), 449–456. doi:10.1016/j.jbusres.2005.10.004

Guindi, M. N., Flaherty, G. T., & Byrne, M. (2018). Every breath you take: How does air pollution affect the international traveller? *Journal of Travel Medicine, 25*(1), tay021. doi:10.1093/jtm/tay021 PMID:29718402

Gulati, R. (1998). Alliances and networks. *Strategic Management Journal, 19*(4), 293–317. doi:10.1002/(SICI)1097-0266(199804)19:4<293::AID-SMJ982>3.0.CO;2-M

Gunn, C. A. (1972). *Vacation scape: Designing Tourist Regions.* Taylor & Francis.

Guttentag, D. A., & Smith, S. L. (2017). Assessing Airbnb as a disruptive innovation relative to hotels: Substitution and comparative performance expectations. *International Journal of Hospitality Management, 64*, 1–10. doi:10.1016/j.ijhm.2017.02.003

Gyimothy, S., Rassing, C., & Wanhill, S. (2000). Marketing Works: A Study of Restaurants on Bornholm. Denmark. *International Journal of Contemporary Hospitality Management, 12*, 371–379.

Hadley, M. B., Vedanthan, R., & Fuster, V. (2018). Air pollution and cardiovascular disease: A window of opportunity. *Nature Reviews. Cardiology, 15*(4), 193–194. doi:10.1038/nrcardio.2017.207 PMID:29297510

Hair, J. F., Black, W. C., Babin, B. J., Anderson, R. E., & Tatham, R. L. (2006). Multivariate Data Analysis (6th ed.). Prentice Hall.

Hair, J. F., Anderson, R. E., Tatham, R. L., & Black, W. C. (2008). *Multivariate data analysis* (7th ed.). Pearson Prentice-Hall.

Halkos, G., & Matsiori, S. (2012). Determinants of willingness to pay for coastal zone quality improvement. *Journal of Socio-Economics*, *41*(4), 391–399. doi:10.1016/j.socec.2012.04.010

Hall, C., & Sharples, L. (2003). The consumption of experiences or the experience of consumption? An introduction to the tourism of taste. *Food Tourism Around the World*, 1–24. . doi:10.1016/B978-0-7506-5503-3.50004-X

Hall, C. M. (2008). *Tourism Planning. Policies, Processes and Relationship*. Pearson-Prentice Hall.

Hall, C. M. (2019). Constructing sustainable tourism development: The 2030 agenda and the managerial ecology of sustainable tourism. *Journal of Sustainable Tourism*, *27*(7), 1044–1060. doi:10.1080/09669582.2018.1560456

Hall, C., M., & Mitchell, R. (2000). Wine Tourism in the Mediterranean: A Tool for Restructuring and Development. *Thunderbird International Business Review*, *42*(4), 445–465.

Hällgren, M. (2010). Groupthink in Temporary Organizations. *International Journal of Managing Projects in Business*, *3*(1), 94–110. doi:10.1108/17538371011014044

Hallowell, R., Schlesinger, L. A., & Zornitsky, J. (1996). Internal Service Quality, Customer and Job Satisfaction: Linkages and Implications for Management. *Human Resource Planning*, *19*, 20–31.

Hall, T., & Barrett, H. (2018). Planning, regeneration and urban policy. In *Urban Geography* (pp. 122–155). Routledge. doi:10.4324/9781315652597-6

Hambrick, M. E., Svensson, P. G., & Kang, S. (2018). Using social network analysis to investigate interorganizational relationships and capacity building within a sport for development coalition. *Sport Management Review*. Advance online publication. doi:10.1016/j.smr.2018.12.002

Hammami, A. M., Stanton, J. L., Periklis, D., Baourakis, G., van Dijk, G., & Mamalis, S. (2018). Tourist's Behaviour Towards Local Cretan Food. *Sustainable Agriculture and Food Security*, 49-61.

Hanna, P., Font, X., Scarles, C., Weeden, C., & Harrison, C. (2018). Tourist destination marketing: From sustainability myopia to memorable experiences. *Journal of Destination Marketing & Management*, *9*, 36–43. doi:10.1016/j.jdmm.2017.10.002

Hartley, J. (2012). *Key Concepts in Creative Industries*. Sage.

Hassan, A. S., & Ali, A. (2019). *Colonial influence on Indian Muslim religious monuments in Penang*. Academic Press.

Hegarty, J. A., & O'Mahony, G. B. (2001). Gastronomy: A phenomenon of cultural expressionism and an aesthetic for living. *International Journal of Hospitality Management*, *20*(1), 3–13. doi:10.1016/S0278-4319(00)00028-1

Heidegger, M. (1996). *Being and Time*. State University of New York Press.

He, M., Li, J., Li, J., & Chen, H. (2018). A comparative study on the effect of soundscape and landscape on tourism experience. *International Journal of Tourism Research*. Advance online publication. doi:10.1002/jtr.2237

Henderson, J. C. (2008). Managing urban ethnic heritage: Little India in Singapore. *International Journal of Heritage Studies*, *14*(4), 332–346. doi:10.1080/13527250802155851

Hennig-Thurau, T., Gwinner, K., Walsh, G., & Gremler, D. (2004). Electronic word-of-mouth via consumer-opinion platforms: What motivates consumers to articulate themselves on the Internet? *Journal of Interactive Marketing*, *18*(1), 38–52. doi:10.1002/dir.10073

Henshaw, V. (2013). *Urban smellscapes: Understanding and designing city smell environments*. Routledge. doi:10.4324/9780203072776

Henshaw, V., Medway, D., Warnaby, G., & Perkins, C. (2015). Marketing the "city of smells.". *Marketing Theory*, *16*(2), 153–170. doi:10.1177/1470593115619970

Herring, H., & Roy, R. (2007). Technological innovation, energy efficient design and the rebound effect. *Technovation*, *27*(4), 194–203. doi:10.1016/j.technovation.2006.11.004

Heskett, J. L., Jones, T. O., Loveman, G. W., Sasser, W. E., & Schlesinger, L. A. (1994). Putting the service–profit chain to work. *Harvard Business Review*, *72*(2), 164–175.

Hess, T., Matt, C., & Wiesböck, F. (2016). Options for formulating a digital transformation strategy. *MISQE*, *105447*. Advance online publication. doi:10.7892/boris

Hieu, V.M., & Rasovska, I. (2017). Craft villages and tourism development, a case study in Phu Quoc island of Vietnam. *Management*, *21*(1), 223–236.

Higgs, M., Plewnia, U., & Ploch, J. (2005). Influence of team composition and task complexity on team performance. *Team Performance Management*, *11*(7/8), 227–250. doi:10.1108/13527590510635134

Hill, M., & Hill, A. (2008). *Investigação por questionário* (2nd ed.). EdiçõesSílabo.

Hinings, B., Gegenhuber, T., & Greenwood, R. (2018). Digital innovation and transformation: An institutional perspective. *Information and Organization*, *28*(1), 52–61. doi:10.1016/j.infoandorg.2018.02.004

Hirsch, A. R. (1995). Effects of ambient odors on slot-machine usage in a las vegas casino. *Psychology and Marketing*, *12*(7), 585–594. doi:10.1002/mar.4220120703

Hong, J. Y., & Jeon, J. Y. (2015). Influence of urban contexts on soundscape perceptions: A structural equation modeling approach. *Landscape and Urban Planning*, *141*, 78–87. doi:10.1016/j.landurbplan.2015.05.004

Hovelynck, J., Craps, M., Dewulf, A., Sips, K., Taillieu, T., & Bouwen, R. (2020). Relational Practices for Generative Multi-Actor Collaboration. The Sage Handbook of Social Constructionist Practice, 258.

Hoven, B. van. (2011). Multi-sensory tourism in the Great Bear Rainforest. *Landabréfið*, 31–49.

Hristov, D. (2015). Investigating DMOs through the lens of social network analysis: Theoretical gaps, methodological challenges and practitioner perspectives. *Advances in Hospitality and Tourism Research*, *3*(1), 18–39.

Hristov, D., & Zehrer, A. (2015). The destination paradigm continuum revisited: DMOs serving as leadership networks. *Tourism Review*, *70*(2), 116–131. doi:10.1108/TR-08-2014-0050

Hrvatski klaster konkurentnosti kreativnih i kulturnih industrija. (2013). Retrieved April 19, 2020, Available at: https://www.udruga-gradova.hr/wordpress/wp-content/uploads/2016/07/5.-HKKKI.pdf

Huang, S., & Hus, K. (2003). Determinants of User Intention toward IT Instruction: An Examination of Internal and External Factors. Knowledge Management and E-Learning. *International Journal (Toronto, Ont.)*, 1.

Huey, R. B., Salisbury, R., Wang, J.-L., & Mao, M. (2007). Effects of age and gender on success and death of mountaineers on Mount Everest. *Biology Letters*, *3*(3), 498–500. doi:10.1098/rsbl.2007.0317 PMID:17698450

Hultén, B. (2011). Sensory marketing: The multi-sensory brand-experience concept. *European Business Review*, *23*(3), 256–273. doi:10.1108/09555341111130245

Hu, W., & Wall, G. (2005). Environmental Management, Environmental Image and the Competitive Tourist Attraction. *Journal of Sustainable Tourism*, *13*(6), 617–635. doi:10.1080/09669580508668584

Hwang, Y. H., Gretzel, U., & Fesenmaier, D. R. (2006). Multicity trip patterns. Tourists to the United States. *Annals of Tourism Research*, *33*(4), 1057–1078. doi:10.1016/j.annals.2006.04.004

Iliopoulou-Georgudaki, J., Kalogeras, A., Konstantinopoulos, P., & Theodoropoulos, C. (2016). Sustainable tourism management and development of a Greek coastal municipality. *International Journal of Sustainable Development and World Ecology*, *23*(2), 143–153. doi:10.1080/13504509.2015.1102780

Imran, S., Alam, K., & Beaumont, N. (2014). Environmental orientations and environmental behaviour: Perceptions of protected area tourism stakeholders. *Tourism Management*, *40*, 290–299. doi:10.1016/j.tourman.2013.07.003

Ineson, E., Benke, E., & László, J. (2013). Employee loyalty in Hungarian hotels. *International Journal of Hospitality Management*, *32*(1), 31–39. doi:10.1016/j.ijhm.2012.04.001

Inman, J. J., & Nikolova, H. (2017). Shopper-facing retail technology: A retailer adoption decision framework incorporating shopper attitudes and privacy concerns. *Journal of Retailing*, *93*(1), 7–28. doi:10.1016/j.jretai.2016.12.006

International Surfing Association. (2014). *ISA 50th Anniversary World Surfing Games*. https://www.isasurf.org/international-surfing-association-celebrate-50-years-success-isa-50th-anniversary-world-surfing-games-punta-rocas-peru/

Ioannides, D., Röslmaier, M., & van der Zee, E. (2018). Airbnb as an instigator of 'tourism bubble' expansion in Utrecht's Lombok neighbourhood. *Tourism Geographies*, 1–19. doi:10.1080/14616688.2017.1409261

Isa, M., Goh, A. T., Abdul Majid, H., Yuen, A., Abdul Rahman, F., Shahidan, S. I., ... Sharuddin, S. N. (2020). *The Kuala Lumpur Heritage Agenda: Reviews and Recommendations for the Kuala Lumpur Draft Structure Plan 2040 & Kuala Lumpur City Plan 2020 by ICOMOS Malaysia*. Academic Press.

Islam, M. W., Ruhanen, L., & Ritchie, B. W. (2018). Adaptive co-management: A novel approach to tourism destination governance? *Journal of Hospitality and Tourism Management*, *37*, 97–106. doi:10.1016/j.jhtm.2017.10.009

ISO/IEC. (2013). *Information technology – Security techniques – Information security management systems – Requirements*. American National Standards Institute.

Ivanov, S. (2019). Ultimate transformation: How will automation technologies disrupt the travel, tourism and hospitality industries? *Zeitschrift fur Tourismuswissenschaft*, *11*(1), 25–43.

Ivona, A., & Privitera, D. (2019). Places and Religious Bands explorations in spiritual tourism. *International Journal of Religious Tourism and Pilgrimage*, *7*(4), 54–63.

Jalilvand, M. R., Ebrahimi, A., & Samiei, N. (2013). Electronic Word of Mouth Effects on Tourists' Attitudes Toward Islamic Destinations and Travel Intention: An Empirical Study in Iran. *Procedia: Social and Behavioral Sciences*, *81*, 484–489. doi:10.1016/j.sbspro.2013.06.465

Jalilvand, M., & Samiei, N. (2012). The effect of electronic word of mouth on brand image and purchase intention.: An empirical study in the automobile industry in Iran. *Marketing Intelligence & Planning*, *30*(4), 460–476. doi:10.1108/02634501211231946

Jamal, T., & Jamrozy, U. (2006). Collaborative networks and partnerships for integrated destination management. In D. Buhalis & C. Costa, C. (Eds.), Tourism management dynamics: Trends, management and tools (pp. 164-172). Elsevier Butterworth-Heinemann. doi:10.1016/B978-0-7506-6378-6.50028-7

Jamal, T. B., & Getz, D. (1995). Collaboration theory and community tourism planning. *Annals of Tourism Research*, *22*(1), 186–204. doi:10.1016/0160-7383(94)00067-3

Jani, D., & Nguni, W. (2016). Pre-trip vs. post-trip destination image variations: A case of inbound tourists to Tanzania. *Original Scientific Paper*, *64*, 27–40.

Jansen-Verbeke, M. (2009). The territoriality paradigm in cultural tourism. *Tourism (Zagreb)*, *19*(1-2), 25–31.

Jia, Y., Ma, H., Kang, J., & Wang, C. (2020). The preservation value of urban soundscape and its determinant factors. *Applied Acoustics*, *168*, 107430. doi:10.1016/j.apacoust.2020.107430

John, L. K., Acquisti, A., & Loewenstein, G. (2010). Strangers on a plane: Context-dependent willingness to divulge sensitive information. *The Journal of Consumer Research*, *37*(5), 858–873.

Jolly, J. (2010, August 28). *Elizabeth Hawley, unrivalled Himalayan record keeper*. Retrieved March 24, 2015, from BBC News: https://www.bbc.com/news/world-south-asia-10268549

Jones, T. E., Apollo, M., & Bui, H. T. (2021). Mountainous Protected Areas & Nature-Based Tourism in Asia. In *Nature-Based Tourism in Asia's Mountainous Protected Areas* (pp. 3–25). Springer. doi:10.1007/978-3-030-76833-1_1

Joppe, M. (2018). Tourism policy and governance: Quo vadis? *Tourism Management Perspectives*, *25*, 201–204. doi:10.1016/j.tmp.2017.11.011

Jöreskog, K. G., & Sörbom, D. (2004). *LISREL 8.7 for Windows* [Computer Software]. Scientific Software International, Inc.

Kabat-Zinn, J. (2014). Tastescape. *Mindfulness*, *5*(1), 102–104. doi:10.100712671-013-0274-y

Kapera, I. (2018). Sustainable tourism development efforts by local governments in Poland. *Sustainable Cities and Society*, *40*, 581–588. doi:10.1016/j.scs.2018.05.001

Kastenholz, E., Davis, D., & Paul, G. (1999). Segmenting Tourism in Rural Areas: The Case of North and Central Portugal. *Journal of Travel Research*, *37*, 353–363.

Katsios, S. (2019). Tourism in 'Yellow Times': The De-formalisation of the Greek Economy and Its Impact on Tourism. In *Yellow Tourism* (pp. 209–225). Springer. doi:10.1007/978-3-319-94664-1_13

Kavaratzis, M., & Hatch, M. J. (2013). The dynamics of place brands: An identity-based approach to place branding theory. *Marketing Theory*, *13*(1), 13. doi:10.1177/1470593112467268

Keller, K. L. (1993). Conceptualizing, Measuring, and Managing Customer-Based Brand Equity. *Journal of Marketing*, *57*(1), 1–22. doi:10.1177/002224299305700101

Kercher, M., Okumus, F., & Okumus, B. (2008). Food Tourism as a Viable Market Segmet: It's All How You Cook the Numbers. *Journal of Travel and Tourism Marketing, Volume*, *25*(2), 137–146.

Khalid, N. S., Othman, R. R., Abdullah, I. C., Marzukhi, M. A., & Nasrudin, N. (2019, November). The Regaining of Traditional Street: Combining Street Network Configuration with Temporary Urban Interventions. *IOP Conference Series. Earth and Environmental Science*, *385*(1), 012019. doi:10.1088/1755-1315/385/1/012019

Khalid, S., Ahmad, M. S., Ramayah, T., Hwang, J., & Kim, I. (2019). Community empowerment and sustainable tourism development: The mediating role of community support for tourism. *Sustainability*, *11*(22), 6248. doi:10.3390u11226248

Khan, S. (2015). Otherness" of ethnic enclave attractions in multicultural cities: A study of Chinatown and Little India. *Asia-Pacific Journal of Innovation in Hospitality and Tourism*, *4*(1), 63–76. doi:10.760340930-015-0004-8

Khoo-Lattimore, C., Mura, P., & Yung, R. (2019). The time has come: A systematic literature review of mixed methods research in tourism. *Current Issues in Tourism*, *22*(13), 1531–1550. doi:10.1080/13683500.2017.1406900

Khoo, S. L. (2021). *Creative City as an Urban Development Strategy: The Case of Selected Malaysian Cities*. Springer Nature. doi:10.1007/978-981-16-1291-6

Khurramov, O. K. (2020). The role of the tourism sector in the digitalization of the service economy. *Economics and Innovative Technologies*, *2020*(1), 6.

Kianicka, S., Buchecker, M., Hunziker, M., & Müller-Böker, U. (2006). Locals and Tourists Sense of Place. *Mountain Research and Development*, *26*(1), 55–63. doi:10.1659/0276-4741(2006)026[0055:LATSOP]2.0.CO;2

Kim, J. J., & Fesenmaier, D. R. (2016). Measuring Human Senses and the Touristic Experience: Methods and Applications. *Analytics in Smart Tourism Design Tourism on the Verge*, 47–63. doi:10.1007/978-3-319-44263-1_4

Kim, H., Rhou, Y., Uysal, M., & Kwon, N. (2017). An examination of the links between corporate social responsibility (CSR) and its internal consequences. *International Journal of Hospitality Management*, *61*, 26–34. doi:10.1016/j.ijhm.2016.10.011

Kim, H., & Richardson, L. (2003). Motion picture impacts on destination images. *Annals of Tourism Research*, *30*(1), 216–237. doi:10.1016/S0160-7383(02)00062-2

Kim, J. (2018). Manhattan's Koreatown as a Transclave: The Emergence of a New Ethnic Enclave in a Global City. *City & Community*, *17*(1), 276–295. doi:10.1111/cico.12276

Kim, J. J., & Fesenmaier, D. R. (2017). Sharing Tourism Experiences: The post-trip experience. *Journal of Travel Research*, *56*(1), 28–40. doi:10.1177/0047287515620491

Kim, S., & Jun, J. (2016). The impact of event advertising on attitudes and visit intentions. *Journal of Hospitality and Tourism Management*, *29*, 1–8. doi:10.1016/j.jhtm.2016.04.002

Kim, S., Park, E., & Lamb, D. (2019). Extraordinary or ordinary? Food tourism motivations of Japanese domestic noodle tourists. *Tourism Management Perspectives*, *29*, 176–186.

Kim, S., Stylidis, D., & Oh, M. (2019). Is perception of destination image stable or does it fluctuate? A measurement of three points in time. *International Journal of Tourism Research*, *21*(4), 447–461. doi:10.1002/jtr.2273

Kim, Y. G., Eves, A., & Scarles, C. (2009). Building a model of local food consumption on trips and holidays: A grounded theory approach. *International Journal of Hospitality Management*, *28*(3), 423–431. doi:10.1016/j.ijhm.2008.11.005

Kinayoglu, G. (2009). Using Audio-Augmented Reality to Assess the Role of Soundscape in Environmental Perception: An Experimental Case Study on the UC Berkeley Campus. *Proceedings: ECAADe 2009 International Conference on Education and Research in Computer Aided Architectural Design in Europe*.

King, R. (2011). *Transforming brickfields: Development and governance in a Malaysian city Houses in Motion: The Experience of Place and the Problem of Belief in Urban Malaysia*. Academic Press.

Kivela, J., & Johns, N. (2002). A mythological approach in analyzing dining out experiences: Implications for remaking of tourist destinations. *An International Tourism Research Conference*.

Klepers, A., & Rozite, M. (2010). Administrative border as a challenge in spatial aspects of tourism competitiveness. In A. Clarke (Ed.), *Constructing Central Europe: Tourism Competitiveness* (pp. 23–33). University Press, University of Pannonia.

Kline, R. (2011). Principles and practice of structural equation modeling (3rd ed.). New York: The Guilford Press.

Knecht, S. (2017). Exploring different levels of stakeholder activity in international institutions: Late bloomers, regular visitors, and overachievers in Arctic Council Working Groups. In *Governing Arctic Change* (pp. 163–185). Palgrave Macmillan. doi:10.1057/978-1-137-50884-3_9

Koch, T., & Windsperger, J. (2017). *Seeing through the network: competitive advantage in the digital economy. J Organ Des.* doi:10.118641469-017-0016-z

Koh, K. Y., & Hatten, T. S. (2002). The tourism entrepreneur. The overlooked player in tourism development studies. *International Journal of Hospitality & Tourism Administration*, *3*(1), 21–48. doi:10.1300/J149v03n01_02

Korsgaard, S., Ferguson, R., & Gaddefors, J. (2015). The best of both worlds: How rural entrepreneurs use placial embeddedness and strategic networks to create opportunities. *Entrepreneurship and Regional Development*, *27*(9-10), 574–598.

Korte, F., & Lames, M. (2018). Characterizing different team sports using network analysis. *Current Issues in Sport Science, 3*(5). doi:10.15203/CISS_2018.005

Koutoulas, D. (2019). *Assignment of study - masterplan for the Cultural Route of Naxos.* Stadium.

Kovačić, S., Jovanović, T., & Šagovnović, I. (2019). Exploring the gap in destination image and destination personality perception between tourists and stakeholders: Case of Roman heritage sites in Serbia. *Turizam*, *23*(3), 133–144. doi:10.5937/turizam23-22773

Kővári, E., Pásztor, J., & Raffay-Danyi, Á. (2021). A pannon egyetem közösségének kultúrafogyasztása, Az érzelmi és kulturális intelligencia összefüggése – veszprém 2023 európa kulturális fővárosa. *Vezetéstudomány,_52*(4), 48-62. doi:10.14267/VEZTUD.2021.04.05

Ko, Y. J., & Pastore, D. L. (2004). Current issues and conceptualizations of service quality in the recreation sport industry. *Sport Marketing Quarterly*, *13*(2), 158–166.

Krakauer, J. (1997). *Into thin air: A personal account of the Mt. Everest disaster*. Villard.

Krishna, A. (2010). *Sensory marketing: Research on the sensuality of products.* Routledge.

Krivacic, D., & Jankovic, S. (2017). Managerial attitudes on environmental reporting: Evidence from Croatia. *Journal of Environmental Accounting and Management*, *5*(4), 327–341. doi:10.5890/JEAM.2017.12.005

Krstinić Nižić, M., Rudan, E., & Trinajstić, M. (2019). The role of creative cities in regional development. *Poslovna Izvrsnost*, *13*(1), 35–55. doi:10.22598/pi-be/2019.13.1.35

Kruger, M., & Saayman, M. (2017). Sand, sea and surf: Segmenting South African surfers. *S.A. Journal for Research in Sport Physical Education and Recreation*, *39*(2), 115–132.

Kucukusta, D., Mak, A., & Chan, X. (2013). Corporate social responsibility practices in four and five-star hotels: Perspectives from Hong Kong visitors. *International Journal of Hospitality Management*, *34*, 19–30. doi:10.1016/j.ijhm.2013.01.010

Kumar, V. (2020). *Morality in the Making of Urban Space* (Doctoral dissertation). State University of New York at Buffalo.

Kumar, S., Valeri, M., & Shekhar. (2021, August 13). Understanding the relationship among factors influencing rural tourism: A hierarchical approach. *Journal of Organizational Change Management.* Advance online publication. doi:10.1108/JOCM-01-2021-0006

Kurtz, J. (2017). *The happy traveler: Unpacking the secrets of better vacations.* Oxford University Press.

Kyrylov, Y., Hranovska, V., Boiko, V., Kwilinski, A., & Boiko, L. (2020). International Tourism Development in the Context of Increasing Globalization Risks: On the Example of Ukraine's Integration into the Global Tourism Industry. *Journal of Risk and Financial Management*, *13*(12), 303. doi:10.3390/jrfm13120303

Lachmann, L. (1977). *The Market as an Economic Process*. Basil Blackwell.

Ladhar, R., & Michaud, M. (2015). EWOM effects on hotel booking intentions, attitudes, trust, and website perceptions. *International Journal of Hospitality Management*, *46*, 36–45. doi:10.1016/j.ijhm.2015.01.010

Lai, P. H., Hsu, Y. C., & Wearing, S. (2016). A social representation approach to facilitating adaptive co-management in mountain destinations managed for conservation and recreation. *Journal of Sustainable Tourism*, *24*(2), 227–244. do i:10.1080/09669582.2015.1062018

Lalicic, L., & Önder, I. (2018). Residents' involvement in urban tourism planning: Opportunities from a smart city perspective. *Sustainability (Switzerland)*, *10*(6), 1852. Advance online publication. doi:10.3390u10061852

Lam, C., & Law, R. (2019). Readiness of upscale and luxury-branded hotels for digital transformation. *International Journal of Hospitality Management*, *79*, 60–69. doi:10.1016/j.ijhm.2018.12.015

Lam, T., Zhang, H., & Baum, T. (2001). An investigation of employee's job satisfaction: The case of hotels in Hong Kong. *Tourism Management*, *22*(2), 157–165. doi:10.1016/S0261-5177(00)00039-X

Larsen, J. (2001). Tourism Mobilities and the Travel Glance: Experiences of Being on the Move. *Scandinavian Journal of Hospitality and Tourism*, *1*(2), 80–98. doi:10.1080/150222501317244010

Lau, C., & Li, Y. (2019). Analyzing the effects of an urban food festival: A place theory approach. *Annals of Tourism Research*, *74*, 43–55. doi:10.1016/j.annals.2018.10.004

Laws, E. (2011). *Tourist destination governance: Practice, theory and issues*. Cabi. doi:10.1079/9781845937942.0000

Lazarow, N., Miller, M. L., & Blackwell, B. (2008). The value of recreational surfing to society. *Tourism in Marine Environments*, *5*(2–3), 145–158. doi:10.3727/154427308787716749

Lee, H. A., Denizci Guillet, B., & Law, R. (2013). An examination of the relationship between online travel agents and hotels: A case study of Choice Hotels International and Expedia.com. *Cornell Hospitality Quarterly*, *54*(1), 95–107. doi:10.1177/1938965512454218

Lee, T. H., Chao, W. H., & Lin, H.-Y. (2018). Cultural inheritance of Hakka cuisine: A perspective from tourists' experiences. *Journal of Destination Marketing & Management*, *7*, 101–111. doi:10.1016/j.jdmm.2016.09.006

Lefebvre, H. (1991). *The Production of Space*. Basil Blackwtl~ Inc.

Leong, Q., Ab. Karim, M., Othman, M., Mohd Adzahan, N., & Sridar, R. (2010). Relationships between Malaysian Food Image, Tourist Satisfaction and Behavioral Intention. *World Applied Sciences Journal*, *10*(10), 164–171.

Levin, D. Z., Walter, J., & Murninghan, J. K. (2011). Dormant Ties: The Value of Reconnecting. *Organization Science*, *22*(4), 923–939. doi:10.1287/orsc.1100.0576

Levy, S., & Park, S.-Y. (2011). An analysis of CSR activities in the lodging industry. *Journal of Hospitality and Tourism Management*, *18*(1), 147–154. doi:10.1375/jhtm.18.1.147

Liberato, D., Ferreira, F. A., & Azevedo, A. F. (2019). Landscape as a Tourism Resource: A Case Study in the Northern Region of Portugal. *Advances in Tourism, Technology and Smart Systems Smart Innovation, Systems and Technologies*, 627–639. . doi:10.1007/978-981-15-2024-2_54

Likert, R. A. (1932). A technique for the measurement of attitudes. *ArchivesofPsychology, 140*, 1–55.

Li, M., Fang, L., Huang, X., & Goh, C. (2015). A spatial–temporal analysis of hotels in urban tourism destination. *International Journal of Hospitality Management, 45*, 34–43. doi:10.1016/j.ijhm.2014.11.005 PMID:32287860

Lima, L. P. (2000). Atitudes: Estrutura e mudança. In J. Vala & M. B. Monteiro (Eds.), *Psicologia social* (4th ed., pp. 187–225). Fundação Calouste Gulbenkian.

Ling, T.P., Noor, S. M., Mustafa, H., & Kiumarsi, S. (2019). *Mindfulness: Exploring Visitor and Communication Factors at Penang Heritage Sites*. Academic Press.

Little India @ Brickfields - VisitKL. (n.d.). Retrieved August 3, 2021, from http://www.visitkl.gov.my/visitklv2/document/files/thingstodo/little_india.pdf

Litvin, S. W., Blose, J. E., & Laird, S. T. (2003). Tourists' use of restaurant webpages: Is the internet a critical marketing tool? *Journal of Vacation Marketing, 11*(2), 155–161. doi:10.1177/1356766705052572

Litvin, S. W., Goldsmith, R. E., & Pan, B. (2008). Electronic word-of mouth in hospitality and tourism management. *Tourism Management, 29*(3), 458–468. doi:10.1016/j.tourman.2007.05.011

Liu, Y. (2016). *Tourism Destination Governance The case study of Hemavan and Tärnaby* (Master Thesis). UMEA Universitet.

Liu, B., Schroeder, A., Pennington-Gray, L., & Farajat, A. (2016). Source market perceptions: How risky is Jordan to travel to? *Journal of Destination Marketing & Management, 5*(4), 294–304. doi:10.1016/j.jdmm.2016.08.005

Liu, J., Kang, J., Behm, H., & Luo, T. (2014). Landscape spatial pattern indices and soundscape perception in a multifunctional urban area, Germany. *Journal of Environmental Engineering and Landscape Management, 22*(3), 1–11. do i:10.3846/16486897.2014.911181

Liu, J., Yang, L., Xiong, Y., & Yang, Y. (2019). Effects of soundscape perception on visiting experience in a renovated historical block. *Building and Environment, 165*, 106375. doi:10.1016/j.buildenv.2019.106375

Liu, X. P. (2012). Perspective on the Problem of the Chinese Urban Space Homogenization and Identity Creation in the Era of Globalization. *Applied Mechanics and Materials, 174-177*, 2226–2231. . doi:10.4028/www.scientific.net/AMM.174-177.2226

Locke, E. A. (1976). The nature and causes of job satisfaction. In M. D. Dunnette (Ed.), *Handbook of industrial and organizational psychology*. Rand McNally.

Loebbecke, C., & Picot, A. (2015). Reflections on societal and business model transformation arising from digitization and big data analytics: A research agenda. *The Journal of Strategic Information Systems, 24*(3), 149–157. doi:10.1016/j.jsis.2015.08.002

Lončarić, D., Prodan, M., & Ribarić, I. (2012). The influence of a visitor's perceptions of a museum's website design on behavioural intentions. *Tourism and Hospitality Management, 18*, 297–312. https://hrcak.srce.hr/ojs/index.php/ekonomski-vjesnik/article/view/3969

Long, M. L. (2004). *Culinary Tourism*. University Press of Kentucky.

Lopes, C. (2005). *Documentos de apoio do módulo 6 (inquérito: amostragem e erros de amostragem, de não-resposta, de cobertura e de ajustamento) do curso de formação em métodos de investigação e análise de dados em ciências sociais*. Textos não publicados. Núcleo de Investigação e Intervenção em Psicologia Social da Faculdade de psicologia e Ciências da Educação da Universidade de Coimbra.

Lopes, S. D. F. (2011). Destination image: Origins, developments and implications. *PASOS. Revista de Turismo y Patrimonio Cultural*, *9*(2), 305–315. doi:10.25145/j.pasos.2011.09.027

Lopes, V., Moreno Pires, S., & Costa, R. (2020). A strategy for a sustainable tourism development of the Greek Island of Chios. *Tourism: An International Interdisciplinary Journal*, *68*(3), 243–260. doi:10.37741/t.68.3.1

Lorenzen, M., & Frederiksen, L. (2008). Why do cultural industries cluster? Localization, urbanization, products and projects. In P. Cooke & L. Lazzeretti (Eds.), *Creative cities, cultural clusters and local economic development* (pp. 155–179). Edward Elgar.

Lőrincz, K. (2017). A fenntarthatóság társadalmi vetületei az Európai Kulturális Főváros projektben. Comitatus különszám, 64-72.

Lőrincz, K., & Raffay, Á. (2019). "Beyond, azaz túllépni saját magunkon" - a turizmus szerepe a Veszprém2023 Európai Kulturális Főváros projektben. Turisztikai és Vidékfejlesztési Tanulmányok, 4(2), 18-38.

Lor, J. J., Kwa, S., & Donaldson, J. A. (2019). Making ethnic tourism good for the poor. *Annals of Tourism Research*, *76*, 140–152. doi:10.1016/j.annals.2019.03.008

Lousã, E. (2013). *Liderança empreendedora e cultura de inovação em organizações de base tecnológica e análise comparativa entre setores de atividade.* Tese de Doutoramento, Universidade de Coimbra.

Lo, W. H., & Cheng, K. L. B. (2020). Does virtual reality attract visitors? The mediating effect of presence on consumer response in virtual reality tourism advertising. *Information Technology & Tourism*, *22*(4), 537–562. doi:10.100740558-020-00190-2

Lowenthal, D. (1998). *The heritage crusade and the spoils of history.* Cambridge University Press. doi:10.1017/CBO9780511523809

Low, K. E. Y. (2015). The sensuous city: Sensory methodologies in urban ethnographic research. *Ethnography*, *16*(3), 295–312. doi:10.1177/1466138114552938

Luecke, R. (2003). *Business communication.* Harvard Business School Press.

Lu, F., Xu, D., Cheng, Y., Dong, S., Guo, C., Jiang, X., & Zheng, X. (2015). Systematic review and meta-analysis of the adverse health effects of ambient PM2.5 and PM10 pollution in the Chinese population. *Environmental Research*, *136*, 196–204. doi:10.1016/j.envres.2014.06.029 PMID:25460637

Lugosi, P. (2014). Mobilising identity and culture in experience co-creation and venue operation. *Tourism Management*, *40*, 165–179. doi:10.1016/j.tourman.2013.06.005

Lundberg, C., Gudmundson, A., & Andersson, T. D. (2009). Herzberg's Two-Factor Theory of work motivation tested empirically on seasonal workers in hospitality and tourism. *Tourism Management*, *30*(6), 890–899. doi:10.1016/j.tourman.2008.12.003

Lusher, D., Robins, G., & Kremer, P. (2010). The Application of Social Network Analysis to Team Sports. *Measurement in Physical Education and Exercise Science*, *14*(4), 211–224. doi:10.1080/1091367X.2010.495559

Lv, X., Li, C., & McCabe, S. (2020). Expanding theory of tourists' destination loyalty: The role of sensory impressions. *Tourism Management*, *77*, 104026. doi:10.1016/j.tourman.2019.104026

Mace, B. L., Bell, P. A., & Loomis, R. J. (2004). Visibility and Natural Quiet in National Parks and Wilderness Areas: Psychological Considerations. *Environment and Behavior*, *36*(1), 5–31. doi:10.1177/0013916503254747

Machado, V., Carrasco, P., Contreiras, J. P., Duarte, A. P., & Gouveia, D. (2018). Governing Locally for Sustainability: Public and Private Organizations' Perspective in Surf Tourism at Aljezur, Costa Vicentina, Portugal. *Tourism Planning & Development, 15*(6), 692–704. doi:10.1080/21568316.2017.1415958

Mach, L., & Ponting, J. (2018). Governmentality and surf tourism destination governance. *Journal of Sustainable Tourism, 26*(11), 1845–1862. doi:10.1080/09669582.2018.1513008

Mach, L., Ponting, J., Brown, J., & Savage, J. (2018). Riding waves of intra-seasonal demand in surf tourism: Analysing the nexus of seasonality and 21st century surf forecasting technology. *Annals of Leisure Research, 5398*, 1–19. doi:10.1080/11745398.2018.1491801

Maguire, K. (2019). Examining the power role of Local Authorities in planning for socio-economic event impacts. *Local Economy, 34*(7), 657–679. doi:10.1177/0269094219889603

Maingi, S. W. (2019). *Sustainable tourism certification, local governance and management in dealing with overtourism in East Africa*. Worldwide Hospitality and Tourism Themes. Retrieved from https://www.emerald.com/insight/content/doi/10.1108/WHATT-06-2019-0034/full/html?casa_token=2HgIS_T2wVQAAAAA:hRTOgr1K-EapXp4i1ADRRIm-J6dkZThwvUQgYHnLBWFUbs7NKRWa-NJw6gIZAbn8gZCdZVdVZqwrbQnFeGZGzMzgetMZsNziu_89rRSzoTQkci98TuTR-

Mair, J., Ritchie, B. W., & Walters, G. (2016). Towards a research agenda for post-disaster and post-crisis recovery strategies for tourist destinations: A narrative review. *Current Issues in Tourism, 19*(1), 1–26. doi:10.1080/13683500.2014.932758

Majewska, J. (2008). Local Governments' Involvement in the Development of Entrepreneurship in Emerging Tourism Destinations. *Zeszyty Naukowe/Akademia Ekonomiczna w Poznaniu*, (99), 127-145.

Makkonen, M., & Rohde, S. (2016). Cross-border regional innovation systems: Conceptual backgrounds, empirical evidence and policy implications. *European Planning Studies, 24*(9), 1623–1642. doi:10.1080/09654313.2016.1184626

Makkonen, T., & Weidenfeld, A. (2016). Knowledge-based urban development of cross-border twin cities. *InternationalJournalKnowledge-BasedDevelopment, 7*(4), 389–406. doi:10.1504/IJKBD.2016.080881

Makridakis, S. (2017). The forthcoming Artificial Intelligence (AI) revolution: Its impact on society and firms. *Futures, 90*, 46–60. doi:10.1016/j.futures.2017.03.006

Malik, M. E. (2012). The Impact of Pay and Promotion on Job Satisfaction. *American Journal of Economics*, 2(4), 6–9. doi:10.5923/j.economics.20120001.02

Malva, F. L., Verso, V. R. L., & Astolfi, A. (2015). Livingscape: A Multi-sensory Approach to Improve the Quality of Urban Spaces. *Energy Procedia, 78*, 37–42. doi:10.1016/j.egypro.2015.11.111

Mangifera, L. (2019). Stakeholders Analysis for Creative Industry Development. *Jurnal Ekonomi Pembangunan: Kajian Masalah Ekonomi dan Pembangunan, 19*(2), 156-162.

Mansor, M. M., Green, D. A., & Metcalfe, A. V. (2019). Detecting Directionality in Time Series. *The American Statistician*, 1–16. doi:10.1080/00031305.2018.1545699

Marasinghe, S., Perera, P., Simpson, G. D., & Newsome, D. (2021). Nature-based tourism development in coastal wetlands of Sri Lanka: An Importance–Performance analysis at Maduganga Mangrove Estuary. *Journal of Outdoor Recreation and Tourism, 33*(May), 100345. doi:10.1016/j.jort.2020.100345

Marchant, B., & Mottiar, Z. (2011). Understanding lifestyle entrepreneurs and digging beneath the issue of profits: Profiling surf tourism lifestyle entrepreneurs in Ireland. *Tourism Planning & Development, 8*(2), 171–183. doi:10.1080/21568316.2011.573917

Marine-Roig, E., Martin-Fuentes, E., & Daries-Ramon, N. (2017). User-generated social media events in tourism. *Sustainability*, *9*, 2250.

Markwell, K., Stevenson, D., & Rowe, D. (2004). Footsteps and memories: Interpreting an Australian urban landscape through thematic walking tours. *International Journal of Heritage Studies*, *10*(5), 457–473. doi:10.1080/1352725042000299063

Marôco, J. (2011). *Análise de equações estruturais: Fundamentos teóricos, software& aplicações*. Pêro Pinheiro: Report Number.

Marôco, J. (2010). *Análise estatística com utilização do SPSS* (3rd ed.). Edições Sílabo.

Marôco, J., & Garcia-Marques, T. (2006). Qual a fiabilidade do alfa de Cronbach? Questões antigas e soluções modernas? *Laboratório de Psicologia*, *4*(1), 65–90.

Marques, L., & Richards, G. (2014). Creative districts around the world (4th ed.). Breda: NHTV.

Martinez, J. M. G., Martín, J. M. M., & Fernández, J. A. S. (2020). Innovation in the measurement of tourism competitiveness. In *Analyzing the relationship between innovation, value creation, and entrepreneurship* (pp. 268–288). IGI Global. doi:10.4018/978-1-7998-1169-5.ch013

Martin, S. A., & Assenov, I. (2012). The genesis of a new body of sport tourism literature: A systematic review of surf tourism research (1997-2011). *Journal of Sport & Tourism*, *17*(4), 257–287. doi:10.1080/14775085.2013.766528

Martin, S. A., & Assenov, I. (2014). Investigating the importance of surf resource sustainability indicators: Stakeholder perspectives for surf tourism planning and development. *Tourism Planning & Development*, *11*(2), 127–148. doi:10.1080/21568316.2013.864990

Martin, S. A., & Assenov, I. (2015). Measuring the conservation aptitude of surf beaches in Phuket, Thailand: An application of the Surf Resource Sustainability Index. *International Journal of Tourism Research*, *17*(2), 105–117. doi:10.1002/jtr.1961

Matt, C., Hess, T., & Benlian, A. (2015). Digital transformation strategies. *Business & Information Systems Engineering*, *57*(5), 339–343. doi:10.100712599-015-0401-5

Matthew, O. A., Ede, C., Osabohien, R., Ejemeyovwi, J., Ayanda, T., & Okunbor, J. (2021). Interaction effect of tourism and foreign exchange earnings on economic growth in Nigeria. *Global Business Review*, *22*(1), 7–22. doi:10.1177/0972150918812985

Matzler, K., & Renzl, B. (2007). Assessing asymmetric effects in the formation of employee satisfaction. *Tourism Management*, *28*(4), 1093–1103. doi:10.1016/j.tourman.2006.07.009

McGehee, N. (2009). Corporate social responsibility within the US lodging industry: An exploratory study. *Journal of Hospitality & Tourism Research (Washington, D.C.)*, *33*(3), 417–437. doi:10.1177/1096348009338532

McKeever, E., Jack, S., & Anderson, A. (2015). Embedded entrepreneurship in the creative re-construction of place. *Journal of Business Venturing*, *30*(1), 50–65.

McKercher, B. (1999). A chaos approach to tourism. *Tourism Management*, *20*(4), 425–434. doi:10.1016/S0261-5177(99)00008-4

McKercher, B., & Ho, S. Y. P. (2006). Assessing the Tourism Potential of Smaller Cultural and Heritage Attractions. *Journal of Sustainable Tourism*, *14*(4), 473–488. doi:10.2167/jost 620.0

McKercher, B., & Koh, E. (2017). Do attractions "attract" tourists? The case of Singapore. *International Journal of Tourism Research*, *19*(6), 661–671. doi:10.1002/jtr.2138

McLean, K. (2017). Smellmap: Amsterdam—Olfactory Art and Smell Visualization. *Leonardo, 50*(1), 92–93. doi:10.1162/LEON_a_01225

Medway, D. (2015). Rethinking Place Branding and the "Other" Senses. *Rethinking Place Branding,* 191–209. . doi:10.1007/978-3-319-12424-7_13

Melina Mercouri Award. (n.d.). https://ec.europa.eu/programmes/creative-europe/actions/capitals-culture_en

Melo, A. I., Ferreira, A. d., Santos, C. A., Augusto, D., Gomes, G., Azevedo, G. M., & Marques, R. (2020). *Projeto PlowDeR: Relatório Final.* UA Editora – Universidade de Aveiro.

Merinero-Rodríguez, R., & Pulido-Fernández, J. I. (2016). Analysing relationships in tourism: A review. *Tourism Management, 54*, 122–135. doi:10.1016/j.tourman.2015.10.010

Micera, R., Presenza, A., Splendiani, S., & Del Chiappa, G. (2013). SMART Destinations: new strategies to manage tourism industry. In G. Schiuma, J. C. Spender, & A. Pulic (Eds.), *IFKAD 2013: 8th International Forum on Knowledge Asset Dynamics. Smart Growth: Organizations, Cities and Communities* (pp. 1405–1422). Zagreb: IKAM - Institute of Knowledge Asset Management.

Michalkó, G., & Rátz, T. (2005). A kulturális turizmus élménygazdaságtani szempontjai. In A magyar városok kulturális gazdasága. MTA Társadalomkutató Központ.

Middleton, V., Fyall, A., Morgan, M., & Ranchhod, A. (2009). *Marketing in travel and tourism.* Routledge Taylor and Francis Group.

Mihali, C. T. (2000). Environmental management of a tourist destination: A factor of tourism competitiveness. *Tourism Management, 21*(1), 65–78. doi:10.1016/S0261-5177(99)00096-5

Milićević, S., Petrović, J., Kostić, M., & Lakićević, M. (2020). Tourism product in the function of improving destination competitiveness: Case of Vrnjačka Banja, Serbia. *Quality - Access to Success, 21*(178), 133–138.

Mill, R. C., & Marrison, A. M. (2002). *The tourism system: An introductory text* (4th ed.). Kemdall/Hunt.

Milman, A., & Pizam, A. (1995). The role of awareness and familiarity with a destination: The central Florida case. *Journal of Travel Research, 33*(3), 21–27. doi:10.1177/004728759503300304

Mimbs, B. P., Boley, B. B., Bowker, J. M., Woosnam, K. M., & Green, G. T. (2020). Importance-performance analysis of residents' and tourists' preferences for water-based recreation in the Southeastern United States. *Journal of Outdoor Recreation and Tourism, 31*(April), 100324. doi:10.1016/j.jort.2020.100324

Minguzzi, A. (2006). Destination competitiveness and the role of destination management organization (DMO): An Italian experience. In Tourism local systems and networking (pp. 211-222). Routledge.

Ministarstvo gospodarstva i održivog ravoja, Središnji državni portal. (2020). Retrieved April 23, 2020, Available at: https://mzoe.gov.hr/o-ministarstvu-1065/djelokrug-4925/vode-i-more/more-i-priobalje-opcenito/1437

Mintzberg, H. (1979). *The structuring of organizations: a synthesis of the research.* Prentice-Hall.

Mintzberg, H. (1983). *Power in and around organizations.* Prentice-Hall.

Mira, M. R., & Breda, Z., (2019a). *Internacionalização de destinos turísticos: Uma abordagem sistémica.* BeauBassin, Mauritius: Novas Edições Acadêmicas, OmniScriptumPublishing.

Mira, M. R., & Breda, Z. (2019b). Internacionalização de destinos turísticos: O papel das DMO. In C. Estevão & C. Costa (Eds.), *Turismo, estratégia, competitividade e desenvolvimento regional: Uma visão estratégica no setor do turismo* (pp. 19–42). Novas Edições Acadêmicas, OmniScriptum Publishing.

Mokhtar, M. R., Yazid, M. F. M., & Shamsudin, M. F. (2020). Sustainability of tourism industry in Malaysia. *Journal of Postgraduate Current Business Research, 1*(1).

Molina, R., Fri`as-Jamilena, D-M., & Castaneda-Garc`ia, A. (2013). The moderating role of past experience in the formation of a tourist destination image and in tourists' behavioral intention. *Current Issues in Tourism, 16*, 107-127. https://www.tandfonline.com/doi/abs/10.1080/13683500.2012.665045

Monks-Husain, A. (2020). "Better Me Than Someone Else": Making Sense of Multiple Publics in Little India. *The American Papers, 2019-2020*, 41.

Moriarty, J. (2016). Business ethics. *Stanford Encyclopedia of Philosophy*. Available at https://plato.stanford.edu/entries/ethics-business/

Morrin, M., & Ratneshwar, S. (2000). The Impact of Ambient Scent on Evaluation, Attention, and Memory for Familiar and Unfamiliar Brands. *Journal of Business Research, 49*(2), 157–165. doi:10.1016/S0148-2963(99)00006-5

Morrin, M., & Ratneshwar, S. (2003). Does it Make Sense to Use Scents to Enhance Brand Memory? *JMR, Journal of Marketing Research, 40*(1), 10–25. doi:10.1509/jmkr.40.1.10.19128

Morrison, M., Gan, S., Dubelaar, C., & Oppewal, H. (2011). In-store music and aroma influences on shopper behavior and satisfaction. *Journal of Business Research, 64*(6), 558–564. doi:10.1016/j.jbusres.2010.06.006

Morvay, S., Rechnitzer, J., & Fekete, D. (2020). Európa Kulturális Fővárosai Kelet-Közép-Európában. *Tér és Társadalom, 34*(1).

Mossberg, L. (2008). Extraordinary Experiences through Storytelling. *Scandinavian Journal of Hospitality and Tourism, 8*(3), 195–210. doi:10.1080/15022250802532443

Moutinho, L., Dionísio, P., & Leal, C. (2007). Surf tribal behaviour: A sports marketing application. *Marketing Intelligence & Planning, 25*(7), 668–690. doi:10.1108/02634500710834160

Mreža kreativni gradovi Hrvatske. (2017). Retrieved May 22, 2020, Available at: https://kreativnigradovi.net/kreativni-gradovi-hrvatske/

Mreža za razvoj i kreativnost. (2013). Retrieved May 18, 2020, Available at: https://hkkkki.eu/

Mulec, I., & Wise, N. (2013). Indicating the competitiveness of Serbia's Vojvodina Region as an emerging tourism destination. *Tourism Management Perspectives, 8*, 68–79. doi:10.1016/j.tmp.2013.07.001

Mumford, M. D. (2003). Where have we been, where are we going? Taking stock in creativity research. *Creativity Research Journal, 15*(2/3), 107–120. doi:10.1207/S15326934CRJ152&3_01

Murali, V., Chaudhuri, S., & Jermaine, C. (2018). *Neural sketch learning for conditional program generation*. International Conference on Learning Representations (ICLR 2018), Vancouver, Canada.

Murtagh, M. J., Minion, J. T., Turner, A., Wilson, R. C., Blell, M., Ochieng, C., Murtagh, B., Roberts, S., Butters, O. W., & Burton, P. R. (2017). The ECOUTER methodology for stakeholder engagement in translational research. *BMC Medical Ethics, 18*(1), 1–12. doi:10.118612910-017-0167-z PMID:28376776

Murugan, I., Perumal, V., & Kamarudin, K. M. (2021). *Challenges in the practice of traditional Kolam among Indian women in the Klang Valley, Malaysia*. Academic Press.

Nadalipour, Z., Khoshkhoo, M. H. I., & Eftekhari, A. R. (2019). An integrated model of destination sustainable competitiveness. *Competitiveness Review*, *29*(4), 314–335. doi:10.1108/CR-12-2017-0086

Najafi, M., & Shariff, M. K. B. M. (2011). The Concept of Place and Sense of Place In Architectural Studies. *World Academy of Science, Engineering and Technology*, 1100–1106.

Naqvi, M. H. A., Jiang, Y., Naqvi, M. H., Miao, M., Liang, C., & Mehmood, S. (2018). The Effect of Cultural Heritage Tourism on Tourist Word of Mouth: The Case of Lok Versa Festival, Pakistan. *Sustainability*, *10*(7), 2391. doi:10.3390u10072391

Nav'ıo-Marco, J., Ruiz-G'omez, L. M., & Sevilla-Sevilla, C. (2018). Progress in information technology and tourism management: 30 years on and 20 years after the internet-Revisiting Buhalis & Law's landmark study about eTourism. *Tourism Management*, *69*, 460–470. doi:10.1016/j.tourman.2018.06.002

Nayan, N. M., Jones, D. S., & Ahmad, S. (2019). Historic Open Space: The Identity of [Padang] Merdeka Square, Kuala Lumpur, Malaysia. *Asian Journal of Behavioural Studies*, *4*(17), 29–39. doi:10.21834/ajbes.v4i17.182

Neto, A. Q., Dimmock, K., Lohmann, G., & Scott, N. (2020). Destination competitiveness: How does travel experience influence choice? *Current Issues in Tourism*, *23*(13), 1673–1687. doi:10.1080/13683500.2019.1641070

Neuhofer, B., Buhalis, D., & Ladkin, A. (2014). A Typology of Technology-Enhanced Tourism Experiences. *International Journal of Tourism Research*, *16*(4), 340–350. doi:10.1002/jtr.1958

Neuman, W. L. (2014). *Social research methods: Qualitative and quantitative approaches* (7th ed.). Pearson.

Neupane, P. C. (2021). Tourism Governance in the aftermath of COVID-19: A Case Study of Nepal. *The Gaze: Journal of Tourism and Hospitality*, *12*(1), 44–69. doi:10.3126/gaze.v12i1.35676

Nikolopoulos, A. (2013). *Alone against all: The negotiation strategy*. Patakis Editions.

Nilsson, J. H. (2019). Urban bicycle tourism: Path dependencies and innovation in Greater Copenhagen. *Journal of Sustainable Tourism*, *27*(11), 1648–1662. doi:10.1080/09669582.2019.1650749

Nilsson, J., Eskilsson, L., & Ek, R. (2010). Creating cross-border destinations: Interreg Programmes and regionalisation in the Baltic Sea Area. *Scandinavian Journal of Hospitality and Tourism*, *10*(2), 153–172. doi:10.1080/15022250903561978

Nor Azureen, R., Shahril, H., Ashraf Siddik Khan, A. R. S., Putri Dahlia, A. R., & Noor Saliza, S. (2016). The Influence of Food Quality on Customer Satisfaction in Fine Dining Restaurant. *Case in Penang*, *2*(2), 45–50.

Novais, M. A., Ruhanen, L., & Arcodia, C. (2018). Destination competitiveness: A phenomenographic study. *Tourism Management*, *64*, 324–334. doi:10.1016/j.tourman.2017.08.014

Novak, J., Purta, M., Marciniak, T., Ignatowicz, K., Rozenbaum, K., & Yarwood, K. (2018). *The rise of digital challengers. How digitization can become the next growth engine for Central and Eastern Europe*. Academic Press.

Nučič, M. (2012). Is sustainability science becoming more interdisciplinary over time? *Acta Geographica Slovenica*, 215–236. doi:10.3986/AGS52109

Nunnally, J. C., & Bernstein, I. H. (1994). Psychometric theory (3rd ed.). New York: McGraw-Hill.

Nunnally, J. C. (1978). *Psychometric theory*. McGraw-Hill.

O'Dell, T. (2005). Experiencescapes: Blurring borders and testing connections. In T. O'Dell & P. Billing (Eds.), *Experiencescapes: Tourism, culture, and economy* (pp. 11–33). Copenhagen Business School Press.

Oakes, T. (2016). Ethnic tourism in China. In *Handbook on ethnic minorities in China.* Edward Elgar Publishing. doi:10.4337/9781784717360.00020

Obermayer, N., Kővári, E., Leinonen, J., Bak, G., & Valeri, M. (2021, August). "How social media practices shape family business performance: The wine industry case study. *European Management Journal.* Advance online publication. doi:10.1016/j.emj.2021.08.003

OECD. (2020). *Operationalisation of Destination Management Organisations in Romania. OECD Tourism Papers, No. 2020/01.* OECD Publishing. doi:10.1787/9074fc18-

Oglesby, L., Künzli, N., Monn, C., Schindler, C., Ackermann-Liebrich, U., & Leuenberger, P. (2000). Validity of Annoyance Scores for Estimation of Long Term Air Pollution Exposure in Epidemiologic Studies: The Swiss Study on Air Pollution and Lung Diseases in Adults (SAPALDIA). *American Journal of Epidemiology, 152*(1), 75–83. doi:10.1093/aje/152.1.75 PMID:10901332

Oliveira, A., & Paulino, F. (2017). European creativity and urban regeneration. *The Journal of Public Space, 2*(2), 127–140. doi:10.5204/jps.v2i2.98

Omar, S. R., & Omar, S. N. (2018). Malaysian heritage food (MHF): a review on its unique food culture, tradition and present lifestyle. *Int J Heritage Art Multimedia, 1*(3), 1-15.

Omar, S. R., & Omar, S. N. (2018). Malaysian Heritage Food (MHF): A Review on its Unique Food Culture, Tradition and Present Lifestyle. *International Journal, 1*(3), 1-15.

Omar, D., Ramlee, M., Yunus, R. M., & Samadi, Z. (2018). Urban Public Space Revitalisation Strategies for George Town World Heritage Site. *Environment-Behaviour Proceedings Journal, 3*(7), 285–291. doi:10.21834/e-bpj.v3i7.1235

Online, N. S. T. (2015). *Najib, Modi jointly launch Torana Gate in Brickfields.* Available at: https://www.nst.com.my/news/2015/11/113341/najib-modi-jointly-launch-torana-gate-brickfields

Online, N. S. T. (2017). *Illegal parking in Brickfields.* Available at: https://www.nst.com.my/actionline/2017/11/304902/illegal-parking-brickfields

Orams, M. B., & Towner, N. (2013). Riding the wave: History, definitions, and a proposed typology of surf-riding tourism. *Tourism in Marine Environments, 8*(4), 173–188. doi:10.3727/154427313X13631129554901

Organisation for Economic Co-operation and Development. (2014). Tourism and the creative economy. Paris: OECD.

Ortner, S. B. (1999). *Life and Death on Mt. Everest: Sherpas and Himalayan Mountaineering.* Princeton University Press. doi:10.1515/9780691211770

Orts, E. W., & Strudler, A. (2009). Putting a stake in stakeholder theory. *Journal of Business Ethics, 88*(4), 605–615. doi:10.100710551-009-0310-y

Ouf, A. M. S. (2001). Authenticity and the Sense of Place in Urban Design. *Journal of Urban Design, 6*(1), 73–86. doi:10.1080/13574800120032914

Ozturk, A. B., Salehi-Esfahani, S., & Bilgihan, A. (2017). Social Media and Destination Marketing. In Advances in Social Media for Travel, Tourism and Hospitality (pp. 107-119). Routledge. doi:10.4324/9781315565736-9

Pagliara, F., Aria, M., Russo, L., Della Corte, V., & Nunkoo, R. (2021). Validating a theoretical model of citizens' trust in tourism development. *Socio-Economic Planning Sciences, 73*, 100922. doi:10.1016/j.seps.2020.100922

Pallis, C., Prasinos, A., & Pallis, P. (2017). The Relation Between Tourism, Local Development and Insularity: Case Study of Southern Aegean, Greece. *Journal of Tourism and Hospitality Management, 5*(6), 223–232.

Palmer, R., & Richards, G. (2004). European Cities and Capitals of Culture. In *Study prepared for the European Commission*. Palmer-Rae Associates.

Pan, S., & Ryan, C. (2009). Tourism Sense-Making: The Role Of The Senses And Travel Journalism. *Journal of Travel & Tourism Marketing*, *26*(7), 625–639. doi:10.1080/10548400903276897

Pant, P., Huynh, W., & Peltier, R. E. (2018). Exposure to air pollutants in Vietnam: Assessing potential risk for tourists. *Journal of Environmental Sciences (China)*, *73*, 147–154. doi:10.1016/j.jes.2018.01.023 PMID:30290863

Panyor, Á., Győriné Kiss, E., & Gulyás, L. (2011). *Bevezetés a desztináció menedzsmentbe*. Keszthely. http://www.tankonyvtar.hu/en/tartalom/tamop425/0034_bevezetes_a__desztinacio_men_jegyzet/bevezetes_a%20_desztinacio_menedzsmentbe_jegyzet.pdf

Papatheodorou, A., Vlassi, E., Gaki, D., Papadopoulou-Kelidou, L., Efthymiou, M., Pappas, D., & Paraschi, P. (2019). The airline–airport–destination authority relationship: the case of Greece. In *Tourist Destination Management* (pp. 27–41). Springer. doi:10.1007/978-3-030-16981-7_3

Paramati, S. R., Shahbaz, M., & Alam, M. S. (2017). Does tourism degrade environmental quality? A comparative study of eastern and western European Union. *Transport Res. D*, *50*, 1–13. doi:10.1016/j.trd.2016.10.034

Parke, K. S., Reisinger, Y., & Noh, E. H. (2010). Luxury shopping in tourism. *International Journal of Tourism Research*, *12*(2), 164–178. doi:10.1002/jtr.744

Park, Y., & Njite, D. (2010). Relationship between destination image and tourists' future behavior: Observations from Jeju Island, Korea. *Asia Pacific Journal of Tourism Research*, *15*(1), 1–20. doi:10.1080/10941660903510024

Patikas, C. (2014), *Participatory destination management and creative tourism: From co-production of tourism products to co-creation of experiences. A reality and stakeholders' check for Thessaloniki, Greece* (Master's thesis). School of Economics, Business Administration & Legal Studies, International Hellenic University, Thessaloniki, Greece.

Patti, S. (2019). Determinants of tourist destination competitiveness in a low-carbon tourism. In *Positioning and Branding Tourism Destinations for Global Competitiveness* (pp. 74–99). IGI Global. doi:10.4018/978-1-5225-7253-4.ch004

Pavlović, D., Avlijaš, G., & Stanić, N. (2016). Tourist perception as key indicator of destination competitiveness. *TEME: Casopis Za Društvene Nauke*, *40*(2).

Pavlovich, K. (2003). The evolution and transformation of a tourism destination network: The Waitomo Caves, New Zealand. *Tourism Management*, *24*(2), 203–216. doi:10.1016/S0261-5177(02)00056-0

Pearce, D. G. (2014). Toward an Integrative Conceptual Framework of Destinations. *Journal of Travel Research*, *53*(2), 141–153. doi:10.1177/0047287513491334

Pearce, D. G., & Schänzel, H. A. (2013). Destinations: Tourists' Perspectives from New Zealand. *International Journal of Tourism Research*. Advance online publication. doi:10.1002/jtr.1955

Pechlaner, H., Beritelli, P., Pichler, S., Peters, M., & Scott, N. (2015). *Contemporary Destination Governance: A Case Study Approach*. Emerald Group Publishing. doi:10.1108/S2042-144320156

Peck, J. (2011). Does touch matter? Insights from haptic research in marketing. In Sensory Marketing (pp. 17–31). Routledge.

Peck, J., & Childers, T. L. (2003). To have and to Hold: The Influence of Haptic Information on Product Judgments. *Journal of Marketing*, *67*(2), 35–48. doi:10.1509/jmkg.67.2.35.18612

Peck, J., & Shu, S. B. (2009). The Effect of Mere Touch on Perceived Ownership. *The Journal of Consumer Research*, *36*(3), 434–447. doi:10.1086/598614

Peck, J., & Wiggins, J. (2006). It Just Feels Good: Customers' Affective Response to Touch and Its Influence on Persuasion. *Journal of Marketing*, *70*(4), 56–69. doi:10.1509/jmkg.70.4.056

Pedersen, E., & Larsman, P. (2008). The impact of visual factors on noise annoyance among people living in the vicinity of wind turbines. *Journal of Environmental Psychology*, *28*(4), 379–389. doi:10.1016/j.jenvp.2008.02.009

Peeters, L. W. J., & Ateljevic, I. (2009). Women empowerment entrepreneurship nexus in tourism: Processes of social innovation. In S. Page & J. Ateljevic (Eds.), *Tourism and entrepreneurship: International perspectives* (pp. 75–89). Routledge.

Pepe, A., & Percoco, A. (2020). *Quel che resta della Legacy di Matera 2019 dopo il Covid-19. Fondazione E.* Mattei.

Pereira, L., Limberger, P., Silva, L., & Pereira, M. (2018). An Empirical Investigation of Destination Branding: The Case of the City of Rio de Janeiro, Brazil. *Sustainability*, 11.

Perinić, L. (2010). Kreativni gradovi: uzroci i posljedice. *Drugost: časopis za kulturalne studije,* (1), 110-115.

Peters, M., & Pikkemaat, B. (2006). Crisis management in Alpine winter sports resorts—The 1999 avalanche disaster in Tyrol. *Journal of Travel & Tourism Marketing*, *19*(2-3), 9–20. doi:10.1300/J073v19n02_02

Petrescu, M. (2013). Marketing research using single-item indicators in structural equation models. *Journal of Marketing Analytics*, *1*(2), 99–117. doi:10.1057/jma.2013.7

Petrevska, B., & Gerasimova, V. M. (2012). Tourism in Regional Development: Empirical Evidence. *Innovative Issues and Approaches in Social Sciences.*, *5*(2), 6–20. doi:10.12959/issn.1855-0541.IIASS-2012-no2-art01

Phillips, J., Wolfe, K., Hodur, N., & Leistritz, L. (2013). Tourist word of mouth and revisit intentions to rural tourism destinations: A case of North Dakota, USA. *International Journal of Tourism Research*, *15*(1), 93–104. doi:10.1002/jtr.879

Phillips, R. (2015). Playful and multi-sensory fieldwork: Seeing, hearing and touching. *New York. Journal of Geography in Higher Education*, *39*(4), 617–629. doi:10.1080/03098265.2015.1084496

Pike, S. (2015). *Destination marketing: essentials*. Routledge. doi:10.4324/9781315691701

Pike, S., Bianchi, C., Keer, G., & Patti, C. (2010). Consumer-based brand equity for Australia as a long-haul tourism destination in an emerging market. *International Marketing Review*, *27*(4), 434–449. doi:10.1108/02651331011058590

Pike, S., & Page, S. J. (2014). Destination Marketing Organizations and destination marketing: A narrative analysis of the literature. *Tourism Management*, *41*, 202–227. doi:10.1016/j.tourman.2013.09.009

Pink, S. (2008). An Urban Tour: The Sensory Sociality of Ethnographic Place-making. *Ethnography*, *9*(2), 175–196. doi:10.1177/1466138108089467

Pizam, A., & Milman, A. (1986). The Social Impacts of Tourism. *Tourism Recreation Research*, *11*(1), 29–33. doi:10.1080/02508281.1986.11014414

Ponting, J. (2009). Projecting paradise: The surf media and the hermeneutic circle in surfing tourism. *Tourism Analysis*, *14*(2), 175–185. doi:10.3727/108354209789116510

Ponting, J., & McDonald, M. G. (2013). Performance, agency and change in surfing tourist space. *Annals of Tourism Research*, *43*, 415–434. doi:10.1016/j.annals.2013.06.006

Porath, C. L. (2015). *Care: How to create an environment where employees and organizations thrive*. Academic Press.

Porteous, J. D. (1985). Smellscape. *Progress in Physical Geography*, *9*(3), 356–378. doi:10.1177/030913338500900303

Porter, B. A., & Usher, L. E. (2019). Sole surfers? Exploring family status and travel behaviour among surf travellers. *Annals of Leisure Research*, *22*(4), 424–443. doi:10.1080/11745398.2018.1484782

Porter, M. (1998). Clusters and the new economics of competition. *Harvard Business Review*, *76*(6), 77–90. PMID:10187248

Portugal, A. C., Campos, F., Martins, F., & Melo, R. (2017). Understanding the relation between serious surfing, surfing profile, surf travel behaviour and destination attributes preferences. *European Journal of Tourism Research*, *16*, 57–73.

Poudyal, N. C., Paudel, B., & Green, G. T. (2013). Estimating the Impact of Impaired Visibility on the Demand for Visits to National Parks. *Tourism Economics*, *19*(2), 433–452. doi:10.5367/te.2013.0204

Prayag, G., & Brittnacher, A. (2014). Environmental Impacts of Tourism on a French Urban Coastal Destination: Perceptions of German and British Visitors. *Tourism Analysis, 19*(4), 461-475. doi:10.3727/108354214X14090817031116

Prayag, G. (2009). Tourists' Evaluation of Destination Image, Satisfaction and Future Behavioral Intentions—The Case of Mauritius. *Journal of Travel & Tourism Marketing*, *26*(8), 836–853. doi:10.1080/10548400903358729

Prayag, G., Hosany, S., & Odeh, K. (2013). The Role of Tourists' Emotional Experiences and Satisfaction in Understanding Behavioural Intentions. *Journal of Destination Marketing & Management*, *2*(2), 118–127. doi:10.1016/j.jdmm.2013.05.001

Prayogo, R., Ketaren, F., & Hati, R. (2016). Electronic Word Of Mouth, Destination Image, And Satisfaction Toward Visit Intention: An Empirical Study In Malioboro Street, Yogyakarta. *Advances in Social Science, Education and Humanities Research*, 81.

Prebensen, N. K., Vittersø, J., & Dahl, T. I. (2013). Value Co-creation significance of tourist resources. *Annals of Tourism Research*, *42*, 240–261. doi:10.1016/j.annals.2013.01.012

Presenza, A., & Cipollina, M. (2010). Analysing tourism stakeholders networks. *Tourism Review*, *65*(4), 17–30. doi:10.1108/16605371011093845

Priskin, J. (2001). Assessment of natural resources for nature-based tourism: The case of the Central coast region of Western Australia. *Tourism Management*, *22*(6), 637–648. doi:10.1016/S0261-5177(01)00039-5

Priya. (2010). *Brickfields business community oppose setting up of kiosks*. Available at: https://www.thestar.com.my/news/community/2010/09/09/brickfields-business-community-oppose-setting-up-of-kiosks

Prokopis, A. C., Farmaki, A., Saveriades, A., & Spanou, E. (2019). The "genius loci" of places that experience intense tourism development. *Tourism Management Perspectives*, *30*, 19–32. doi:10.1016/j.tmp.2019.01.002

PutriviN. (2020). *How Digital Innovation and Dynamic Organization Concept Can Help Creative Industries to Survive COVID-19 Global Pandemic*. https://ssrn.com/abstract=3590926 doi:10.2139/ssrn.3590926

Qiu, M., Zhang, J., & Zheng, C. (2018). Exploring tourists' soundscape emotion and its impact on sustainable tourism development. *Asia Pacific Journal of Tourism Research*, *23*(9), 862–879. Advance online publication. doi:10.1080/10941665.2018.1494614

Quercia, D., Schifanella, R., Aiello, L. M., & McLean, K. (2015). Smelly Maps: The Digital Life of Urban Smellscapes. *9th International Conference on Web and Social Media, ICWSM 2015*, 327-336.

R. (2019, September 4). *KL 5th world's most-visited city, Bangkok takes top spot - MasterCard*. Retrieved August 1, 2021, from https://www.nst.com.my/news/nation/2019/09/518721/kl-5th-worlds-most-visited-city-bangkok-takes-top-spot-mastercard

Rae, D. (2020). Intercultural Entrepreneurship in Creative Place-Making. In *Value Construction in the Creative Economy*. Palgrave Macmillan. doi:10.1007/978-3-030-37035-0_7

Rahman, S. (2018). Emerging Built Heritage Commodification of Boutique Hotels In World Heritage Site: Evidence From George Town, Penang, Malaysia. *Planning Malaysia Journal, 16*(8).

Rahman, N. A., Halim, N., & Zakariya, K. (2018, September). Architectural value for urban tourism placemaking to rejuvenate the cityscape in Johor Bahru. *IOP Conference Series. Materials Science and Engineering, 401*(1), 012010. doi:10.1088/1757-899X/401/1/012010

Raihan, T., & Al Karim, R. (2017, April). CSR and employee job satisfaction: A case from MNCS Bangaldesh. *Global Journal of Human Resource Management, 5*(3), 26–39.

Raisi, H., Baggio, R., Barratt-Pugh, L., & Willson, G. (2020). A network perspective of knowledge transfer in tourism. *Annals of Tourism Research, 80*(March), 102817. doi:10.1016/j.annals.2019.102817

Raisi, H., Baggio, R., Barratt-Pugh, L., & Willson, G. (2018). Hyperlink Network Analysis of a Tourism Destination. *Journal of Travel Research, 57*(5), 671–686. doi:10.1177/0047287517708256

Rajan, S. I., & Percot, M. (Eds.). (2020). *Dynamics of Indian migration: Historical and current perspectives.* Taylor & Francis. doi:10.4324/9780367818043

Rajantheran, M. (2009). *Siddha philosophy in the Malaysian Indian Culture: A historical outlook.* Academic Press.

Rajantheran, M., Gill, S. S., Muniapan, B., Silllalee, K., & Manimaran, S. (2014). A critical analysis of siddha tradition in the context of Malaysian Hindu culture. *Life Science Journal, 11*(7).

Ramlee, M., Omar, D., Mohd Yunus, R., & Samadi, Z. (2018). Attributes of successful public spaces through users perception. *Asian Journal of Quality of Life, 3*(11), 21–28. doi:10.21834/ajqol.v3i11.118

Ramos, G. M., & Fernandes, J. L. (2016). Tourism territories in low density areas: The case of Naturtejo geopark in Portugal. *Journal of Tourism. Heritage & Services Marketing, 2*(1), 14–21.

Rasdi, A. L. M., Som, A. P. M., & Usolludin, H. H., & Nordin, N. (2019). Responsible Tourism and Quality of Life: Perspective of Local Community. Journal of Tourism. *Hospitality and Environment Management, 4*(13), 49–56.

Rasoolimanesh, S. M., Ringle, C. M., Jaafar, M., & Ramayah, T. (2017). Urban vs. rural destinations: Residents' perceptions, community participation and support for tourism development. *Tourism Management, 60*, 147–158. doi:10.1016/j.tourman.2016.11.019

Ratten, V. (2018). Entrepreneurial intentions of surf tourists. *Tourism Review, 73*(2), 262–276. doi:10.1108/TR-05-2017-0095

Rátz, T. (2014). Az EKF-turizmus mint lehetséges niche termék. A kulturális turizmus egy különleges területének vizsgálata. Turizmus Bulletin, 15(2), 24-33.

Rauch, S., Marzolo, M., Dal Cappello, T., Ströhle, M., Mair, P., Pietsch, U., ... Strapazzon, G. (2021). Severe traumatic brain injury and hypotension is a frequent and lethal combination in multiple trauma patients in mountain areas–an analysis of the prospective international Alpine Trauma Registry. *Scandinavian Journal of Trauma, Resuscitation and Emergency Medicine, 29*(1), 1–10. doi:10.118613049-021-00879-1 PMID:33407690

Ravindran, S. (2017). *DBKL: No more double-parking in Brickfield's Little India*. Available at: https://www.thestar.com.my/metro/metro-news/2017/10/14/dbkl-no-excuse-for-doubleparking-city-hall-to-conduct-more-enforcement-activities-in-brickfields-to/

Raykov, T., & Marcoulides, G. A. (2006). *A first course in structural equation modeling* (2nd ed.). Lawrence Erlbaum Associates Publishers.

Raymond, C. (2007). Creative Tourism New Zealand: The practical challenges of developing creative tourism. In G. Richards & J. Wilson (Eds.), *Tourism, Creativity and Development* (pp. 145–157). Routledge.

Razvojna agencija Grada Dubrovnika. (2014). Retrieved November 14, 2020, Available at: https://dura.hr/news/dani_kreativnih_kulturnih_industrija/index.html

Reagan, B. (2011). *Community Benefits and Accountable Development.* IRLE Job Creation and Local Communities Speaker Series.

Reinartz, W., Wiegand, N., & Imschloss, M. (2019). The impact of digital transformation on the retailing value Chain. *International Journal of Research in Marketing*, *36*(3), 350–366. doi:10.1016/j.ijresmar.2018.12.002

Reinhold, S., Beritelli, P., & Grünig, R. (2019). A business model typology for destination management organizations. *Tourism Review*, *74*(6), 1135–1152. doi:10.1108/TR-03-2017-0065

Remoaldo, P., Matos, O., Freitas, I., Gôja, R., Araújo Alves, J., Ribeiro, V., Pereira, M., & Xavier, C. (2020). An international overview of certified practices in creative tourism in rural and urban territories. *Journal of Hospitality & Tourism Research (Washington, D.C.)*, 1–29. doi:10.1177/1096348020950792

Ren, G., & Hong, T. (2017). Investigating Online Destination Images Using a Topic-Based Sentiment Analysis Approach. *Sustainability*, *9*(10), 9. doi:10.3390u9101765

Ren, T., Can, M., Paramati, S. R., Fang, J., & Wu, W. (2019). The impact of tourism quality on economic development and environment: Evidence from Mediterranean countries. *Sustainability*, *11*(8), 2296. doi:10.3390u11082296

Richards, G. & Raymond, C. (2000). Creative tourism. *Atlas News*, *23*(8), 16-20.

Richards, G. (2020). Designing creative places: The role of creative tourism. *Annals of Tourism Research*, *85*, 1–11. doi:10.1016/j.annals.2020.102922

Richards, G., & Wilson, J. (2006). Developing creativity in tourist experiences: A solution to the serial reproduction of culture? *Tourism Management*, *27*(6), 1209–1223. doi:10.1016/j.tourman.2005.06.002

Richter, A., & Riemer, K. (2013). Malleable end-user software. *Business & Information Systems Engineering*, *5*(3), 195–197. doi:10.100712599-013-0260-x

Rijeka. (2020). *Europska prijestolnica kulture.* Retrieved November 16, 2020, Available at: https://rijeka2020.eu/en/

Ritchie, B. W., & Jiang, Y. (2019). A review of research on tourism risk, crisis and disaster management: Launching the annals of tourism research curated collection on tourism risk, crisis and disaster management. *Annals of Tourism Research*, *79*, 102812. doi:10.1016/j.annals.2019.102812

Ritchie, J. R. B., & Crouch, G. I. (2003). *The Competitive Destination: A Sustainable Tourism Perspective.* CABI Publishing. doi:10.1079/9780851996646.0000

Robaina, M., Madaleno, M., Silva, S., Eusébio, C., Carneiro, M. J., Gama, C., Oliveira, K., Russo, M. A., & Monteiro, A. (2020). The relationship between tourism and air quality in five European countries. *Economic Analysis and Policy*, *67*, 261–272. doi:10.1016/j.eap.2020.07.012

Rodan, S., & Galunic, C. (2003). More than Network Structure: How Knowledge Heterogeinity Influences Managerial Performance and Innovativeness. *Strategic Management Journal.*

Rodaway, P. (1994). *Sensuous geographies: body, sense and place*. Routledge. doi:10.4324/9780203306864

Rodríguez-Antón, J., & Alonso-Almeida, M. M. (2011). Quality certification systems and their impact on employee satisfaction in services with high levels of customer contact. *Total Quality Management & Business Excellence*, *22*(2), 145–157. doi:10.1080/14783363.2010.529640

Roman, M., Roman, M., Prus, P., & Szczepanek, M. (2020). Tourism competitiveness of rural areas: Evidence from a region in Poland. *Agriculture*, *10*(11), 569. doi:10.3390/agriculture10110569

Romero-García, L. E., Aguilar-Gallegos, N., Morales-Matamoros, O., Badillo-Piña, I., & Tejeida-Padilla, R. (2019). Urban tourism: A systems approach – state of the art. *Tourism Review*, *74*(3), 679–693. doi:10.1108/TR-06-2018-0085

Rosso, B.D., Dekas, K. H., & Wrzesniewski, A. (2010). On the meaning of work: A theoretical integration and review. Research in Organizational Behavior, 30, 91-127. doi:10.1016/j.riob.2010.09.001

Rovira, N. (2016). ¿Ciudades en el mapa o en la guía turística? Venta de la ciudad y sentido del lugar. *Revista CIDOB d'AfersInternacionals*, *113*(113), 89–105. doi:10.24241/rcai.2016.113.2.89

Rowlands, J. (1997). *Questioning empowerment: working with women in Honduras. Academic Press.*

Rudan, E. (2012). Razvojne perspektive kreativnoga turizma Hrvatske. *Ekonomska Misao i Praksa*, (2), 713–730.

Rueger, J., Dolfsma, W., & Aalbers, R. (2020). Perception of Peer Advice in Online Health Communities: Access to Lay Expertise. *Social Science & Medicine*. Advance online publication. doi:10.1016/j.socscimed.2020.113117 PMID:33865092

Rumere, N., Wiranatha, A. S., & Pujaastawa, I. B. G. (2020). Planning of Country Border Area as A Tourism Destination in Jayapura City. *E-Journal of Tourism*, 384-398.

Russell, R., & Faulkner, B. (1999). Movers and shakers: Chaos makers in tourism development. *Tourism Management*, *20*(4), 411–423. doi:10.1016/S0261-5177(99)00014-X

Russell, R., & Faulkner, B. (2004). Entrepreneurship, chaos and the tourism area lifecycle. *Annals of Tourism Research*, *31*(3), 556–579. doi:10.1016/j.annals.2004.01.008

Russo, M. A., Relvas, H., Gama, C., Lopes, M., Borrego, C., Rodrigues, V., Robaina, M., Madaleno, M., Carneiro, M. J., Eusébio, C., & Monteiro, A. (2020). Estimating emissions from tourism activities. *Atmospheric Environment, 220*. doi:10.1016/j.atmosenv.2019.117048

Saenz-de Miera, O., & Rosselló, J. (2014). Modelling tourism impacts on air pollution: The case study of PM 10 in Mallorca. *Tourism Management*, *40*, 273–281. doi:10.1016/j.tourman.2013.06.012

Saldanha, A. (2002). Music tourism and factions of bodies in Goa. *Tourist Studies*, *2*(1), 43–62. doi:10.1177/1468797602002001096

Salisbury, R., & Hawley, E. (2004). *The Himalayan Database - The Expedition Archives of Elizabeth Hawley*. The American Alpine Club Press.

Sallent, O., Palau, R., & Guia, J. (2011). Exploring the legacy of sport events on sport tourism networks. *European Sport Management Quarterly*, *11*(4), 397–421. doi:10.1080/16184742.2011.599208

Salpeteur, M., Calvet-Mir, L., Diaz-Reviriego, I., & Reyes-García, V. (2017). Networking the environment: Social network analysis in environmental management and local ecological knowledge studies. *Ecology and Society*, *22*(1), art41. doi:10.5751/ES-08790-220141

Salvatore, R., Chiodo, E., & Fantini, A. (2018). Tourism transition in peripheral rural areas: Theories, issues and strategies. *Annals of Tourism Research, 68*, 41–51. doi:10.1016/j.annals.2017.11.003

Samad, N. S. A., Abdul-Rahim, A. S., Afandi, S. H. M., & Johari, M. M. (2021, May). Factors influencing the public park use in Kuala Lumpur, Malaysia. *IOP Conference Series. Earth and Environmental Science, 756*(1), 012085. doi:10.1088/1755-1315/756/1/012085

Samaraweera, D., Hamid, J. A., Khatibi, A. A., Azam, S. F., & Dharmaratne, I. (2018). Measuring Burnout of Human Service Workers: Testing the Copenhagen Burnout Inventory in Sri Lankan Context. *European Journal of Education Studies.*

Santich, B. (2004). The study of gastronomy and its relevance to hospitality education and training. *International Journal of Hospitality Management, 23*(1), 15–24. doi:10.1016/S0278-4319(03)00069-0

Santos, C., Melo, A., Ferreira, A., Augusto, D., Gomes, G., Azevedo, G., . . . Marques, R. (2019). PLowDeR - Framework para Análise do Impacto Económico e Social das Atividades Turísticas nos Territórios de Baixa Densidade: o Caso das Aldeias Históricas de Portugal. Aveiro.

Santos, C. A., & Yan, G. (2008). Representational politics in Chinatown: The ethnic other. *Annals of Tourism Research, 35*(4), 879–899. doi:10.1016/j.annals.2008.06.006

Santos, V., Ramos, P., Sousa, B., Almeida, N., & Valeri, M. (2021a). Factors influencing touristic consumer behavior. *Journal of Organizational Change Management*. Advance online publication. doi:10.1108/JOCM-02-2021-0032

Santos, V., Sousa, B., Ramos, P., & Valeri, M. (2021b). Emotions and Involvement in Tourism Settings. *Current Issues in Tourism*, 1–6. Advance online publication. doi:10.1080/13683500.2021.1932769

Sanz-Ibáñez, C., & Clavé, S. (2014). The evolution of destinations: Towards an evolutionary and relational economic geography approach. *Tourism Geographies, 16*(4), 563–579. doi:10.1080/14616688.2014.925965

Saraniemi, S. (2011). From destination image building to identity-based branding. *International Journal of Culture, Tourism and Hospitality Research, 5*(3), 247–254. doi:10.1108/17506181111156943

Saraniemi, S., & Komppula, R. (2019). The development of a destination brand identity: A story of stakeholder collaboration. *Current Issues in Tourism, 22*(9), 1116–1132. doi:10.1080/13683500.2017.1369496

Saraniemi, S., & Kylänen, M. (2011). Problematizing the Concept of Tourism Destination: An Analysis of Different Theoretical Approaches. *Journal of Travel Research, 50*(2), 133–143. doi:10.1177/0047287510362775

Sarantakou, E., & Terkenli, T. S. (2019). Non-institutionalized forms of tourism accommodation and overtourism impacts on the landscape: The case of Santorini, Greece. *Tourism Planning & Development, 16*(4), 411–433. doi:10.1080/21568316.2019.1569122

Sarasa, J. (2015). Mito y estrategias para un destino romántico: La ciudad de Teruel. *Cuadernos de Turismo, 36*(36), 39–54. doi:10.6018/turismo.36.230871

Scarpato, R. (2002). Gastronomy Studies in Search of Hospitality. *Journal of Hospitality and Tourism Management, 9*(2), 152–163.

Schafer, R. M. (1977). *The Tuning of the World.* Knopf.

Scheske, C., Arroyo Rodriguez, M., Buttazzoni, J. E., Strong-Cvetich, N., Gelcich, S., Monteferri, B., Rodríguez, L. F., & Ruiz, M. (2019). Surfing and marine conservation: Exploring surf-break protection as IUCN protected area categories and other effective area-based conservation measures. *Aquatic Conservation, 29*(S2), 195–211. doi:10.1002/aqc.3054

Schmitt, B. (1999). Experiential Marketing. *Journal of Marketing Management, 15*(1-3), 53–67. doi:10.1362/026725799784870496

Schmitt, B. (2010). Key concepts of experience marketing. *Foundations and Trends in Marketing, 5*(2), 55–112. doi:10.1561/1700000027

Schneider, M., Bernasch, D., Weymann, J., Holle, R., & Bärtsch, P. (2002). Acute mountain sickness: Influence of susceptibility, preexposure, and ascent rate. *Medicine and Science in Sports and Exercise, 34*(12), 1886–1891. doi:10.1097/00005768-200212000-00005 PMID:12471292

Schumacker, R. E., & Lomax, R. G. (2012). *A beginner's guide to structural equation modeling*. Routledge Academic. doi:10.4324/9780203851319

Schumann, J. H., Von Wangenheim, F., & Groene, N. (2014). Targeted online advertising Using reciprocity appeals to increase acceptance among users of free web services. *Journal of Marketing, 78*(1), 59–75. doi:10.1509/jm.11.0316

Schwartz, M., & Carroll, A. B. (2003). Corporate social responsibility: A three domain approach. *Business Ethics Quarterly, 13*(4), 503–530. doi:10.5840/beq200313435

Scott, N., Baggio, R., & Cooper, C. (2008). *Network analysis and tourism*. Channel View Publications. doi:10.21832/9781845410896

Scuttari, A., Volgger, M., & Pechlaner, H. (2016). Transition management towards sustainable mobility in Alpine destinations: Realities and realpolitik in Italy's South Tyrol region. *Journal of Sustainable Tourism, 24*(3), 463–483. doi:10.1080/09669582.2015.1136634

See-To, W. K., & Ho, K. (2014). Value co-creation and purchase intention in social network sites: The role of electronic Word-of-Mouth and trust – A theoretical analysis. *Computers in Human Behavior, 31*, 182–189. doi:10.1016/j.chb.2013.10.013

Sen, S., & Lerman, D. (2007). Why are you telling me this? An examination into negative consumer reviews on the Web. *Journal of Interactive Marketing, 12*(4), 76–94. doi:10.1002/dir.20090

Sertakova, E., Koptseva, N., Kolesnik, M., Libakova, N., Luzan, V., & Sergeeva, N. (2016). Brand-management of Siberian cities: Krasnoyarsk as a case study. *International Review of Management and Marketing, 6*(5), 185–191.

SETE. (2010). *The Gastronomy in the Greek Tourism Marketing*. Available in Http://www.sete.gr/files/ Media/Ebook/2010/100222%20gastronomy%20f.pdf

Shahriar, A. Z. M. (2018). Gender differences in entrepreneurial propensity: Evidence from matrilineal and patriarchal societies. *Journal of Business Venturing, 33*(6), 762–779.

Sharma, P. (2012). Tourism in Nepal 2030. Nepal, 2030, 73-92.

Shaw, G., & Williams, A. (2009). Knowledge transfer and management in tourism organisations: An emerging research agenda. *Tourism Management, 30*(3), 325–335. doi:10.1016/j.tourman.2008.02.023

Sheehan, L., Vargas-Sánchez, A., Presenza, A., & Abbate, T. (2016). The Use of Intelligence in Tourism Destination Management: An Emerging Role for DMOs. *International Journal of Tourism Research, 18*(6), 549–557. doi:10.1002/jtr.2072

Sheresheva, M. Y. (2020). Coronavirus and tourism. *Population and Economics, 4*(2), 72–76. doi:10.3897/popecon.4.e53574

Shi, B., Zhao, J., & Chen, P. (2017). Exploring urban tourism crowding in Shanghai via crowdsourcing geospatial data. *Current Issues in Tourism, 20*(11), 1186–1209. doi:10.1080/13683500.2016.1224820

Shih, H. Y. (2006). Network characteristics of drive tourism destinations: An application of network analysis in tourism. *Tourism Management, 27*(5), 1029–1039. doi:10.1016/j.tourman.2005.08.002

Silva, S. F., & Ferreira, J. C. (2014). The social and economic value of waves: An analysis of Costa de Caparica, Portugal. *Ocean and Coastal Management, 102*, 58–64. doi:10.1016/j.ocecoaman.2014.09.012

Singh, A. (2011). India–Malaysia Strategic Relations. *Maritime Affairs: Journal of the National Maritime Foundation of India, 7*(1), 85–105.

Skånberg, A., & Öhrström, E. (2002). Adverse Health Effects In Relation To Urban Residential Soundscapes. *Journal of Sound and Vibration, 250*(1), 151–155. doi:10.1006/jsvi.2001.3894

Smith, D., Menon, S., & Sivakumar, K. (2005). Online Peer and Editorial Recommendations, Trust, and Choice in Virtual Markets. *Journal of Interactive Marketing, 19*(3), 15–37. doi:10.1002/dir.20041

Soares, J., Baidal, J., & Gândar, J. (2015). La evolución de destinos turísticos litorales consolidados: Análisis comparado de balneario Camboriú (Brasil) y Benidorm (España). *Annales de Géographie, 35*(2), 143–166. doi:10.5209/rev_AGUC.2015.v35.n2.50118

Soda, G., Usai, A., & Zaheer, A. (2004). Network Memory: The Influence of Past and Current Networks on Performance. *Academy of Management Journal, 47*(6), 893–906.

Sofronov, B. (2018). The development of the travel and tourism industry in the world. *Annals of Spiru Haret University. Economic Series, 18*(4), 123–137.

Soisalon-Soininen, T., & Lindroth, K. (2006). Regional Tourism Co-operation in Progress. In Tourism Local Systems And Networking. Elsevier Science & Technology.

Solikhah, N. (2020, April). Ethnic tourism and sustainable of vernacular settlement in Cibeo Village, Baduy Dalam. *IOP Conference Series. Earth and Environmental Science, 452*(1), 012021. doi:10.1088/1755-1315/452/1/012021

Soltani, A. (2019). Abusability testing: Considering the ways your technology might be used for harm. Enigma 2019, Burlingame, CA.

Son, A., & Pearce, P. (2005). Multi-Faceted Image Assessment. *Journal of Travel & Tourism Marketing, 18*(4), 21–35. doi:10.1300/J073v18n04_02

Sotomayor, S., & Barbieri, C. (2016). An exploratory examination of serious surfers: Implications for the surf tourism industry. *International Journal of Tourism Research, 18*(1), 62–73. doi:10.1002/jtr.2033

South Aegean Region. (2019). http://www.pnai.gov.gr/

Spangenberg, E. R., Crowley, A. E., & Henderson, P. W. (1996). Improving the Store Environment: Do Olfactory Cues Affect Evaluations and Behaviors? *Journal of Marketing, 60*(2), 67–80. doi:10.1177/002224299606000205

Spiller, R. (2000). Ethical Business and Investment: AModel for Business and Society. *Journal of Business Ethics, 27*(1/2), 149–160. doi:10.1023/A:1006445915026

Spinelli, M. A., & Canavos, G. C. (2000). Investigating the relationship between employees satisfaction and guest satisfaction. *The Cornell Hotel and Restaurant Administration Quarterly, 41*(6), 29–33.

Spyriadis, T., Fletcher, J., & Fyall, A. (2013). Destination management organisational structures. In C. Costa, E. Panyik, & D. Buhalis (Eds.), *Trends in European tourism planning and organization* (pp. 77–91). Channel View Publications. doi:10.21832/9781845414122-010

Stankova, M., & Amoiradis, C. (2019). Effects Of Promotion Practices On The Sustainable Development Of Tourist Destinations. *Entrepreneurship*, *7*(1), 84–96.

Steinbachné Hajmásy, G. (2020). *A dolgozói elégedettség vizsgálata a belső CSR tevékenységek tükrében a Balaton régió négy- és ötcsillagos szállodáinak bevonásával* (PhD thesis).

Stipanović, C., Rudan, E., & Zubović, V. (2019). Cultural and creative industries in urban tourism innovation – the example of city of Rijeka. *Tourism in South East Europe – ToSEE*, (5), 655-666.

Stipanović, C., Rudan, E., & Zadel, Z. (2018). Possibilities of entrepreneurial development in cultural tourism of Croatian cities. In *24th Biennial International Congress: Tourism and hospitality industry*. Opatija: Faculty of Tourism and Hospitality Management Opatija, University of Rijeka.

Stokes, R. (2004). Relationships and networks for shaping events tourism: An Australian study. *Event Management*, *10*(2/3), 145–158.

Strahilevitz, M., & Myers, J. G. (1998). Donations to charity as purchase incentives: How well they work may depend on what you are trying to sell. *The Journal of Consumer Research*, *24*(4), 434–446. doi:10.1086/209519

Strange, I. (2016). Urban planning, architecture and the making of creative spaces. In P. Long & N. D. Morpeth (Eds.), *Tourism and the creative industries: theories, policies and practice* (pp. 100–113). Routledge. doi:10.4324/9781315735870-6

Strategic cooperation of Veszprém-Balaton2023 Plc. with the University of Pannonia. (n.d.). https://veszprembalaton2023.hu/hu/news/veszprem-balaton-2023-egyuttmukodes-felsofokon

Strobl, A., & Peters, M. (2013). Entrepreneurial reputation in destination networks. *Annals of Tourism Research*, *40*, 59–82. doi:10.1016/j.annals.2012.08.005

Su, L., & Fan, X. A. (2011). *Study on the Relationships between Service Quality, Satisfaction, Trust and Loyalty among Rural Tourism*. Paper presented at 8th International Conference on Service Systems and Service Management, Tianjin. https://ieeexplore.ieee.org/document/5959418

Suárez Velasco, J. E., Huaraca Vera, L. E., & Paladines, G. V. (2017). Desarrollo turístico de los mercados municipales en centros históricos. Mercado San Francisco. Quito-Ecuador. *Revista Publicando*, *11*(2), 331–339.

Sudipta, K. S., & Sarat, L. (2010). Cultural tourism in Malaysia in the perspective of Indian tourists: A study. *Revista de turism-studii si cercetari in turism*, (10), 48-53.

Sulaiman, M. Z., & Wilson, R. (2019). Translating Tourism Landscapes: From Nature to Urban. In *Translation and Tourism* (pp. 89–127). Springer. doi:10.1007/978-981-13-6343-6_6

Su, M. M., Wall, G., & Ma, Z. (2019). A multi-stakeholder examination of destination image: Nanluoguxiang heritage street, Beijing, China. *Tourism Geographies*, *21*(1), 2–23. doi:10.1080/14616688.2017.1385031

Szyszkowicz, M., Kousha, T., Kingsbury, M., & Colman, I. (2016). Air Pollution and Emergency Department Visits for Depression: A Multicity Case-Crossover Study. *Environmental Health Insights*, *10*, 155–161. doi:10.4137/EHI.S40493 PMID:27597809

Tabachnick, B. G., & Fidell, L. S. (2013). *Using multivariate statistics* (6th ed.). Pearson Education.

Tang, J., Yuan, X., Ramos, V., & Sriboonchitta, S. (2019). Does air pollution decrease inbound tourist arrivals? The case of Beijing. *Asia Pacific Journal of Tourism Research*, *24*(6), 597–605. doi:10.1080/10941665.2019.1610004

Tavitiyaman, P., & Qu, H. (2013). Destination Image and Behavior Intention of Travelers to Thailand: The Moderating Effect of Perceived Risk. *Journal of Travel and Tourism Management*, *3*(3), 169–185. doi:10.1080/10548408.2013.774911

Tawil, M., Tawil, Y. P., Rahmarini, G. M., & Salmon, I. P. (2021). *Collaborative Governance as a Strategy for Optimizing the Potential of Coastal Areas and Local Tourism: A Study in Tangerang Regency*. Academic Press.

Taylor, J. P. (2001). Authenticity and sincerity in tourism. *Annals of Tourism Research*, *28*(1), 7–26. doi:10.1016/S0160-7383(00)00004-9

Teddlie, C., & Yu, F. (2007). Mixed methods sampling: A typology with examples. *Journal of Mixed Methods Research*, *1*(1), 77–100. doi:10.1177/1558689806292430

Telegraph. (2013, July 24). *The world's most dangerous mountains*. Retrieved March 27, 2015, from Telegraph: https://www.telegraph.co.uk/travel/activityandadventure/10197830/The-worlds-most-dangerous-mountains.html?frame=2625884

Telfer, D. J. (2001). Strategic alliances along the Niagara Wine Route. *Tourism Management*, *22*(1), 21–30. doi:10.1016/S0261-5177(00)00033-9

Telfer, D., & Wall, G. (1996). Linkages between tourism and food production. *Annals of Tourism Research*, *23*(3), 635–653.

The Star. (2015). *Police launch pilot project to 'clean up' Brickfields*. Available at: https://www.thestar.com.my/Metro/Community/2015/09/28/Focused-sweep-in-Brickfields

Thees, H., Pechlaner, H., Olbrich, N., & Schuhbert, A. (2020). The Living Lab as a Tool to Promote Residents' Participation in Destination Governance. *Sustainability*, *12*(3), 1–27. doi:10.3390u12031120

Thiumsak, T., & Ruangkanjanases, A. (2016). Factors influencing international visitors to revisit Bangkok, Thailand. *Journal of Economics, Business and Management*, *4*, 220-230. http://www.joebm.com/index.php?m=content&c=index&a=show&catid=57&id=709

Thorun, C. (2018). Corporate Digital Responsibility: Unternehmerische Verantwortung in der digitalen Welt. In C. Gärtner & C. Heinrich (Eds.), *Fallstudien zur Digitalen Transformation* (pp. 173–191). Springer Gabler. doi:10.1007/978-3-658-18745-3_9

Thuraisingham, J., & Valu, V. V. (2020). 10. The financial vulnerability of Indian micro and small entrepreneurs in Malaysia. In *Vulnerable Groups in Malaysia* (pp. 140–151). De Gruyter Oldenbourg. doi:10.1515/9783110610970-010

Tikkanen, I. (2007). Maslow's hierarchy and food tourism in Finland: Five cases. *British Food Journal*, *109*(3), 635–653.

Toanoglou, M., Chemli, S., & Valeri, M. (2021). The organizational impact of COVID-19 crisis on travel perceived risk across four continents. *Journal of Organizational Change Management*. Advance online publication. doi:10.1108/JOCM-12-2020-0369

Tomor, Z., Meijer, A., Michels, A., & Geertman, S. (2019). Smart Governance For Sustainable Cities: Findings from a Systematic Literature Review. *Journal of Urban Technology*, *26*(4), 3–27. doi:10.1080/10630732.2019.1651178

Torija, A. J., Ruiz, D. P., & Ramos-Ridao, Á. (2011). Required stabilization time, short-term variability and impulsiveness of the sound pressure level to characterize the temporal composition of urban soundscapes. *Applied Acoustics*, *72*(2-3), 89–99. doi:10.1016/j.apacoust.2010.09.011

Towner, N. (2018). Surfing tourism and local stakeholder collaboration. *Journal of Ecotourism*, *17*(3), 268–286. doi:10.1080/14724049.2018.1503503

Trang, N. T., & Lee, G. (2018). Senses in leisure tourism: Scale development and its relationship with emotions. *International Journal of Tourism and Hospitality Research*, *32*(4), 21–39. doi:10.21298/IJTHR.2018.4.32.4.21

Trasande, L., & Thurston, G. D. (2005). The role of air pollution in asthma and other pediatric morbidities. *The Journal of Allergy and Clinical Immunology*, *115*(4), 689–699. doi:10.1016/j.jaci.2005.01.056 PMID:15805986

Travel and tourism economic impact. (2018). *World – WTTC.* https://www.wttc.org/-/media/files/reports/economic-impact-research/regions-2018/world2018. pdf

Trinchini, L., Kolodii, N. A., Goncharova, N. A., & Baggio, R. (2019). Creativity, innovation and smartness in destination branding. *International Journal of Tourism Cities, 5*(4), 529–543. doi:10.1108/IJTC-08-2019-0116

Trivellas, P., Kakkos, N., Vasiliadis, L., & Belias, D. (2016). Residents' perceptions toward cultural, social and economic benefits and costs of tourism industry: An empirical survey. In *Tourism and Culture in the Age of Innovation* (pp. 171–181). Springer. doi:10.1007/978-3-319-27528-4_12

Trivellas, P., Kakkos, N., Vasiliadis, L., & Belias, D. (2017). Sustainability, social marketing and host attitudes about Tourism in the city of Larissa. In *Strategic Innovative Marketing* (pp. 89–95). Springer. doi:10.1007/978-3-319-33865-1_11

Trunfio, M., & Della Lucia, M. (2019). Engaging Destination Stakeholders in the Digital Era: The Best Practice of Italian Regional DMOs. *Journal of Hospitality & Tourism Research (Washington, D.C.), 43*(3), 349–373. doi:10.1177/1096348018807293

Tsai, W., & Ghoshal, S. (1998). Social Capital and Value Creation: The Role of Intrafirm Networks. *Academy of Management Journal, 41*(4), 464–476.

Tsekos, T. N., & Triantafyllopoulou, A. (2016). From municipal socialism to the sovereign debt crisis: Local services in Greece 1980–2015. In *Public and Social Services in Europe* (pp. 135–149). Palgrave Macmillan. doi:10.1057/978-1-137-57499-2_10

Tsiotas, D., Belias, D., & Niavis, S. (2020). Internal Marketing on the Tourism Sector: Examination of How the Staff Can Develop on Greek Hotels. *Strategic Innovative Marketing and Tourism*, 963-970.

Tucker, H., & Boonabaana, B. (2011). A critical analysis of tourism, gender and poverty reduction. *Journal of Sustainable Tourism.*

Turismo de Portugal. (2019). https://travelbi.turismodeportugal.pt/pt-pt/Documents/Turismo%20em%20Portugal/turismo-em-numeros-2019.pdf

Turistička zajednica grada Biograd na Moru. (2020). Retrieved November 16, 2020, Available at: https://discover-biograd.com

Turistička zajednica općine Bale. (2020). Retrieved November 16, 2020, Available at: https://www.bale-valle.hr/

Turok, I. (2009). The distinctive city: Pitfalls in the pursuit of differential advantage. *Environment & Planning A, 41*(1), 13–30. doi:10.1068/a37379

Ullah, A. A., Akram Hossain, M., & Huque, A. S. (2021). Non-Conventional Migration: An Evolving Pattern in South Asia. *Journal of Asian and African Studies.* doi:10.1177/00219096211008468

Umbelino, J. (1998). Turismo em espaço rural: da utopia à realidade. In Turismo Horizontes Alternativos (pp. 175-188). Lisbon: Colibri.

Umjetnička galerija Dubrovnik. (2020). Retrieved November 15, 2020, Available at: https://www.ugdubrovnik.hr/?file=naslovna

United Nations Conference on Trade and Development. (2020). Retrieved November 28, 2020, Available at: https://unctad.org/webflyer/world-investment-report-2020

United Nation's Global Compact. (2000). United Nations.

UNWTO. (2007). *A practical guide to tourism destination management.* World Tourism Organization.

UNWTO. (2010). *Survey on destination governance: Evaluation report.* World Tourism Organization.

UNWTO. (2016). *Highlights of the 1st UNWTO Conference on Accessible Tourism in Europe (San Marino, 19-20 November 2014).* 10.18111/9789284417902

Upadhya, A., & Vij, M. (2020). Creative tourist experience: Role of destination management organizations. In *Destination Management and Marketing: Breakthroughs in Research and Practice* (pp. 763–783). IGI Global. doi:10.4018/978-1-7998-2469-5.ch042

Urbina, S. (2014). *Essentials of psychological testing.* Wiley.

Urbinati, A., Landoni, P., Cococcioni, F., & De Giudici, L. (2020). Stakeholder management in open innovation projects: A multiple case study analysis. *European Journal of Innovation Management.*

Valenzuela-Garcia, H., Parella, S., & Güell, B. (2017). *Revisiting the 'ethnic enclave economy': Resilient adaptation of small businesses.* Academic Press.

Valeri, M. (2021b). *Organization Theories. Implications for the Strategic Management.* Springer.

Valeri, M. (Ed.). (2021). Tourism risk. Crisis and recovery management. Emerald Publishing.

Valeri, M. (Ed.). (2021a). Tourism risk. Crisis and recovery management. Emerald Publishing.

Valeri, M., & Katsoni, V. (Eds.). (2021), Gender and tourism: challenges and entrepreneurial opportunities. Emerald Publishing. doi:10.1108/9781801173223

Valeri, M. (2015). Sustainability development and competitiveness of Rome as a tourist destination. *Tourism and Hospitality Management, 21*(2), 203–217. doi:10.20867/thm.21.2.7

Valeri, M. (2016). Networking and cooperation practices in the Italian tourism business. *Journal of Tourism, Heritage & Services Marketing, 2*(1), 30–35. doi:10.5281/zenodo.376333

Valeri, M. (2021b). *Organization Theories. Implications for the Strategic Management.* Springe.

Valeri, M., & Baggio, R. (2020a). Social network analysis: Organizational implications in tourism management. *The International Journal of Organizational Analysis.* Advance online publication. doi:10.1108/IJOA-12-2019-1971

Valeri, M., & Baggio, R. (2020b). Italian tourism intermediaries: A social network analysis exploration. *Current Issues in Tourism, 24*(9), 1270–1283. doi:10.1080/13683500.2020.1777950

Valeri, M., & Baggio, R. (2020c). A critical reflection on the adoption of blockchain in tourism. *Journal Information Technology and Tourism, 23*(2), 121–132. doi:10.100740558-020-00183-1

Valeri, M., & Baggio, R. (2021). Increasing the efficiency of knowledge transfer in an Italian tourism system: A network approach. *Current Issues in Tourism,* 1–16. Advance online publication. doi:10.1080/13683500.2021.1937960

Valeri, M., & Fadlon, L. (2018). Co-evolutionary prospects in tourism. *International Journal of Business and Management, 13*(9), 95–105. doi:10.5539/ijbm.v13n9p95

Valeri, M., & Katsoni, V. (Eds.). (2021). *Gender and tourism: challenges and entrepreneurial opportunities.* Emerald Publishing.

Vallbona, M. C., & Pérez, M. D. (2016). Los mercados de abastos y las ciudades turisticas. *PASOS Revista De Turismo y Patrimonio Cultural, 14*(2), 401–416. doi:10.25145/j.pasos.2016.14.026

van der Zee, E., Gerrets, A. M., & Vanneste, D. (2017). Complexity in the governance of tourism networks: Balancing between external pressure and internal expectations. *Journal of Destination Marketing & Management*, *6*(4), 296–308. doi:10.1016/j.jdmm.2017.07.003

Van der Zee, E., & Vanneste, D. (2015). Tourism networks unravelled; a review of the literature on networks in tourism management studies. *Tourism Management Perspectives*, *15*, 46–56. doi:10.1016/j.tmp.2015.03.006

Vannini, M.C., Nasciuti, G., Suzzani, V., & Tavecchio, G. (2019). Brand building analysis report Villa Bernasconi, Internal report. *Proceedings of Progetto LIBERTYTUTTI – Fondazione Cariplo and Progetto Maison Bernasconi – Regione Lombardia.*

Vareiro, L., Ribeiro, J. C., & Remoaldo, P. C. (2019). What influences a tourist to return to a cultural destination? *International Journal of Tourism Research*, *21*(2), 280–290. doi:10.1002/jtr.2260

Vasiliadis, L., Trivellas, P., Belias, D., Meleas, J., Kyriakou, D., & Koustelios, A. (2016). Cultural tourism revisited: The case of Thessaly. In *Tourism and culture in the age of innovation* (pp. 69–78). Springer. doi:10.1007/978-3-319-27528-4_4

Vassakis, K., Petrakis, E., Kopanakis, I., Makridis, J., & Mastorakis, G. (2019). Location-based social network data for tourism destinations. In *Big data and innovation in tourism, travel, and hospitality* (pp. 105–114). Springer. doi:10.1007/978-981-13-6339-9_7

Veal, A. J. (2006). *Research methods for leisure and tourism: A practical guide* (3rd ed.). Financial Times - Prentice Hall/Pearson Education.

Ventura, F. (1994). Women in Italian Agriculture: New Roles, New Problems. In L. van der Plas & M. Fonte (Eds.), *Rural Gender Studies in Europe* (pp. 80–90). Van Gorcum.

Verčič, A. T., & Ćorić, D. S. (2018). The relationship between reputation, employer branding and corporate social responsibility. *Public Relations Review*, *44*, 442–452.

Vermeulen, R. (2015). Pursuing the peripheral path? A path-dependent analysis of the Frankfurt and Munich fairs. *European Planning Studies*, *23*(2), 332–348. doi:10.1080/09654313.2013.867318

Vernon, J., Essex, S., Pinder, D., & Curry, K. (2005). Collaborative policymaking: Local sustainable projects. *Annals of Tourism Research*, *32*(2), 325–345. doi:10.1016/j.annals.2004.06.005

Veszprém-Balaton2023 Image Manual. (n.d.). http://archiv.veszprembalaton2023.hu/arculat

Veszprém-Balaton2023 official website. (n.d.). www.veszprembalaton2023.hu

Vial, G. (2019). Understanding digital transformation: A review and a research agenda. *The Journal of Strategic Information Systems*, *28*(2), 118–144. doi:10.1016/j.jsis.2019.01.003

Vieira, I. C. (2014). *Desenvolvimento turístico sustentável em pequenas cidades históricas: O caso de Lamego* (Tese de Doutoramento). Vila Real: UTAD.

Viollon, S., Lavandier, C., & Drake, C. (2002). Influence of visual setting on sound ratings in an urban environment. *Applied Acoustics*, *63*(5), 493–511. doi:10.1016/S0003-682X(01)00053-6

Virág, Á. (2016). A turisztikai együttműködések gazdasági szerepe. Studia Mundi – Economica, 3(1), 115-124.

Virfinija, J. (2016). Interaction between Cultural/Creative Tourism and Tourism/Cultural Heritage Industries. In L. Butowski (Ed.), *Tourism – From Empirical Research Towards Practical Application* (pp. 137–156). Tech.

Vivant, E. (2013). Creatives in the city: Urban contradictions of the creative city. *City. Cultura e Scuola*, *4*(2), 57–63. doi:10.1016/j.ccs.2013.02.003

Vodeb, K., & Rudež, H. (2016). Possibilities for inter-destination collaboration in tourism in the case of Opatija and Portorož: A managerial perspective. *RevijaZaSociologiju*, *46*(2), 205–227. doi:10.5613/rzs.46.2.4

Volgger, M., Pechlaner, H., & Pichler, S. (2017). *The practice of destination governance: A comparative analysis of key dimensions and underlying concepts*. Academic Press.

Volgger, M., & Pechlaner, H. (2014). Requirements for destination management organizations in destination governance: Understanding DMO success. *Tourism Management*, *41*, 64–75. doi:10.1016/j.tourman.2013.09.001

Volgger, M., & Pechlaner, H. (2015). Governing networks in tourism: What have we achieved, what is still to be done and learned? *Tourism Review*, *70*(4), 298–312. doi:10.1108/TR-04-2015-0013

Volo, S. (2020). Tourism statistics, indicators and big data: A perspective article. *Tourism Review*, *75*(1), 304–309. doi:10.1108/TR-06-2019-0262

Vosoughi, S., Roy, D., & Aral, S. (2018). The spread of true and false news online. *Science*, *359*(6380), 1146–1151. doi:10.1126cience.aap9559 PMID:29590045

Vostrá, H., Jindrová, A., & Dömeová, L. (2011). The Position of the CR among the EU States Based on Selected Measures of the Lisbon Strategy. *Journal of Competitiveness*, *11*(3), 50–57.

Vu, H. M., & Ngo, V. M. (2019). *Strategy Development from Triangulated Viewpoints for a Fast Growing Destination Toward Sustainable Tourism Development–A Case Of Phu Quoc Islands in Vietnam*. Academic Press.

Vu, H. M., & Ngo, V. M. (2019). Strategy Development from Triangulated Viewpoints for a Fast Growing Destination Toward Sustainable Tourism Development–A Case Of Phu Quoc Islands in Vietnam. *Journal of Tourism and Services*, *10*(18), 117–140. doi:10.29036/jots.v10i18.86

Vukovic, D. B., Petrovic, M., Maiti, M., & Vujko, A. (2021). Tourism development, entrepreneurship and women's empowerment–Focus on Serbian countryside. *Journal of Tourism Futures*.

Wachsmuth, D., & Weisler, A. (2018). Airbnb and the rent gap: Gentrification through the sharing economy. *Environment and Planning A. Economy and Space*, *50*(6), 1147–1170.

Waitt, G., & Duffy, M. (2010). Listening and tourism studies. *Annals of Tourism Research*, *37*(2), 457–477. doi:10.1016/j.annals.2009.10.017

Wangenheim, F., & Bayón, T. (2004). The effect of word of mouth on services switching: Measurement and moderating variables. *European Journal of Marketing*, *38*, 1173–1185. doi:10.1108/03090560410548924

Wang, L., Fang, B., & Law, R. (2018). Effect of air quality in the place of origin on outbound tourism demand: Disposable income as a moderator. *Tourism Management*, *68*, 152–161. doi:10.1016/j.tourman.2018.03.007

Wang, M. C., & Wang, C. S. (2018). Tourism, the environment, and energy policies. *Tourism Economics*, *24*(7), 821–838. doi:10.1177/1354816618781458

Wang, S., & Gu, K. (2020). Pingyao: The historic urban landscape and planning for heritage-led urban changes. *Cities (London, England)*, *97*, 102489. doi:10.1016/j.cities.2019.102489

Wang, Y. (2008). Collaborative destination marketing: Understanding the dynamic process. *Journal of Travel Research*, *47*(2), 151–166. doi:10.1177/0047287508321194

Wang, Y., & Krakover, S. (2008). Destination marketing: Competition, cooperation or coopetition? *International Journal of Contemporary Hospitality Management, 20*(2), 126–141. doi:10.1108/09596110810852122

Wang, Y., Shen, H., Ye, S., & zhou, L. (2020). Being rational and emotional: An integrated model of residents' support of ethnic tourism development. *Journal of Hospitality and Tourism Management, 44*, 112–121. doi:10.1016/j.jhtm.2020.05.008

Warner, S., Bowers, M. T., & Dixon, M. A. (2012). Team dynamics: A social network perspective. *Journal of Sport Management, 26*(1), 53–66. doi:10.1123/jsm.26.1.53

Wäsche, H., Dickson, G., Woll, A., & Brandes, U. (2017). Social network analysis in sport research: An emerging paradigm. *European Journal of Sport Science, 14*(2), 138–165. doi:10.1080/16138171.2017.1318198

Wearing, S. L., & Foley, C. (2017). Understanding the tourist experience of cities. *Annals of Tourism Research, 65*, 97–107. doi:10.1016/j.annals.2017.05.007

Webster, C., & Ivanov, S. (2020). Robotics, artificial intelligence, and the evolving nature of work. In B. George & J. Paul (Eds.), *Digital Transformation in Business and Society Theory and Cases* (pp. 127–143). Palgrave-MacMillan. doi:10.1007/978-3-030-08277-2_8

Weidenfeld, A. (2013). Tourism and cross border regional innovation systems. *Annals of Tourism Research, 42*, 191–213. doi:10.1016/j.annals.2013.01.003

Welter, F. (2011). Contextualizing entrepreneurship-conceptual challenges and ways forward. *Entrepreneurship Theory and Practice, 35*(1), 165–184.

West, J. B. (1984). Human Physiology at Extreme Altitudes on Mount Everest. *Science, 223*(4638), 784–788. doi:10.1126cience.6364351 PMID:6364351

Whang, H., Yong, S., & Ko, E. (2016). Pop culture, destination images, and visit intentions: Theory and research on travel motivations of Chinese and Russian tourists. *Journal of Business Research, 69*(2), 631–641. doi:10.1016/j.jbusres.2015.06.020

Whatmore, S. (1988). *Farming Women: Gender, Work and Family Enterprise*. Macmillan.

Whelan, E., Parise, S., De Valk, J., & Aalbers, H. L. (2011). Creating Employee Networks That Deliver Open Innovation. *MIT Sloan Management Review, 53*(1), 37–44.

Willford, A. (2020). 8. The Figure of the Tamil in Modern Malaysia. In *Clio/Anthropos* (pp. 223–272). Stanford University Press. doi:10.1515/9780804772402-009

Williams, C., & Buswell, J. (2003). Service Quality in Leisure and Tourism. Cromwell Press.

Williams, N. L., Inversini, A., Ferdinand, N., & Buhalis, D. (2017). Destination eWOM: A macro and meso network approach? *Annals of Tourism Research, 64*, 87–101. doi:10.1016/j.annals.2017.02.007

Wirtz, J., Patterson, P., Kunz, W., Gruber, T., Lu, V. N., Paluch, S., & Martins, A. (2018). Brave new world: Service robots in the frontline. *Journal of Service Management, 29*(5), 907–931. doi:10.1108/JOSM-04-2018-0119

Wirtz, J., So, K. K. F., Mody, M., Chun, E. H., Liu, S., & Chun, H. (2019). Platforms in the peer-to-peer sharing economy. *Journal of Service Management, 30*(4), 452–483. doi:10.1108/JOSM-11-2018-0369

Wissmann, T. (2014). *Geographies of urban sound*. Routledge.

Wolf, E. (2002). *Culinary tourism: A tasty economic proposition*. http//: www.culinarytourism.org/faq.php

Wondirad, A., Tolkach, D., & King, B. (2020). Stakeholder collaboration as a major factor for sustainable ecotourism development in developing countries. *Tourism Management, 78*, 104024. doi:10.1016/j.tourman.2019.104024

Wong, N. L., & Lau, K. L. (2016). Voices of Third Generation Malaysians: Malaysian or Malaysian Indian'. *Sarjana, 31*(1), 45–54.

Woosnam, K. M., Maruyama, N. U., Ribeiro, M. A., & Joo, D. (2019). Explaining minority residents' attitudes of ethnic enclave tourism from general perceptions of tourism impacts. *Journal of Tourism and Cultural Change, 17*(4), 467–484. doi:10.1080/14766825.2019.1601407

World Tourism Organization (UNWTO). (2000). *Public-Private Sector Cooperation-Enhancing Tourism Competitiveness*. World Tourism Organization.

World Tourism Organization (UNWTO). (2007). *A Practical Guide to Tourism Destination Management*. World Tourism Organization.

World Tourism Organization (UNWTO). (2019). *UNWTO Guidelines for Institutional Strengthening of Destination Management Organizations (DMOs) – Preparing DMOs for new challenges*. World Tourism Organization.

World Tourism Organization. (2017). *Affiliate Members Report. Volume sixteen. In Second Global Report on Gastronomy Tourism*. UNWTO.

World Travel & Tourism Council. (n.d.). *Travel & Tourism Economic Impact 2018-Greece*. Retrieved from https://www.wttc.org/-/media/files/reports/economic-impact-research/countries-2018/greece2018.pdf

WTO. (2003). *Proceedings of the 1st International Conference on Climate Change and Tourism Djerba, Tunisia*, 9-11 April 2003. WTO.

Xiang, Z. (2018). From digitization to the age of acceleration: On information technology and tourism. *Tourism Management Perspectives, 25*, 147–150. doi:10.1016/j.tmp.2017.11.023

Xiang, Z., & Fesenmaier, D. R. (2017). Big Data Analytics, Tourism Design and Smart Tourism. In F. Zheng (Ed.), *Analytics in Smart Tourism Design, Concepts and Methods*. doi:10.1007/978-3-319-44263-1_17

Xiao, J., Tait, M., & Kang, J. (2018). A perceptual model of smellscape pleasantness. *Cities (London, England), 76*, 105–115. doi:10.1016/j.cities.2018.01.013

Xiaomei, W., & Daming, X. (2018). The mismatches between minority language practices and national language policy in Malaysia: A linguistic landscape approach. *Kajian Malaysia: Journal of Malaysian Studies, 36*(1).

Xiong, J., Hashim, N. H., & Murphy, J. (2015). Multisensory image as a component of destination image. *Tourism Management Perspectives, 14*, 34–41. doi:10.1016/j.tmp.2015.03.001

Xiong, L., King, C., & Piehler, R. (2013). "That's not my job": Exploring the employee perspective in the development of brand ambassadors. *International Journal of Hospitality Management, 35*, 348–359. doi:10.1016/j.ijhm.2013.07.009

Xu, B., & Dong, A. (2020). Evaluating the Impact of Air Pollution on China's Inbound Tourism: A Gravity Model Approach. *Sustainability, 12*(4), 1456. doi:10.3390u12041456

Xu, F., Lin, X., Li, S., & Niu, W. (2018). Is Southern Xinjiang Really Unsafe? *Sustainability, 10*(12), 4639. Advance online publication. doi:10.3390u10124639

Xu, Z., & Zhang, J. (2016). Antecedents and consequences of place attachment: A comparison of Chinese and Western urban tourists in Hangzhou, China. *Journal of Destination Marketing & Management, 5*(2), 86–96. doi:10.1016/j.jdmm.2015.11.003

Yahya, M. S. S., Safian, E. E. M., & Burhan, B. (2020). The Development and Distribution Pattern of Railway Network for Urban Public Transport Using GIS from 1990 Until 2019 in The Klang Valley and Kuala Lumpur, Malaysia. *Journal of Social Transformation and Regional Development, 2*(2), 1–10.

Yang, W., & Kang, J. (2005a). Acoustic comfort evaluation in urban open public spaces. *Applied Acoustics, 66*(2), 211–229. . 011 . doi:10.1016/j.apacoust.2004.07

Yan, G., Watanabe, N. M., Shapiro, S. L., Naraine, M. L., & Hull, K. (2018). Unfolding the Twitter scene of the 2017 UEFA Champions League Final: Social media networks and power dynamics. *European Sport Management Quarterly*, 1–18.

Yanga, Y., Chenb, M.-H., Sud, C.-H., & Yu-Xia, L. (2020). Asymmetric effects of tourist arrivals on the hospitality industry. *International Journal of Hospitality Management, 84*, 102323. doi:10.1016/j.ijhm.2019.102323

Yang, L., & Wall, G. (2016). *Planning for ethnic tourism*. Routledge. doi:10.4324/9781315600741

Yang, X., & Rivers, C. (2009). Antecedents of CSR practices in MNCs' subsidiaries: A stakeholder and institutional perspective. *Journal of Business Ethics, 86*(2), 155–169. doi:10.100710551-009-0191-0

Yap, J. B. H., Chua, C. Y., & Skitmore, M. (2021). Towards Sustainable Mobility with Transit-Oriented Development (TOD): Understanding Greater Kuala Lumpur. *Planning Practice and Research, 36*(3), 314–336. doi:10.1080/02697459.2021.1883249

Yázigi, E., Carlos, A. F. A., & Cruz, R. C. A. (1996). *Turismo: Espaço, paisagem e cultura*. Turismo cultural. Impacto ambiental. Geografia. Politica social. Conservacion de la naturaleza. São Paulo Hucitec.

Yeh, C. M. (2013). Tourism involvement, work engagement and job satisfaction among frontline hotel employees. *Annals of Tourism Research, 42*, 214–239. doi:10.1016/j.annals.2013.02.002

Ye, S., Wu, J. S., & Zheng, C. J. (2019). Are tourists with higher expectation more sensitive to service performance? Evidence from urban tourism. *Journal of Destination Marketing & Management, 12*, 64–73. doi:10.1016/j.jdmm.2019.01.002

Yin, I., & Abdullah, J. (2020). The development control of urban centre in Kuala Lumpur. *Planning Malaysia, 18*(13).

Yoon, M., & Suh, J. (2003). Organizational citizenship behavior and service quality as external effectiveness of contact employees. *Journal of Business Research, 56*(8), 597–611. doi:10.1016/S0148-2963(01)00290-9

Youn, H., Lee, K., & Lee, S. (2018). Effects of corporate social responsibility on employees in the casino industry. *Tourism Management, 68*, 328–335. doi:10.1016/j.tourman.2018.03.018

Yuan, J., Deng, J., Pierskalla, C., & King, B. (2018). Urban tourism attributes and overall satisfaction: An asymmetric impact-performance analysis. *Urban Forestry & Urban Greening, 30*, 169–181. doi:10.1016/j.ufug.2018.02.006

Yucel. (2012). Examining the Relationships among Job Satisfaction, Organizational Commitment, and Turnover Intention: An Empirical Study. *International Journal of Business and Management, 7*(20), 44–58.

Yuen, K., Loh, H. S., Zhou, Q., & Wong, Y. D. (2018). Determinants of job satisfaction and performance of seafarers. *Transportation Research Part A, Policy and Practice, 110*, 1–12. doi:10.1016/j.tra.2018.02.006

Yunus, R. M., Samadi, Z., & Omar, D. (2018). Community's Participation: A strategic approach in revitalizing heritage streets. *Journal of ASIAN Behavioural Studies, 3*(10), 29–37. doi:10.21834/jabs.v3i10.301

Yüzbaşıoğlu, N., & Doğan, O. (2021). Strategic governance in the marina sector in the context of marine tourism cluster. *Almatourism-Journal of Tourism, Culture and Territorial Development, 12*(23), 59–82.

Zaheer, A., & Bell, G. G. (2005). Benefiting from Network Position: Firm Capabilities, Structural Holes, and Performance. *Strategic Management Journal*, *26*(9), 809–825. doi:10.1002mj.482

Zaheer, A., & Soda, G. (2009). Network Evolution: The Origins of Structural Holes. *Administrative Science Quarterly*, *54*(1), 1–31. doi:10.2189/asqu.2009.54.1.1

Zahra, A. (2011). Rethinking regional tourism governance: The principle of subsidiarity. *Journal of Sustainable Tourism*, *19*(4), 535-552. doi:.576764 doi:10.1080/09669582.2011

Zainal, A., Nizan, A., & Nizam, M. (2010). *Malaysian Gastronomy Routes as a Tourist*. Academic Press.

Zainol, N. R., Al Mamun, A., Hassan, H., Rajennd, A., & Muniady, L. (2017). Examining the effectiveness of micro-enterprise development programs in Malaysia. *Journal of International Students*, *10*(2).

Zainuddin, Z., Zahari, M. S. M., Radzi, S. M., Hanafiah, M. H. M., & Ishak, N. (2018). Tourism images and the international tourists response behaviour: A case of Langkawi Island, Malaysia. *International Journal of Academic Research in Business & Social Sciences*, *8*(16), 60–82.

Zainudin, H., Bachek, S. H., & Haron, N. A. (2016). Preservation of Culture and Built Heritage in New Urban Development: A Case Study on Little India Brickfields. Academic Press.

Zamri, N. A. (2020). Walkways for visually impaired person in Brickfields. Academic Press.

Zeithaml, A. (1988). Consumer Perceptions of Price, Quality, and Value: A Means-End Model and Synthesis of Evidence. *Journal of Marketing*, *52*(3), 2–22. doi:10.1177/002224298805200302

Zeithaml, A., & Bitner, J. (1996). *Services marketing*. McGraw-Hill.

Zhang, A., Zhong, L., Xu, Y., Wang, H., & Dang, L. (2015). Tourists' Perception of Haze Pollution and the Potential Impacts on Travel: Reshaping the Features of Tourism Seasonality in Beijing, China. *Sustainability*, *7*(3), 2397–2414. doi:10.3390u7032397

Zhang, T., Chen, J., & Hu, B. (2019). Authenticity, Quality, and Loyalty: Local Food and Sustainable Tourism Experience. *Sustainability*, *11*(12), 3437. doi:10.3390u11123437

Zhao, X., Ghiselli, R., Law, R., & Ma, J. (2016). Motivating frontline employees: Role of job characteristics in work and life satisfaction. *Journal of Hospitality and Tourism Management*, *27*, 27–38. doi:10.1016/j.jhtm.2016.01.010

Ziakas, V., & Costa, C. A. (2010). Explicating inter-organizational linkages of a host community's events network. *International Journal of Event and Festival Management*, *1*(2), 132–147. doi:10.1108/17852951011056919

Zopiatis, A., Constanti, P., & Theocharous, A. L. (2014). Job involvement, commitment, satisfaction and turnover: Evidence from hotel employees in Cyprus. *Tourism Management*, *41*, 129–140. doi:10.1016/j.tourman.2013.09.013

About the Contributors

Marco Valeri is a Senior Lecturer of Organizational Behavior (Master Degree and Doctoral Course), Faculty of Economics, Niccolò Cusano University, Rome (Italy). He is Visiting Professor at Faculty of Social Sciences and Leisure Management, School of Hospitality, Tourism and Events, Taylor's University (Malaysia). His research fields include competitive advantage, sustainability and green practices, strategy implementation, knowledge management, family business and tourism hotel/lodging management, crisis management, destination marketing, information technology and developing countries, network analysis. He was a visiting professor in several Universities: University of Eichastaett – Ingolstadt (Germa-ny), Universitè Jean Moulin, Lyon (France), Holy Spirit University of Kaslik (USEK), Lebanon, University of Pannonia (Hungary), University of West Attica (Greece), TBS Business School (Toulouse). He is member of several editorial board of international tour-ism journals, reviewer and editor of several handbooks on entrepreneurship, tourism and hospitality management (Emerald Publishing, Springer and IGI Global).

* * *

Hendrik Aalbers is an Associate Professor (tenured) in Strategy and Innovation at the department of Business Administration of the Radboud University Nijmegen, The Netherlands. He has been a visiting scholar at Imperial College London, National University of Ireland at Galway and The Tokyo Institute of Technology. His research focuses on organizational restructuring and innovation and revolves around the following research lines: Collaboration and intra-firm network intervention, Organizational restructuring and downsizing and Advanced technology business model development. when it comes to alpine tourism, Rick has been a snowboarding and downhill biking adept, scaling alpine regions on four of the continents, and as a waterman always in keen pursuit of such activity whenever such extreme sports touristic opportunity arises.

Albattat Ahmad is an Assistant Professor and Postgraduate Coordinator in Post Graduate Centre, Management and Science University, Shah Alam, Selangor, Malaysia. He is a visiting Professor and external examiner in Medan Academy of Tourism (Akpar Medan). He holds a doctoral degree in Hospitality Management "Disaster and Emergency Planning and Preparedness" from University Sains Malaysia (USM). He worked as an Assistant Professor, Ammon Applied University College, Amman, Jordan, and Researcher at Sustainable Tourism Research Cluster (STRC), Penang, Malaysia. He was working for the Jordanian hospitality industry for 17 years. He has participated and presented research papers in several International academic conferences He is an active member of Scientific and Editorial Review

Board for various SCOPUS journals. His latest works have been published in the refereed international journals, conference proceedings, books, and book chapters.

Saleh Mohamad Ali is a full-time Ph.D. graduating candidate at the University of Pannonia Faculty of Management Science and Business Administration. Mohamad holds a master degree in Business administration, two bachelor degrees in Management information system and Business administration from Arts, Sciences, and Technology University in Lebanon. After joining the University of Pannonia Ph.D. school of management, he taught undergraduate and graduate classes. He also participated in many international research groups launched by the university. Research interest: International management, Strategic management, Information technology, Multinational companies, Corporate social responsibility, Stakeholders' satisfaction, National and organizational culture, Globalization

Vincenzo Asero is Assistant Professor at University of Catania, Department of Political and Social Sciences, where he currently teaches Economics, Regional Economics, and Heritage and Tourism Development. He is Academic Member of the TEMA+ European Territories: Heritage and Development, Erasmus Mundus Joint Master Course, and Academic Member of the Tourism, Leisure & Recreation Unit of ATINER. His research interests cover Tourism, Local Development, Regional Economics, and Business Networks. In his area of expertise, he is actively involved as consultant and published articles, book chapters and edited books. He is an active reviewer for several international journals on tourism and currently co-Editor of the Athens Journal of Tourism.

Graça Azevedo has a PhD in Management (scientific field: Accounting) at the Institute of Higher Labor and Enterprise – ISCTE. She has published and presented several articles in national and international journals and conferences as well. She is a member of the Editorial Board of international journal "Accounting and Finance Research". She currently teaches courses in scientific field of Financial Accounting at the Institute of Higher Learning in Accounting and Administration of the University of Aveiro. Her research interests are related to financial reporting, impression management strategies and international accounting standards.

Angela Besana is Associate Professor of Economics at IULM University. Angela Besana has been teaching microeconomics, macroeconomics, economics of culture and tourism, economics of arts and fundraising for more than two decades. Her research concerns economics of creativity and music, economics of nonprofit organizations, microeconomics, and macroeconomics.

Sandro Billi is an expert in Destination Development and Marketing, with more than 30 years' experience in research and tourism strategies. He has a wealth of experience in producing studies and marketing plans for government organisations, private companies, trade associations, and public authorities.

Ricardo Biscaia is an associate researcher at CIPES and an invited assistant professor at Universidade de Aveiro and Instituto Politécnico do Porto. He holds a PhD in Economics since January 2014, with a thesis on Location Theory. His main research interests are in the fields of Industrial Economics, Regional and Urban Economics and Higher Education Policies, the latter being focused on Higher Education funding, graduate job markets and bibliometrics. Some of his work has been published in several Journals, such as Public Administration Review, Papers in Regional Science, Applied Economics, Higher

Education, Studies in Higher Education or Cambridge Journal of Economics. Ricardo joined the Faculty of Economics at the Universidade do Porto as an assistant professor in September 2021.

Zélia Breda holds a PhD in Tourism, a Master's degree in Chinese Studies (Business and International Relations) and a degree in Tourism Management and Planning from the University of Aveiro, where she is Assistant Professor, at the Department of Economics, Management, Industrial Engineering, and Tourism, holding the position of Director of the Master in Tourism Management and Planning. She is a full member of the Research Unit' Governance, Competitiveness and Public Policies' of the University of Aveiro; a founding member and vice-president of the Observatory of China (www.observatoriodachina.org), and a senior researcher of the spin-off company Idtour (www.idtour.pt). She is also a member of the editorial and scientific boards of a few academic national and international journals, as well as a member of the organizing and scientific committees of international tourism conferences. She has authored and co-authored several national and international papers and communications on tourism, and has been taking part in several research projects in the tourism field, both as a member of the team and as a consultant.

Maria João Carneiro is Assistant Professor of Tourism at the University of Aveiro (UA) (in Portugal) and researcher at the Governance, Competitiveness, and Public Policy (GOVCOPP) research unit at this University. She holds a Degree in Tourism Management and Planning from UA, an MBA from New University of Lisbon and a PhD in Tourism from UA. She is the Director of the degree program (licenciatura) in Tourism of the UA. She has published papers in several journals and conference proceedings, and has also presented several works in conferences and seminars. Her research interests are competitiveness of tourism destinations, consumer behaviour in tourism and destination marketing.

Alexandra Correia, Ph.D in Tourism, is an Associate Professor of Tourism Studies at Polytechnic Institute of Viana do Castelo. She is the Head of the Master in Innovative Tourism Development, and her main interests include wine tourism, inter-business cooperation, tourism experiences, outdoor tourism, and digital marketing. She presented papers at national and international conferences and has been a member of scientific committees. She has been involved in funded I&D national and international projects, along with projects with local private and public organisations.

Rúben Duarte has a degree in New Communication Technologies (2014), and a Master in Communication and Multimedia (2016), both from the University of Aveiro. In his Master's dissertation, he developed a dashboard model for presenting information to citizens, applied to the context of Águeda, where he is from. He is currently integrated in the TheoFrameAccountability Project, at ISCA-UA, responsible for the development and management of the technological platform.

Celeste Eusébio is an Associate Professor of Tourism and a researcher at the GOVCOPP Research Unit at the University of Aveiro (Portugal). She holds a Degree in Tourism Management and Planning from the University of Aveiro (1995), a Master in Economics from the University of Coimbra (1998) and a Doctoral Degree in Tourism, also from the University of Aveiro (2006). Her research interests include tourism economics, tourism impacts, tourism forecasts and consumer behavior in tourism.

Augusta Ferreira completed his first degree at the Higher Institute of Accounting and Administration of Aveiro; she got his Masters Degree in Business Management Specialization in Corporate Finance at Minho University and she holds a PhD in Accounting at Aveiro University and Minho University. Currently she teaches Accounting within the accounting scientific area at Higher Institute Accounting and Administration of Aveiro (Aveiro University) and she is interested in researching public sector accounting, accounting information systems, disclosure of financial information and accountability. She has presented and published many articles in national and international journals and conferences.

Tomáš Gajdošík is associate professor at the Department of Tourism, Faculty of Economics, Matej Bel University, Banská Bystrica. In his research he focuses on smart tourism and tourism information technologies. He also deals with the issues of destination management, governance and leadershisp. To date he has contributed to several monographs, textbooks and scientific journals such as Tourism Review, Tourism Economics, Tourism Management Perspectives, Quaestiones Geographicae, Czech Journal of Tourism, Economic Review of Tourism and others.

Gonçalo Gomes is the Investment Support's Head of Division of Tourism Center of Portugal Regional Entity (TCP) since 2014. Previously he coordinated this area since 2007 at TCP, where he worked since 1999. He holds a degree in Public Administration Management from ISCSP (Lisbon University). He was the creator of José Manuel Alves Prize, a tourism entrepreneurship award in Center of Portugal as well as of the Academic Thesis Prize, that elects the best new tourism project and the best tourism related thesis, both held annually by TCP since 2015. He was the co-creator of the Center of Portugal Sustainable Tourism Observatory. His main interests as fields of work and/or study are tourism entrepreneurship, digital skills in tourism enterprises, tourism activities monitoring, sustainable tourism and legal and financial issues for tourism SME.

Gyöngyi Steinbachné Hajmásy is a junior lecturer at the Tourism Department, Faculty of Business and Economics, University of Pannonia. She graduated as an Economist specializing in Tourism and Hospitality in 2010 at the University of Pannonia. She got her master's degree in 2013 as an Economist specializing in Tourism Management at the University of Pannonia. Her PhD research focuses on corporate social responsibility in the hospitality industry. She has gained practical experience working in the hospitality industry for several years and also as a project manager at the Balaton Riviéra DMO.

Edit Kővári is an associate professor at the University of Pannonia Faculty of Business and Economics. Her PhD thesis which was completed at the University of Derby Business School focusing on hotel managers' emotional intelligence and its relation to task and contextual performance within organisational culture. She also wrote 3 books and several articles, took part in various national and international projects as a professional expert. In 2015 she founded the Faculty's mentor program (Pannon Pentor Program) for Hungarian and international students. Apart from being a professor she is a practising coach and trainer of emotional intelligence, communication and resilience. She also works with Hungarian settlements' decision makers and influencers to develop liveable city environments. Edit was presentation member for European Capital of Culture (ECOC) bid for Veszprém 2023 and after winning the title she became the University of Pannonia representative of ECOC.

Katalin Lőrincz is an associate professor and vice dean for general affair at Faculty of Business and Economics, University of Pannonia, Hungary. Her main research area is the relationship between tourism, mobility, quality of life and sustainability. As a lecturer, expert and leader of a recent circular economy research program, she focuses on the sustainable tourism, the Lake Balaton region and the geographical elements of various destinations. Involved in the local community, she takes an active role in the preparation of the Veszprém-Balaton 2023 European Capital of Culture programme.

Mara Madaleno is a Ph.D. in Economics from the University of Aveiro. Director of the Master in Economics from DEGEIT, University of Aveiro. Assistant Professor in Finance and Economics. Vice-coordinator of the research line Systems for Decision Support at GOVCOPP. Vice-director of the Master in Data Science for Social Sciences.

Rui Pedro Figueiredo Marques received the PhD degree in Computer Science in 2014 from the universities of Minho, Aveiro and Porto, three of the top universities in the north of Portugal. In 2008 he concluded his Masters degree in Electronics and Telecommunications Engineering, at the University of Aveiro, and in 2005 he graduated in the same area, also from the University of Aveiro. His main research interests are related to the integration of Information Systems into organizational topics, contributing with solutions to improve the risk management and the organizational efficiency. He is Professor of Information Systems and Technology since 2007 at the Higher Institute of Accounting and Administration, University of Aveiro, Portugal.

Maria Manuela Santos Natário is Coordinating Professor at Polytechnics of Guarda, Portugal. She holds a PhD in Economics of Innovation and Competitiveness by University of Évora, and as the Master in Management by the University of Beira Interior. She teaches Innovation and Creativity, Innovation and Competitiveness, International Management and Economy. She is also Research Fellow at Research Unit for Inland Development (Portugal). Her main interests include Economy of Innovation, Innovation and Competitiveness. She has published articles in international journals such as European Journal of Innovation Management, Tékhne Studies Review, Journal of Business Economics and Management, Journal of the American Academy of Business, Journal of Comparative International Management, Journal of Knowledge-based Innovation in China. She has published articles in international books.

Christos Papadimitriou is an Assistant Professor in Management - Human Resources Management, Neapolis University Pafos.

Panoraia Poulaki is teaching and research staff, University of the Aegean, Greece.

Susana Andreia Salgueiro Rachão is a Ph.D. in Tourism from the University of Aveiro, Portugal and has a Msc in Tourism, Innovation and Development from the Polytechnic Institute of Viana do Castelo (IPVC) and a Bachelor's degree in Tourism from the same institution. She is a research member at CETRAD (Centre of Transdisciplinary Studies for Development, UTAD). Her areas of interest are creative and food-and-wine tourism, innovation and product development and tourism market analysis.

Margarita Matias Robaina holds a PhD in Economics, a Master's in Business Economics and a degree in Economics from the University of Aveiro. She is a member of the Governance, Competitiveness and Public Policy Research Unit (GOVCOOP), in the Competitiveness, Innovation and Sustainability group. Its areas of interest are Energy Economics, Environmental and Natural Resources Economics, Energy and Environmental Policy and General Equilibrium Models. She is Assistant Professor in the

Department of Economics, Management, Industrial Engineering and Tourism (DEGEIT) at the University of Aveiro since 2011. Margarita has several publications in books and international journals with scientific arbitration, and has participated in several conferences, both in the Organizing Committee, in the Scientific Committee.

Vitor Rodrigues is a research fellow at the Department of Economics, Management, Industrial Engineering and Tourism, University of Aveiro (UA). With a Master in Tourism Management and Planning and a Degree in Tourism from the UA, his research work has been focused on Chinese Outbound Tourism, Meetings Industry and Tourism Governance, counting several scientific works as author and co-author.

Ioannis F. Rossidis holds a bachelor degree in Business Administration from University of Piraeus, a MBA from University of the Aegean, a Msc in Health Management from the National School of Public Health and a Phd in Administrative Science from Panteion University, Public Administration Department (national scholarship). Dr. Rossidis has been teaching over the last decade in undergraduate and post graduate programs in various Greek and Cypriot Universities. Dr. Rossidis has also worked as Financial Director in the Hellenic Quality Assurance& Accreditation Agency, as Ministry of Finance, Piraeus Bank and ATE Bank executive and as manager in Leivatho Hotel (4star). Dr. Rossidis, has contributed to various conferences while he has authored numerous published papers (80 in total). His research interests are administrative science, change management, public management, knowledge management.

Elena Rudan, PhD, is an associate professor at the Faculty of Tourism and Hospitality Management, University of Rijeka, Croatia. She is the Head of the Department of International Economics and Development and the Head of the undergraduate study "Business Economics in Tourism and Hospitality" at the Faculty of Tourism and Hospitality Management, University of Rijeka. Her areas of interest are cultural and creative tourism, entrepreneurship in tourism, circular economy on the sustainability of cultural tourism, strategic development and new strategic directions in tourism. She is involved in scientific and professional projects at national and international levels mainly in the field of cultural tourism, heritage tourism and strategic management. Elena Rudan is the author of scientific papers, book chapters, books and professional contributions.

Carlos Alberto Lourenço dos Santos completed his first degree in Electrical Engineering at the University of Coimbra he got his masters degree in Electrical Engineering Sciences and he holds a PhD in Information Systems and Computer Engineering from the Instituto Superior Técnico (Technical University of Lisbon). Currently he teaches Computer Sciences within the area of scientific computing at Higher Institute for Accountancy and Administration of Aveiro (Aveiro University) and he is interested in researching information systems auditing, accounting information systems, XBRL language and Business Intelligence. He as presented and published about 50 articles in national and international journals and conferences.

Goretti Silva is Associate Professor in Tourism Studies at the Polytechnic Institute of Viana do Castelo, Portugal. She obtained her PhD degree at Bournemouth University (UK), about the role of relationships, conceptualized as social capital and social support, in the creation and management of Tourism related small family businesses in rural areas. Her research interests include tourism as a strategy for rural development, particularly in peripheral regions, SME's in Tourism, tourism destinations' sustainable development, and tourism planning at regional, local and project level. Prior to being a member of academia she has collaborated with several organisations as tourism consultant, namely the Peneda Gerês National Park in the North of Portugal. She is currently involved in several research projects, both at national and international level, namely Project INCOME: Innovative Cooperation Business – HEI learning model for Tourism, co-funded by Erasmus + (KA2); and Project TURNOUT aiming at understanding the development of Outdoor Tourism in the North region (Portugal), co-financed by FEDER.

Luís Soares holds a degree in Tourism from Instituto Politécnico de Viana do Castelo (IPVC). Former research of the funded Project, TURNOUT - Development of Outdoor Tourism in the North Region of Portugal. Currently working at Your Tours, in Braga, Portugal, cultural touring and customized trips in Portugal. His main interests include cultural tourism, tourism experiences, and outdoor tourism.

Ramon Spildooren is a member of the Radboud Centre for Organization Restructuring, focusing on network analytics and network change at large.

Christian Stipanović, PhD, is a full professor with tenure at the Faculty of Tourism and Hospitality Management, University of Rijeka, Croatia. His research interest relate to entrepreneurship in tourism, strategic management in tourism, audio management, business intelligence, event management, cultural and creative industries, logistic processes in tourism destinations. He is the author of two university books, eight chapters in books and 80 scientific papers. He leads national and international scientific and professional projects mainly in the field of strategic management, concept development and audio management.

Maria Cristina Vannini, archaeologist by background, museologist and expert in museum studies and cultural projects, founded in 2004 soluzionimuseali-ims, a consulting agency for museums. Her expertise is hybrid, having worked in archaeology and in the field of marketing research. Former Director General of INSULA-UNESCO, National Secretary of ICOM Italy and member of the Board of ICOFOM - ICOM International Committee of Museology, she was Trustee of the Board of European Museum Forum / EMYA from 2014 to 2018 and presently she is National Correspondent for Italy. Lecturer at various Italian universities and international institutions, she has been a contracting professor at IULM University, Milan, since 2016 and, since 2019, in the Master: Tourism Management. She is a consultant at YAS Management Reform in Dubai for the development of cultural projects. Currently she is a member of Europa Nostra International Council and President of the jury of the European Heritage Awards / Europa Nostra Awards in the Education, Training and Awareness Raising section.

Vedran Zubović, MSc, is working as an Assistant at the Faculty of Tourism and Hospitality Management in Opatija, University of Rijeka, Croatia. His areas of interest are strategic management in tourism and entrepreneurial orientation in selective forms of tourism. He is attending postgraduate doctoral studies "Sustainable Development Management" at the same university.

Index

V

W

www.ingramcontent.com/pod-product-compliance
Lightning Source LLC
LaVergne TN
LVHW060303011026
849161LV00011B/2764